Proceedings
of the
STEKLOV INSTITUTE
OF MATHEMATICS

1984, ISSUE 1

Quasilinear Degenerate and Nonuniformly Elliptic and Parabolic Equations of Second Order

by

A. V. Ivanov

Translation of
ТРУДЫ
ордена Ленина
МАТЕМАТИЧЕСКОГО ИНСТИТУТА
имени В. А. СТЕКЛОВА

Том 160 (1982)

АКАДЕМИЯ НАУК

СОЮЗА СОВЕТСКИХ СОЦИАЛИСТИЧЕСКИХ РЕСПУБЛИК

ТРУДЫ

ордена Ленина

МАТЕМАТИЧЕСКОГО ИНСТИТУТА

имени В. А. СТЕКЛОВА

CLX

А. В. ИВАНОВ

КВАЗИЛИНЕЙНЫЕ ВЫРОЖДАЮЩИЕСЯ И НЕРАВНОМЕРНО ЭЛЛИПТИЧЕСКИЕ И ПАРАБОЛИЧЕСКИЕ УРАВНЕНИЯ ВТОРОГО ПОРЯДКА

Ответственный редактор (Editor-in-chief)
академик С. М. НИКОЛЬСКИЙ (S. M. Nikol'skiĭ)

Заместитель ответственного редактора (Assistant to the editor-in-chief)
профессор Е. А. ВОЛКОВ (E. A. Volkov)

издательство "наука"
Ленинград 1982

Translated by J. R. SCHULENBERGER

Library of Congress Cataloging in Publication Data

Ivanov, A. V.

Quasilinear degenerate and nonuniformly elliptic and parabolic equations of second order.
(Proceedings of the Steklov Institute of Mathematics; 1984, issue 1)
Translation of: Kvazilineĭnye vyrozhdaiushchiesia i naravnomerno ellipticheskie i parabolicheskie uravneniia vtorogo poriadka.
Bibliography: p.
1. Differential equations, Elliptic—Numerical solutions. 2. Differential equations, Parabolic—Numerical solutions. 3. Boundary value problems—Numerical solutions. I. Title. II. Series: Trudy ordena Lenina Matematicheskogo instituta imeni V. A. Steklova. English; 1984, issue 1.
QA1.A413 1984, issue 1 [QA377] 510s [515.3'53] 84-12386
ISBN 0-8218-3080-5

March 1984

PROCEEDINGS OF THE STEKLOV INSTITUTE OF MATHEMATICS IN THE ACADEMY OF SCIENCES OF THE USSR

TABLE OF CONTENTS

RUSSIAN TABLE OF CONTENTS*

* The American Mathematical Society scheme for transliteration of Cyrillic may be found at the end of index issues of *Mathematical Reviews.*

ABSTRACT. This volume is devoted to questions of solvability of basic boundary value problems for quasilinear degenerate and nonuniformly elliptic and parabolic equations of second order, and also to the investigation of differential and certain qualitative properties of solutions of such equations. A theory of generalized solvability of boundary value problems is constructed for quasilinear equations with specific degeneracy of ellipticity or parabolicity. Regularity of generalized solutions of quasilinear degenerate parabolic equations is studied. Existence theorems for a classical solution of the first boundary value problem are established for large classes of quasilinear nonuniformly elliptic and parabolic equations.

1980 *Mathematics Subject Classifications.* Primary 35J65, 35J70, 35K60, 35K65, 35D05; Secondary 35D10.

Correspondence between *Trudy Mat. Inst. Steklov.* and *Proc. Steklov Inst. Math.*

Russian imprint	Vol.	English imprint
1966	74	1967
1965	75	1967
1965	76	1967
1965	77	1967
1965	78	1967
1965	79	1966
1965	80	1968
1966	81	1968
1966	82	1967
1965	83	1967
1965	84	1968
1966	85	1967
1965	86	1967
1966	87	1967
1967	88	1969
1967	89	1968
1967	90	1969
1967	91	1969
1966	92	1968
1967	93	1970
1968	94	1969
1968	95	1971
1968	96	1970
1968	97	1969
1968	98	1971
1967	99	1968
1971	100	1974
1972	101	1975
1967	102	1970
1968	103	1970
1968	104	1971
1969	105	1971
1969	106	1972

Russian imprint	Vol.	English imprint
1971	107	1976
1968	108	1971
1971	109	1974
1970	110	1972
1970	111	1972
1971	112	1973
1970	113	1972
1970	114	1974
1971	115	1974
1971	116	1973
1972	117	1974
1972	118	1976
1973	119	1976
1974	120	1976
1972	121	1974
1973	122	1975
1973	123	1975
1976	124	1978, Issue 3
1973	125	1975
1973	126	1975
1975	127	1977
1972	128	1974
1973	129	1976
1978	130	1979, Issue 4
1974	131	1975
1973	132	1975
1973	133	1977
1975	134	1977
1975	135	1978, Issue 1
1975	136	1978, Issue 2
1976	137	1978, Issue 4
1975	138	1977

Russian imprint	Vol.	English imprint Year	Issue
1980	139	1982	1
1976	140	1979	1
1976	141	1979	2
1976	142	1979	3
1977	143	1980	1
1979	144	1980	2
1980	145	1981	1
1978	146	1980	3
1980	147	1981	2
1978	148	1980	4
1978	149	1981	3
1979	150	1981	4
1980	151	1982	2
1980	152	1982	3
1981	153	1982	4
1983	154	1984	4
1981	155	1983	1
1980	156	1983	2
1981	157	1983	3
1981	158	1983	4
1983	159	1984	2
1982	160	1984	1
1983	161	1984	3

PREFACE

This monograph is devoted to the study of questions of solvability of main boundary value problems for degenerate and nonuniformly elliptic and parabolic equations of second order and to the investigation of differential and certain qualitative properties of the solutions of such equations. The study of various questions of variational calculus, differential geometry, and the mechanics of continuous media leads to quasilinear degenerate or nonuniformly elliptic and parabolic equations. For example, some nonlinear problems of heat conduction, diffusion, filtration, the theory of capillarity, elasticity theory, etc. lead to such equations. The equations determining the mean curvature of a hypersurface in Euclidean and Riemannian spaces, including the equation of minimal surfaces, belong to the class of nonuniformly elliptic equations. The Euler equations for many variational problems are quasilinear, degenerate or nonuniformly elliptic equations.

With regard to the character of the methods applied, this monograph is organically bound with the monograph of O. A. Ladyzhenskaya and N. N. Ural′tseva, *Linear and quasilinear equations of elliptic type*, and with the monograph of O. A. Ladyzhenskaya, V. A. Solonnikov, and N. N. Ural′tseva, *Linear and quasilinear equations of parabolic type*. In particular, a theory of solvability of the basic boundary value problems for quasilinear, nondegenerate and uniformly elliptic and parabolic equations was constructed in those monographs.

The monograph consists of four parts. In Part I the principal object of investigation is the question of classical solvability of the first boundary value problem for quasilinear, nonuniformly elliptic and parabolic equations of nondivergence form. A priori estimates of the gradients of solutions in a closed domain are established for large classes of such equations; these estimates lead to theorems on the existence of solutions of the problem in question on the basis of the well-known results of Ladyzhenskaya and Ural′tseva. In this same part qualified local estimates of the gradients of solutions are also established, and they are used, in particular, to establish two-sided and one-sided Liouville theorems. A characteristic feature of the a priori estimates for gradients of solutions obtained in Part I is that these estimates are independent of any minorant for the least eigenvalue of the matrix of coefficients of the second derivatives on the solution in question of the equation. This circumstance predetermines the possibility of using these and similar estimates to study quasilinear degenerate elliptic and parabolic equations.

Parts II and III are devoted to the construction of a theory of solvability of the main boundary value problems for large classes of quasilinear equations with a nonnegative characteristic form. In Part II the class of quasilinear, so-called $(A, \mathbf{b})$-elliptic equations is introduced. Special cases of this class are the classical elliptic and parabolic quasilinear equations, and also linear equations with an arbitrary nonnegative characteristic form. The general boundary value problem (in particular, the first, second, and third boundary value problems) is formulated for $(A, \mathbf{b})$-elliptic equations, and the question of existence and uniqueness of a generalized solution of energy type to such a problem for the class of $(A, \mathbf{b}, m, \mathbf{m})$-elliptic equations is investigated. Theorems on the existence and uniqueness of regular generalized solutions of the first boundary value problem for $(A, \mathbf{b})$-elliptic equations are also established in this part.

In Part III questions of the solvability of the main boundary value problems are studied in detail for important special cases of $(A, \mathbf{b})$-elliptic equations—$(A, \mathbf{0})$-elliptic and so-called $(A, \mathbf{0})$-parabolic equations, which are more immediate generalizations of classical elliptic and parabolic quasilinear equations. All the conditions under which theorems on the existence and uniqueness of a generalized solution (of energy type) of the general boundary value problem are established for $(A, \mathbf{0}, m, \mathbf{m})$-elliptic and $(A, \mathbf{0}, m, \mathbf{m})$-parabolic equations are of easily verifiable character. Theorems on the existence and uniqueness of so-called A-regular generalized solutions of the first boundary value problem are also established for $(A, \mathbf{0})$-elliptic equations. Examples are presented which show that for equations of this structure the investigation of A-regularity of their solutions (in place of ordinary regularity) is natural. These results are applied to the study of a certain class of nonregular variational problems.

Some of the results in Parts II and III are also new for the case of linear equations with an arbitrary nonnegative characteristic form. For these equations a theory of boundary value problems has been constructed in the works of G. Fichera, O. A. Oleĭnik, J. J. Kohn and L. Nirenberg, M. I. Freĭdlin, and others.

Part IV is devoted to the study of properties of generalized solutions of quasilinear, weakly degenerate parabolic equations. From the results obtained it is evident how the properties of generalized solutions of the equation improve as the regularity of the functions forming the equation improves. This improvement, however, is not without limit as in the case of nongenerate parabolic equations, since the presence of the weak degeneracy poses an obstacle to the improvement of the differential properties of the functions forming the equation.

This monograph is not a survey of the theory of quasilinear elliptic and parabolic equations, and for this reason many directions of this theory are not reflected here. The same pertains to the bibliography.

The author expresses his gratitude to Ol'ga Aleksandrovna Ladyzhenskaya for a useful discussion of the results presented here. The very idea of writing this monograph is due to her.

BASIC NOTATION

We denote n-dimensional real space by $\mathbb{R}^n$; $x = (x_1, \ldots, x_n)$ is a point of $\mathbb{R}^n$, and Ω is a domain (an open, connected set) in $\mathbb{R}^n$; the boundary of Ω is denoted by $\partial\Omega$. All functions considered are assumed to be real.

Let G be a Lebesgue-measurable set $\mathbb{R}^n$. Functions equivalent on G, i.e., having equal values for almost all (a.a.) $x \in G$ are assumed to be indistinguishable (coincident).

$L^p(G)$, $1 \leqslant p < +\infty$, denotes the Banach space obtained by introducing the norm

$$\|u\|_{p,G} \equiv \|u\|_{L^p(G)} = \left(\int_G |u(x)|^p \, dx\right)^{1/p}$$

on the set of all Lebesgue-measurable functions u on the set g with finite Lebesgue integral $\int_G |u(x)|^p \, dx$.

$L^\infty(G)$ is the Banach space obtained by introducing the norm

$$\|u\|_{\infty,G} \equiv \|u\|_{L^\infty(G)} = \operatorname*{ess\,sup}_{x \in G} |u(x)|$$

on the set of all measurable and essentially bounded functions on G.

$L^p_{\text{loc}}(G)$, $1 \leqslant p \leqslant +\infty$, denotes the set of functions belonging to $L^p(G')$ for any subdomain G' strictly interior to G (i.e., G' such that $\overline{G'} \subset G$).

$W^l_p(G)$ is the familiar Sobolev space obtained by introducing (on the set of all functions u which with all their partial derivatives through order l belong to the space $L^p(\Omega)$, $p \geqslant 1$) the norm

$$\|u\|_{W^l_p(\Omega)} = \sum_{k=0}^{l} \|D^k u\|_{p,\Omega},$$

where

$$D^k u \equiv \frac{\partial^k u}{\partial x_1^{k_1} \cdots \partial x_n^{k_n}}, \qquad |k| = k_1 + \cdots + k_n.$$

$C^m(\Omega)$ ($C^\infty(\Omega)$) denotes the class of all functions continuously differentiable m times on Ω (all infinitely differentiable functions on Ω), and $C^m(\overline{\Omega})$ ($C^\infty(\overline{\Omega})$), where $\overline{\Omega}$ is the closure of Ω, is the set of those functions in $C^m(\Omega)$ ($C^\infty(\Omega)$) for which all partial derivatives through order m (all partial derivatives) can be extended to continuous functions on $\overline{\Omega}$. The set of all continuous functions on Ω (on $\overline{\Omega}$) is denoted simply by $C(\Omega)$ ($C(\overline{\Omega})$).

The support of a function $u \in C(\Omega)$ is the set $\operatorname{sup} u \equiv \overline{\{x \in \Omega : u(x) \not\equiv 0\}}$. $C_0^m(\Omega)$ ($C_0^\infty(\Omega)$) denotes the set of all functions in $C^m(\Omega)$ ($C^\infty(\Omega)$) having compact support in Ω.

Let K be a compact set in $\mathbb{R}^n$. A function u defined on K is said to belong to the class $C^\alpha(K)$, where $\alpha \in (0,1)$, if there exists a constant c such that

$$|u(x) - u(x')| \leqslant c|x - x'|^\alpha, \quad \forall x, x' \in K, x \neq x'.$$

In this case it is also said that the function u is Hölder continuous with exponent α on the set K. The least constant c for which this inequality holds is called the Hölder constant of the function u on the set K and is denoted by $\langle u \rangle_K^{(\alpha)}$. In particular, if $u \in C^\alpha(\overline{\Omega})$, then

$$\langle u \rangle_{\overline{\Omega}}^{(\alpha)} = \sup_{x, x \in \Omega, x \neq x'} \frac{|u(x) - u(x')|}{|x - x'|^\alpha}.$$

If on the set $C^\alpha(\Omega)$ we introduce the norm

$$\|u\|_\Omega^{(\alpha)} = \sup_\Omega |u(x)| + \langle u \rangle_{\overline{\Omega}}^{(\alpha)},$$

then we obtain a Banach space which is also denoted by $C^\alpha(\overline{\Omega})$.

Functions u satisfying the condition

$$|u(x) - u(x')| \leqslant c|x - x'|, \quad \forall x, x' \in \overline{\Omega}, x \neq x',$$

are called Lipschitz continuous on $\overline{\Omega}$. Such functions form a Banach space $\mathrm{Lip}(\overline{\Omega})$ with norm defined by

$$|u|_{\mathrm{Lip}(\Omega)} = \sup_{\overline{\Omega}} |u(x)| + \langle u \rangle_{\overline{\Omega}}^{(1)},$$

where the Lipschitz constant $\langle u \rangle_{\overline{\Omega}}^{(1)}$ is defined in the same way as $\langle u \rangle_{\overline{\Omega}}^{(\alpha)}$ but with $\alpha = 1$. $\mathrm{Lip}(\Omega)$ denotes the collection of functions continuous in Ω and belonging to $\mathrm{Lip}(\overline{\Omega}')$ for all $\overline{\Omega}' \subset \Omega$.

$C^{m+\alpha}(\overline{\Omega})$ denotes the Banach space with elements which are functions of the class $C^m(\overline{\Omega})$ having derivatives of mth order belonging to the class $C^\alpha(\overline{\Omega})$; the norm is given by

$$\|u\|_\Omega^{(m+\alpha)} = \sum_{|k|=0}^{m} \sup_\Omega |D^k u(x)| + \sum_{|k|=m} \langle D^k u(x) \rangle_{\overline{\Omega}}^{(\alpha)}.$$

$C^{m+\alpha}(\Omega)$ denotes the set of functions belonging to $C^{m+\alpha}(\overline{\Omega}')$ for all Ω' such that $\overline{\Omega}' \subset \Omega$.

We denote by $\tilde{C}^k(\overline{\Omega})$ the set of all functions of the class $C^{k-1}(\overline{\Omega})$ such that all their partial derivatives of order k are piecewise continuous in $\overline{\Omega}$ (and are hence bounded in Ω). In particular, $\tilde{C}^1(\overline{\Omega})$ denotes the set of all continuous and piecewise differentiable functions in $\overline{\Omega}$.

Let Γ be a fixed subset of $\partial\Omega$. We denote by $\tilde{C}^1_{0,\Gamma}(\overline{\Omega})$ the set of all functions in the class $\tilde{C}_1(\overline{\Omega})$ which are equal to zero outside some (depending on the function) n-dimensional neighborhood of Γ. In the case $\Gamma = \partial\Omega$ we denote the corresponding set by $\tilde{C}^1_0(\Omega)$.

A domain Ω is called strongly Lipschitz if there exist constants $R > 0$ and $L > 0$ such that for any point $x_0 \in \partial\Omega$ it is possible to construct a (orthogonal) Cartesian coordinate system $y_1, \ldots, y_n$ with center at x_0 such that the intersection of $\partial\Omega$ with the cylinder $C_{R,L} \equiv \{y \in \mathbb{R}^n: \sum_{i=1}^{n-1} y_i^2 < R^2, |y_n| < 2LR\}$ is given by the equation

$$y_n = \varphi(y'), \qquad y' \equiv (y_1, \ldots, y_{n-1}),$$

where $\varphi(y')$ is a Lipschitz function on the domain $\{|y'| \leqslant R\}$, with Lipschitz constant $\langle\varphi\rangle^{(1)}_{\{|y'| \leqslant R\}}$ not exceeding L, and

$$\overline{\Omega} \cap C_{R,L} \equiv \{y \in \mathbb{R}^n: |y'| \leqslant R, \varphi(y') \leqslant y_n \leqslant 2LR\}.$$

It is known that any convex domain is strongly Lipschitz.

A domain Ω with boundary $\partial\Omega$ is called a domain of class C^k, $k \geqslant 1$, if for any point of $\partial\Omega$ there is a neighborhood ω such that $\partial\Omega \cap \omega$ can be represented in the form

$$x_l = \varphi_l(x_1, \ldots, x_{l-1}, x_{l+1}, \ldots, x_n) \qquad (*)$$

for some $l \in \{1, \ldots, n\}$, and the function φ_l belongs to the class $C^k(\omega_l)$, where ω_l is the projection of $\omega \cap \partial\Omega$ onto the plane $x_l = 0$.

We further introduce the classes $\tilde{C}^{(k)}$, $k \geqslant 1$, of domains with piecewise smooth boundary (see the definition of the classes $B^{(k)}$ in [102]). It is convenient to introduce the class of such domains by induction on the dimension of the domain. An interval is a one-dimensional domain of class $\tilde{C}^{(k)}$. A domain $\Omega \subset \mathbb{R}^n$ with boundary $\partial\Omega$ belongs to the class $\tilde{C}^{(k)}$ if its boundary coincides with the boundary of the closure $\overline{\Omega}$ and it can be decomposed into a finite number of pieces S^j, $j = 1, \ldots, N$, homeomorphic to the $(n-1)$-dimensional ball which possibly intersect only at boundary points and are such that each piece S^j can be represented in the form $(*)$ for some $l \in \{1, \ldots, n\}$, where the function φ_l is defined in an $(n-1)$-dimensional closed domain $\bar{\sigma}$ of class $\tilde{C}^{(k)}$ on the plane $x_l = 0$ and $\varphi_l \in \tilde{C}^k(\bar{\sigma})$.

If $\Omega \in \tilde{C}^{(k)}$, $k \geqslant 1$, then the formula for integration by parts can be applied to $\Omega \cup \partial\Omega$; it transforms an n-fold integral over Ω into an $(n-1)$-fold integral over $\partial\Omega$.

Let B_1 and B_2 be any Banach spaces. Following [96], we write $B_1 \to B_2$ to denote the continuous imbedding of B_1 in B_2. In other words, this notation means $B_1 \subset B_2$ (each element of B_1 belongs to B_2) and there exists a constant $c > 0$ such that

$$\|u\|_{B_2} \leqslant c\|u\|_{B_1}, \quad \forall u \in B_1.$$

Above we have presented only the notation and definitions which will be most frequently encountered in the text. Some other commonly used notation and terms will be used without special clarification. Much notation and many definitions will be introduced during the course of the exposition.

In the monograph the familiar summation convention over twice repeated indices is often used. For example, $a^{ij}u_{x_i x_j}$ means the sum $\sum_{i,j=1}^n a^{ij}u_{x_i x_j}$, etc.

Within each chapter formulas are numbered to reflect the number of the section and the number of the formula in that section. For example, in the notation (2.8) the first number indicates the number of the section in the given chapter, while the second number indicates the number of the formula in that particular section. A three-component notation is used when it is necessary to refer to a formula of another chapter. For example, in Chapter 7 the notation (5.1.2) is used to refer to formula (1.2) of Chapter 5. Reference to numbers of theorems and sections is made similarly. The formulas in the introductions to the first, second, and third parts of the monograph are numbered in special fashion. Here the numbering reflects only the number of the formula within the given introduction. There are no references to these formulas outside the particular introduction.

PART I

QUASILINEAR, NONUNIFORMLY ELLIPTIC AND PARABOLIC EQUATIONS OF NONDIVERGENCE TYPE

Boundary value problems for linear and quasilinear elliptic and parabolic equations have been the object of study of an enormous number of works. The work of O. A. Ladyzhenskaya and N. N. Ural'tseva, the results of which are consolidated in the monographs [83] and [80], made a major contribution to this area. In these monographs the genesis of previous work is illuminated, results of other mathematicians are presented, and a detailed bibliography is given. In addition to this work we note that contributions to the development of the theory of boundary value problems for quasilinear elliptic and parabolic equations were made by S. N. Bernstein, J. Schauder, J. Leray, S. L. Sobolev, L. Nirenberg, C. Morrey, O. A. Oleĭnik, M. I. Vishik, J. L. Lions, E. M. Landis, A. Friedman, A. I. Koshelev, V. A. Solonnikov, F. Browder, E. DeGiorgi, J. Nash, J. Moser, D. Gilbarg, J. Serrin, I. V. Skrypnik, S. N. Kruzhkov, Yu. A. Dubinskiĭ, N. S. Trudinger, and many other mathematicians.

The so-called uniformly elliptic and parabolic equations formed the main object of study in the monographs [83] and [80]. Uniform ellipticity to the equation

$$\sum_{i,j=1}^{n} a^{ij}(x, u, \nabla u) u_{x_i x_j} = a(x, u, \nabla u) \tag{1}$$

in a domain $\Omega \subset \mathbb{R}^n$, $n \geqslant 2$ (uniform parabolicity of the equation

$$-u_t + \sum_{i,j=1}^{n} a^{ij}(x, t, u, \nabla u) u_{x_i x_j} = a(x, t, u, \nabla u) \tag{2}$$

in the cylinder $Q = \Omega \times (0, T] \subset \mathbb{R}^{n+1}$, $n \geqslant 1$) means that for this equation not only the condition of ellipticity $a^{ij}(x, u, p)\xi_i\xi_j > 0$ for all $\xi \in \mathbb{R}^n$, $\xi \neq 0$ (the condition of parabolicity $a^{ij}(x, t, u, p)\xi_i\xi_j > 0$ for all $\xi \in \mathbb{R}^n$, $\xi \neq 0$) is satisfied, but also the following condition: for all $(x, u, p) \in \overline{\Omega} \times \{|u| \leqslant m\} \times \mathbb{R}^n$ ($(x, t, u, p) \in \overline{Q} \times \{|u| \leqslant m\} \times \mathbb{R}^n$)

$$\Lambda(x, u, p) \leqslant c\lambda(x, u, p) \quad (\Lambda(x, t, u, p) \leqslant c\lambda(x, t, u, p)), \tag{3}$$

where λ and Λ are respectively the least and largest eigenvalues of the matrix of coefficients of the leading derivatives, and c is a constant depending on the parameter m. In view of the results of Ladyzhenskaya and Ural'tseva the problem of solvability of a boundary value problem for a nonuniformly elliptic or parabolic equation reduces to the question of constructing a priori estimates of the maximum moduli of the gradients of solutions for a suitable one-parameter family of similar equations. Much of Part I of the present monograph is devoted to this question. The question of the validity of a priori estimates of the maximum moduli of the gradients of solutions for quasilinear elliptic and parabolic equations is the key question, since the basic restrictions on the structure of such equations arise precisely at this stage.

Nonuniformly elliptic equations of the form (1) are considered in Chapter 1. It is known (see [83] and [163]) that to be able to ensure the existence of a classical solution of the Dirichlet problem for an equation of the form (1) for any sufficiently smooth boundary function it is necessary to coordinate the behavior as $p \to \infty$ of the right side $a(x, u, p)$ of the equation with the behavior as $p \to \infty$ of a certain characteristic of the equation determined by its leading terms $a^{ij}(x, u, p)$, $i, j = 1, \dots, n$. Serrin [163] proved that growth of $a(x, u, p)$ as $p \to \infty$ faster than the growth of each of the functions $\mathscr{E}_1(x, u, p)\psi(|p|)$ and $\mathscr{E}_2(x, u, p)$ as $p \to \infty$, where

$$\mathscr{E}_1(x, u, p) \equiv a^{ij}(x, u, p)p_i p_j, \quad \int^{+\infty} \frac{d\rho}{\rho\psi(\rho)} = +\infty,$$

$$\mathscr{E}_2(x, u, p) = \mathrm{Tr}\|a^{ij}(x, u, p)\|\,|p|,$$

leads to the nonexistence of a classical solution of the Dirichlet problem for a certain choice of infinitely differentiable boundary functions. On the other hand, sufficient conditions for classical solvability of the Dirichlet problem for any sufficiently smooth boundary function obtained in [127], [77], [79], [163], [29], [81], [31], [34], [35], [83] and [165] for various classes of uniformly and nonuniformly elliptic equations afford the possibility of considering as right sides of (1) functions $a(x, u, p)$ growing as $p \to \infty$ no faster than $\mathscr{E}_1$. Sufficient conditions for this solvability of the Dirichlet problem obtained in [29], [31], [34] and [35] for rather large classes of nonuniformly elliptic equations and in [165] for equations with special structure make it possible to consider as right sides of (1) functions $a(x, u, p)$ growing as $p \to \infty$ no faster than $\mathscr{E}_2$.

Thus, the functions (or majorants, as we call them) $\mathscr{E}_1$ and $\mathscr{E}_2$ control the admissible growth of the right side of the equation. In connection with this, one of the first questions of the general theory of boundary value problems for quasilinear elliptic equations of the form (1) is the question of distinguishing classes of equations for which the conditions for solvability of the Dirichlet problem provide the possibility of natural growth of the right side $a(x, u, p)$ as $p \to \infty$, i.e. growth not exceeding the growth of at least one of the majorants $\mathscr{E}_1$ and $\mathscr{E}_2$. Just such classes are distinguished in [127], [77]–[79], [163], [29], [31], [34], [35], [165] and [83].

The author's papers [29], [31], [34] and [35], on which Chapter 1 of this monograph is based, distinguish large classes of nonuniformly elliptic equations of this sort. For them a characteristic circumstance is that the conditions imposed on the leading coefficients of the equation are formulated in terms of the majorants $\mathscr{E}_1$ and $\mathscr{E}_2$ and refer not to the individual coefficients a^{ij} but rather to aggregates of the form $A^\tau \equiv a^{ij}(x, u, p)\tau_i\tau_j$, where $\tau = (\tau_1, \dots, \tau_n)$ is an arbitrary unit vector in $\mathbb{R}^n$. It is important that under these conditions the established a priori estimates of the gradients of solutions do not depend on any minorant for the least eigenvalue λ of the matrix $\|a^{ij}\|$. This circumstance predetermines the possibility of using the results obtained here also in the study of boundary value problems for quasilinear degenerate elliptic equations, and this is done in Parts II and III.

As in the case of uniformly elliptic equations, the establishment of an a priori estimate of $\max_{\bar{\Omega}}|\nabla u|$ breaks down into two steps: 1) obtaining $\max_{\partial\Omega}|\nabla u|$ in terms of $\max_{\Omega}|u|$, and 2) obtaining an estimate of $\max_{\Omega}|\nabla u|$ in terms of $\max_{\partial\Omega}|\nabla u|$ and $\max_{\Omega}|u|$. The estimates of $\max_{\partial\Omega}|\nabla u|$ are first established by means of the technique of global barriers developed by Serrin (see [163]).

In particular, the modifications of Serrin's results obtained in this way are found to be useful in studying quasilinear degenerate equations. We then establish local estimates of the gradients of solutions of equations of the form (1) by combining the use of certain methods characteristic of the technique of global barriers with constructions applied by Ladyzhenskaya and Ural'tseva (see [83]). The results obtained in this way constitute a certain strengthening (for the case of nonuniformly elliptic equations) of the corresponding results of [77], [142] and [83] on local estimates of $|\nabla u|$ on the boundary of a domain.

Further on in Chapter 1 a priori estimates of $\max_{\Omega}|\nabla u|$ in terms of $\max_{\partial\Omega}|\nabla u|$ are established. The method of proof of such estimates is based on applying the maximum principle for elliptic equations. This circumstance relates it to the classical methods of estimating gradients of solutions that took shape in the work of S. N. Bernstein (for $n = 2$) and Ladyzhenskaya (for $n \geqslant 2$) and applied in [77]–[79], [163], [127], [11] and elsewhere. Comparison of the results of [77]–[79], [163] and [127] with those of [29], [31], [34] and [35] shows, however, that these methods have different limits of applicability. The estimate of $\max_{\Omega}|\nabla u|$ is first established for a class of equations with structure described in terms of the majorant $\mathscr{E}_1$ (see (1.6.4)). This class contains as a special case the class of quasilinear uniformly elliptic equations considered in [83]. An estimate of $\max_{\Omega}|\nabla u|$ is then obtained for a class of equations with structure described in terms of the majorant $\mathscr{E}_2$ (see (1.7.1)). This class contains, in particular, the equation with principal part which coincides with the principal part of the equation of minimal surfaces (1.7.13). The latter is also contained in the third class of equations for which an estimate of $\max_{\Omega}|\nabla u|$ is established. The structure of this class has a more special character (see (1.8.1)). This class is distinguished as a separate class in the interest of a detailed study of the equations of surfaces with a given mean curvature. We note that the conditions on the right side of an equation of the form (1.7.13) which follow from (1.8.1) do not coincide with conditions following from (1.7.1). The class of equations determined by conditions (1.8.1) contains as special cases some classes of equations of the type of equations of surfaces with prescribed mean curvature which have been distinguished by various authors (see [163], [4] and [83]).

We note that the works [171], [103], [104], [54], [30] and [55], in which the so-called divergence method of estimating $\max_{\Omega}|\nabla u|$ developed by Ladyzhenskaya and Ural'tseva for uniformly elliptic equations is used, are also devoted to estimating $\max_{\Omega}|\nabla u|$ for solutions of nonuniformly elliptic equations of the form (1). In these works the structure of equation (1) is not characterized in terms of the majorants $\mathscr{E}_1$ and $\mathscr{E}_2$.

With the help of a fundamental result of Ladyzhenskaya and Ural'tseva on estimating the norm $\|u\|_{C^{1+\alpha}(\Omega)}$ in terms of $\|u\|_{C^1(\Omega)}$ for solutions of arbitrary elliptic equations of the form (1), at the end of Chapter 1 theorems on the existence of classical solutions of the Dirichlet problem are derived from the a priori estimates obtained earlier. Analogous results on the solvability of the Dirichlet problem have also been established by the author for certain classes of nonuniformly elliptic systems [37]. Due to the limited length of this monograph, however, these results are not presented.

The first boundary value problem for nonuniformly parabolic equations of the form (2) is studied in Chapter 2. As in the case of elliptic equations of the form (1), the leading coefficients of (2) determine the admissible growth of the right side $a(x, t, u, p)$ as $p \to \infty$, since growth of $a(x, t, u, p)$ as $p \to \infty$ faster than the growth of each of the functions $\mathscr{E}_1\psi(|p|)$ and $\mathscr{E}_2$ as $p \to \infty$, where

$$\mathscr{E}_1 \equiv a^{ij}(x, t, u, p)p_i p_j \quad \text{and} \quad \mathscr{E}_2 = \operatorname{Tr}\|a^{ij}(x, t, u, p)\|\,|p|,$$

$$\int^{+\infty} \frac{d\rho}{\psi(\rho)\rho} = +\infty,$$

leads to the nonexistence of a classical solution of the first boundary value problem for certain infinitely differentiable boundary functions (see, for example, [136]). Sufficient conditions for classical solvability of the first boundary value problem for any sufficiently smooth boundary function obtained in [78], [83] and [98] for uniformly parabolic equations, in [136] for a certain special class of nonuniformly parabolic equations, and in [38] for a large class of nonuniformly parabolic equations of the form (2) make it possible to consider functions growing no faster than $\mathscr{E}_1$ as right sides of the equation. In [38], on the basis of which Chapter 2 of the monograph is written, sufficient conditions are also obtained for classical solvability of the first boundary value problem which admit right sides $a(x, t, u, p)$ growing as $p \to \infty$ no faster than the function $\mathscr{E}_2$. (We remark that, as in the case of elliptic equations, the meaning of the expression "growth of a function as $p \to \infty$" has relative character.)

Thus, the majorants $\mathscr{E}_1$ and $\mathscr{E}_2$ control the admissible growth of the right side of the equation also in the case of parabolic equations. However, the presence of the term u_t in (2) alters the picture somewhat. The situation is that among the sufficient conditions ensuring the solvability of the first boundary value problem for any sufficiently smooth boundary functions and under natural conditions on the behavior of $a(x, t, u, p)$ as $p \to \infty$ there is the condition

$$\mathscr{E}_1\psi(|p|) + \mathscr{E}_2 \to \infty \quad \text{as } p \to \infty. \tag{4}$$

When condition (4) is violated we establish the existence of a classical solution of the first boundary value problem under natural conditions on growth of $a(x, t, u, p)$ in the case of an arbitrary sufficiently smooth boundary function depending only on the space variables. This assumption (the independence of the boundary function of t when condition (4) is violated) is due, however, to an inherent feature of the problem. We prove a nonexistence theorem (see Theorem 2.5.2) which asserts that if conditions which are in a certain sense the negation of condition (4) are satisfied there exist infinitely differentiable boundary functions depending in an essential way on the variable t for which the first boundary value problem has no classical solution.

Returning to the discussion of sufficient conditions for solvability of the first boundary value problem given in Chapter 2, we note that, as in the elliptic case, the conditions on the leading coefficients $a^{ij}(x, t, u, p)$ of the equation pertain to the summed quantities $A^\tau \equiv a^{ij}(x, t, u, p)\tau_i\tau_j$, $\tau \in \mathbb{R}^n$, $|\tau| = 1$, and are formulated in terms of the majorants $\mathscr{E}_1$ and $\mathscr{E}_2$. Here it is also important to note that the structure of these conditions and the character of the basic a priori estimates established for solutions of (2) do not depend on the "parabolicity constant" of the equation. This determines at the outset the possibility of using the results obtained to study in addition boundary value problems for quasilinear degenerate parabolic equations. In view of the results of Ladyzhenskaya and Ural′tseva (see [80]), the proof of classical solvability of the first boundary value problem for equations of the form (2) can be reduced to establishing an a priori estimate of $\max_Q|\nabla u|$, where ∇u is the spatial gradient, for solutions of a one-parameter family of equations (2) having the same structure as the original equation (see §2.1).

To obtain such an estimate we first find an a priori estimate of ∇u on the parabolic boundary Γ of the cylinder Q on the basis of the technique of global barriers. We then establish a priori estimates of $\max_Q|\nabla u|$ in terms of $\max_\Gamma|\nabla u|$ and $\max_Q|u|$. Sufficient conditions for the validity of this estimate are formulated in terms of both the majorant $\mathscr{E}_1$ and the majorant $\mathscr{E}_2$. The first class of equations of the form (2) for the solutions of which this estimate is established (see (2.3.2)) contains as a special case the class of quasilinear uniformly parabolic equations considered in [83]. The second class of equations distinguished in this connection and characterized by conditions formulated in terms of the majorant $\mathscr{E}_2$ contains, in particular, the parabolic analogue of the equation of given mean curvature (see (2.3.25)). Such equations find application in the mechanics of continuous media. From our estimates the proof of existence of a classical solution of the first boundary value problem is assembled with the help of a well-known result of Ladyzhenskaya and Ural′tseva on estimating the norm $\|u\|_{C^{1+\alpha}(\bar{Q})}$ in terms of $\|u\|_{C^1(\bar{Q})}$ for solutions of arbitrary parabolic equations of the form (2).

Chapter 3, which concludes Part I, is devoted to obtaining local estimates of the gradients of solutions of quasilinear elliptic equations of the form (1) and their application to the proof of certain qualitative properties of solutions of these equations. In the case of uniformly elliptic equations local estimates of the gradients of solutions of equations of the form (1) have been established by Ladyzhenskaya and Ural′tseva (see [83]). In [142], [166], [26] and [83] these estimates are extended to certain classes of nonuniformly elliptic equations of the form (1). In these works the modulus $|\nabla u(x_0)|$ of the gradient of a solution u at an arbitrary interior point x_0 of Ω is estimated in terms of $\max_{K_\rho(x_0)}|u|$, where $K_\rho(x_0)$ is a ball of radius ρ with center at x_0. The author's results [26] obtained in connection with this estimate find reflection at the beginning of Chapter 3. The estimate in question is established here under conditions formulated in terms of the majorant $\mathscr{E}_1$ (see (3.1.1)–(3.1.6)). An important feature of these conditions and of the estimate of $|\nabla u(x_0)|$ is that they are independent of the ellipticity constant of equation (1), i.e., of any minorant for the least eigenvalue of the matrix $\|a^{ij}(x, u, \nabla u)\|$ at the solution in question of this equation. Therefore, the result is meaningful even for the case of uniformly elliptic equations. Moreover, this affords the possibility of using the estimate to study quasilinear degenerate elliptic equations. We remark also that in place of a condition on the degree of elliptic nonuniformity of the equation (see [83]) conditions (3.1.1)–(3.1.6) express a restriction on characteristics of elliptic nonuniformity which are more general than this degree.

More special classes of equations of the form (1) for which $|\nabla u(x_0)|$ can be expressed in terms of $\max_{K_\rho(x_0)} u$ or $\min_{K\rho(x_0)} u$ alone or, generally, in terms of structural characteristics alone of the equation are distinguished in the work of L. A. Peletier and J. Serrin [157]. The author's paper [48] is devoted to analogous questions. Estimates of $|\nabla u(x_0)|$ of this sort are also presented in Chapter 3. The local estimates of the gradients are used in this chapter to obtain theorems of Liouville type and (in a special case) to prove a Harnack inequality. Theorems of Liouville type for quasilinear elliptic equations of nondivergence form were the subject of study in [166], [26], [157] and [48]. Two-sided Liouville theorems, consisting in the assertion that any solution that is bounded in modulus or does not have too rapid growth in modulus as $p \to \infty$ is identically constant, are established in [157] for the nonlinear Poisson equation $\Delta u = f(u, \nabla u)$ and in [26] for quasilinear elliptic equations of the form $a^{ij}(\nabla u)u_{x_ix_j} = a(u, \nabla u)$ admitting particular elliptic nonuniformity. In particular, the results of [26] give a limiting two-sided Liouville theorem for the Euler equation of the variational problem on a minimum of the integral $\int_\Omega (1 + |\nabla u|^2)^{m/2}\, dx$, $m > 1$, i.e, for the equation

$$\left[(1 + |\nabla u|^2)\delta_i^j + (m - 2)u_{x_i}u_{x_j}\right]u_{x_ix_j} = 0. \tag{5}$$

Namely, it follows from Theorem 3.1.1 that for any sufficiently smooth solution u in $\mathbb{R}^n$ of (5) for any $x_0 \in \mathbb{R}^n$ there is the estimate $|\nabla u(x_0)| \leqslant c \operatorname{osc}_{K_\rho(x_0)} u \cdot \rho^{-1}$, where the constant c depends only on n and m. This easily implies that any sufficiently smooth solution of (5) in $\mathbb{R}^n$ which grows as $|x| \to \infty$ like $o(|x|)$ is identically constant. This result cannot be strengthened, since a linear function is a solution of (5) in $\mathbb{R}^n$.

For certain classes of uniformly elliptic equations, in [157] one-sided Liouville theorems are proved in which the a priori condition on the solution has a one-sided character: only bounded growth as $\rho \to \infty$ of the quantity $\sup_{|x| \leqslant \rho} u$ or $\inf_{|x| \leqslant \rho} u$ is assumed. In [48] one-sided Liouville theorems are established for other classes of equations of the form (1). So-called weak Liouville theorems in which aside from a priori boundedness of the growth of the function itself at infinity bounded growth of the gradient is also required were proved in [166], [26], [157] and [48]. The exposition of theorems of Liouville type in Chapter 3 is based on the author's papers [26] and [48]. In [39] and [44] two-sided theorems of Liouville type were established for certain classes of elliptic systems of nondivergence form, but in the present monograph Liouville theorems for elliptic systems are not discussed. We do not mention here the large cycle of works on Liouville theorems for linear elliptic equations and systems and for quasilinear elliptic equations of divergence form in which the results are obtained by a different method.

CHAPTER 1
THE DIRICHLET PROBLEM FOR QUASILINEAR, NONUNIFORMLY ELLIPTIC EQUATIONS

§1. The basic characteristics of a quasilinear elliptic equation

In a bounded domain $\Omega \subset \mathbb{R}^n$, $n \geqslant 2$, we consider the quasilinear equation

$$\mathscr{L}u \equiv a^{ij}(x, u, \nabla u)u_{x_ix_j} - a(x, u, \nabla u) = 0, \tag{1.1}$$

where $a^{ij} = a^{ji}$, $i, j = 1, \ldots, n$, which satisfies the ellipticity condition

$$a^{ij}(x, u, p)\xi_i\xi_j > 0 \quad \text{for all } \xi \in \mathbb{R}^n,\ \xi \neq 0, \text{ and all } (x, u, p) \in \overline{\Omega} \times \mathbb{R} \times \mathbb{R}^n. \tag{1.2}$$

Regarding the functions $a^{ij}(x, u, p)$, $i, j = 1, \ldots, n$ and $a(x, u, p)$ in this chapter it is always assumed that they are at least continuous in $\overline{\Omega} \times \mathbb{R} \times \mathbb{R}^n$. In the investigation of conditions for solvability of the Dirichlet problem for equation (1.1), i.e., the problem

$$\mathscr{L}u = 0 \quad \text{in } \Omega, \qquad u = \varphi \quad \text{on } \partial\Omega, \tag{1.3}$$

where φ is a given function, the first question is naturally that of the admissible structure of this equation, i.e., the question of under what conditions on the functions $a^{ij}(x, u, p)$, $i, j = 1, \ldots, n$, and $a(x, u, p)$ a problem of the type (1.3) has a classical solution in any (at least strictly convex) domain Ω with a sufficiently smooth boundary $\partial\Omega$ and for any sufficiently smooth boundary function φ. Here a classical solution of problem (1.3) is understood to be any function $u \in C^2(\Omega) \cap C(\overline{\Omega})$ satisfying (1.1) in Ω and coinciding with φ on $\partial\Omega$. While the admissible structure of linear elliptic equations is determined mainly by the condition of sufficient smoothness of the coefficients, in the study of quasilinear elliptic equations the foremost conditions are those of admissible growth of $a(x, u, p)$ as $p \to \infty$, depending on the behavior as $p \to \infty$ of certain characteristics of equation (1.1) determined by the leading coefficients $a^{ij}(x, u, p)$, $i, j = 1, \ldots, n$. The first of these characteristics, the function

$$\mathscr{E}_1 \equiv \mathscr{E}_1(x, u, p) = a^{ij}(x, u, p)p_ip_j \equiv Ap \cdot p, \tag{1.4}$$

where $A \equiv \|a^{ij}(x, u, p)\|$, was known as far back as the early work of S. N. Bernstein. The second such characteristic is the function

$$\mathscr{E}_2 \equiv \mathscr{E}_2(x, u, p) = \operatorname{Tr} A|p|. \tag{1.5}$$

The growth of the right side $a(x, u, p)$ of the equation as $p \to \infty$ cannot simultaneously exceed the growth of the functions $\mathscr{E}_1$ and $\mathscr{E}_2$ in the following sense.

THEOREM 1.1 (J. SERRIN). *Let Ω be a bounded domain in $\mathbb{R}^n$ whose boundary $\partial\Omega$ contains at least one point x_0 at which there is tangent a ball K from the exterior of the domain (so that $K \cap \overline{\Omega} = \{x_0\}$). Suppose that*

$$|a(x, u, p)| \geqslant \mathscr{E}_1(x, u, p)\psi(|p|) \quad \textit{for } x \in \overline{\Omega}, |u| \geqslant m, |p| \geqslant l, \tag{1.6}$$

where m and l are positive constants and the function $\psi(\rho)$, $0 \leqslant \rho < +\infty$, satisfies the condition

$$\int^{+\infty} \frac{d\rho}{\rho\psi(\rho)} < +\infty, \tag{1.7}$$

and suppose that

$$\frac{|a(x, u, p)|}{\mathscr{E}_2(x, u, p)} \to +\infty \quad \text{as } p \to \infty \text{ uniformity in } x \in \overline{\Omega},\ u \in \mathbb{R}. \tag{1.8}$$

Then there exists an infinitely differentiable function $\varphi(x)$ *in* $\overline{\Omega}$ *for which the Dirichlet problem* (1.3) *has no classical solution.*

Theorem 1.1 is proved in [163] by means of the technique of global barriers that is developed there.

For concrete elliptic equations of the form (1.1) a decisive role is usually played by one of the functions $\mathscr{E}_1$ or $\mathscr{E}_2$, namely, the one with greater growth as $p \to \infty$. Thus, for uniformly elliptic equations characterized by the condition

$$\Lambda(x, u, p)\lambda^{-1}(x, u, p) \leqslant \text{const}, \qquad x \in \overline{\Omega},\ u \in \mathbb{R},\ p \in \mathbb{R}^n, \tag{1.9}$$

where $\Lambda = \Lambda(x, u, p)$ and $\lambda = \lambda(x, u, p)$ are respectively the greatest and least eigenvalues of the matrix $A(x, u, p)$, the function $\mathscr{E}_1$ always plays the decisive role, since in this case $\mathscr{E}_1 \sim \lambda|p|^2$ and $\mathscr{E}_2 \sim \lambda|p|$ as $p \to \infty$. For the (normalized) equation of a surface of given mean curvature

$$\left(\frac{1 + |\nabla u|^2}{|\nabla u|^2}\delta_i^j - \frac{u_{x_i}u_{x_j}}{|\nabla u|^2}\right)u_{x_ix_j} = n\mathscr{H}(x, u, \nabla u)\frac{(1 + |\nabla u|^2)^{3/2}}{|\nabla u|^2}, \tag{1.10}$$

on the other hand, the decisive role is played by $\mathscr{E}_2$, since in this case $\mathscr{E}_1 \equiv 1$ and $\mathscr{E}_2 > (n - 1)|p|$. Below the functions $\mathscr{E}_1$ and $\mathscr{E}_2$ are also called *majorants*.

The "positive role" of the majorant $\mathscr{E}_1$, consisting in the admissibility, for the solvability of problem (1.3), of growth of $a(x, u, p)$ as $p \to \infty$ no faster than the growth of $\mathscr{E}_1$ as $p \to \infty$ (when, of course, certain conditions of another type are also satisfied), was originally justified in the case $n = 2$ by Bernstein [127] and in the case $n \geqslant 2$ by Ladyzhenskaya and Ural'tseva [77] within the confines of the class of uniformly elliptic equations. These same authors presented examples showing that in a particular sense growth of the right side of the equation as $p \to \infty$ considerably more rapid than the growth of $\mathscr{E}_1$ as $p \to \infty$, generally speaking, is already not admissible for the solvability of the Dirichlet problem even in the case of a strictly convex domain Ω. This is the "negative role" of the majorant $\mathscr{E}_1$. The "positive role" of $\mathscr{E}_1$ was then confirmed in [79], [163], [29], [81], [31], [34], [35] and [83] for various classes of nonuniformly elliptic equations.

Serrin's Theorem 1.1, which was proved after the work of Ladyzhenskaya and Ural'tseva on uniformly elliptic equations, makes precise the fact that within the general framework of nonuniformly elliptic equations the limit of inadmissible growth is already determined by the two functions $\mathscr{E}_1$ and $\mathscr{E}_2$ and shows that the "negative role" of the majorant's $\mathscr{E}_1$ and $\mathscr{E}_2$ has universal character. Apparently, Serrin's paper [163] is the first in which the "negative role" of $\mathscr{E}_2$ is exposed. We remark that in [163] the "positive role" of $\mathscr{E}_2$ was demonstrated only at the stage of obtaining an estimate of $\max_{\partial\Omega}|\nabla u|$, and this for equations of special structure. The "positive role" of $\mathscr{E}_2$ was subsequently justified in the author's papers [29], [31], [34] and [35] for fairly large classes of nonuniformly elliptic equations, and in [165] for equations of special structure.

§2. A conditional existence theorem

We first present two known fundamental results which play a basic role in the reduction of the proof of classical solvability of problem (1.3) to the problem of constructing an a priori estimate of $\max_{\overline{\Omega}}(|u| + |\nabla u|)$ for solutions of a one-parameter family of Dirichlet problems related to problem (1.3). We present these results within a framework sufficient for our purposes in this monograph.

SCHAUDER'S THEOREM. *Suppose that the coefficients of the linear equation*

$$a^{ij}(x)u_{x_ix_j} = a(x) \tag{2.1}$$

belong to the class $C^{l-2+\alpha}(\Omega)$, where $l \geqslant 2$ is an integer, $\alpha \in (0,1)$ and Ω is a bounded domain in $\mathbb{R}^n$, $n \geqslant 2$, and suppose that the following ellipticity condition is satisfied:

$$a^{ij}(x)\xi_i\xi_j \geqslant \nu\xi^2, \qquad \nu = \text{const} > 0, \xi \in \mathbb{R}^n, x \in \overline{\Omega}. \tag{2.2}$$

Then any solution $u \in C^2(\Omega)$ of (2.1) *belongs to the class $C^{l+\alpha}(\overline{\Omega}')$, and*

$$\|u\|_{C^{l+\alpha}(\overline{\Omega}')} \leqslant c_1, \tag{2.3}$$

where $\overline{\Omega}' \subset \overline{\Omega}'' \subset \Omega$ and the constant c_1 depends only on $\|u\|_{C(\overline{\Omega}'')}, n, \nu, \alpha, l$, $\|a^{ij}\|_{C^{l-2+\alpha}(\overline{\Omega})}$, $\|a\|_{C^{l-2+\alpha}(\overline{\Omega})}$, and the distance of Ω' from $\partial\Omega$. If the domain Ω belongs to the class $C^{l+\alpha}$ and the boundary function $\varphi \in C^{l+\alpha}(\overline{\Omega})$, then the Dirichlet problem

$$a^{ij}(x)u_{x_ix_j} = a(x) \quad \text{in } \Omega, \qquad u = \varphi \quad \text{on } \partial\Omega \tag{2.4}$$

has precisely one classical solution u with $u \in C^{l+\alpha}(\overline{\Omega})$, and

$$\|u\|_{C^{l+\alpha}(\overline{\Omega})} \leqslant c_2,$$

where c_2 depends only on n, ν, α, l, $\|a^{ij}\|_{C^{l-2+\alpha}(\overline{\Omega})}$, $\|a\|_{C^{l-2+\alpha}(\overline{\Omega})}$, $\|\varphi\|_{C^{l+\alpha}(\overline{\Omega})}$ and on the $C^{l+\alpha}$-norms of the functions describing the boundary $\partial\Omega$.

Schauder's theorem is a well-known classical result.

THEOREM OF LADYZHENSKAYA AND URAL'TSEVA [83]. *Suppose that a function $u \in C^2(\overline{\Omega})$ satisfying the condition*

$$\max_{\overline{\Omega}}|u| \leqslant m, \quad \max_{\overline{\Omega}}|\nabla u| \leqslant M, \tag{2.5}$$

is a solution of (1.1) *in a bounded domain $\Omega \subset \mathbb{R}^n$, $n \geqslant 2$, and that equation* (1.1) *is elliptic at this solution in the sense that*

$$a^{ij}(x, u(x), \nabla u(x))\xi_i\xi_j \geqslant \nu\xi^2, \qquad \nu = \text{const} > 0, \xi \in \mathbb{R}^n, x \in \overline{\Omega}. \tag{2.6}$$

Suppose that on the set $\mathscr{F}_{\Omega,m,M} \equiv \overline{\Omega} \times \{|u| \leqslant m\} \times \{|p| \leqslant M\}$ the functions $a^{ij}(x, u, p)$, $i, j = 1, \ldots, n$, and $a(x, u, p)$ satisfy the condition

$$|a^{ij}| + \left|\frac{\partial a^{ij}}{\partial x}\right| + \left|\frac{\partial a^{ij}}{\partial u}\right| + \left|\frac{\partial a^{ij}}{\partial p}\right| + |a| \leqslant M_1 \equiv \text{const} > 0 \quad \text{on } \mathscr{F}_{\Omega,m,M}. \tag{2.7}$$

Then there exists a number $\gamma \in (0,1)$, depending only on n, ν, M and M_1, such that for any subdomain Ω', $\overline{\Omega}' \subset \Omega$,

$$\|\nabla u\|_{C^{\gamma}(\overline{\Omega}')} \leqslant c_1, \tag{2.8}$$

where c_1 depends only on n, ν, M, M_1, and the distance from Ω' to $\partial\Omega$. If the domain Ω belongs to the class C^2 and $u = \varphi(x)$ on $\partial\Omega$, where $\varphi \in C^2(\overline{\Omega})$, then

$$\|\nabla u\|_{C^{\gamma}(\overline{\Omega})} \leqslant c_2, \tag{2.9}$$

where c_2 and γ depend on the same quantities as the constants c_1 and γ in (2.8) *and also on $\|\varphi\|_{C^2(\bar{\Omega})}$ and the C^2-norms of the functions describing $\partial\Omega$.*

The following conditional existence theorem for a classical solution of the Dirichlet problem is established by means of the theorems of Schauder and Ladyzhenskaya-Ural′tseva and the familiar topological principle of Leray-Schauder (in Schäffer's form) for the existence of a fixed point of a compact operator in a Banach space.

THEOREM 2.1. *Suppose that the functions $a^{ij}(x, u, p)$, $i, j = 1, \dots, n$, and $a(x, u, p)$ belong to the class $C^2(\bar{\Omega} \times \mathbb{R} \times \mathbb{R}^n)$, where Ω is a bounded domain in $\mathbb{R}^n$, $n \geqslant 2$, and suppose that for any $(x, u, p) \in \bar{\Omega} \times \mathbb{R} \times \mathbb{R}^n$*

$$a^{ij}(x, u, p)\xi_i\xi_j \geqslant \nu|\xi|^2, \qquad \nu = \text{const} > 0, \forall \xi \in \mathbb{R}^n. \tag{2.10}$$

Assume also that the domain Ω belongs to the class C^3 and the function $\varphi \in C^3(\bar{\Omega})$. Finally, suppose that for any solution $v \in C^2(\bar{\Omega})$ of the problem

$$\mathcal{L}_\tau v \equiv a^{ij}(x, v, \nabla v)v_{x_ix_j} - \tau a(x, v, \nabla v) = 0 \quad \text{in } \Omega, \qquad v = \tau\varphi \quad \text{on } \partial\Omega, \tau \in [0, 1] \tag{2.11}$$

there is the estimate

$$\max_{\bar{\Omega}} (|v| + |\nabla v|) \leqslant c_0, \tag{2.12}$$

where $c_0 = \text{const} > 0$ does not depend on either v or τ. Then the Dirichlet problem (1.3) *has at least one classical solution. Moreover, this solution belongs to the class $C^2(\bar{\Omega})$.*

A proof of Theorem 2.1 is given, for example, in [163]. We remark that other, more general one-parameter families of problems can be considered in place of the family of problems (2.11) (see [83] and [163]). Theorem 2.2 determines the program for investigating the solvabilty of the Dirichlet problem for a general elliptic equation of the form (1.1). It reduces to the successive proof of the a priori estimates of $\max_{\bar{\Omega}}|u|$ and $\max_{\bar{\Omega}}|\nabla u|$. The estimate of $\max_{\bar{\Omega}}|\nabla u|$ is usually carried out in two steps: 1) an estimate of $\max_{\partial\Omega}|\nabla u|$ in terms of $\max_{\Omega}|u|$, and 2) an estimate of $\max_{\bar{\Omega}}|\nabla u|$ in terms of $\max_{\partial\Omega}|\nabla u|$ and $\max_{\Omega}|u|$. Many sufficient conditions for obtaining an a priori estimate of $\max_{\Omega}|u|$ are presently known (see [83], [82], [163] and others). In connection with this, in our monograph the estimates of $\max_{\bar{\Omega}}|v|$ for solutions of problems (2.11) which do not depend on τ are usually postulated in formulations of conditions for the solvability of problem (1.3). The subsequent considerations in Part I are mainly devoted to constructing various methods of estimating $\max_{\bar{\Omega}}|\nabla u|$.

§3. Some facts about the barrier technique

LEMMA 3.1 (SERRIN). *Suppose that in a bounded domain $\Omega \subset \mathbb{R}^n$, $n \geqslant 2$, a function $u \in C^2(\Omega) \cap C^1(\bar{\Omega})$ satisfies* (1.1), *where it is assumed that condition* (1.2) *is satisfied and that $a^{ij}(x, u, p)$, $i, j = 1, \dots, n$, and $a(x, u, p)$ are continuous functions of their arguments in $\bar{\Omega} \times \mathbb{R} \times \mathbb{R}^n$. Suppose that for any constant $c \geqslant 0$ the* (*barrier*) *function $\omega \in C^2(\Omega) \cap C^1(\bar{\Omega})$ satisfies in Ω the inequality*

$$\mathcal{L}(\omega + c) \equiv a^{ij}(x, \omega + c, \nabla\omega)\omega_{x_ix_j} - a(x, \omega + c, \nabla\omega) \leqslant 0. \tag{3.1}$$

If $u \leqslant \omega$ on $\partial\Omega$, then $u \leqslant \omega$ throughout $\bar{\Omega}$.

Lemma 3.1 is proved in [163]. The proof is based on applying the strong maximum principle for linear elliptic equations.

Suppose that Ω belongs to the class C^3. In a subdomain $D_0 \subset \Omega$ abutting $\partial\Omega$ it is possible to define a function $d = d(x)$ as the distance from the point $x \in D_0$ to $\partial\Omega$ (i.e., $d(x) = \operatorname{dist}(x, \partial\Omega)$). The domain D_0 is characterized by the condition

$$D_0 \equiv \{x \in \Omega\colon d(x) < \delta_0\}, \tag{3.2}$$

and the number $\delta_0 > 0$ is determined solely by the boundary $\partial\Omega$. We shall henceforth always asume that $\delta_0 \leqslant K^{-1}$, where K is the supremum of the absolute values of the normal curvatures on $\partial\Omega$. It is proved in [163] that if this condition is satisfied the function $d(x)$ belongs to the class $C^2(D_0)$.

LEMMA 3.2 (SERRIN). *Let $\Omega \subset \mathbb{R}^n$, $n \geqslant 2$, be a bounded domain of class C^3, and let*

$$\begin{gathered} \omega(x) = \varphi(x) + h(d), \qquad d = d(x) \equiv \operatorname{dist}(x, \partial\Omega), \\ x \in D \equiv \{x \in \Omega\colon d(x) < \delta\}, \end{gathered} \tag{3.3}$$

where $\varphi \in C^3(\overline{D})$, $h(d) \in C^2((0, \delta)) \cap C([0, \delta])$, $0 < \delta < \delta_0$, δ_0 is the number from condition (3.2), *and $h'(d) > 0$ on* $[0, \delta]$. *In the domain D the expression*

$$\mathscr{L}(\omega + c) \equiv a^{ij}(x, \omega + c, \nabla\omega)\omega_{x_ix_j} - a(x, \omega + c, \nabla\omega),$$

$c = \text{const} > 0$, *can then be bounded above by the expression*

$$\mathscr{F}(h''/h'^2) + K\operatorname{Tr} Ah' + a^{ij}\varphi_{x_ix_j} - a, \tag{3.4}$$

where $\mathscr{F} = A(p - p_0)\cdot(p - p_0)$, $A \equiv \|a^{ij}(x, \omega(x) + c, \nabla\omega(x))\|$, $p = \nabla\omega(x)$, $p_0 = \nabla\varphi(x)$, $p = p_0 + \nu h'$, *ν is the unit inner normal to $\partial\Omega$ at the point $y = y(x) \in \partial\Omega$ closest to x on $\partial\Omega$*, $h' = h'(d(x))$, $K = \sup_{i=1,\dots,n-1, y\in\partial\Omega}|k_i(y)|$, *the* $k_i(y)$, $i = 1, \dots, n-1$, *are the principal curvatures of the surface $\partial\Omega$ at the point y, and* $a = a(x, \omega(x) + c, \nabla\omega(x))$. *If the domain Ω is convex, then the expression* $\mathscr{L}(\omega + c)$, $c = \text{const} > 0$, *can also be bounded above in the domain D by the expression*

$$\mathscr{F}((h'' + Kh')/h'^2) - k\operatorname{Tr} Ah' + a^{ij}\varphi_{x_ix_j} - a, \tag{3.5}$$

where

$$k = \inf_{i=1,\dots,n-1,\, y\in\partial\Omega} k_i(y) \quad \textit{and} \quad k \geqslant 0.$$

PROOF. The results of Lemma 3.2 follow in an obvious way from the formulas obtained in [163] (see pp. 422, 423).

We now present a simple proposition related to the investigation of a single ordinary differential equation.

LEMMA 3.3. *Suppose a positive, continuous function $\Phi(\rho)$, $0 \leqslant \rho < +\infty$, satisfies the condition*

$$\int^{+\infty} \frac{d\rho}{\rho^2\Phi(\rho)} = +\infty,$$

and suppose the number $\delta_0 > 0$ is fixed. Then for an $q > 0$ and $\alpha > 0$ there exist a number $\delta \in (0, \delta_0)$ and a function $h = h(d) \in C^2((0, \delta)) \cap C([0, \delta])$ satisfying the *conditions*

$$h''/h'^3 + \Phi(h') = 0 \quad on\ (0, \delta),$$
$$h(0) = 0, \quad h(\delta) = q, \quad h'(d) \geqslant \alpha \quad on\ (0, \delta), \tag{3.6}$$

where δ depends only on q, α, and $\Phi(\rho)$.

PROOF. Because of the first condition, there exists a number β such that

$$q = \int_{\bar{\alpha}}^{\beta} \frac{d\rho}{\rho^2 \Phi(\rho)}, \tag{3.7}$$

where $\bar{\alpha} = \max(\alpha, q\delta_0^{\delta - 1})$, so that $\beta > \bar{\alpha} \geqslant \alpha$. Let

$$\delta = \int_{\bar{\alpha}}^{\beta} \frac{d\rho}{\rho^3 \Phi(\rho)}. \tag{3.8}$$

Obviously, $\delta \leqslant q/\bar{\alpha} \leqslant \delta_0$. We now define $h(d)$ on $[0, \delta]$ by the following parametric equations:

$$h = \int_{\rho}^{\beta} \frac{d\rho}{\rho^2 \Phi(\rho)}, \quad d = \int_{\rho}^{\beta} \frac{d\rho}{\rho^3 \Phi(\rho)}, \qquad \bar{\alpha} \leqslant \rho \leqslant \beta. \tag{3.9}$$

It is easy to see by direct verification that $h''/h'^3 + \Phi(h') = 0$ on $(0, \delta)$. It is also obvious that $h' = \rho \geqslant \bar{\alpha} \geqslant \alpha$ on $[0, \delta]$, $h(0) = 0$, and $h(\delta) = q$. The lemma is proved.

§4. Estimates of $|\nabla u|$ on the boundary $\partial\Omega$ by means of global barriers

In this section estimates of the normal derivative of a solution of problem (4.3), and thereby of the entire gradient of this solution on $\partial\Omega$, are established by means of the technique of global barriers developed by Serrin. The results presented here are a modification of the corresponding results of [163]. Below D_0 denotes the subdomain of Ω defined by (3.2). We also assume that condition (1.2) is satisfied for equation (1.1).

THEOREM 4.1. *Suppose that on the set $\mathfrak{N}_{m,l} \equiv \overline{D}_0 \times \{|u| \leqslant m\} \times \{|p| > l\}$ (m and l are nonnegative constants) the functions a^{ij}, $\partial a^{ij}/\partial p_k$, a and $\partial a/\partial p_k$, $i, j, k = 1, \ldots, n$, are continuous and satisfy the condition*

$$|a(x, u, p)| \leqslant \psi(|p|)\mathscr{E}_1(x, u, p) + \delta(|p|)\mathscr{E}_2(x, u, p), \tag{4.1}$$

where $\mathscr{E}_1$ and $\mathscr{E}_2$ are defined by (1.4) and (1.5), $\psi(\rho)$, $0 \leqslant \rho < +\infty$, is a positive, monotone, continuous function such that $\lim_{\rho \to +\infty} \psi(\rho)/\rho = 0$ and for all $c = \text{const} \geqslant 0$ the function $\rho\psi(\rho \pm c)$ is montonic and

$$\int^{+\infty} \frac{d\rho}{\rho\psi(\rho \pm c)} = +\infty, \tag{4.2}$$

and $\delta(\rho)$, $0 < \rho < +\infty$, is a nonnegative, nonincreasing function such that

$$\lim_{\rho \to \infty} \delta(\rho) = 0. \tag{4.3}$$

Let $u \in C^2(D_0) \cap C^1(\overline{D}_0)$ be an arbitrary function satisfying (1.1) in the domain D_0 which coincides on $\partial\Omega$ with a function $\varphi \in C^3(\overline{D}_0)$ and is such that $|u(x)| \leqslant m$ in D_0. Suppose also that the domain Ω is strictly convex and belongs to the class C^3. Then

$$|\partial u(y)/\partial \nu| \leqslant M_0, \qquad y \in \partial\Omega, \tag{4.4}$$

where $\partial u/\partial\nu$ is the derivative in the direction of the inner normal to $\partial\Omega$ at the point $y \in \partial\Omega$ and M_0 depends only on m, l, the number δ_0 from condition (3.2), *the functions $\psi(\rho)$ and $\delta(\rho)$ from* (4.1), $\|\varphi\|_{C^2(\bar{D}^0)}$, *and also on k^{-1} and K where*

$$k = \inf_{i=1,\dots,n-1,\, y\in\partial\Omega} k_i(y)$$

and the $k_i(y)$, $i = 1,\dots,n-1$, are the principal curvatures of $\partial\Omega$ at $y \in \partial\Omega$. If it is additionally assumed that on the set $\mathfrak{N}_{m,l}$

$$\psi(|p|)\mathscr{E}_1(x, u, p) \geqslant \mathscr{E}_2(x, u, p), \tag{4.5}$$

then the estimate (4.4) *holds without the assumption of strict convexity of Ω. In this case the constant M_0 in* (4.4) *depends on the same quantities as previously with the exception of k^{-1}.*([1])

PROOF. We first assume that conditions (4.1) and (4.5) are satisfied for any u (more precisely, on the set $\mathfrak{N}_{m,l} \equiv \bar{D}_0 \times \mathbb{R} \times \{|p| > l\}$). This assumption will be eliminated at the end of the proof. We first prove the first part of the theorem assuming that there is no condition (4.5) Suppose that the function ω is defined by (3.3) where $\varphi(x)$ is the function in the hypotheses of the theorem; the choice of the function $h(d) \in C^2((0, \delta)) \cap C([0, \delta])$, $0 < \delta \leqslant \delta_0$, and, in particular, of the number δ characterizing the domain of $h(d)$ we specify below. Applying Lemma 3.2, we obtain for any constant c the inequality

$$\mathscr{L}(\omega + c) \leqslant \mathscr{F}\left((h'' + Kh')/h'^2\right) - k\operatorname{Tr} A h' + a^{ij}\varphi_{x_i x_j} - a, \quad x \in D, \tag{4.6}$$

where in correspondence with the notation adopted in Lemma 3.2 $\mathscr{F} = A(p - p_0)\cdot(p - p_0)$, $A = \|a^{ij}(x, \omega(x) + c, \nabla\omega(x))\|$, $p = \nabla\omega(x)$, $p_0 = \nabla\varphi(x)$, $p = p_0 + \nu h'$, $K = \sup_{i=1,\dots,n-1}|k_i(y)|$ and $a = a(x, \omega(x) + c, p)$. We suppose that $h'(d) \geqslant c_\varphi + l + 1$ on $[0, \delta]$, where

$$c_\varphi \equiv \|\varphi\|_{C(\bar{D})} + \sum_{i=1}^{n} \|\varphi_{x_i}\|_{C(\bar{D})} + \sum_{i,j=1}^{n} \|\varphi_{x_i x_j}\|_{C(\bar{D})}.$$

Then $|p| > l$. Applying (4.1), estimating $|a^{ij}\varphi_{x_i x_j}| \leqslant c_\varphi \operatorname{Tr} A$, and taking also into account that $h' - c_\varphi \leqslant |p| \leqslant h' + c_\varphi \leqslant 2h'$, we obtain the estimate

$$a^{ij}\varphi_{x_i x_j} - a \leqslant c_\varphi \operatorname{Tr} A + \psi(h' \pm c_\varphi)\mathscr{E}_1 + 2h'\delta(h' - c_\varphi)\operatorname{Tr} A, \qquad x \in D. \tag{4.7}$$

We shall now prove that

$$\mathscr{E}_1 \leqslant 2\mathscr{F} + 4c_\varphi^2 \operatorname{Tr} A \quad \text{on } D. \tag{4.8}$$

Indeed, noting that $|Ap\cdot p_0| \leqslant |Ap||p_0| \leqslant c_\varphi|Ap|$ and $|Ap| \leqslant (\operatorname{Tr} A)^{1/2}(Ap\cdot p)^{1/2}$, we obtain the inequality

$$|Ap\cdot p_0| \leqslant c_\varphi(\operatorname{Tr} A)^{1/2}(Ap\cdot p)^{1/2}.$$

Then

$$\begin{aligned}\mathscr{F} &= Ap\cdot p - 2Ap\cdot p_0 + Ap_0\cdot p_0 \geqslant Ap\cdot p - 2Ap\cdot p_0 \\ &\geqslant Ap\cdot p - 2c_\varphi(\operatorname{Tr} A)^{1/2}(Ap\cdot p)^{1/2} \geqslant \tfrac{1}{2}\mathscr{E}_1 - 2c_\varphi^2 \operatorname{Tr} A,\end{aligned}$$

([1]) The second part of Theorem 4.1 is a result of Serrin [163] (see pp. 432 and 433).

whence we obtain (4.8). It follows from (4.6)–(4.8) that for the function ω on the set D we have

$$\mathscr{L}(\omega + c) \leqslant \mathscr{F} h' \left\{ \frac{h''}{h'^3} + \frac{K}{h'^2} + \frac{2\psi(h' \pm c_\varphi)}{h'} \right\}$$
$$+ h' \left\{ -k + \frac{c_\varphi}{h'} + \frac{4c_\varphi^2 \psi(h' \pm c_\varphi)}{h'} + 2\delta(h' - c_\varphi) \right\} \operatorname{Tr} A, \tag{4.9}$$

where $c = \text{const} > 0$.

Because of conditions (4.2) and (4.3),

$$\frac{c_\varphi}{\rho} + \frac{4c_\varphi^2 \psi(\rho \pm c_\varphi)}{\rho} + 2\delta(\rho - c_\varphi) \to 0$$

as $\rho \to +\infty$. Therefore, assuming that $h' \geqslant \alpha_0$, where $\alpha_0 > 0$ is a sufficiently large number depending only on k, c_φ, $\psi(\rho)$, and $\delta(\rho)$, we can relax inequality (4.9) by dropping the nonpositive second pair of braces in it. Then

$$\mathscr{L}(\omega + c) \leqslant \mathscr{F} h' \{ h''/h'^3 + \Phi(h') \}, \qquad x \in D, \tag{4.10}$$

where

$$\Phi(\rho) = K/\rho^2 + 2\psi(\rho \pm c_\varphi)/\rho \quad \text{and} \quad \int^{+\infty} \frac{d\rho}{\rho^2 \Phi(\rho)} = +\infty.$$

We shall now specify the choice of the function $h(d)$ and its domain. Let δ be defined by (3.8) for $\bar{\alpha} = \max(\alpha, q\delta_0^{-1})$, $q = c_\varphi + m$, and $\alpha = \max(\alpha_0, c_\varphi + l + 1)$, and suppose that on $[0, \delta]$ the function $h(d)$ is defined by (3.9). From Lemma 3.3 it then follows that $\delta \in (0, \delta_0)$ and

$$\mathscr{L}(\omega + c) \leqslant 0, \qquad x \in D. \tag{4.11}$$

Moreover, $u \leqslant \omega$ on ∂D. Indeed, since $h(0) = 0$, it follows that $\omega = \varphi = u$ on $\partial\Omega$. On the set $\{x \in \Omega : d(x) = \delta\}$, however, the inequalities $u \leqslant m = (m + c_\varphi) - c_\varphi = h(\delta) - c_\varphi \leqslant \omega$ hold. It then follows from Lemma 3.1 that $u \leqslant \omega$ in D. In view of the fact that $u = \omega$ on $\partial\Omega$, from the last inequality we obviously obtain

$$\partial u/\partial \nu \leqslant |\partial \omega/\partial \nu| \quad \text{on } \partial\Omega. \tag{4.12}$$

Since the function $\tilde{u} \equiv -u$ is a solution of an equation having precisely the same structure as the original equation, from what has been proved we also obtain

$$-\partial u/\partial \nu \leqslant |\partial \omega/\partial \nu| \quad \text{on } \partial\Omega. \tag{4.13}$$

The estimate (4.4) obviously follows from (4.12) and (4.13).

We now proceed to consider the second part of the theorem (i.e., we assume that both conditions (4.1) and (4.5) are satisfied). This part of the theorem was proved by Serrin [163]. For completeness we repeat Serrin's proof here. Because of condition (4.5) and the fact that $|p|\psi^{-1}(|p|) \to \infty$, there exists a constant α_1, depending only on $\psi(\rho)$, such that

$$\mathscr{E}_1 \geqslant 8c_\varphi^2 \operatorname{Tr} A \quad \text{on } \mathfrak{N}_{\infty, \max(l, \alpha_1)}. \tag{4.14}$$

Let $h' \geqslant \alpha = c_\varphi + l + \alpha_1 + 1$. Then $|p| > l$, and hence from (4.8) and (4.14) we obtain

$$\mathscr{F} \geqslant \tfrac{1}{2}\mathscr{E}_1 - 2c_\varphi^2 \operatorname{Tr} A \geqslant \tfrac{1}{4}\mathscr{E}_1 \quad \text{on } D. \tag{4.15}$$

Estimating $\mathscr{L}(\omega + c)$ above by (3.4) and taking into account that under condition (4.5) it is possible to set $\delta(\rho) \equiv 0$ in (4.1) with no loss of generality, we obtain

$$\mathscr{L}(\omega + c) \leqslant \mathscr{F}h'\left\{\frac{h''}{h'^3} + \frac{(K + c_\varphi)\operatorname{Tr} A}{\mathscr{F}} + \frac{\psi(|p|)\mathscr{E}_1}{h'\mathscr{F}}\right\}, \qquad x \in D. \quad (4.16)$$

Recalling (4.5), (4.15) and also the inequalities $h' - c_\varphi \leqslant |p| \leqslant h' + c_\varphi$, from (4.16) we deduce the inequality

$$\mathscr{L}(\omega + c) \leqslant \mathscr{F}h'\{h''/h'^3 + \Phi(h')\}, \qquad x \in D, \qquad (4.17)$$

where

$$\Phi(\rho) = 4(c_\varphi + K + 1)\big(\psi(\rho \pm c_\varphi)/(\rho - c_\varphi)\big).$$

It is obvious that $\int^{+\infty}(d\rho/\rho^2\Phi(\rho)) = +\infty$. Since (4.17) coincides in form with (4.10), the remainder of the proof does not differ from the proof of the preceding case. Thus, an estimate of the form (4.4) has also been established in this case.

We shall now eliminate the assumption that conditions (4.1) and (4.5) are satisfied for all u. Suppose that these conditions are satisfied on the set $\mathfrak{N}_{m,l}$, where $m \geqslant \max_{D_0}|u|$. We consider a new equation of the form (1.1) with a matrix of leading coefficients defined by

$$\hat{A}(x, u, p) = \begin{cases} A(x, -m, p) & \text{for } u < -m, \\ A(x, u, p) & \text{for } -m \leqslant u \leqslant m, \\ A(x, m, p) & \text{for } u > m, \end{cases} \qquad (4.18)$$

and with a similarly defined lower-order term $\hat{a}(x, u, p)$. It is obvious that in D_0 the function u also satisfies the new equation $\hat{a}^{ij}u_{x_ix_j} - \hat{a} = 0$ for which conditions of the form (4.1) and (4.5) are satisfied for all $u \in \mathbb{R}$. The validity of an estimate of the form (4.4) then follows from what has been proved above. Theorem 4.1 is proved.

It is useful to record also the following version of Theorem 4.1.

THEOREM 4.1′. *Suppose that the functions a^{ij}, $\partial a^{ij}/\partial p_k$, a and $\partial a/\partial p_k$, $i, j, k = 1, \ldots, n$, are continuous on the set $\mathfrak{N}_{m,l}$, and suppose that for all $x \in D_0$, all $u \in [-m, m]$ and any $\rho \geqslant l = \mathrm{const} \geqslant 0$*

$$|a(x, u, p\boldsymbol{\nu})| \leqslant \psi(\rho)\mathscr{E}_1(x, u, \rho\boldsymbol{\nu}) + \delta(\rho)\operatorname{Tr} A(x, u, \rho\boldsymbol{\nu})\rho, \qquad (4.19)$$

where $A = \|a^{ij}\|$, $\boldsymbol{\nu} = \boldsymbol{\nu}(y(x))$ is the unit inner normal to $\partial\Omega$ at the point $y(x)$ closest to $x \in D_0$ on $\partial\Omega$, and the functions $\mathscr{E}_1$, ψ, and δ are the same as in Theorem 4.1. Suppose that a function $u \in C^2(D_0) \cap C(\overline{D}_0)$ satisfies (1.1) in D_0, is equal to 0 on $\partial\Omega$ (i.e., $\varphi \equiv 0$), and $|u(x)| \leqslant m$ in $\overline{D}_0$. Assume that the domain Ω is strictly convex and belongs to the class C^3. Then the estimate (4.4) holds, where the constant M_0 depends only on m, l, δ_0, $\psi(\rho)$, $\delta(\rho)$, k^{-1}, and K. If it is additionally assumed that for all $x \in \overline{D}_0$, $u \in [-m, m]$ and $\rho \geqslant l$

$$\psi(\rho)\mathscr{E}_1(x, u, \rho\boldsymbol{\nu}) \geqslant \operatorname{Tr} A(x, u, \rho\boldsymbol{\nu})\rho, \qquad (4.20)$$

then (4.4) holds without the assumption of strict convexity of Ω. In this case the constant M_0 does not depend on k.

PROOF. This theorem follows directly from the proof of Theorem 4.1.

REMARK 4.1. In [163] certain classes of nonuniformly elliptic equations of the form (4.1) are distinguished which are beyond the framework of condition (4.5) but for which an estimate of the form (4.4) can nevertheless be established not only for convex domains. Here, however, conditions arise on the curvature of the boundary surface which depend on the behavior of the right side of the equation. It is shown in [163] that these conditions are inherent in the nature of the problem.

§5. Estimates of $|\nabla u|$ on the boundary by means of local barriers

In obtaining an estimate of $|\nabla u|$ at a fixed point $y_0 \in \partial\Omega$ it is not always expedient to impose conditions on the entire boundary $\partial\Omega$ and the entire boundary function φ as must be done in using the method of global barriers presented in §4. In the present section the estimate of $|\nabla u(y_0)|$ is based on the construction of local barriers which cause constraints only on the part of the boundary near the point y_0 and on the restriction of φ to a neighborhood of y_0. To construct local barriers we use certain methods characteristic of Serrin's technique of global barriers, and also constructions applied by Ladyzhenskaya and Ural'tseva [83]. The results obtained here are a strengthening (for the case of nonuniformly elliptic equations) of the corresponding results of [83] regarding local estimates of $|\nabla u|$ on the boundary of the domain.

Let Ω be a bounded domain in $\mathbb{R}^n$, $n \geqslant 2$, with boundary $\partial\Omega$. We consider an open part Γ of $\partial\Omega$ containing the point $y_0 \in \partial\Omega$. We assume that Γ belongs to the class C^3. Suppose there exists a number $\delta_0 > 0$ such that for each point x in the domain

$$D_\Gamma^0 \equiv \{x \in \Omega : x = y + \tau\nu(y),\ y \in \Gamma,\ \tau \in (0, \delta_0)\} \tag{5.1}$$

there is a unique point $y = y(x) \in \Gamma$ such that $\operatorname{dist}(x, \Gamma) = \operatorname{dist}(x, \partial\Omega) = \operatorname{dist}(x, y(x))$. It can be proved precisely as in the case $\Gamma = \partial\Omega$ that the function $d = d(x) = \operatorname{dist}(x, \Gamma)$ defined on D_Γ^0 belongs to the class C^2 if δ_0 is sufficiently small (see [163]).

Let $K_r(y_0)$ be a ball of radius $r > 0$ with center at y_0. We set

$$\Omega_r = K_r(y_0) \cap \Omega, \quad S_r = K_r(y_0) \cap \Gamma, \tag{5.2}$$

where the number $r > 0$ is assumed to be so small that $K_r(y_0) \cap \Gamma = K_r(y_0) \cap \partial\Omega$ and $\Omega_r \subset D_\Gamma^0$. We also suppose that condition (1.2) is satisfied for equation (1.1).

THEOREM 5.1. *Suppose that on the set* $\mathfrak{N}_{\Gamma,m,l} \equiv \overline{D_\Gamma^0} \times \{|u| \leqslant m\} \times \{|p| > l\}$ *the functions* a^{ij}, $\partial a^{ij}/\partial p_k$, a *and* $\partial a/\partial p_k$, $i, j, k = 1, \dots, n$, *are continuous and satisfy inequality* (4.1), *where* $\mathscr{E}_1$, $\mathscr{E}_2$, $\psi(\rho)$ *and* $\delta(\rho)$ *are the same as in Theorem* 4.1. *Suppose that the function u satisfies the conditions*

$$\begin{gathered} u \in C^2(\Omega_r) \cap C^1(\overline{\Omega}_r), \qquad |u| \leqslant m \quad \text{in } \Omega_r, \\ \mathscr{L}u \equiv a^{ij}(x, u, \nabla u)u_{x_ix_j} - a(x, u, \nabla u) = 0 \quad \text{in } \Omega_r, \\ u = \varphi \quad \text{on } S_r, \qquad \varphi \in C^3(\Omega_r). \end{gathered} \tag{5.3}$$

Suppose also that the following conditions are satisfied:

Ω_r is contained in a ball $K_R(x_0)$ of radius $R > 0$ with center at the point x_0 lying on the axis defined by the vector $\mathbf{v}$ of the inner normal to $\partial\Omega$ at the point y_0, where $K_R(x_0)$ is tangent to $\partial\Omega$ at y_0, (5.4)

and

$$k = \inf_{i=1,\dots,n-1,\, y\in\Gamma} k_i(y) > 0, \tag{5.5}$$

where $k_1(y),\dots,k_{n-1}(y)$ *are the principal curvatures of the surface* $\partial\Omega$ *at the point* $y \in \Gamma$.[2] *Then*

$$|\partial u(y_0)/\partial\nu| \leqslant M_0, \tag{5.6}$$

where M_0 *depends only on* m, l, $\|\varphi\|_{C^2(\Omega_r)}$, $K \equiv \sup_{y\in\Gamma, i=1,\dots,n-1} k_i(y)$, k^{-1}, Rr^{-2}, *the surface* Γ, *and also on the functions* $\psi(\rho)$ *and* $\delta(\rho)$ *from condition* (4.1). *If it is additionally assumed that condition* (4.5) *is satisfied on the set* $\mathfrak{N}_{\Gamma,m,l}$, *then the estimate* (5.6) *remains valid if in place of* (5.4) *and* (5.5) *the following condition is satisfied*:

> *there exists an open ball* $K_R(x_*)$ *of radius* $R > 0$ *with center at the point* x_* *which has no common points with* Ω_r *and contains the point* y_0 *on its boundary*.[3] (5.7)

In this case the constant M_0 *in* (5.6) *depends only on* m, l, $\|\varphi\|_{C^2(\Omega_r)}$, K, R, r^{-1}, Γ *and the function* $\psi(\rho)$.

PROOF. We assume with no loss of generality that (4.1) and (4.5) are valid on $\mathfrak{N}_{\Gamma,\infty,l}$ (i.e., for any values of the variable u; see the end of the proof of Theorem 4.1). We set $D_\Gamma \equiv \{x \in D_\Gamma^0 : d(x) < \delta\}$, where $0 < \delta < \delta_0$ and δ_0 is the number in (5.1). On D_Γ we consider the function

$$\omega(x) = f(x) + h(d(x)), \qquad f(x) = \varphi(x) + \mu\rho(x), \quad \mu = \text{const} > 0, \tag{5.8}$$

where $\rho(x)$ is defined on D_Γ by the formula $\rho(x) = \operatorname{dist}(x, P_{y_0})$ and P_{y_0} denotes the tangent plane to $\partial\Omega$ at y_0; $h(d) \in C^2((0,\delta)) \cap C([0,\delta])$, and $d(x) = \operatorname{dist}(x,\Gamma)$. The number $\delta \in (0,\delta_0)$, the function $h(d)$, $d \in [0,\delta]$, and also the constant $\mu > 0$ will be chosen below. Since Lemma 3.2 has local character, its results can also be used in the case of local barriers of the form (5.8) (in other words, the estimates of $\mathscr{L}(\omega + c)$ in terms of the expression (3.5) or (3.4) are valid under the conditions of Theorem 5.1 on the set D_Γ). We first prove the first part of the theorem assuming that there is no condition (4.5). In view of Lemma 3.2, for any constant $c \geqslant 0$ we have

$$\mathscr{L}(\omega + c) \leqslant \mathscr{F}((h'' + Kh')/h'^2) - k\operatorname{Tr} Ah' + a^{ij}f_{x_ix_j} - a, \qquad x \in D_\Gamma, \tag{5.9}$$

where the notation adopted in Lemma 3.2 has been used. Because of the linearity of the function $\rho(x)$ on D_Γ we have $a^{ij}\rho_{x_ix_j} = 0$, and hence (5.9) coincides in form with (4.6) although the arguments of the functions a^{ij}, a and $\mathscr{F}$ in (5.9) and (4.6) are different, since $f(x) = \varphi(x) + \mu\rho(x)$.

Suppose the condition $h'(d) \geqslant c_\varphi + \mu + l + 1$ is satisfied on $[0,\delta]$, where $c_\varphi \equiv \|\varphi\|_{C^2(D_\Gamma)}$. From this condition we obtain the inequality $h'(d) \geqslant \max_{D_\Gamma}|\nabla f| + l + 1$ on $[0,\delta]$, since $|\nabla f| \leqslant |\nabla\varphi| + \mu \leqslant c_\varphi + \mu$ on D_Γ in view of the fact that $|\nabla\rho| = 1$. For $p \equiv |\nabla\omega(x)|$ the condition $|p| > l$ is then satisfied. Applying precisely the same

---

(2) It is not hard to see that condition (5.4) follows from (5.5). However, for the proof of the theorem it is convenient to write (5.4) separately.

(3) It is obvious that condition (5.7) is more general than (5.4) (i.e., (5.4) implies (5.7)).

arguments as in the proof of Theorem 4.1 with c_φ replaced by $c_\varphi + \mu$, we establish the inequality

$$\mathscr{L}(\omega + c) \leqslant \mathscr{F}h'\{h''/h'^3 + \Phi(h')\}, \qquad x \in D_\Gamma, \tag{5.10}$$

where

$$\Phi(\rho) = K/\rho^2 + 2\psi(\rho \pm c_\varphi \pm \mu)/\rho$$

(it is obvious that $\int^{+\infty}(d\rho/\rho^2\Phi(\rho)) = +\infty$).

We now specify the choice of $h(d)$ and $\delta \in (0, \delta_0)$. Let δ be defined by (3.8) with $\bar{\alpha} = \max(\alpha, q\delta_0^{-1})$, $q = c_\varphi + m$ and $\alpha = \max(\alpha_0, c_\varphi + \mu + l + 1)$, where α_0 depends only on k, c_φ, μ, and the functions $\psi(\rho)$ and $\delta(\rho)$ and is such that the second pair of braces in (4.9) is negative. Let $h(d)$ be defined on $[0, \delta]$ by (3.9) so that $h(0) = 0$, $h(\delta) = c_\varphi + m$ and $\alpha \leqslant h' \leqslant \beta$ on $[0, \delta]$, where β is determined from (3.7) for the values of $\bar{\alpha}$ and q indicated above. From Lemma 3.3 it then follows that

$$\mathscr{L}(\omega + c) \leqslant 0, \qquad x \in D_\Gamma \cap \Omega_r. \tag{5.11}$$

We shall prove that for a suitable choice of the constant $\mu > 0$ the inequality $u \leqslant \omega$ holds on the boundary $\partial(D_\Gamma \cap \Omega_r)$. It is obvious that $\partial(D_\Gamma \cap \Omega_r) = S_r \cup S_r' \cup S_r''$, where $S_r' = \Omega_r \cap \{x \in D_\Gamma\colon d(x) = \delta\}$, $S_r'' = \partial(D_\Gamma \cap \Omega_r)\backslash(S_r \cup S_r')$, and the set S_r'' is that part of the boundary surface of the ball K_r cut out by surfaces Γ and $\{x \in D_\Gamma\colon d(x) = \delta\}$. Indeed, taking into account the form of the function ω and the properties of $h(d)$ and $\rho(x)$, it is easy to see that $u \leqslant \omega$ on $S_r \cup S_r'$. To prove the inequality on S_r'' we note that from geometric considerations we have

$$\inf_{\partial\Omega_r\backslash S_r} \rho(x) \geqslant r^2/2R, \tag{5.12}$$

where R is the number in condition (5.4), and we choose $\mu = (c_\varphi + m)(2R/r^2)$. Then on S_r'' we have

$$\begin{aligned}\omega(x) &\geqslant -\max_{\Omega_r}|\varphi| + (c_\varphi + m)\frac{2R}{r^2}\inf_{S_r''}\rho(x)\\ &\geqslant -c_\varphi + (c_\varphi + m)\frac{2R}{r^2}\inf_{\partial\Omega_r\backslash S_r}\rho(x) \geqslant m \geqslant u(x).\end{aligned}$$

Thus,

$$\mathscr{L}(\omega + c) \leqslant 0 \quad \text{in } D_\Gamma \cap \Omega_r, \qquad u \leqslant \omega \quad \text{on } \partial(D_\Gamma \cap \Omega_r). \tag{5.13}$$

Applying Lemma 3.1, we deduce from (5.13) that $u \leqslant \omega$ throughout $\Omega_r \cap D_\Gamma$. Taking into account that $u(y_0) = \omega(y_0) = \varphi(y_0)$, we obtain

$$\partial u(y_0)/\partial\nu \leqslant |\partial\omega(y_0)/\partial\nu| \leqslant \beta + c_\varphi + \mu.$$

An upper bound for $-\partial u(y_0)/\partial\nu$ is obtained in a similar way,so that $|\partial u(y_0)/\partial\nu| \leqslant \beta + c_\varphi + \mu$, whence the first part of the theorem follows.

We now prove the second part of the theorem. In this case the proof of (5.6) proceeds in two steps. At the first step we establish (5.6) by replacing condition (5.7) by the stronger condition (5.4). At the second step the assumption regarding this replacement of conditions is eliminated. Thus, we first assume that conditions (4.1), (4.5), (5.3), and (5.4) are satisfied. In view of (4.5) we shall assume with no loss of generality that in condition (4.1) $\delta(\rho) \equiv 0$. On D_Γ we again consider the function

$\omega(x)$ defined by (5.8). Using the upper bound for $\mathscr{L}(\omega + c)$ in terms of an expression of the form (3.4) on the basis of Lemma 3.2 and taking also into account that $a^{ij}\rho_{x_i x_j} = 0$ on D_Γ, we obtain

$$\mathscr{L}(\omega + c) \leqslant \mathscr{F}h''/h'^2 + K\operatorname{Tr} Ah' + a^{ij}\varphi_{x_i x_j} - a, \qquad x \in D_\Gamma. \tag{5.14}$$

Assuming that $h' \geqslant \alpha = c_\varphi + \mu + l + \alpha_1 + 1$, where $\alpha_1 > 0$ is determined by (4.14) and depends only on the function $\psi(\rho)$ just as in the proof of the second part of Theorem 4.1, we establish that

$$\mathscr{L}(\omega + c) \leqslant \mathscr{F}h'\{h''/h'^3 + \Phi(h')\}, \qquad x \in D_\Gamma, \tag{5.15}$$

where

$$\Phi(\rho) = 4(c_\varphi + \mu + K + 1)\frac{\psi(\rho \pm c_\varphi \pm \mu)}{\rho - c_\varphi - \mu}$$

satisfies the condition $\int^{+\infty}(d\rho/\rho^2\Phi(\rho)) = +\infty$.

We define the number δ and the function $h(d)$ by the same formulas as in the proof of the first part of the theorem. The following conditions are then satisfied: $\mathscr{L}(\omega + c) \leqslant 0$ in $D_\Gamma \cap \Omega_r$ and $u \leqslant \omega$ on $S_r \cup S_r'$. Choosing μ in exactly the same way as in the proof of the first part of the theorem and taking account of inequality (5.12), we find that $u \leqslant \omega$ on $\partial(D_\Gamma \cap \Omega_r)$. The estimate (5.6) is deduced from what has been proved in the same way as above.

Finally, we show how to eliminate the assumption regarding the replacement of condition (5.7) by (5.4). Suppose that conditions (4.1), (4.5), and (5.7) are satisfied. We make a change of variables

$$\tilde{x} = \tilde{x}(x), \tag{5.16}$$

which realizes the transformation of inversion relative to the sphere $\partial K_R(x_*)$ (see (5.7)). Under this transformation the domain Ω_r goes over into a domain $\tilde{\Omega}_r$, and $\tilde{\Omega}_r$ is contained in a ball $K_R(x_*)$ of radius R with center at the point x_* lying on the axis defined by the vector of the inner (relative to the new domain $\tilde{\Omega}_r$) normal $\boldsymbol{\nu}$ to $\partial\Omega$ at the point $\tilde{y}_0 = y_0$ (it is obvious that y_0 is a fixed point of the transformation (5.16)). Thus, a condition of the form (5.4) is satisfied for the new domain $\tilde{\Omega}_r$.

It is obvious that the transformation (5.16) defined above realizes a diffeomorphism of class C^∞ between $\overline{\Omega}_r$ and $\tilde{\overline{\Omega}}_r$. In particular, the function $\tilde{x} = \tilde{x}(x)$, $x \in \Omega_r$, and the inverse function $x = x(\tilde{x})$, $\tilde{x} \in \tilde{\Omega}_r$, are bounded together with their partial derivatives of first and second orders by a constant depending only on R and the diameter of Ω_r. Equation (1.1) is thus transformed into an equation of the form

$$\tilde{a}^{kl}(\tilde{x}, u, \tilde{\nabla} u)u_{\tilde{x}_k\tilde{x}_l} - \tilde{a}(\tilde{x}, u, \tilde{\nabla} u) = 0, \tag{5.17}$$

where

$$\tilde{a}^{kl} = a^{ij}\frac{\partial \tilde{x}_k}{\partial x_i}\frac{\partial \tilde{x}_l}{\partial x_j}, \qquad \tilde{a} = a - a^{ij}\frac{\partial^2 \tilde{x}_k}{\partial x_i \partial x_j}u_{\tilde{x}_k}.$$

Setting $\nabla u(x) = p$ and $\tilde{\nabla} u(\tilde{x}) = \tilde{p}$, noting that

$$\begin{gathered} c_1|p| \leqslant |\tilde{p}| \leqslant c_2|p|, \quad c_3\operatorname{Tr} A \leqslant \operatorname{Tr}\tilde{A} \leqslant c_4 \operatorname{Tr} A, \\ \mathscr{E}_1 \equiv Ap\cdot p = \tilde{A}\tilde{p}\cdot\tilde{p} \equiv \tilde{\mathscr{E}}_1, \end{gathered} \tag{5.18}$$

where c_1, c_2, c_3, c_4 are positive constants, $A \equiv \|a^{ij}(x, u, p)\|$ and $\tilde{A} \equiv \|\tilde{a}^{ij}(\tilde{x}, u, \tilde{p})\|$, and observing that

$$\tilde{A} = CAC^*, \qquad C = \|\partial \tilde{x}_k/\partial x_i\|, \tag{5.19}$$

from conditions (4.1) and (4.5) we deduce the inequalities

$$\tilde{\psi}(|\tilde{p}|)\tilde{\mathscr{E}}_1 \geqslant c_5 \operatorname{Tr} \tilde{A}|\tilde{p}|, \qquad c_5 = \text{const} > 0, \tag{5.20}$$

and

$$|\tilde{a}(\tilde{x}, u, \tilde{p})| \leqslant c_6 \tilde{\psi}(|\tilde{p}|)\tilde{\mathscr{E}}_1, \qquad c_6 = \text{const} > 0, \tag{5.21}$$

where $\tilde{\psi}(\tilde{\rho}) = c_7\psi(c_8\tilde{\rho})$, c_7, $c_8 = \text{const} > 0$, obviously satisfies the condition $\int^{+\infty}(d\tilde{\rho}/\tilde{\rho}\tilde{\psi}(\tilde{\rho})) = +\infty$. Thus, for (5.17) and the domain $\tilde{\Omega}_r$ all the conditions are satisfied under which (in the proof of the second part of the theorem) an estimate of the form (5.6) was established, i.e.,

$$|\partial \tilde{u}(y_0)/\partial \tilde{\nu}| \leqslant M_0. \tag{5.22}$$

Returning to the old variables, from (5.22) we deduce (5.6). The proof of Theorem 5.1 is completed.

In the sequel we shall use the following version of the second part of Theorem 5.1.

THEOREM 5.1′. *Suppose that the functions* a^{ij}, $\partial a^{ij}/\partial p_k$, a *and* $\partial a/\partial p_k$, $i, j, k = 1,\dots,n$, *are continuous on the set* $\mathfrak{N}_{\Gamma,m,l}$, *and suppose that for all* $x \in D_\Gamma^0$, *all* $u \in [-m, m]$ *and any numbers* μ *and* t *such that* $\mu > 0$ *and* $t \geqslant \mu + l$, *where* $l = \text{const} > 0$, *the inequalities*

$$|a(x, u, \mu\nabla\rho + t\nu)| \leqslant \psi(|\mu\nabla\rho + t\nu|)\mathscr{E}_1(x, u, \mu\nabla\rho + t\nu) \tag{5.23}$$

and

$$\psi(|\mu\nabla\rho + t\nu)|\mathscr{E}_1(x, u, \mu\nabla\rho + t\nu) \geqslant |\mu\nabla\rho + t\nu| \operatorname{Tr} A(x, u, \mu\nabla\rho + t\nu) \tag{5.24}$$

hold, where $A \equiv \|a^{ij}\|$, $\rho = \rho(x) = \operatorname{dist}(x, P_{y_0})$, P_{y_0} *is the tangent plane to* $\partial\Omega$ *at the point* y_0, $\nu = \nu(y(x))$ *is the interior unit normal to* $\partial\Omega$ *at the point* $y(x)$ *closest to* $x \in D_\Gamma^0$ *on* Γ, *and the functions* $\mathscr{E}_1$ *and* ψ *are the same as in Theorem* 5.1. *Suppose that the function* u *satisfies condition* (5.3) *in the case* $\varphi \equiv 0$. *Suppose also that condition* (5.4) *is satisfied. Then the estimate* (5.6) *holds, in which the constant* M_0 *depends only on* m, l, K, R, r^{-1}, δ_0 *and the function* ψ.

PROOF. Theorem 5.1′ follows directly from the proof of the second part of Theorem 5.1. An analogous modification of the formulation could also be made for the first part of the theorem. However, we shall omit this, since it is not used anywhere in the sequel.

REMARK 5.1. If under the conditions of Theorem 5.1 (5.1′) the function φ is identically 0 on $\overline{\Omega}$, then the estimate (4.4) ((5.6)) has the form

$$|\partial u(y)/\partial\nu| \leqslant \beta \qquad (|\partial u(y_0)/\partial\nu| \leqslant \beta + \mu), \tag{5.25}$$

where β is determined from (3.7) with

$$\bar{\alpha} = \max(\alpha, m\delta_0^{-1}), \qquad \alpha = \max(\alpha_0, l, + 1)$$

$$\left(\bar{\alpha} = \max(\alpha, m\delta_0^{-1}), \qquad \alpha = \max(\alpha_0, \mu + l + 1), \qquad \mu = m2R/r^2\right)$$

and α_0 depends only on the functions $\psi(\rho)$ and $\delta(\rho)$, and on $\partial\Omega$.

§6. Estimates of $\max_\Omega|\nabla u|$ for equations with structure described in terms of the majorant $\mathscr{E}_1$

Suppose that the functions $a^{ij}(x, u, p)$, $i, j = 1,\dots,n$, and $a(x, u, p)$ forming equation (1.1) belong to the class $C^1(\mathfrak{M}_{\Omega,m,L})$, where $\mathfrak{M}_{\Omega,m,L} \equiv \overline{\Omega} \times \{|u| \leqslant m\} \times \{|p| > L\}$, m and L being positive constants. We suppose also that on $\mathfrak{M}_{\Omega,m,L}$

$$a^{ij}(x, u, p)\xi_i\xi_j \geqslant 0, \qquad \xi \in \mathbb{R}^n. \tag{6.1}$$

Let $\tau = (\tau_1,\dots,\tau_n)$ be an arbitrary fixed vector with $|\tau| = 1$. We set

$$A^\tau \equiv a^{ij}(x, u, p)\tau_i\tau_j \equiv A\tau\cdot\tau. \tag{6.2}$$

We further introduce the following notation for an arbitrary function $\Phi \equiv \Phi(x, u, p)$:

$$\delta\Phi \equiv \sum_{k=1}^n \frac{p_k}{|p|}\frac{\partial\Phi}{\partial x_k} + |p|\frac{\partial\Phi}{\partial u}, \qquad p\Phi_p \equiv \sum_{k=1}^n p_k\frac{\partial\Phi}{\partial p_k}. \tag{6.3}$$

THEOREM 6.1. *Suppose that on the set $\mathfrak{M}_{\Omega,m,L}$ for any τ, $|\tau| = 1$*

$$\begin{gathered} |pA_p^\tau| \leqslant \sqrt{\mu_1 A^\tau\mathscr{E}_1\omega(|p|)}\,|p|^{-1}, \quad |\delta A^\tau| \leqslant \sqrt{\sigma_1 A^\tau\mathscr{E}_1/\omega(|p|)}, \\ |a - pa_p| \leqslant \mu_2\mathscr{E}_1, \quad \delta a \geqslant -\sigma_2\mathscr{E}_1|p|\omega^{-1}(|p|), \qquad \mathscr{E}_1 > 0, \end{gathered} \tag{6.4}$$

where μ_1 and μ_2 are arbitrary nonnegative constants, σ_1 and σ_2 are nonnegative constants which are sufficiently small, depending on n, μ_1, μ_2 and m,(4) *and $\omega(\rho) > 0$, $0 \leqslant \rho < +\infty$, is an arbitrary nondecreasing, continuous function. Then for any solution $u \in C^3(\Omega) \cap C^1(\overline{\Omega})$ of* (1.1) *satisfying the condition*

$$\max_{\overline{\Omega}}|u| \leqslant m, \tag{6.5}$$

the quantity $\max_{\overline{\Omega}}|\nabla u|$ *can be estimated in terms of m, $M_1 \equiv \max_{\partial\Omega}|\nabla u|$, n, L, μ_1 and μ_2 alone.*

PROOF. Applying to (1.1) the operator $u_{x_k}(\partial/\partial x_k)$ and setting $v = \sum_1^n u_{x_k}^2$, we obtain the identity

$$\tfrac{1}{2}a^{ij}v_{ij} = a^{ij}u_{ki}u_{kj} + \tfrac{1}{2}\left[a_{p_l} - a^{ij}_{p_l}u_{ij}\right]v_l + \sqrt{v}\left(\delta a - \delta a^{ij}u_{ij}\right), \quad x \in \Omega. \tag{6.6}$$

Here and henceforth we use the abbreviated notation for derviatives: $v_l \equiv v_{x_l}$, $u_{ij} \equiv u_{x_ix_j}$, etc. We multiply both sides of (6.6) by $f(v(x))$, where $f(v) > 0$ and $f'(v) \geqslant 0$ and $v > 0$, and we introduce the function

$$w = \int_0^v f(t)\,dt. \tag{6.7}$$

Taking into account that $w_i = fv_i$ and $w_{ij} = fv_{ij} + f'v_iv_j$, from (6.7) we deduce the identity

$$\begin{gathered} \tfrac{1}{2}a^{ij}w_{ij} = \tfrac{1}{2}f'a^{ij}v_iv_j + fa^{ij}u_{ki}u_{kj} + \tfrac{1}{2}\left[a_{p_l} - a^{ij}_{p_l}u_{ij}\right]w_l \\ + f\sqrt{v}\left(\delta a - \delta a^{ij}u_{ij}\right), \qquad x \in \Omega. \end{gathered} \tag{6.8}$$

Let $z = z(u)$ be a twice differentiable, positive function defined on $[-m, m]$. We consider the function $\overline{w}$ defined by

$$w = z(u)\overline{w}. \tag{6.9}$$

(4) This dependence will be specified in the proof of the theorem.

Taking into account that $w_i = z'u_i\overline{w} + z\overline{w}_i$ and $w_{ij} = z''u_iu_j\overline{w} + z'u_{ij}\overline{w} + z'u_i\overline{w}_j + z'u_j\overline{w}_i + z\overline{w}_{ij}$, from (6.8) we deduce the identity

$$za^{ij}\overline{w}_{ij} + \sum_{k=1}^{n} b^k\overline{w}_k = -z''\mathscr{E}_1\overline{w} + \frac{f'}{f^2}z'^2\mathscr{E}_1\overline{w}^2 + 2fa^{ij}u_{ki}u_{kj} + z'(pa_p - a)\overline{w}$$
$$-z'(a^{ij})_{p_l}u_lu_{ij}\overline{w} + 2f\sqrt{v}\left(\delta a - \delta a^{ij}u_{ij}\right), \tag{6.10}$$

where $\mathscr{E}_1 \equiv a^{ij}(x, u, \nabla u)u_{x_i}u_{x_j}$, and the form of the functions b^k is irrelevant for subsequent considerations. In deriving (6.10) it was also taken into account that $a^{ij}u_{ij} = a$ in Ω. The identity (6.10) will subsequently be used only on the set $\Omega_L \equiv \{x \in \Omega: |\nabla u| > L\}$.

We consider the matrix $\|u_{ij}\|$ at a fixed point of Ω_L. Let $T \equiv \|t_{ij}\|$ be the orthogonal matrix reducing $\|u_{ij}\|$ to the diagonal matrix $\|\tilde{u}_{ij}\|$, where $\tilde{u}_{ij} = 0$ for $i \neq j$. Then

$$\|u_{ij}\| = T\|\tilde{u}_{ij}\|T^*. \tag{6.11}$$

We denote by $\tilde{A}$ the matrix

$$\tilde{A} = T^*AT. \tag{6.12}$$

We shall prove the inequalities

$$|z'(a^{ij})_{p_l}u_lu_{ij}\overline{w}| \leq \frac{1}{2}fa^{ij}u_{ki}u_{kj} + c_0\frac{z'^2}{f}\frac{\omega}{v}\mathscr{E}_1\overline{w}^2,$$
$$|2\sqrt{v}\,f\delta a^{ij}u_{ij}| \leq \frac{1}{2}fa^{ij}u_{ki}u_{kj} + 2\sigma_1 n\frac{fv}{\omega}\mathscr{E}_1, \tag{6.13}$$

where $c_0 = \mu_1 n/2$ and μ_1, σ_1 are the constants in (6.4). Indeed, we have

$$\begin{aligned} z'(a^{ij})_{p_l}u_lu_{ij}\overline{w} &= z'(\tilde{a}^{ii})_{p_l}u_l\tilde{u}_{ii}\overline{w} = z'(A^{\tau_i})_{p_l}u_l\tilde{u}_{ii}\overline{w}, \\ 2\sqrt{v}\,f\delta a^{ij}u_{ij} &= 2\sqrt{v}\,f\delta\tilde{a}^{ii}\tilde{u}_{ii} = 2\sqrt{v}\,f\delta A^{\tau_i}\tilde{u}_{ii}, \end{aligned} \tag{6.14}$$

where $p_l = u_l$, τ_i is the ith column of the orthogonal matrix T (so that $|\tau_i| = 1$), and $\tilde{a}^{ii} = t^*_{ik}a^{kl}t_{li} = A\tau^i \cdot \tau^i \equiv A^{\tau_i}$. If $\tilde{a}^{ii} = 0$ for some value of $i \in \{1,\dots,n\}$, condition (6.4) implies that for this index i also

$$(\tilde{a}^{ii})_p p = 0, \quad \delta\tilde{a}^{ii} = 0. \tag{6.15}$$

If $\tilde{a}^{ii} \neq 0$, then, applying the Cauchy inequality and taking account of (6.4), we can estimate the corresponding terms in (6.14) as follows:

$$\begin{aligned} |z'(\tilde{a}^{ii})_{p_l}u_l\tilde{u}_{ii}\overline{w}| &\leq \frac{1}{2}f\tilde{a}^{ii}\tilde{u}_{ii}^2 + \frac{1}{2}\frac{z'^2}{f}\mu_1\frac{\omega}{v}\mathscr{E}_1\overline{w}^2, \\ |2f\sqrt{v}\,\delta\tilde{a}^{ii}\tilde{u}_{ii}| &\leq \frac{1}{2}f\tilde{a}^{ii}\tilde{u}_{ii}^2 + 2\sigma_1 f\frac{v}{\omega}\mathscr{E}_1. \end{aligned} \tag{6.16}$$

We note that there is no summation over the index i in (6.16). The inequalities (6.16) have been established only for those indiced i for which $\tilde{a}^{ii} \neq 0$ at the point $x \in \Omega_L$ under consideration. However, in view of (6.15) the inequalities (6.16) are trivially valid for indices i for which $\tilde{a}^{ii} = 0$. Therefore, summing (6.16) over all $i = 1,\dots,n$ and taking into account that $\tilde{a}^{ii}\tilde{u}_{ii}^2 = a^{ij}u_{ik}u_{kj}$, we obtain (6.13).

We shall assume that the function f has the form

$$f(v) = \exp\left\{2c_0\int_1^v \frac{\omega(t)}{t}\,dt\right\}$$

for $v \geqslant 1$. This function satisfies the conditions

$$f'/f = 2c_0(\omega/v), \quad f(v) \geqslant 1 \quad \text{for } v \geqslant 1. \tag{6.17}$$

From (6.10), (6.13), and (6.17) we then obtain

$$\begin{aligned} za^{ij}\overline{w}_{ij} + \sum_{k=1}^n b^k w_k \geqslant \left(-z'' + c_0\frac{z'^2}{z}\frac{w}{vf}\omega\right)\mathscr{E}_1\overline{w} + fa^{ij}u_{ki}u_{kj} \\ + z'(pa_p - a)\overline{w} + 2f\sqrt{v}\,\delta a - 2n\sigma_1\frac{fv}{\omega}\mathscr{E}_1, \qquad x \in \Omega_L. \end{aligned} \tag{6.18}$$

Assuming, with no loss of generality, that $c_0 \geqslant 1$ and $\omega(\rho) \geqslant 1$, $0 \leqslant \rho < +\infty$, we see that

$$f \leqslant 8c_0(\omega/v)w \quad \text{for } v \geqslant L^2. \tag{6.19}$$

Indeed, since ω, which may be considered a function of v, is assumed to be nondecreasing, it follows that

$$f(v) - 1 = \int_1^v f'(t)\,dt = 2c_0\int_1^v \frac{\omega(t)}{t}f(t)\,dt \leqslant 2c_0\frac{\omega(v)}{v}F(v), \tag{6.20}$$

where $F(v) \equiv v\int_1^v t^{-1}f(t)\,dt$. The function $t^{-1}f(t)$, $t \geqslant 1$, increases, since in view of (6.17)

$$\begin{aligned} \left(\frac{f(t)}{t}\right)' &= \frac{f'(t)t - f(t)}{t^2} = \frac{f(t)}{t^2}\left(\frac{f'(t)t}{f(t)} - 1\right) \\ &= \frac{f(t)}{t^2}(2c_0\omega(t) - 1) \geqslant \frac{f(t)}{t^2} > 0 \end{aligned} \tag{6.21}$$

for $t \geqslant 1$. Therefore,

$$\begin{aligned} F(v) &= \int_1^v F'(t)\,dt = \int_1^v f(t)\,dt + \int_1^v dt\int_1^t \frac{f(\tau)}{\tau}\,d\tau \\ &\leqslant \int_1^t f(t)\,dt + \int_1^v \frac{f(t)}{t}\int_1^t d\tau \leqslant 2w. \end{aligned} \tag{6.22}$$

It is obvious that (6.19) follows from (6.20) and (6.22). Now taking condition (6.4) into account in regard to $pa_p - a$ and δa, and also (6.19), we deduce from (6.18) that

$$za^{ij}\overline{w}_{ij} + \sum_{k=1}^n b^k\overline{w}_k \geqslant \left[-z'' + \frac{1}{8}\frac{z'^2}{z} - \mu_2|z'| - 16c_0(n\sigma_1 + \sigma_2)z\right]\mathscr{E}_1\overline{w}. \tag{6.23}$$

We now choose some function $z \in C^2([-m, m])$ satisfying the conditions

$$\begin{aligned} -z'' + (1/8)(z'^2/z) - \mu_2|z'| - 16c_0(n\sigma_1 + \sigma_2)z > 0 \quad \text{on } [-m, m], \\ z(u) \geqslant c_1 = \text{const} > 0 \quad \text{on } [-m, m]. \end{aligned} \tag{6.24}$$

To prove the existence of such a function it is necessary to require that the constants σ_1 and σ_2 be sufficiently small in dependence on m, μ_2, and c_0. We consider, for example, the function $z(u) = 1 + e^{\alpha m} - e^{\alpha u}$. Taking into account that $z' = -\alpha e^{\alpha u}$ and $z'' = -\alpha^2 e^{\alpha u}$, we have

$$-z'' - \mu_2|z'| - 16c_0(n\sigma_1 + \sigma_2)z \geqslant \alpha(\alpha - \mu_2)e^{\alpha u} - 16c_0(n\sigma_1 + \sigma_2)(1 + e^{\alpha m}).$$

It is then obvious that condition (6.24) will certainly be satisfied for this choice of the function z if it is assumed that $\alpha = \mu_2 + 1$ and required that

$$(\mu_2 + 1)e^{-(\mu_2+1)m} > 16c_0(n\sigma_1 + \sigma_2)(1 + e^{(\mu_2+1)m}). \tag{6.25}$$

Condition (6.25) is the smallness condition for the quantities σ_1 and σ_2 in (6.4).

In view of (6.24) and the assumption of positivity of $\mathscr{E}_1$ on $\mathfrak{M}_{\Omega,m,L}$ it follows from (6.23) that $\overline{w}$ cannot achieve a maximum on Ω_L. Therefore,

$$\max_{\overline{\Omega}} \overline{w} \leqslant \max\left\{ \max_{\partial\Omega}(w/z), \max_{\Omega\setminus\Omega_L}(w/z)\right\}. \tag{6.26}$$

Taking into account that $w = \int_0^v f(t)\,dt$, we deduce from (6.26) that

$$\max_{\overline{\Omega}} w \leqslant \frac{\max z}{\min z} \max\left\{ \int_0^{M_1^2} f(t)\,dt, \int_0^{L^2} f(t)\,dt \right\}, \tag{6.27}$$

where $M_1 = \max_{\partial\Omega}|\nabla u|$. Noting (6.7), the form of the function z, and the fact that f is nondecreasing, from (6.27) we easily obtain

$$\int_0^{M^2} f(t)\,dt \leqslant \max\left\{ \int_0^{(aM_1)^2} f(t)\,dt, \int_0^{(aL)^2} f(t)\,dt \right\}, \tag{6.28}$$

where we have used the notation $M \equiv \max_{\overline{\Omega}}|\nabla u|$ and $a(1 + e^{(\mu_2+1)m})^{1/2}$. Then

$$M \leqslant a^{1/2}\max\{M_1, L\}. \tag{6.29}$$

Theorem 6.1 is proved.

REMARK 6.1. It is obvious that in condition (6.4) the constants σ_1 and σ_2 satisfying (6.25) can be replaced by functions $\sigma_1(|p|)$ and $\sigma_2(|p|)$ assuming that $\sigma_1(\rho)$, $\sigma_2(\rho) \to 0$ as $\rho \to +\infty$. Assuming that the parameter L is sufficiently large, we then reduce this case to that considered in Theorem 6.1.

REMARK 6.2. We note that the conditions of Theorem 6.1 admit degeneracy of the matrix $A \equiv \|a^{ij}(x, u, p)\|$ characterized by the condition $\mathscr{E}_1(x, u, p) > 0$ on the set $\mathfrak{M}_{\Omega,m,L}$. We shall distinguish an especially important special case of Theorem 6.1 obtained by the assumption that $\omega(\rho) = \text{const} \geqslant 1$.

THEOREM 6.1′. *Suppose that on the set* $\mathfrak{M}_{\Omega,m,L}$

$$\begin{gathered} |pA_p^\tau| \leqslant \sqrt{\mu_1 A^\tau \mathscr{E}_1}\,|p|^{-1}, \quad |\delta A^\tau| \leqslant \sqrt{\sigma_1 A^\tau \mathscr{E}_1}, \quad |pa_p - a| \leqslant \mu_2\mathscr{E}_1, \\ \delta a \geqslant -\sigma_2\mathscr{E}_1|p|, \qquad \mathscr{E}_1 > 0, \end{gathered} \tag{6.30}$$

where $\mu_1, \mu_2, \sigma_1, \sigma_2 = \text{const} \geqslant 0$, *and that condition* (6.25) *is satisfied, where* $c_0 = \mu_1 n/2$. *Then for any solution* $u \in C^3(\Omega) \cap C^1(\overline{\Omega})$ *satisfying* (6.5) *there is an estimate of the form* (6.29), *where* $M_1 = \max_{\partial\Omega}|\nabla u|$ *and* $a = (1 + \exp((\mu_2 + 1)m)^{1/2}$.

Theorem 6.1′ contains as a special case the well-known result of Ladyzhenskaya and Ural′tseva on the estimate of $\max_{\overline{\Omega}}|\nabla u|$ in terms of $\max_{\overline{\Omega}}|u|$ and $\max_{\partial\Omega}|\nabla u|$ for solutions of quasilinear uniformly elliptic equations [83]. The conditions of the corresponding theorem of Ladyzhenskaya and Ural'tseva can be written in the form

$$\begin{gathered}\Lambda\lambda^{-1} \leqslant c, \quad |pa_p^{ij}| \leqslant \tilde{\mu}_1\lambda, \quad |\delta a^{ij}| \leqslant \tilde{\sigma}_1\lambda|p|, \qquad i, j = 1,\dots,n, \\ |pa_p - p| \leqslant \tilde{\mu}_2\lambda|p|^2, \quad \delta a \geqslant -\tilde{\sigma}_2\lambda|p|^3 \quad \text{on } \mathfrak{M}_{\Omega,m,L},\end{gathered} \tag{6.31}$$

where Λ and λ are respectively the largest and least eigenvalues of the matrix A, $\tilde{\mu}_1$ and $\tilde{\mu}_2$ are arbitrary constants, and $\tilde{\sigma}_1$ and $\tilde{\sigma}_2$ are sufficiently small constants. It is easy to see that the conditions (6.30) follow from (6.31), since for any τ with $|\tau| = 1$ we have

$$\sqrt{\lambda} \leqslant c\sqrt{A^\tau}, \qquad \sqrt{\lambda} \leqslant c\sqrt{\mathscr{E}_1}\,|p|^{-1}, \qquad \lambda|p|^2 \leqslant c\mathscr{E}_1, \tag{6.32}$$

where the constant c in (6.32) does not depend on τ. We note that very recently [84] Ladyzhenskaya and Ural'tseva have strengthened their result by replacing the sufficiently small constants σ_1 and σ_2 in (6.31) by arbitrary constants. This is accomplished using the proof of an a priori estimate of the Hölder norm $\|u\|_{C^\alpha(\bar{\Omega})}$ for solutions of uniformly elliptic equations.

§7. The estimate of $\max_\Omega|\nabla u|$ for equations with structure described in terms of the majorant $\mathscr{E}_2$

Suppose that the functions $a^{ij}(x, u, p)$, $i, j = 1,\dots,n$, and $a(x, u, p)$ belong to the class $C^1(\mathfrak{M}_{\Omega,m,L})$, and suppose that condition (6.1) is satisfied on $\mathfrak{M}_{\Omega,m,L}$.

THEOREM 7.1. *Suppose that on the set $\mathfrak{M}_{\Omega,m,L}$ for any τ, $|\tau| = 1$, the conditions*

$$\begin{gathered}|A_p^\tau|\,|p| \leqslant \sqrt{\mu_1 A^\tau \operatorname{Tr} A}, \quad |\delta A^\tau| \leqslant \sqrt{\sigma_1 A^\tau \operatorname{Tr} A}, \\ |a_p|\,|p| \leqslant \sigma_2\mathscr{E}_2, \quad \delta a \geqslant -\sigma_3\mathscr{E}_2, \quad \operatorname{Tr} A > 0\end{gathered} \tag{7.1}$$

are satisfied, where μ_1 is an arbitrary nonnegative constant, and σ_1, σ_2, σ_3 are nonnegative constants which are sufficiently small, depending on n, μ_1, and the diameter of the domain Ω. Then for any solution $u \in C^3(\Omega) \cap C^1(\bar{\Omega})$ of (1.1) satisfying condition (6.5) the quantity $\max_{\bar{\Omega}}|\nabla u|$ can be estimated in terms of m, $M_1 \equiv \max_{\partial\Omega}|\nabla u|$, L, n, μ_1, and the diameter d of Ω, alone. Under the additional condition $\sum_{i,j=1}^n a^{ij} \geqslant \varepsilon_0 \operatorname{Tr} A$ on $\mathfrak{M}_{\Omega,m,L}$, where $\varepsilon_0 =$ const > 0, in (7.1) in place of σ_2 it is possible to admit an arbitrary constant $\mu_2 \geqslant 0$.

PROOF. Applying to (1.1) the operator $u_k(\partial/\partial x_k)$ and setting $v = \sum_1^n u_k^2$, we obtain (6.6). This identity will be considered below only on the set $\Omega_L = \{x \in \Omega\colon |\nabla u| > L\}$. Arguing exactly as in the derivation of (6.13) in the proof of Theorem 6.1, we shall prove that at each point of Ω_L

$$\begin{gathered}|(a^{ij})_{p_l}u_{ij}v_l| \leqslant \frac{1}{2}a^{ij}u_{ki}u_{kj} + \frac{\mu_1 n}{2}\frac{|\nabla v|^2}{v}\operatorname{Tr} A, \\ |\sqrt{v}\,\delta a^{ij}u_{ij}| \leqslant \frac{1}{2}a^{ij}u_{ki}u_{kj} + \frac{\sigma_1 n}{2}\operatorname{Tr} Av.\end{gathered} \tag{7.2}$$

Indeed, using (6.11) and (6.12) at a fixed point $x \in \Omega_L$ and taking into account that $\tilde{a}^{ii} = A^{\tau_i}$, $i = 1,\dots,n$, where $\tau_i = (t_{1i},\dots,t_{ni})$, by means of the first two conditions of (1.1) we obtain

$$\begin{aligned} |a^{ij}_{p_l} u_{ij} v_l| &= |\tilde{a}^{ii}_{p_l} \tilde{u}_{ii} v_l| \leqslant \frac{1}{2}\tilde{a}^{ii}\tilde{u}^2_{ii} + \frac{\mu_1}{2}\operatorname{Tr}\tilde{A}\frac{|\nabla v|^2}{v}, \\ |\sqrt{v}\,\delta a^{ij} u_{ij}| &= |\sqrt{v}\,\delta \tilde{a}^{ii}\tilde{u}_{ii}| \leqslant \frac{1}{2}\tilde{a}^{ii}\tilde{u}^2_{ii} + \frac{\sigma_1}{2}\operatorname{Tr}\tilde{A}v, \end{aligned} \tag{7.3}$$

where there is no summation on the index i and only those values of this index are taken for which $\tilde{a}^{ii} \neq 0$. However, for those values of the index $i \in \{1,\dots,n\}$ for which $\tilde{a}^{ii} = 0$ at the point $x_0 \in \Omega_L$ in question the inequalities (7.3) are satisfied trivially, since in (7.1) for $\tau = \tau_i$ it follows in this case that $\tilde{a}^{ii}_p = 0$ and $\delta\tilde{a}^{ii} = 0$. Summing (7.3) over $i = 1,\dots,n$, we obtain (7.2). From (6.6) and (7.2) we get

$$\begin{aligned} a^{ij}v_{ij} \geqslant a^{ij}u_{ki}u_{kj} + a_{p_l}v_l + 2\sqrt{v}\,\delta a - \mu_1 n \operatorname{Tr} A\frac{|\nabla v|^2}{v} \\ -\sigma_1 n \operatorname{Tr} A v, \qquad x \in \Omega_L. \end{aligned} \tag{7.4}$$

Let $z = z(x)$ be a positive function in $\bar{\Omega}$ belonging to $C^2(\bar{\Omega})$. We introduce the function $\bar{v}$ defined by

$$v = z\bar{v}. \tag{7.5}$$

Taking into account that $v_i = z_i\bar{v} + z\bar{v}_i$ and $v_{ij} = z_{ij}\bar{v} + z_i\bar{v}_j + z_j\bar{v}_i + z\bar{v}_{ij}$, we deduce from (7.4) that

$$\begin{aligned} za^{ij}\bar{v}_{ij} + b^k\bar{v}_k \geqslant -a^{ij}z_{ij}\bar{v} + a^{ij}u_{ki}u_{kj} + a_{pl}z_l\bar{v} \\ +2\sqrt{v}\,\delta a - \mu_1 n \operatorname{Tr} A\frac{|\nabla z|^2}{z}\bar{v} - \sigma_1 nz \operatorname{Tr} A\bar{v}, \end{aligned} \tag{7.6}$$

where the form of the b^k is irrelevant for what follows. Taking account of conditions (7.1) on $|a_p|$ and δa, we deduce from (7.6) that

$$za^{ij}\bar{v}_{ij} + b^k\bar{v}_k \geqslant \left[-a^{ij}z_{ij} - \left(\mu_1 n\frac{|\nabla z|^2}{z} + \sigma_2|\nabla z| + (n\sigma_1 + 2\sigma_3)z\right)\operatorname{Tr} A\right]\bar{v}, \qquad x \in \Omega_L. \tag{7.7}$$

We choose some function $z = z(x)$ satisfying the conditions

$$\begin{aligned} -a^{ij}z_{ij} - \left(\mu_1 n\frac{|\nabla z|^2}{z} + \sigma_2|\nabla z| + (n\sigma_1 + 2\sigma_3)z\right)\operatorname{Tr} A > 0 \quad \text{on } \Omega_L, \\ z \geqslant c_1 = \text{const} > 0 \quad \text{on } \Omega_L. \end{aligned} \tag{7.8}$$

Suppose, for example, $z(x) = \alpha + d^2 - |x|^2$, where $\alpha = \text{const} > 1$, d is the diameter of Ω, and we assume with no loss of generality that the origin is contained in Ω. Taking into account that $z_i = -2x_i$ and $z_{ij} = -2\delta_i^j$, where δ_i^j is the Kronecker symbol, we easily see that conditions (7.8) are satisfied if we first choose the constant $\alpha > 1$ so that $\mu_1 n(4d^2/\alpha) \leqslant 1/2$ and then impose on σ_1, σ_2 and σ_3 the condition

$$2\sigma_2 d + (n\sigma_1 + 2\sigma_3)(\alpha + d^2) < 1/2. \tag{7.9}$$

This is the smallness condition on the quantities σ_1, σ_2 and σ_3 in (7.1). In view of (7.8) and the condition of positivity of $\mathscr{E}_2$ on Ω_L it follows from (7.7) that the function $\bar{v}$ cannot achieve a maximum on Ω_L. Therefore,

$$\max_{\bar{\Omega}} \bar{v} \leqslant \max\Big\{ \max_{\partial\Omega}(v/z), \max_{\{|\nabla u|<L\}}(v/z)\Big\}. \tag{7.10}$$

Taking the form of the function z into account, we then find that

$$\max_{\bar{\Omega}} v \leqslant \big((\alpha + d^2)/d^2\big)\max\{M_1^2, L^2\}, \tag{7.11}$$

where $M_1 = \max_{\partial\Omega}|\nabla u|$, whence we obtain the desired estimate

$$\max_{\bar{\Omega}}|\nabla u| \leqslant \big(1 + \sqrt{\alpha}/d\big)\max\{M_1, L\}. \tag{7.12}$$

We now suppose that on $\mathfrak{M}_{\Omega,m,L}$ we have $\Sigma_{i,j=1}^n a^{ij} \geqslant \varepsilon_0 \operatorname{Tr} A$, $\varepsilon_0 = \text{const} > 0$. In this case in place of the function $z = \alpha + d^2 - |x|^2$ we consider $z = \alpha + e^{\beta n d} - e^{-\beta\Sigma_1^n x_k}$, where α and β are positive constants, d is the diameter of Ω, and, as above, we suppose that the origin is contained in Ω. Taking account of the relations $z_i = \beta e^{-\beta\Sigma_1^n x_k}$ and $z_{ij} = -\beta^2 e^{-\beta\Sigma_1^n x_k}$, we see that (7.8) will be satisfied if we first choose $\beta > 0$ such that $\beta^2\varepsilon_0 - \sigma_2 n\beta = \frac{1}{2}\beta^2\varepsilon_0$ and then choose $\alpha > 1$ so that $\frac{1}{2}\beta^2 e^{-\beta n d}\varepsilon_0 - \mu_1 n\beta^2 e^{2\beta n d}\alpha^{-1} \geqslant \frac{1}{4}\beta^2\varepsilon_0 e^{-\beta n d}$ and assume that σ_1 and σ_2 are so small that

$$(n\sigma_1 + 2\sigma_3)(\alpha + e^{\beta n d}) < \tfrac{1}{4}\beta\varepsilon_0 e^{-\beta n d}.$$

Then, arguing in exactly the same way, we obtain the estimate

$$\max_{\bar{\Omega}}|\nabla u| \leqslant \sqrt{(\alpha + e^{\beta n d})/\alpha}\,\max(M_1, L),$$

and in deriving it no condition was imposed on σ_2. Theorem 7.1 is proved.

REMARK 7.1. Theorem 7.1 continues to hold if in (7.1) the constants σ_1, σ_2 and σ_3 satisfying (7.9) are replaced by functions $\sigma_1(|p|)$, $\sigma_2(|p|)$ and $\sigma_3(|p|)$, assuming that $\sigma_1(\rho), \sigma_2(\rho), \sigma_3(\rho) \to 0$ as $\rho \to +\infty$.

REMARK 7.2. We note that the conditions of Theorem 7.1 admit degeneracy of the matrix $A \equiv \|a^{ij}(x, u, p)\|$ characterized by the condition $\operatorname{Tr} A > 0$ on $\mathfrak{M}_{\Omega,m,L}$.

As an example we consider an equation with principal part which coincides with the principal part of the normalized equation of minimal surfaces, i.e,. an equation of the form

$$\left(\frac{1 + |\nabla u|^2}{|\nabla u|^2}\delta_i^j - \frac{u_{x_i}u_{x_j}}{|\nabla u|^2}\right)u_{x_i x_j} = a(x, u, \nabla u). \tag{7.13}$$

In this case

$$a^{ij} = \frac{1 - |p|^2}{|p|^2}\delta_i^j - \frac{p_i p_j}{|p|^2}, \qquad A^\tau = \left(\frac{1 + |p|^2}{|p|^2}\delta_i^j - \frac{p_i p_j}{|p|^2}\right)\tau_i\tau_j,$$

$$\mathscr{E}_1 \equiv 1, \qquad \operatorname{Tr} A = n - 1 + \frac{n}{|p|^2}, \qquad \mathscr{E}_2 = (n-1)|p| + \frac{n}{|p|},$$

$$\frac{\partial A^\tau}{\partial p_k} = -\frac{2p_k + 2\tau_k p\cdot\tau|p|^2 - 2(p\cdot\tau)^2 p_k}{|p|^4},$$

$\tau \in \mathbb{R}^n$, $|\tau| = 1$. Any vector τ, $|\tau| = 1$, can be represented in the form $\tau = \alpha\xi + \beta\sigma$, when $\xi\cdot\sigma = 0$, $|\xi| = 1$, $\sigma = p/|p|$ and $\alpha^2 + \beta^2 = 1$. Then obviously $A^\tau = \alpha^2 + |p|^{-2}$

and $|\partial A^\tau/\partial p_k||p| \leqslant 2|p|^{-2} + 2\alpha$, where in deriving the last inequality it is noted that $|\beta| \leqslant 1$ and $|\xi_k| \leqslant 1$. From what has been proved it is evident that

$$\sqrt{A^\tau \operatorname{Tr} A} \geqslant \sqrt{\frac{n-1}{2}}\,(\alpha + |p|^{-1}), \quad |A_p^\tau||p| \leqslant 2\sqrt{n}\,(\alpha + |p|^{-1}), \tag{7.14}$$

whence we easily obtain the inequality $|A_p^\tau||p| \leqslant 2\sqrt{2n/(n-1)}\,\sqrt{A^\tau \operatorname{Tr} A}$. In addition, taking into account that $\delta A^\tau \equiv 0$ and $\operatorname{Tr} A > n - 1$, we conclude that the first two and the last conditions in (7.1) are satisfied for (7.13) with $\mu_1 = 8n/(n-1)$ and $\sigma_1 = 0$.

For the possibility of applying Theorem 7.1 to equation (7.13) it is necessary to require that on $\mathfrak{M}_{\Omega,m,L}$ (for some known $L \geqslant 0$) the conditions

$$|a_p| \leqslant \sigma_2, \qquad \delta a \geqslant -\sigma_3|p| \tag{7.15}$$

are satisfied, where σ_2 and σ_3 are sufficiently small, depending on n and d (see (7.9)). We consider, in particular, the normalized equation of a surface of given mean curvature of the form (1.10) for which

$$a(x, u, p) = \mathscr{H}(x, u, p)\frac{(1 + |p|^2)^{3/2}}{|p|^2}.$$

We suppose that

$$\mathscr{H}(x, u, p) = H_0 + \hat{H}(x, u, p),$$

where H_0 is an arbitrary constant. In this case (7.15) is satisfied if we require that

$$|\mathscr{H}| + |p||\hat{H}_p| \leqslant \sigma_2, \qquad \delta\hat{H} \geqslant -\sigma_3 \quad \text{on } \mathfrak{M}_{\Omega,m,L} \tag{7.16}$$

for sufficiently smalll σ_2 and σ_3.

§8. The estimate of $\max_\Omega|\nabla u|$ for a special class of equations

In the preceding section, in particular, conditions were indicated on the right side of equation (7.13) which ensure an estimate of $\max_\Omega|\nabla u|$ in terms of $\max_\Omega|u|$ and $\max_{\partial\Omega}|\nabla u|$. (Equation (7.13) is related to the study of various questions of geometry and continuum mechanics.) In this section we distinguish a class of equations containing (7.13), for which this estimate is constructed by another method that imposes different restrictions on their structure. In particular, in the case of (7.13) conditions are imposed on the right side $a(x, u, p)$ which, generally speaking, are not contained in conditions (7.15) and which do not contain them.

THEOREM 8.1. *Suppose that on the set $\mathfrak{M}_{\Omega,m,L}$ for any τ, $|\tau| = 1$, $\tau \cdot p = 0$, the conditions*

$$\begin{gathered} |pA_p^\tau| \leqslant \sqrt{\sigma_1 A^\tau \mathscr{E}_1}\,|p|^{-1}, \qquad |\delta A^\tau| \leqslant \sqrt{\mu_1 A^\tau \mathscr{E}_1}, \qquad \mathscr{E}_1 > 0, \\ \mathscr{E}_1 - p(\mathscr{E}_1)_p > \mu_2 \mathscr{E}_1, \qquad |\delta\mathscr{E}_1| \leqslant \mu_3 \mathscr{E}_1|p|, \\ |a - pa_p| \leqslant \mu_4 \mathscr{E}_1, \qquad \delta a \geqslant -\mu_5 \mathscr{E}_1|p| \end{gathered} \tag{8.1}$$

are satisfied, where $\mu_1, \dots, \mu_5$ are arbitrary nonnegative constants, and the constant $\sigma_1 \geqslant 0$ is sufficiently small in dependence on $\mu_2 > 0$. Then for any solution $u \in C^3(\Omega) \cap C^1(\overline{\Omega})$ of (1.1) satisfying condition (6.5) the quantity $\max_\Omega|\nabla u|$ can be estimated in terms of only m, $M_1 \equiv \max_{\partial\Omega}|\nabla u|$, L, n and $\mu_1, \dots, \mu_5$.

PROOF. We introduce the new unknown function $\bar u$ defined from the equality

$$u = K^{-1}\ln \bar u, \qquad K = \text{const.} \tag{8.2}$$

Taking into account $u_i = K^{-1}\bar u^{-1}\bar u_i$ and $u_{ij} = K^{-1}\bar u^{-1}\bar u_{ij} - Ku_iu_j$, we conclude that $\bar u$ satisfies the equation

$$\bar a^{ij}\bar u_{ij} - \bar a = 0, \tag{8.3}$$

where

$$\bar a^{ij} = a^{ij}\left(x, K^{-1}\ln \bar u, K^{-1}\bar u^{-1}\bar p\right),$$
$$\bar a = K\bar u a\left(x, K^{-1}\ln \bar u, K^{-1}\bar u^{-1}\bar p\right) + K^2\bar u\mathscr{E}_1\left(x, K^{-1}\ln \bar u, K^{-1}\bar u^{-1}\bar p\right).$$

Applying to (8.3) the operator $\bar u_k(\partial/\partial \bar x_k)$ and setting $\bar v = \sum_1^n \bar u_k^2$, we obtain

$$\tfrac12 a^{ij}\bar v_{ij} = a^{ij}\bar u_{ki}\bar u_{kj} + \tfrac12\left(\bar a_{\bar p l} - \bar a^{ij}_{\bar p l}\bar u_{ij}\right)\bar v_l + \sqrt{\bar v}\left(\bar\delta\bar a - \bar\delta a^{ij}\bar u_{ij}\right), \tag{8.4}$$

where

$$\bar\delta \equiv \frac{\bar p_k}{|p|}\frac{\partial}{\partial x_k} + |\bar p|\frac{\partial}{\partial u}.$$

If $\Phi = \Phi(x, u, p) \equiv \Phi(x, K^{-1}\ln \bar u, K^{-1}\bar u^{-1}\bar p)$, then

$$\bar\delta\Phi = (p_k/|p|)\Phi_{x_k} + |p|\Phi_u - K|p|\Phi_p p \equiv \delta\Phi - K|p|p\Phi_p. \tag{8.5}$$

If $\Phi = \Phi(\bar u)$, then

$$\bar\delta\Phi = K\bar u|p|\Phi_{\bar u}. \tag{8.6}$$

Using (8.5) and (8.6), and taking into account that $v = |p|^2 = \sum_1^n u_k^2 = K^{-2}\bar u^{-2}(\sum_1^n \bar u_k^2) \equiv K^{-2}\bar u^{-2}\bar v$, we obtain

$$\begin{aligned}\sqrt{\bar v}\,\bar\delta\bar a &= \bar v\left[Ka + K^2\mathscr{E}_1 + \delta a/|p| - Kpa_p + K(\delta\mathscr{E}_1/|p|) - K^2p(\mathscr{E}_1)_p\right],\\ \sqrt{\bar v}\,\bar\delta a^{ij}\bar u_{ij} &= \sqrt{\bar v}\left[\delta a^{ij}u_{ij} - K|p|(a^{ij})_p p\bar u_{ij}\right].\end{aligned} \tag{8.7}$$

It follows from (8.4) and (8.7) that if $\bar v$ assumes a maximum value at an interior point x_0 of Ω, then at this point the following equality holds:

$$\begin{aligned}\frac12 a^{ij}\bar v_{ij} &= a^{ij}\bar u_{ki}\bar u_{kj} + \bar v\left[K^2\left(\mathscr{E}_1 - p(\mathscr{E}_1)_p\right) + K\frac{\delta\mathscr{E}_1}{|p|} + K(a - pa_p) + \frac{\delta a}{|p|}\right]\\ &\quad + \sqrt{\bar v}\left[\delta a^{ij}\bar u_{ij} - K|p|p(a^{ij})_p\bar u_{ij}\right].\end{aligned} \tag{8.8}$$

We estimate the terms contained in the last square brackets in (8.8). Let $T = \|t_{ij}\|$ be the orthogonal matrix reducing the matrix $\|\bar u_{ij}\|$ at x_0 to the diagonal matrix $\|\tilde u_{ij}\|$ with $\tilde u_{ij} = 0$ for $i \neq j$, so that an equality of the form (6.11) holds. We set $\tilde A = T^*AT$. Noting then that $\tilde a^{ii} = A^{\tau_i} \equiv A\tau_i \cdot \tau_i$, $i = 1,\dots,n$, where $\tau_i = (t_{1i},\dots,t_{ni})$ is the ith column of the matrix T with $|\tau_i| = 1$, we obtain

$$\begin{aligned}K|p|p(a^{ij})_p\bar u_{ij} &= K|p|p(\tilde a^{ii})_p\tilde u_{ii} = K|p|p(A^{\tau_i})_p\tilde u_{ii},\\ \delta a^{ij}\bar u_{ij} &= \delta\tilde a^{ii}\tilde u_{ii} = \delta A^{\tau_i}\tilde u_{ii}.\end{aligned} \tag{8.9}$$

The definition of the vector τ_i implies the vector equality $\|\bar u_{ij}\|\tau_i = \tilde u_{ii}\tau_i$ for all $i = 1,\dots,n$. Taking the inner product of both sides of this equality with the vector $\nabla\bar u$, we obtain for any $i = 1,\dots,n$ the equality

$$\left(\|\bar u_{ij}\|\tau_i\right)\cdot\nabla\bar u = \tilde u_{ii}\left(\tau_i\cdot\nabla\bar u\right). \tag{8.10}$$

Writing the left side of (8.10) as a sum (over $k, l = 1,\dots,n$) $\bar{u}_{kl}t_{li}\bar{u}_k \equiv \frac{1}{2}\bar{v}_l t_{li}$ and noting that at x_0 the derivatives $\bar{v}_l = 0$, $l = 1,\dots,n$ (the necessary condition for an extreme), we obtain

$$\tilde{u}_{ii}(\tau_i \cdot \nabla\bar{u}) = 0, \qquad i = 1,\dots,n. \tag{8.11}$$

It follows from (8.11) that for any $i = 1,\dots,n$ either $\tilde{u}_{ii} = 0$ or $\tau_i \cdot p = 0$, since $\nabla\bar{u} \equiv \bar{p} = K\bar{u}p$, and $\bar{u} \neq 0$, since $\bar{u} = e^{Ku}$.

We now suppose that the point x_0 of the maximum in $\overline{\Omega}$ of $\bar{v}$ belongs to the domain Ω_L (i.e., $x_0 \in \Omega$ and $|\nabla u(x_0)| > L$). We shall prove that then

$$\begin{aligned} |\sqrt{\bar{v}}\,K|p|p(a^{ij})_p\bar{u}_{ij}| &\leqslant \tfrac{1}{2}a^{ij}\bar{u}_{ki}\bar{u}_{kj} + \tfrac{1}{2}nK^2\sigma_1\mathscr{E}_1\bar{v}, \\ |\sqrt{\bar{v}}\,\delta a^{ij}\bar{u}_{ij}| &\leqslant \tfrac{1}{2}a^{ij}\bar{u}_{ki}\bar{u}_{kj} + \tfrac{1}{2}n\mu_1\mathscr{E}_1\bar{v}. \end{aligned} \tag{8.12}$$

Suppose that for some index $i \in \{1,\dots,n\}$ at the point $x_0 \in \Omega_L$ in question the conditions $\tilde{a}^{ii} > 0$ and $\tilde{u}^{ii} \neq 0$ hold. In this case $\tau_i \cdot p = 0$, and therefore we can use the corresponding conditions of (8.1) to estimate $p(A^{\tau_i})_p$ and δA^{τ_i}. Applying the Cauchy inequality, we then obtain, for these values of i,

$$\begin{aligned} |\sqrt{\bar{v}}\,K|p|p(\tilde{a}^{ii})_p\tilde{u}_{ii}| &\leqslant \tfrac{1}{2}\tilde{a}^{ii}\tilde{u}_{ii}^2 + \tfrac{1}{2}K^2\sigma_1\mathscr{E}_1\bar{v}, \\ |\sqrt{\bar{v}}\,\delta\tilde{a}^{ii}\tilde{u}_{ii}| &\leqslant \tfrac{1}{2}\tilde{a}^{ii}\tilde{u}_{ii}^2 + \tfrac{1}{2}\mu_1\mathscr{E}_1\bar{v}. \end{aligned} \tag{8.13}$$

If for some index $i \in \{1,\dots,n\}$ at the point x_0 at least one of the numbers $\tilde{a}^{ii}$ or $\tilde{u}_{ii}$ is equal to 0, then the left sides in (8.13) are equal to 0, since for $\tilde{u}_{ii} = 0$ this is obvious, while for $\tilde{a}^{ii} = 0$ from the first two inequalities in (8.1) for $\tau = \tau_i$ it follows that $(\tilde{a}^{ii})_p p = 0$ and $\delta\tilde{a}^{ii} = 0$. Thus, in this case inequalities (8.13) are trivially satisfied. Summing (8.13) over all indices $i = 1,\dots,n$, we obtain (8.12). From (8.8) and (8.12) it follows that, under our assumption, at x_0 the following inequality holds:

$$\frac{1}{2}a^{ij}\bar{v}_{ij} \geqslant \left\{K^2\left[\mathscr{E}_1 - p(\mathscr{E}_1)_p - \frac{n\sigma_1}{2}\mathscr{E}_1\right] + K\left[\frac{\delta\mathscr{E}_1}{|p|} + a - pa_p\right] + \frac{\delta a}{|p|} - \frac{n\mu_1}{2}\mathscr{E}_1\right\}\bar{v}. \tag{8.14}$$

We suppose that $\sigma_1 \leqslant \mu_2/n$ and assume that the constant K is so large that $\frac{1}{2}\mu_2K^2 - (\mu_3 + \mu_4)K - (\mu_5 + n\mu_1/2) \geqslant (\mu_2/4)K$. From (8.14) it then follows that $\frac{1}{2}a^{ij}\bar{v}_{ij} > 0$ at x_0, which contradicts our assumption that at the point $x_0 \in \Omega_L$ the function $\bar{v}$ has its maximum value in $\overline{\Omega}$. Hence,

$$\max_{\overline{\Omega}} \bar{v} \leqslant \max\left\{\max_{\partial\Omega} \bar{v},\ \max_{\Omega\setminus\Omega_L} \bar{v}\right\}. \tag{8.15}$$

Recalling that $v = K^{-2}\bar{u}^{-2}\bar{v}$, we deduce from (8.15) that

$$\max_{\overline{\Omega}} v \leqslant \left(\max_{\overline{\Omega}} \bar{u}^2 / \min_{\overline{\Omega}} \bar{u}^2\right) \max\left\{\max_{\partial\Omega} v,\ L^2\right\}, \tag{8.16}$$

whence it easily follows that

$$\max_{\overline{\Omega}} |\nabla u| \leqslant e^{Km} \max\left\{\max_{\partial\Omega} |\nabla u|,\ L\right\}. \tag{8.17}$$

Theorem 8.1 is proved.

REMARK 8.1. As the proof shows, the result of Theorem 8.1 remains valid if in place of (8.1) we require that the following inequalities are satisfied on $\mathfrak{M}_{\Omega,m,L}$ for all τ, $|\tau| = 1$, $\tau \cdot p = 0$:

$$\begin{aligned} &\mathscr{E}_1 - p(\mathscr{E}_1)_p - \frac{n}{2}\frac{\left(pA_p^\tau\right)^2|p|^2}{A^\tau} \geqslant c_0 = \text{const} > 0, \\ &\frac{\delta\mathscr{E}_1 + \delta a}{\mathscr{E}_1|p|} - \frac{a - pa_p}{\mathscr{E}_1} - \frac{n}{2}\frac{|\delta A^\tau|^2}{A^\tau} \leqslant c_1 = \text{const} > 0, \qquad \mathscr{E}_1 > 0. \end{aligned} \tag{8.18}$$

It is obvious that both (8.1) and (8.18) admit the particular degeneracy of ellipticity of equation (1.1) characterized by the condition $\mathscr{E}_1 > 0$ on $\mathfrak{M}_{\Omega,m,L}$.

As an example of Theorem 8.1 we consider equation (7.13). Since for this equation for all τ, $|\tau| = 1$, $\tau \cdot p = 0$, the relations $A^\tau = 1 + |p|^{-2}$, $|pA_p^\tau| = 2|p|^{-2}$, $\mathscr{E}_1 = 1$ and $\sqrt{A^\tau\mathscr{E}_1}/|p| \geqslant |p|^{-1}$ hold, for $|p| > L$ the first condition in (8.1) is satisfied with $\sigma_1 = 4L^{-2}$, the second and fourth are satisfied with $\mu_1 = \mu_3 = 0$, and the third with $\mu_2 > 1$. It is then obvious that the smallness condition for σ_1 expressed by the inequality $\sigma_1 \leqslant \mu_2/n$ in the proof of Theorem 8.1 will be satisfied for (7.13) provided that $L > 2n$. Thus, in order that it be possible to apply Theorem 8.1 to (7.13) it is necessary that for some fixed $L > 2n$ the condition

$$|a - pa_p| \leqslant \mu_4, \quad \delta a \geqslant -\mu_5|p| \quad \text{on } \mathfrak{M}_{\Omega,m,L} \tag{8.19}$$

be satisfied for any constants μ_4 and μ_5.

Comparing (8.19) and (7.15), we observe that the second condition on the growth of the right side as $p \to \infty$ in (8.19) is weaker than the corresponding condition in (7.15), while the first condition in (8.19) is, generally speaking, stronger than the first condition in (7.15), but it may be weaker in the case of special structure of the function $a(x, u, p)$. In particular, we consider an equation of the form (1.10) for which

$$a(x, u, p) = \mathscr{H}(x, u, p)\frac{\left(1 + |p|^2\right)^{3/2}}{|p|^2}.$$

Taking into account

$$\frac{\left(1 + |p|^2\right)^{3/2}}{|p|^2} = |p| + \varphi(|p|),$$

where $\varphi(|p|) \sim (3/2)|p|^{-1}$ as $|p| \to \infty$, we write

$$\begin{aligned} a - pa_p &= \mathscr{H}\{[|p| + \varphi(|p|)] - p[|p| + \varphi(|p|)]_p\} - p\mathscr{H}_p[|p| + \varphi(|p|)] \\ &= \mathscr{H}(\varphi - p\varphi_p) - p\mathscr{H}_p[|p| + \varphi(|p|)], \quad \delta a = \delta\mathscr{H}[|p| + \varphi(|p|)]. \end{aligned} \tag{8.20}$$

Since $\varphi - p\varphi_p = O(|p|^{-1})$ and $|p| + \varphi(|p|) = O(|p|)$ as $p \to \infty$, (8.19) and (8.20) imply the following conditions on the growth of the function $\mathscr{H}(x, u, p)$:

$$|\mathscr{H}| \leqslant \mu_0|p|, \quad |p\mathscr{H}_p| \leqslant \mu_1/|p|, \quad \delta\mathscr{H} \geqslant -\mu_3 \quad \text{on } \mathfrak{M}_{\Omega,m,L}, \tag{8.21}$$

where μ_0, μ_1 and μ_3 are arbitrary constants.

Comparing (8.21) and (7.16), we note that the conditio generally speaking, a whole order stronger than the condition However, if we consider the case $\mathscr{H}(x, u, p) = h(x, u, p/|p|)$; then homogeneity of degree zero of the function h in p we have $ph_p = 0$, so special case to the conditions (8.21) there correspond the conditions

$$|h| \leqslant \mu_0, \qquad \delta h \geqslant -\mu_3 \quad \text{on } \mathfrak{M}_{\Omega,m,L} \tag{8.22}$$

with arbitrary constants μ_0 and μ_3, while conditions (7.16) for $\hat{H} = h(x, u, p/|p|)$ retain their form, i.e.,

$$|h| + |p|\,|h_p| \leqslant \sigma_2, \qquad \delta h \geqslant -\sigma_3 \quad \text{on } \mathfrak{M}_{\Omega,m,L} \tag{8.23}$$

with sufficiently small constants σ_2 and σ_3. It is clear that in the latter case conditions (8.22) are somewhat weaker than (8.23).

REMARK 8.2. The class of equations of the form (1.1) distinguished in Theorem 8.1 contains classes of equations such as the equation of a surface of given mean curvature, which have been singled out by various authors (see [163], [4] and [83]). In particular, in the monograph [83] a class of equations defined by the following conditions on $\mathfrak{M}_{\Omega,m,L}$ was distinguished:

$$\begin{gathered}\left(\sum_{i,j=1}^{n} (a^{ij})^2\right)^{1/2} \leqslant c_0\nu, \quad \nu\xi^2 \leqslant a^{ij}\xi_i\xi_j \leqslant c_1\nu\xi^2 \quad \text{for } \xi \in \mathbb{R}^n,\ \xi\cdot p = 0,\\ \nu = \nu(x, u, p) > 0, \quad |(a^{ij})_p p| \leqslant \sigma_1\sqrt{\mathscr{E}_1\nu}\,|p|^{-1}, \quad |\delta a^{ij}| \leqslant \mu_1\sqrt{\mathscr{E}_1\nu}\,,\\ \mathscr{E}_1 - p(\mathscr{E}_1)_p \geqslant \mu_2\mathscr{E}_1, \quad |\delta\mathscr{E}_1| \leqslant \mu_3\mathscr{E}_1|p|, \quad |a - pa_p| \leqslant \mu_4\mathscr{E}_1, \quad \delta a \geqslant -\mu_5\mathscr{E}_1|p|.\end{gathered} \tag{8.24}$$

Since for $\xi = \tau$, where $|\tau| = 1$ and $\tau \cdot p = 0$, (8.24) implies the inequality $\nu \leqslant A^\tau$, it follows from the condition $|p(a^{ij})_p| \leqslant \sigma_1\sqrt{\mathscr{E}_1\nu}\,|p|^{-1}$ that

$$|pA^\tau_p| \leqslant \sigma_1\sqrt{\mathscr{E}_1 A^\tau}\,|p|^{-1},$$

and the condition $|\delta a^{ij}| \leqslant \mu_1\sqrt{\mathscr{E}_1\nu}$ implies that

$$|\delta A^\tau| \leqslant \tilde{\mu}_1\sqrt{A^\tau\mathscr{E}_1}\,.$$

Thus, (8.24) implies (8.1). We remark also that the classes of equations of the form (1.1) distinguished in this connection in [163] and [4] are defined by certain assumptions which imply conditions (8.24).

§9. The existence theorem for a solution of the Dirichlet problem in the case of an arbitrary domain Ω with a sufficiently smooth boundary

In Theorem 2.1 the estimate (2.12) is postulated for solutions of problems of the form (2.11) under the assumption that these solutions belong to the class $C^2(\overline{\Omega})$ while in Theorems 6.1, 7.1 and 8.1 the estimates of $\max_{\overline{\Omega}}|\nabla u|$ are obtained for solutions of the class $C^3(\Omega) \cap C^1(\overline{\Omega})$. Because of this and also in connection with the desire to relax somewhat the conditions of Theorem 2.1 on the boundary $\partial\Omega$ and the boundary function φ, to be able to obtain existence theorems for the Dirichlet problem (1.3) on the basis of the results obtained above it is necessary to carry out certain supplementary arguments. These form the content of this and the next section.

THEOREM 9.1. *Let Ω be a bounded domain in $\mathbb{R}^n$, $n \geqslant 2$, of class C^2, and suppose that the functions $a^{ij}(x, u, p)$, $i, j = 1, \ldots, n$, and $a(x, u, p)$ are continuous in $\overline{\Omega} \times \mathbb{R} \times \mathbb{R}^n$ and have partial derivatives $\partial a^{ij}/\partial p_k$, $\partial a^{ij}/\partial u$, $\partial a^{ij}/\partial x_k$, $\partial a/\partial p_k$, $\partial a/\partial u$ and $\partial a/\partial x_k$ which are bounded on any compact set in $\overline{\Omega} \times \mathbb{R} \times \mathbb{R}^n$. Suppose condition* (1.2) *is satisfied, and assume that for any solution $v \in C^2(\overline{\Omega})$ of problem* (2.11) *for all $\tau \in [0, 1]$ the estimate*

$$\max_{\Omega} |v| \leqslant m \tag{9.1}$$

holds. Suppose that on $\mathfrak{N}_{m,l} \equiv \overline{D_0} \times \{|u| \leqslant m\} \times \{|p| > l\}$ (where D_0 is the domain defined by condition (3.2), *m is the constant in* (9.1), *and l is a nonnegative constant) the inequalities* (4.1) *and* (4.5) *are satisfied, while on the set $\mathfrak{M}_{\Omega,m,L} \equiv \overline{\Omega} \times \{|u| \leqslant m\} \times \{|p| > L\}$, where m is the constant in* (9.1) *and L is a nonnegative constant, conditions* (6.4) *and* (6.25) *hold. Suppose also that the function $\varphi \in C^2(\overline{\Omega})$. Then problem* (1.3) *has at least one classical solution u, and $u \in C^{1+\gamma}(\overline{\Omega}) \cap C^{2+\gamma}(\Omega)$ for some $\gamma \in (0, 1)$.*

PROOF. We first suppose that $\Omega \in C^3$, $\varphi \in C^3$, and the functions $a^{ij}(x, u, p)$, $i, j = 1, \ldots, n$, and $a(x, u, p)$ belong to the class $C^2(\overline{\Omega} \times \mathbb{R} \times \mathbb{R}^n)$. Suppose that a function $v \in C^2(\overline{\Omega})$ is a solution of problem (2.11) for some $\tau \in [0, 1]$. From the hypothesis of the theorem it follows that (9.1) holds for v. Since the functions $\tilde{a} = \tau a$ and $\tilde{\varphi} = \tau\varphi$, $\tau \in [0, 1]$, satisfy exactly the same conditions as a and φ, from Theorem 4.1 we obtain

$$\max_{\partial\Omega} |\nabla v| \leqslant M_1, \tag{9.2}$$

where M_1 does not depend on either v or τ. In order to use Theorem 6.1 now, it is first necessary to verify that actually $v \in C^3(\Omega)$. Indeed, since the functions

$$x \to a^{ij}(x, v(x), \nabla v(x)), \quad i, j = 1, \ldots, n, \qquad x \to a(x, v(x), \nabla v(x))$$

belong to $C^1(\overline{\Omega})$ while $\tau\varphi \in C^3(\Omega)$ and $\Omega \in C^3$, it follows from Schauder's theorem that $v(x) \in C^{2+\alpha}(\overline{\Omega})$, $\alpha \in (0, 1)$. Then the functions

$$x \to a^{ij}(x, v(x), \nabla v(x)), \quad i, j = 1, \ldots, n, \qquad x \to a(x, v(x), \nabla v(x))$$

belong to the class $C^{1+\alpha}(\overline{\Omega})$, $\alpha \in (0, 1)$. Again applying Schauder's theorem, we conclude that $v \in C^{3+\alpha}(\Omega)$. Since for (2.11) conditions (6.4) are satisfied with the same constants as for the original equation (1.1), from Theorem 6.1 we obtain the estimate

$$\max_{\Omega} |\nabla v| \leqslant \overline{M_1}, \tag{9.3}$$

where $\overline{M_1}$ does not depend either on v or τ. In view of (9.1), (9.3) it follows from Theorem 2.1 that (1.3) has at least one solution $u \in C^2(\overline{\Omega})$.

We shall now eliminate the superfluous assumptions of smoothness of Ω, φ, a^{ij}, and a. Let Ω, φ, a^{ij} and a be as in the formulation of Theorem 9.1. We use the standard method of approximating Ω, φ, a^{ij}, and a by the respective objects possessing the degree of smoothness used above, taking into account the compactness of the family of solutions of problems of the form (1.3) so obtained (see, for example, [163], p. 453). It may be assumed that the approximating domains $\tilde{\Omega}$ expand and are contained in Ω. We note that the constants M_1 and $\overline{M_1}$ in (9.2) and

(9.3) depend only on the C^2-norms of $\partial\Omega$ and φ and on the known quantities in those conditions on the structure of (1.1) stipulated in the formulation of Theorem 9.1. Therefore, for the solutions $\tilde{u}$ of the approximating problems we obtain the uniform estimate

$$\max_{\tilde{\Omega}} (|\tilde{u}| + |\nabla \tilde{u}|) \leqslant c.$$

Applying the theorem of Ladyzhenskaya and Ural'tseva, for these solutions we establish the uniform estimate

$$\|\tilde{u}\|_{C^{1+\gamma}(\overline{\tilde{\Omega}})} \leqslant c_1 \tag{9.4}$$

for some fixed $\gamma \in (0,1)$. From (9.4) and Schauder's theorem we obtain the uniform estimates

$$\|\tilde{u}\|_{C^{2+\gamma}(\overline{\Omega'})} \leqslant c_2(\Omega') \tag{9.5}$$

for each Ω', $\overline{\Omega}' \subset \tilde{\Omega}$. Applying now the classical Arzelà-Ascoli theorem, we find a sequence of solutions of the approximating problems which converges in $C^2_{\mathrm{loc}}(\Omega)$ to a function $u \in C^2(\Omega) \cap C(\overline{\Omega})$. It is obvious that this function is a solution of (1.3). From (9.4) it follows easily that this solution u belongs to $C^{1+\gamma}(\overline{\Omega})$. From Schauder's theorem it then follows that $u \in C^{2+\gamma}(\Omega) \cap C^{1+\gamma}(\overline{\Omega})$. Theorem 9.1 is proved.

REMARK 9.1. If all the conditions of Theorem 9.1 are satisfied and it is additionally required that $\Omega \in C^{2+\beta}$ and $\varphi \in C^{2+\beta}(\overline{\Omega})$, $\beta \in (0,1)$, then problem (1.3) has a solution $u \in C^{2+\gamma}(\overline{\Omega})$, $\gamma \in (0, \beta]$.

§10. Existence theorem for a solution of the Dirichlet problem in the case of a strictly convex domain Ω

The following results are established in exactly the same way as Theorem 9.1.

THEOREM 10.1. *Suppose that all the hypotheses of Theorem* 9.1 *are satisfied with the exception of condition* (4.5). *Suppose, moreover, that the domain* Ω *is strictly convex. Then problem* (1.3) *has at least one solution* $u \in C^{1+\gamma}(\overline{\Omega}) \cap C^{2+\gamma}(\Omega)$ *for some* $\gamma \in (0,1)$. *If, however,* $\Omega \in C^{2+\beta}$ *and* $\varphi \in C^{2+\beta}(\overline{\Omega})$, $\beta \in (0,1)$, *then* $u \in C^{2+\gamma}(\overline{\Omega})$ *for some* $\gamma \in (0, \beta]$.

THEOREM 10.2. *Let* Ω *be a strictly convex bounded domain in* $\mathbb{R}^n$, $n \geqslant 2$, *belonging to the class* C^2, *and suppose that the functions* $a^{ij}(x, u, p)$, $i, j = 1, \dots, n$, *and* $a(x, u, p)$ *are continuous in* $\overline{\Omega} \times \mathbb{R} \times \mathbb{R}^n$ *and have all their partial derivatives of first order bounded on any compact set in* $\overline{\Omega} \times \mathbb{R} \times \mathbb{R}^n$. *Let condition* (1.2) *be satisfied, and assume that for any solution* $v \in C^2(\overline{\Omega})$ *of problem* (2.11) *for any* $\tau \in [0,1]$ *the estimate* (9.1) *holds. Suppose that condition* (4.1) *is satisfied on the set* $\mathfrak{N}_{m,l}$, *while on the set* $\mathfrak{M}_{\Omega,m,L}$ *either conditions* (7.1) *and* (7.9) *or condition* (8.1) *is satisfied with* $\sigma_1 \leqslant \mu_2/n$. *Suppose, finally, that* $\varphi \in C^2(\overline{\Omega})$. *Then problem* (1.3) *has at least one solution* $u \in C^{1+\gamma}(\overline{\Omega}) \cap C^{2+\gamma}(\Omega)$ *for some* $\gamma \in (0,1)$. *If, however,* $\Omega \in C^{2+\beta}$ *and* $\varphi \in C^{2+\beta}(\overline{\Omega})$, $\beta \in (0,1)$, *then* $u \in C^{2+\gamma}(\overline{\Omega})$ *for some* $\gamma \in (0, \beta]$.

Before formulating the next theorem, we prove a lemma in which an a priori estimate of $\max_{\overline{\Omega}}|u|$ is established for solutions of problem (1.3). This lemma is also of independent interest.

LEMMA 10.1. *Let Ω be an arbitrary bounded domain in $\mathbb{R}^n$, $n \geqslant 2$, and suppose that the functions $a^{ij}(x, u, p)$, $i, j = 1, \dots, n$, and $a(x, u, p)$ are continuous in $\overline{\Omega} \times \mathbb{R} \times \mathbb{R}^n$ and satisfy condition* (1.2). *Assume that on the set $\mathscr{F}_{\Omega, m_0, l_0} \equiv \overline{\Omega} \times \{|u| \geqslant m_0\} \times \{|p| \geqslant l_0\}$, where m_0 and l_0 are some fixed positive numbers, the inequality*

$$a(x, u, p) \geqslant -(\delta_1|p| + \delta_2|u|)\operatorname{Tr} A \tag{10.1}$$

is satisfied; where δ_1 and δ_2 are nonnegative constants satisfying the condition $2\delta_1 + (1 + \hat{d}^2)\delta_2 < 2$, where $\hat{d}$ is a geometric characteristic of the domain Ω. Then for any classical solution u of problem (1.3) $\max_{\overline{\Omega}}|u|$ *can be bounded by a quantity depending only on* $\max_{\overline{\Omega}}|\varphi|$ *and on the structure of equation* (1.1). *If additionally on the set $\mathscr{F}_{\Omega, m_0, l_0}$ the condition*

$$\sum_{i,j=1}^{n} a^{ij} \geqslant \varepsilon_0 \operatorname{Tr} A, \qquad \varepsilon_0 = \text{const} > 0, \tag{10.2}$$

is satisfied, then the estimate of $\max_{\overline{\Omega}}|u|$ *indicated above is preserved when the constant δ_1 in condition* (10.1) *is replaced by an arbitrary constant $\mu \geqslant 0$; here the constant δ_2 must be sufficiently small, depending on ε_0, μ and the size of Ω.*

PROOF. Let $z = z(x)$ be a function of class $C^2(\overline{\Omega})$ such that $z > 0$ and $|\nabla z| > 0$ in $\overline{\Omega}$. We introduce a new unknown function $\bar{u}$ by setting $u = z\bar{u}$. It is obvious that in Ω the function $\bar{u}$ satisfies the identity

$$z a^{ij}\bar{u}_{ij} + b^k \bar{u}_k + a^{ij} z_{ij} \bar{u} = a, \tag{10.3}$$

where the form of b^k is irrelevant for subsequent arguments. We suppose that $\bar{u}$ achieves a greatest value at a point $x_0 \in \Omega$. Using the necessary conditions for an extremum, we then conclude that at this point

$$a - a^{ij} z_{ij} \bar{u} \leqslant 0, \tag{10.4}$$

where $a = a(x_0, z_0\bar{u}_0, (\nabla z)_0 \bar{u}_0)$, $z_0 = z(x_0)$, $\bar{u}_0 = \bar{u}(x_0)$ and $(\nabla z)_0 = \nabla z(x_0)$.

We fix a number m_1 satisfying the condition

$$m_1 = \max\left\{m_0, \frac{\max_{\overline{\Omega}} z}{\min_{\overline{\Omega}}|\nabla z|} l_0\right\}. \tag{10.5}$$

We suppose that $\bar{u}_0 \geqslant m_1/z_0$. Then $u_0 \geqslant m_1 \geqslant m_0$, and in view of (10.5)

$$|(\nabla z)_0 \bar{u}_0| = |(\nabla z)_0| u_0/z_0 \geqslant l_0 m_1/m_1 = l_0.$$

At x_0 it is therefore possible to apply condition (10.1). It follows that at x_0

$$a \geqslant -(\delta_1|(\nabla z)_0| + \delta_2 z_0)\bar{u}_0 \operatorname{Tr} A. \tag{10.6}$$

Suppose that the function z satisfies the conditions

$$\begin{gathered} -a^{ij} z_{ij} - (\delta_1|\nabla z| + \delta_2 z)\operatorname{Tr} A > 0 \quad \text{in } \Omega, \\ z > 0 \quad \text{in } \overline{\Omega}, \qquad |\nabla z| > 0 \quad \text{in } \overline{\Omega}. \end{gathered} \tag{10.7}$$

If (10.2) is not satisfied, we set $z = 1 + \hat{d}^2 - |x|^2$, where $\hat{d} = \sup_{\overline{\Omega}}|x|$, and we may assume with no loss of generality that $\inf_{\overline{\Omega}}|x| \geqslant 1$. Then, taking into account that on $\overline{\Omega}$ we have the relations $z \geqslant 1$, $z \leqslant 1 + \hat{d}^2$, $z_i = -2x_i$, $|\nabla z| \geqslant 2$ and $z_{ij} = -2\delta_i^j$, it is easy to see that all the conditions (10.7) are satisfied if

$$2\delta_1 + (1 + \hat{d}^2)\delta_2 < 2. \tag{10.8}$$

Now if (10.8) is satisfied it follows from (10.4), (10.6), and (10.7) that the assumption made above that $\bar{u}$ achieves its greatest value at a point $x_0 \in \Omega$ where $\bar{u}_0 \geqslant m_1/z_0$ is impossible. Therefore,

$$\max_{\bar{\Omega}} \bar{u} \leqslant \max\Big\{ \max_{\partial\Omega} \bar{u},\ m_1/z_0 \Big\}. \tag{10.9}$$

We then have

$$\max_{\bar{\Omega}} u \leqslant \Big(\max_{\bar{\Omega}} z / \min_{\bar{\Omega}} z \Big) \max\Big\{ \max_{\partial\Omega} |\varphi|,\ m_1 \Big\}. \tag{10.10}$$

Taking the form of the function z and (10.5) into account, we deduce from (10.10) that

$$\max_{\bar{\Omega}} u \leqslant (1 + \hat{d}^2) \max\Big\{ \max_{\partial\Omega} |\varphi|,\ m_0, (1 + \hat{d}^2) l_0 \Big\}. \tag{10.11}$$

Since $\tilde{u} = -u$ is a solution of an equation having precisely the same structure as the original equation, it follows that $\max_{\bar{\Omega}}(-u)$ can also be bounded by the right side of (10.11). Thus,

$$\max_{\bar{\Omega}} |u| \leqslant (1 + \hat{d}^2) \max\Big\{ \max_{\partial\Omega} |\varphi|,\ m_0, (1 + \hat{d}^2) l_0 \Big\}. \tag{10.12}$$

We now suppose that (10.2) is also satisfied. Setting $z = 1 + e^{\beta n \hat{d}} - e^{-\beta \Sigma_1^n x_k}$, where $\beta = \text{const} > 0$ and $\hat{d} = \max_{\bar{\Omega}} |x|$, and taking into account that on $\bar{\Omega}$ the relations $z \geqslant 1$, $z \leqslant 1 + e^{\beta n \hat{d}}$, $z_i = \beta e^{-\beta \Sigma_1^n x_k}$, $|\nabla z| \geqslant \beta e^{-\beta \Sigma_1^n x_k}$ and $z_{ij} = -\beta^2 e^{-\beta \Sigma_1^n x_k}$ are then satisfied, we easily see that all the conditions (10.7) are satisfied if

$$\varepsilon_0 \beta > \delta_1, \quad \delta_2 (1 + e^{\beta n \hat{d}}) < \varepsilon_0 \beta - \delta_1. \tag{10.13}$$

But if conditions (10.13) are satisfied, it follows from (10.4), (10.6), and (10.7) that our assumption that $\max_{\bar{\Omega}} \bar{u}$ is achieved at a point $x_0 \in \Omega$ where $\bar{u}_0 \geqslant m_1/z_0$ is impossible. From this, as in the proof of the first part of the theorem, we deduce (10.10). Now, taking the form of z and (10.5) into account, we obtain

$$\max_{\bar{\Omega}} |u| \leqslant (1 + e^{\beta n \hat{d}}) \max\Big\{ \max_{\partial\Omega} |\varphi|,\ m_0, (1 + e^{\beta n \hat{d}}) l_0 \Big\}. \tag{10.14}$$

Lemma 10.1 is proved.

THEOREM 10.3. *Let $\Omega \subset \mathbb{R}^n$, $n \geqslant 2$, be a bounded, strictly, convex domain of class C^2, and suppose that the functions $a^{ij}(x, u, p)$, $i, j = 1, \dots, n$, and $a(x, u, p)$ are continuous in $\bar{\Omega} \times \mathbb{R} \times \mathbb{R}^n$ and have all their partial derivatives of first order bounded on any compact set in $\bar{\Omega} \times \mathbb{R} \times \mathbb{R}^n$. Suppose that condition* (1.2) *is satisfied, and that on $\mathscr{F}_{\Omega, m_0, l_0}$ either conditions* (10.1) *and* (10.8) *or conditions* (10.1), (10.2) *and* (10.13) *for any constant $\varepsilon_0 > 0$ are satisfied. Suppose further that condition* (4.1) *is satisfied on $\mathfrak{N}_{m,l}$, and conditions* (7.1) *and* (7.9) *are satisfied on $\mathfrak{M}_{\Omega, m, L}$. Then the result of Theorem* 10.2 *holds.*

PROOF. In view of Theorem 10.2, to prove Theorem 10.3 it suffices to establish an a priori estimate of the form (9.1) for any solution $v \in C^2(\bar{\Omega})$ of problem (2.11) and any $\tau \in [0, 1]$ (an estimate that is independent of both v and τ). Now such an estimate obviously follows from Lemma 10.1. Theorem 10.3 is proved.

REMARK 10.1. In condition (10.1), which is part of the hypothesis of Theorem 10.3, the constant δ_2 which is subject to a smallness condition cannot be replaced by an arbitrary constant $\mu \geqslant 0$ while preserving the result of Theorem 10.3, since it is known that there exist values $\lambda \in \mathbb{R}$ for which the Dirichlet problem

$$\Delta u + \lambda u = 0 \quad \text{in } \Omega, \qquad u = \varphi \quad \text{on } \partial\Omega$$

has no classical solution for certain $\varphi \in C^\infty(\overline{\Omega})$ in the disk $\Omega \equiv \{|x| \leqslant 1\}$.

CHAPTER 2

THE FIRST BOUNDARY VALUE PROBLEM FOR QUASILINEAR, NONUNIFORMLY PARABOLIC EQUATIONS

§1. A conditional existence theorem

Let $Q = \Omega \times (0, T]$, where Ω is a bounded domain in $\mathbb{R}^n$, $n \geqslant 1$, and $T > 0$. In Q we consider the quasilinear equation

$$\mathscr{L}u \equiv -u_t + a^{ij}(x, t, u, \nabla u)u_{x_i x_j} - a(x, t, u, \nabla u) = 0, \tag{1.1}$$

where $a^{ij} = a^{ji}$, $x = (x_1, \dots, x_n)$ and $\nabla u = (u_{x_1}, \dots, u_{x_n})$, which satisfies the parabolicity condition

$$a^{ij}(x, t, u, p)\xi_i\xi_j > 0 \ \ \forall \xi \in \mathbb{R}^n, \xi \neq 0, \forall (x, t, u, p) \in \overline{Q} \times \mathbb{R} \times \mathbb{R}^n. \tag{1.2}$$

Regarding the functions $a^{ij}(x, t, u, p)$, $i, j = 1, \dots, n$, and $a(x, t, u, p)$, it is henceforth always assumed that they are at least continuous in $\overline{Q} \times \mathbb{R} \times \mathbb{R}^n$.

We denote by Γ the parabolic boundary of the cylinder Q, $\Gamma = (\partial\Omega \times [0, T]) \cup (\Omega \times \{t = 0\})$, and by Γ' the part of Γ consisting of points not belonging to the set $\partial\Omega \times \{t = 0\}$, i.e.,

$$\Gamma' = (\partial\Omega \times (0, T]) \cup (\Omega \times \{t = 0\}).$$

Let $C^{2,1}(Q)$ ($C^{2,1}(\overline{Q})$) denote the set of all functions $u(x, t)$ continuous in Q ($\overline{Q}$) together with u_t, u_{x_i} and $u_{x_i x_j}$, $i, j = 1, \dots, n$. Similarly, $C^{2,1}(Q \cup \Gamma')$ denotes the set of all functions $u(x, t)$ continuous together with u_t, u_{x_i} and $u_{x_i x_j}$ in $Q \cup \Gamma'$. Let Q' be an arbitrary compact set contained in $\overline{Q}$. We denote by $C_\alpha(Q')$ the set of all functions $u(x, t)$ satisfying the inequality

$$\|u\|_{\alpha, Q'} \equiv \max_{Q'} |u(x, t)| + \max_{(x,t),(x',t') \in Q'} \frac{|u(x, t) - u(x', t')|}{(|x - x'|^2 + |t - t'|)^\alpha} \leqslant K.$$

where $K = \text{const} \geqslant 0$ and $\alpha = \text{const} \in (0, 1)$.

Let $C_{1+\alpha}(Q')$ ($C_{2+\alpha}(Q')$), $\alpha \in (0, 1)$, be the set of all functions $u(x, t)$ for which u, $u_{x_i} \in C^\alpha(Q')$, $i = 1, \dots, n$ ($u, u_{x_i}, u_{x_i x_j}, u_t \in C^\alpha(Q')$, $i, j = 1, \dots, n$). $C_{2+\alpha}(Q \cup \Gamma')$ denotes the set of all $u(x, t)$ belonging to $C_{2+\alpha}(Q')$ for any compact set $Q' \subset Q \cup \Gamma$.

Finally, for $\alpha \in (0,1)$ let

$$\|u\|_{1+\alpha,Q'} \equiv \|u\|_{\alpha,Q'} + \sum_{i=1}^{n} \|u_{x_i}\|_{\alpha,Q'};$$

$$\|u\|_{2+\alpha,Q'} \equiv \|u\|_{1+\alpha,Q'} + \sum_{i,j=1}^{n} \|u_{x_i x_j}\|_{\alpha,Q'} + \|u_t\|_{\alpha,Q'}.$$

In this chapter we study the question of classical solvability of the first boundary value problem for equation (1.1), i.e., the problem

$$\mathscr{L}u = 0 \quad \text{in } Q, \qquad u = \varphi \quad \text{on } \Gamma, \tag{1.3}$$

where $\varphi = \varphi(x, t)$ is a given function. Here a classical solution of (1.3) is understood to be any function $u \in C^{2,1}(Q) \cap C(\overline{Q})$ satisfying (1.1) in Q and coinciding with φ on the parabolic boundary Γ. We shall set forth the basic results which play the crucial role in reducing the proof of classical solvability of problem (1.3) to the problem of constructing an a priori estimate of $\max_{\overline{Q}}(|u| + |\nabla u|)$ for solutions of a suitable family of one-parameter boundary value problems related to (1.3).

THEOREM 1.1 (A. FRIEDMAN AND V. A. SOLONNIKOV). *Let $\Omega \in C^{2+\alpha}$, $\varphi \in C_{2+\alpha}(\overline{Q})$, $A^{ij}(x,t) \in C_\alpha(\overline{Q})$ and $f(x,t) \in C_\alpha(\overline{Q})$, where $\alpha \in (0,1)$. Then the linear problem*

$$-W_t + A^{ij}(x,t)W_{x_i x_j} - f(x,t) = 0 \quad \text{in } Q, \qquad W = \varphi \quad \text{on } \Gamma, \tag{1.4}$$

where $A^{ij}(x,t)\xi_i\xi_j \geqslant \nu|\xi|^2$, $\nu = \text{const} > 0$ and $(x,t) \in Q$, has a unique solution $W \in C_{2+\alpha}(Q \cup \Gamma') \cap C_{1+\beta}(\overline{Q})$ for any $\beta \in (0,1)$; and

$$\|W\|_{2+\alpha,\overline{Q}'} \leqslant c_1 \tag{1.5}$$

and

$$\|W\|_{1+\beta,\overline{Q}} \leqslant c_2, \tag{1.6}$$

where $\overline{Q}'$ is any compact set contained in $Q \cup \Gamma'$,

$$c_1 = c_1\big(n, \nu, \|a^{ij}\|_{\alpha,\overline{\omega}}, \|f\|_{\alpha,\overline{Q}}, \|\varphi\|_{2+\alpha,\overline{Q}}, d\big),$$

d is the distance from Q' to $\partial\Omega \times \{t = 0\}$, and

$$c_2 = c_2\big(n, \nu, \|a^{ij}\|_{\alpha,\overline{Q}}, \|f\|_{\alpha,\overline{Q}}, \|\varphi\|_{2+\alpha,\overline{Q}}, \beta\big).$$

If the boundary function φ satisfies the compatibility condition on $\partial\Omega \times \{t = 0\}$

$$-\varphi_t + A^{ij}(x,t)\varphi_{x_i x_j} - f(x,t) = 0, \tag{1.7}$$

then the solution $W \in C_{2+\alpha}(\overline{Q})$.

PROOF. Theorem 1.1 is a combination of well-known results of Friedman [123] and Solonnikov [116] (see also [80], Russian pp. 260–261 and 388–389, English pp. 223–224 and 341–342). In particular, the estimate (1.5) was obtained by Friedman, and (1.6) by Solonnikov.

THEOREM 1.2 (LADYZHENSKAYA AND URAL'TSEVA). *Suppose that a function $u \in C^{2,1}(Q \cup \Gamma') \cap C(\overline{Q})$ satisfies (1.1) in Q and coincides on Γ with a function $\varphi(x,t) \in C^{2,1}(\overline{Q})$. Suppose that at this solution*

$$\nu|\xi|^2 \leqslant a^{ij}(x,t,u(x,t),\nabla u(x,t))\xi_i\xi_j \leqslant \mu|\xi|^2, \quad \forall \xi \in \mathbb{R}^n;\ \nu, \mu = \text{const} > 0,$$

and the coefficients $a^{ij}(x, t, u, p)$, $i, j = 1, \ldots, n$, *and* $a(x, t, u, p)$ *are continuous together with all their partial derivatives of first order in the region*

$$\{(x, t) \in \overline{Q}\} \times \{|u| \leqslant m\} \times \{|p| \leqslant M\},$$

where $m = \max_{\overline{Q}}|u(x, t)|$, $M = \max_{\overline{Q}}|\nabla u(x, t)|$. *Finally, suppose that* $\partial\Omega \in C^2$. *Then there exists* $\alpha \in (0, 1)$ *for which* $u \in C_{1+\alpha}(\overline{Q})$, *and*

$$\|u\|_{1+\alpha,\overline{Q}} \leqslant c_3, \tag{1.8}$$

where the constant c_3 *depends only on* n, ν, μ, m, M, *the upper bounds in* $\overline{Q}$ *of the moduli of the functions* a^{ij}, $\partial a^{ij}/\partial p_k$, $\partial a^{ij}/\partial u$, $\partial a^{ij}/\partial x_k$, $\partial a^{ij}/\partial t$, a, $\partial a/\partial p_k$, $\partial a/\partial u$, $\partial a/\partial x_k$ *and* $\partial a/\partial t$, $i, j, k = 1, \ldots, n$, *computed at the solution in question, on* $\|\varphi\|_{C^{2,1}(\overline{Q})}$, *and on the* C^2-*norms of the functions describing the boundary* $\partial\Omega$. *The exponent* α *is determined by these same quantities.*

A proof of Theorem 1.2 is contained, in particular, in the monograph [80] (see Russian pp. 608–609 and 505, English pp. 532–533 and 446).

We now present a well-known theorem on a fixed point of a compact operator in a Banach space which will be used below.

LERAY - SCHAUDER THEOREM (in Schäffer's form). *Let* T *be a compact operator taking the Banach space* B *into itself. If for any elements* $v \in B$ *satisfying the equation*

$$v = \tau T v, \qquad \tau \in [0, 1],$$

the inequality $\|v\|_B \leqslant c$ *holds with a constant* c *not depending either on* v *or on* $\tau \in [0, 1]$, *then the operator* T *has at least one fixed point in* B, *i.e., there exists* $v \in B$ *for which* $v = Tv$.

We now proceed to the proof of the main theorem of this section.

THEOREM 1.3. *Suppose that the functions* $a^{ij}(x, t, u, p)$, $i, j = 1, \ldots, n$, *and* $a(x, t, u, p)$ *belong to the class* C^1 *on the set* $\overline{Q} \times \{|u| \leqslant m\} \times \{|p| \leqslant M\}$ *and satisfy the condition* (1.2), *and that* $\Omega \in C_{2+\gamma}$ *and* $\varphi \in C_{2+\gamma}(\overline{Q})$, $\gamma \in (0, 1)$. *If an arbitrary solution* $v \in C^{2,1}(\overline{Q})$ *of the problem*

$$-v_t + a^{ij}(x, t, v, \nabla v) v_{x_i x_j} - \tau a(x, t, v, \nabla v) = 0 \quad \text{in } Q, \quad v = \tau\varphi \quad \text{on } \Gamma, \tag{1.9}$$

where $\tau \in [0, 1]$, *satisfies the inequalities*

$$\max_{\overline{Q}}|v| \leqslant m, \quad \max_{\overline{Q}}|\nabla v| \leqslant M, \tag{1.10}$$

where m *and* M *do not depend either on* v *or* $\tau \in [0, 1]$, *and on* $\partial\Omega \times \{t = 0\}$ *the conditions* $\varphi = 0$, $\nabla\varphi = 0$, *and*

$$-\varphi_t + a^{ij}(x, t, \varphi, \nabla\varphi)\varphi_{x_i x_j} - a(x, t, \varphi, \nabla\varphi) = 0 \tag{1.11}$$

are satisfied, then problem (1.3) *has at least one solution* $u \in C_{2+\alpha}(\overline{Q})$, $\alpha \in (0, \gamma]$.

PROOF. Let $w \in C_{1+\alpha}(\overline{Q})$, where the choice of the exponent $\alpha \in (0, \gamma)$ will be specified below. We consider the linear problem

$$\begin{gathered} -W_t + a^{ij}(x, t, w, \nabla w) W_{x_i x_j} - a(x, t, w, \nabla w) = 0 \quad \text{in } Q, \\ W = \varphi \quad \text{on } \Gamma, \end{gathered} \tag{1.12}$$

i.e., a problem of the form (1.4) with $A^{ij}(x, t) = a^{ij}(x, t, w, \nabla w)$, $i, j = 1, \ldots, n$, and $f(x, t) = a(x, t, w, \nabla w)$. It is obvious that for problem (1.12) all the conditions of Theorem 1.1 are satisfied. Therefore, (1.12) has a unique solution

$$W \in C_{2+\alpha}(Q \cup \Gamma') \cap C_{1+\beta}(\overline{Q}) \quad \text{for all } \beta \in (0,1)$$

which defines an operator T mapping the Banach space $C_{1+\alpha}(\overline{Q})$ into itself. It is easy to see that this operator is compact. Indeed, by Solonnikov's estimate (1.6) any bounded set in $C_{1+\alpha}(\overline{Q})$ is mapped by T into a set bounded in $C_{1+\beta}(\overline{Q})$, where $\beta > \alpha$. Since the imbedding of $C_{1+\beta}(\overline{Q})$ in $C_{1+\alpha}(\overline{Q})$ is compact, this implies the compactness of T.

We shall now prove that the set of all fixed points of the operators τT, $\tau \in [0,1]$, is bounded in $C_{1+\alpha}(\overline{Q})$ for an appropriate choice of $\alpha \in (0,1)$. Let $v = \tau T v$, $\tau \in [0,1]$, and $v \in C_{2+\alpha}(Q \cup \Gamma') \cap C_{1+\beta}(\overline{Q})$, for all $\beta \in (0,1)$. From the definition of T it then follows that such a function v is a solution of (1.9). In view of (1.11) and Theorem 1.1, $v \in C^{2+\alpha}(\overline{Q}) \subset C^{2,1}(\overline{Q})$. According to Theorem 1.3, inequalities (1.10) hold for this function. Applying Theorem 1.2, we conclude that (1.8) holds for v for some particular value of $\alpha \in (0, \gamma]$. We take this value of α as the selected value in considering the space $C_{1+\alpha}(\overline{Q})$. Applying the Leray-Schauder theorem in Schäffer's form, we conclude that problem (1.3) has at least one solution $u \in C_{2+\alpha}(\overline{Q})$. Theorem 1.3 is proved.

§2. Estimates of $|\nabla u|$ on Γ

LEMMA 2.1 (D. E. EDMUND AND L. A. PELETIER). *Suppose that a function $u \in C^{2,1}(Q) \cap C^{1,0}(\overline{Q})$, where $C^{1,0}(\overline{Q})$ is the set of all functions $u(x,t)$ which are continuous in $\overline{Q}$, together with the derivatives u_{x_i}, $i = 1, \ldots, n$, satisfies an equation of the form* (1.1) *in the cylinder Q, where it is assumed that condition* (1.2) *is satisfied and that a^{ij}, $\partial a^{ij}/\partial p_k$, a and $\partial a/\partial p_k$, $i, j, k = 1, \ldots, n$, are continuous functions of their arguments in $\overline{Q} \times \mathbb{R} \times \mathbb{R}^n$. Suppose that the* (*barrier*) *function $\omega(x,t) \in C^{2,1}(Q) \cap C(\overline{Q})$ for any constant $c \geqslant 0$ satisfies in $\Omega \times (0, T]$ the inequality*

$$\mathscr{L}(\omega + c) \equiv -\omega_t + a^{ij}(x, t, \omega + c, \nabla\omega)\omega_{x_i x_j} - a(x, t, \omega + c, \nabla\omega) \leqslant 0. \quad (2.1)$$

If $u(x,t) \leqslant \omega(x,t)$ on the parabolic boundary Γ of Q, then $u(x,t) \leqslant \omega(x,t)$ also throughout Q.

A proof is given in [136]. It is based on applying the strict maximum principle for linear parabolic equations.

THEOREM 2.1. *Let $\Omega \in C^3$, and suppose that the functions a^{ij}, $\partial a^{ij}/\partial p_k$, a and $\partial a/\partial p_k$, $i, j, k = 1, \ldots, n$, are continuous and satisfy condition* (1.2) *in $\overline{Q} \times \mathbb{R} \times \mathbb{R}^n$. Suppose that on the set $\{D_0 \times (0, T]\} \times \{|u| \leqslant m\} \times \{|p| > l\}$, where D_0 is defined by condition* (1.3.2), $m = \text{const} \geqslant 0$, $l = \text{const} \geqslant 0$, *the inequality*

$$|a(x, t, u, p)| \leqslant \psi(|p|)\mathscr{E}_1 + \delta(|p|)\mathscr{E}_2 \quad (2.2)$$

holds, where $\mathscr{E}_1 \equiv a^{ij}(x, t, u, p)p_i p_j$, $\mathscr{E}_2 = \operatorname{Tr} A|p|$, $A \equiv \|a^{ij}(x, t, u, p)\|$, $\psi(\rho)$, $0 \leqslant \rho < +\infty$, is a positive, monotonic, continuous function satisfying condition (1.4.2), *and $\delta(\rho)$, $0 < \rho < +\infty$, is a nonnegative nonincreasing function satisfying* (1.4.3). *Let $u \in C^{2,1}(Q) \cap C^{1,0}(\overline{Q})$ be an arbitrary solution in $D_0 \times (0, T)$ of* (1.1) *satisfying the condition $u = \varphi$ on Γ for $\varphi \in C^{3,1}(\overline{Q})$, where $C^{3,1}(\overline{Q})$ is the set of functions $u(x,t)$*

continuous in $\overline{Q}$ together with their derivatives u_{x_i}, $u_{x_ix_j}$, $u_{x_ix_jx_k}$, and u_t, and the inequality $|u| \leqslant m$ in $D_0 \times (0, T)$. Further, assume that the condition

$$\mathscr{E}_1\psi(|p|) + \mathscr{E}_2 \to \infty \quad \text{as } p \to \infty \ (\text{uniformly in } (x,t) \in \overline{Q},\, u \in [-m, m]) \tag{2.3}$$

is satisfied. Then the following assertions are true:

1) *If the domain Ω is strictly convex, then*

$$|\partial u(y,t)/\partial \nu| \leqslant M_1 \quad \text{on } \partial\Omega \times (0, T), \tag{2.4}$$

where $\partial u/\partial \nu$ is the derivative in the direction of the inner normal to $\partial\Omega \times (0, T]$ at the point $(y, t) \in \partial\Omega \times (0, T]$ and the constant M_1 depends only on m, l, $\psi(\rho)$, $\delta(\rho)$, $\|\varphi\|_{C^{2,1}(\overline{Q})}$, k^{-1} and K, where

$$k = \inf_{\substack{i=1,\dots,n-1\\ y\in\partial\Omega}} k_i(y), \qquad K = \sup_{\substack{i=1,\dots,n-1\\ y\in\partial\Omega}} k_i(y),$$

and the $k_i(y)$ are the principal curvatures of the surface $\partial\Omega$ at the point y.

2) *If on the set $\{D_0 \times (0, T]\} \times \{|u| \leqslant m\} \times \{|p| > l\}$ the inequality*

$$\mathscr{E}_2 \leqslant \psi(|p|)\mathscr{E}_1 \tag{2.5}$$

holds, where $\psi(\rho)$, $0 \leqslant \rho < +\infty$, is the same function as in (2.2), then (2.4) holds with a constant M_1 depending only on m, l, $\psi(\rho)$, $\|\varphi\|_{C^{2,1}(\overline{Q})}$, and

$$K \equiv \sup_{\substack{i=1,\dots,n-1\\ y\in\partial\Omega}} |k_i(y)|.$$

PROOF. We start by proving the first part of the theorem. We shall assume that condition (2.2) is satisfied for all $u \in \mathbb{R}$ (see the proof of Theorem 1.4.1). In $D_0(0, T]$ we consider the barrier function

$$\omega(x,t) = \varphi(x,t) + h(d), \tag{2.6}$$

where $h(d) \equiv h(d(x))$, $d(x) = \operatorname{dist}(x, \partial\Omega)$, $x \in D_0$, and $h(d) \in C^2(0, \delta) \cap C([0, \delta])$, $0 < \delta < \delta_0$; we shall specify the choice of the function $h(d)$ and the number δ below. Applying Lemma 1.3.2, for all $c = \text{const} > 0$ we obtain

$$\mathscr{L}(\omega + c) \leqslant \mathscr{F}\frac{h'' + Kh'}{h'^2} - k \operatorname{Tr} A h' + \left(-\varphi_t + a^{ij}\varphi_{x_ix_j} - a\right), \tag{2.7}$$

where

$$\mathscr{F} = A(p - p_0)(p - p_0), \qquad A = \|a^{ij}(x, t, \omega(x,t) + c, \nabla\omega(x,t))\|,$$
$$p = \nabla\omega(x,t), \qquad p_0 = \nabla\varphi(x,t), \qquad p = p_0 + \nu h',$$
$$a = a(x, t, \omega(x,t) + c, \nabla\omega(x,t)).$$

Here and everywhere in this chapter, ∇ denotes the spatial gradient.

Inequality (2.7) differs from (1.4.6) only by the term $-\varphi_t$ on the right side. Since conditions (2.2) and (1.4.1) are completely analogous, just as in the proof of the right side of Theorem 1.4.1, we obtain the estimate

$$\mathscr{L}(\omega + c) \leqslant |\varphi_t| + Fh'\left\{\frac{h''}{h'^3} + \frac{K}{h'^2} + \frac{2\psi(h' \pm c_\varphi)}{h'}\right\}$$
$$+ h'\left\{-k + \frac{c_\varphi}{h'} + \frac{4c_\varphi^2\psi(h' \pm c_\varphi)}{h'} + 2\delta(h' - c_\varphi)\right\} \operatorname{Tr} A, \tag{2.8}$$

where

$$c_\varphi = \max_{\overline{\Omega}} \left(|\varphi| + |\nabla \varphi| + |D^2\varphi| \right), \qquad |D^2\varphi|^2 = \sum_{i,j=1}^{n} \varphi_{x_i x_j}^2,$$

provided that $h'(d) \geqslant c_\varphi + l + 1$ on $[0, \delta]$. Taking the properties of the functions $\psi(\rho)$ and $\delta(\rho)$ into account and assuming that $h'(d) \geqslant \alpha$, where $\alpha = \max\{c_\varphi + l + 1, \alpha_0, \alpha_1\}$, α_0 is chosen so that

$$\frac{c_\varphi}{\rho} + \frac{4c_\varphi^2 \psi(\rho \pm c_\varphi)}{\rho} + 2\delta(\rho - c_\varphi) \leqslant \frac{k}{2},$$

and $\alpha_1 > 0$ will be chosen below, we deduce from (2.8) that

$$\mathscr{L}(\omega + c) \leqslant |\varphi_t| + \mathscr{F} h' \left\{ \frac{h''}{h'^2} + \Phi(h') \right\} - \frac{k}{2} \operatorname{Tr} A h' - \frac{1}{2} \mathscr{F} h' \Phi(h'), \qquad (2.9)$$

and (2.9) is valid for all $(x, t) \in D \times (0, T]$, where $D \equiv \{x \in D_0\colon \operatorname{dist}(x, \partial\Omega) < \delta\}$ and $\Phi(\rho) = 2K(\rho^{-2} + 2\psi(\rho + c_\varphi)\rho^{-1})$.

We now specify the choice of $h(d)$ and δ. Let δ be defined by (1.3.8) with $\bar{\alpha} = \max(\alpha, q\delta_0^{-1})$, $q = c_\varphi + m$, and suppose that $h(d)$ is defined on $[0, \delta]$ by (1.3.9). From Lemma 1.3.3 it then follows that $\delta \in (0, \delta_0)$ and

$$\mathscr{L}(\omega + c) \leqslant \max_{\overline{Q}} |\varphi_t| - \tfrac{1}{2} \mathscr{F} h' \Phi(h') - \tfrac{k}{2} \operatorname{Tr} A h' \quad \text{on } D \times (0, T). \qquad (2.10)$$

Taking into account (1.4.8) and the fact that $\Phi(\rho)$ may be assumed to be decreasing, we write in place of (2.10)

$$\mathscr{L}(\omega + c) \leqslant \max_{\overline{Q}} |\varphi_t| - \tfrac{1}{4} \mathscr{E}_1 h' \Phi(h') - \tfrac{k}{4} \operatorname{Tr} A h' \quad \text{on } D \times (0, T], \qquad (2.11)$$

provided that α_1 is sufficiently large. Taking (2.3) and the relations $h' \geqslant |p| - c_\varphi \geqslant \frac{1}{2}|p|$ into account, we can choose α_1 so large that the right side of (2.11) is certainly negative, i.e.,

$$\mathscr{L}(\omega + c) \leqslant 0 \quad \text{on } D \times (0, T]. \qquad (2.12)$$

Here we have also used the fact that $\frac{1}{4} h' \Phi(h') \geqslant \psi(h' \pm c_\varphi) \geqslant \psi(|p|)$. It is obvious that $u \leqslant \omega$ on $\partial\Omega \times (0, T]$ and on $D \times \{t = 0\}$. On the set $\{x \in \Omega\colon d(x) = \delta\} \times (0, T]$, however, we have the inequalities $u \leqslant m = (m + c_\varphi) - c_\varphi = h(\delta) - c_\varphi \leqslant \omega$. Thus, $u \leqslant \omega$ on the parabolic boundary of the cylinder $D \times (0, T]$. Applying Lemma 2.1, we conclude that $u \leqslant \omega$ also in $D \times (0, T]$. From this (see (1.4.12) and (1.4.13)) we obtain (2.4).

The first part of Theorem 2.1 has thus been proved. We now prove the second part assuming again that condition (2.2) is satisfied for all $u \in \mathbb{R}$. Because of condition (2.5) it may be assumed that $\delta(\rho) \equiv 0$ in (2.2). Exactly as in the proof of the second part of Theorem 1.4.1, we obtain

$$\mathscr{L}(\omega + c) \leqslant |\varphi_t| + \mathscr{F} h' \left\{ \frac{h''}{h'^2} + \Phi(h') \right\} - \frac{1}{2} \mathscr{F} h' \Phi(h'), \qquad (2.13)$$

where

$$\Phi(\rho) = 8(c_\varphi + K + 1) \frac{\psi(\rho \pm c_\varphi)}{\rho - c_\varphi},$$

provided that $h' \geqslant c_\varphi + l + 1 + \alpha_1$ and $\alpha_1 > 0$ is chosen so that an inequality of the form (1.4.14) is satisfied on $\{D \times (0, T]\} \times \mathbb{R} \times \{|p| > \max(l, \alpha_1)\}$.

We shall now specify the choice of the function $h(d)$ and the number δ. Let δ be determined from (1.3.8) with $\bar{\alpha} = \max(\alpha, q\delta_0^{-1})$, $\alpha = c_\varphi + l + 1 + \alpha_1 + \alpha_2$, where the choice of $\alpha_2 > 0$ will be specified below, $q = c_\varphi + m$, and $h(d)$ is defined on $[0, \delta]$ by (1.3.9). From Lemma 1.3.3 and relations (1.4.8) and (1.4.14) it then follows that $\delta \in (0, \delta_0)$ and

$$\begin{aligned}\mathscr{L}(\omega + c) &\leqslant \max_{\bar{Q}}|\varphi_t| - \tfrac{1}{2}\mathscr{F}h'\Phi(h') \leqslant \max_{\bar{Q}}|\varphi_t| - \tfrac{1}{4}\mathscr{E}_1 h'\Phi(h') + \tfrac{1}{8}\mathscr{E}_1 h'\Phi(h') \\ &\leqslant \max_{\bar{Q}}|\varphi_t| - \tfrac{1}{8}\mathscr{E}_1 h'\Phi(h'). \end{aligned} \tag{2.14}$$

Taking (2.3) and the relations $h' - c_\varphi \leqslant |p| \leqslant h' + c_\varphi$ into account, we can choose α_2 so large that the right side of (2.14) is negative, so that (2.12) holds. The remainder of the proof of (2.4) is precisely the same as in the proof of the first part of the theorem. Theorem 2.1 is proved.

We remark that the validity of the estimate (2.4) when in the second part of Theorem 2.1 the conditions (2.2) and (2.5) are replaced by the stronger condition

$$|p| + \operatorname{Tr} A \cdot |p| + |a| \leqslant \psi(|p|)\mathscr{E}_1,$$

where $\psi(\rho)$, $0 \leqslant \rho < +\infty$, is the same sort of function as in (2.2) and (2.5), was established in [136]. This result of [136] is thus a special case of the second part of Theorem 2.1.

THEOREM 2.2. *Let $\Omega \in C^3$, and suppose that the functions $a^{ij}(x, t, u, p)$, $i, j = 1, \ldots, n$, and $a(x, t, u, p)$ are continuous and satisfy condition* (1.2) *in $\bar{Q} \times \mathbb{R} \times \mathbb{R}^n$. Assume that on the set $\{D_0 \times (0, T]\} \times \{|u| \leqslant m\} \times \{|p| > l\}$ the inequality*

$$|a(x, t, u, p)| \leqslant \psi(|p|)\mathscr{E}_1 + \delta(|p|)\mathscr{E}_2 \tag{2.15}$$

holds, where the functions $\mathscr{E}_1$, $\mathscr{E}_2$, $\psi(\rho)$ and $\delta(\rho)$, $0 \leqslant \rho < +\infty$, are the same as in Theorem 2.1. *Let $u \in C^{2,1}(Q) \cap C^{1,0}(\bar{Q})$ be an arbitrary solution of* (1.1) *in $D_0 \times (0, T)$ which satisfies the conditions $u = \varphi$ on Γ, where $\varphi = \varphi(x) \in C^2(\bar{\Omega})$, an $|u| \leqslant m$ in $D_0 \times (0, T]$. Then the following assertions are true*:

1) *If the domain Ω is strictly convex, then the estimate* (2.4) *holds with a constant M_1 depending only on m, l, $\psi(\rho)$, $\delta(\rho)$, $\|\varphi\|_{C^2(\bar{\Omega})}$, k^{-1}, and K.*

2) *If on $\{D_0 \times (0, T]\} \times \{|u| \leqslant m\} \times \{|p| > l\}$ inequality* (2.5) *is satisfied, where $\psi(\rho)$ is the same function as in* (2.13), *then the estimate* (2.4) *holds with a constant M_1 depending only on m, l, $\psi(\rho)$, $\|\varphi\|_{C^2(\bar{\Omega})}$, and K.*

PROOF. Theorem 2.2 is proved in exactly the same way as in Theorem 2.1, where now condition (2.3) is not used, since $\varphi_t \equiv 0$ in (2.9) and (2.13). The remainder of the proof of Theorem 2.1 is unchanged. Theorem 2.2 is proved.

§3. Estimates of $\max_{\bar{Q}}|\nabla u|$

In this section we assume that the functions $a^{ij}(x, t, u, p)$, $i, j = 1, \ldots, n$, and $a(x, t, u, p)$ belong to the class $C^1(\mathfrak{M}_{Q,m,L})$, where $\mathfrak{M}_{Q,m,L} \equiv \bar{Q} \times \{|u| \leqslant m\} \times \{|p| > L\}$, $m, L = \text{const} \geqslant 0$. We suppose also that

$$a^{ij}(x, t, u, p)\xi_i\xi_j \geqslant 0, \quad \forall \xi \in \mathbb{R}^n, \text{ on } \mathfrak{M}_{Q,m,L}. \tag{3.1}$$

Let $\tau = (\tau_1, \dots, \tau_n)$ be an arbitrary vector with $|\tau| = 1$. We set $A^\tau \equiv a^{ij}(x, t, u, p)\tau_i\tau_j \equiv A\tau \cdot \tau$ and for an arbitrary function $\Phi(x, t, u, p)$ we introduce the notation

$$\partial\Phi \equiv \sum_{k=1}^{n} \frac{p_k}{|p|}\frac{\partial\Phi}{\partial x_k} + |p|\frac{\partial\Phi}{\partial u}, \quad p\Phi_p \equiv \sum_{k=1}^{n} p_k \frac{\partial\varphi}{\partial p_k}.$$

THEOREM 3.1. *Suppose that on the set $\mathfrak{M}_{Q,m,L}$ for any τ, $|\tau| = 1$, the inequalities*

$$|pA_p^\tau| \leqslant \sqrt{\mu_1 A^\tau \mathscr{E}_1 \omega(|p|)}\,|p|^{-1}, \quad |\delta A^\tau| \leqslant \sqrt{\frac{\sigma_1 A^\tau \mathscr{E}_1}{\omega(|p|)}}, \tag{3.2}$$
$$|a - pa_p| \leqslant \mu_2\mathscr{E}_1 + \tilde{\mu}_2, \quad \delta a \geqslant -\sigma_2\mathscr{E}_1|p|[\omega(|p|)]^{-1} - \tilde{\mu}_3|p|, \quad \mathscr{E}_1 > 0$$

are satisfied, where $\mathscr{E}_1 \equiv \mathscr{E}_1(x, t, u, p) \equiv a^{ij}(x, t, u, p)p_ip_j$, μ_1, μ_2, $\tilde{\mu}_2$ and $\tilde{\mu}_3$ are arbitrary nonnegative constants, σ_1 and σ_2 are nonnegative constants which are sufficiently small in dependence on n, μ_1, μ_2, T, $\tilde{\mu}_3$, m,(¹) and $\omega(\rho)$, $0 \leqslant \rho < +\infty$, is an arbitrary nondecreasing continuous function. Then for any solution $u \in C^{3,1,1}(Q) \cap C^{1,0}(\overline{Q})$(²) of (1.1) such that $\max_{\overline{Q}}|u| \leqslant m$ the estimate

$$\max_{\overline{Q}}|\nabla u| \leqslant \overline{M}_1 \tag{3.3}$$

holds, where $\overline{M}_1$ depends only on n, m, $M_1 \equiv \max_\Gamma|\nabla u|$, L, μ_1, μ_2, $\tilde{\mu}_2$, and T.

PROOF. Applying the operator $u_k(\partial/\partial x_k)$ to (1.1) and setting $v = \sum_1^n u_k^2$, where $u_k \equiv u_{x_k}$, we obtain the identity in Q

$$\tfrac{1}{2}\left(-v_t + a^{ij}v_{ij}\right) = a^{ij}u_{ki}u_{kj} + \tfrac{1}{2}\left[a_{p_l} - a_{p_l}^{ij}u_{ij}\right]v_l + \sqrt{v}\left(\delta a - \delta a^{ij}u_{ij}\right), \tag{3.4}$$

where in (3.4) and below we use the abbreviated notation for partial derivatives $v_l \equiv v_{x_l}$, $u_{ij} \equiv u_{x_ix_j}$, etc. Setting $v = e^{\kappa t}\bar{v}$, $\kappa = \text{const} > 0$ in (3.4), we obtain the identity

$$-\bar{v}_t + a^{ij}\bar{v}_{ij} = 2e^{-\kappa t}a^{ij}u_{ki}u_{kj} + \left[a_{p_l} - a_{p_l}^{ij}u_{ij}\right]\bar{v}_l$$
$$+2\bar{v}\left(\frac{\kappa}{2} + \frac{\delta a}{|p|} - \frac{\delta a^{ij}}{|p|}u_{ij}\right), \tag{3.5}$$

which we henceforth consider only on the set $Q_L \equiv \{(x, t) \in Q: |\nabla u| > L\}$. Taking account of the condition on δa in (3.2) and setting $\kappa = 2\tilde{\mu}_3$, on Q_L we obtain

$$-\bar{v}_t + a^{ij}\bar{v}_{ij} \geqslant 2e^{-\kappa T}a^{ij}u_{ki}u_{kj} + \left[a_{p_l} - a_{p_l}^{ij}u_{ij}\right]\bar{v}_l + 2\bar{v}\left(\frac{\tilde{\delta} a}{|p|} - \frac{\delta a^{ij}}{|p|}u_{ij}\right), \tag{3.6}$$

where $\tilde{\delta} a \equiv \delta a + \mu_3|p|$, so that by (3.2)

$$\tilde{\delta} a \geqslant -\sigma_2\mathscr{E}_1(|p|/\omega).$$

We multiply both sides of (3.6) by $f(\bar{v}(x))$, where $f(\bar{v})$, $f'(\bar{v}) \geqslant 0$ for $\bar{v} > 0$, and we set $w = \int_0^{\bar{v}} f(t)\,dt$. Let $z = z(u)$ be a positive, twice differentiable function on

(¹) This dependence will be specified in the proof of the theorem.

(²) $C^{3,1,1}(Q)$ denotes the set of functions u which are continuous in Q together with u_t, u_{tx_i}, u_{x_i}, $u_{x_ix_j}$ and $u_{x_ix_jx_k}$, $i, j, k = 1, \dots, n$.

$[-m, m]$. We denote by $\overline{w}$ the function defined by $w = z(u)\overline{w}$. Exactly as in the proof of Theorem 1.6.1, we obtain (see the derivation of (1.6.18))

$$z\left(-\overline{w}_t + a^{ij}\overline{w}_{ij}\right) + b^k\overline{w}_k \geqslant \left(-z'' + c_0\frac{z'^2}{z}\frac{w\omega}{\overline{v}f}\right)\mathscr{E}_1\overline{w} + e^{-\kappa T}fa^{ij}u_{ki}u_{kj}$$
$$+z'(pa_p - a)\overline{w} + 2f\overline{v}\frac{\tilde{\delta}a}{|p|} - 2\sigma_1 ne^{\kappa T}\frac{f\overline{v}}{\omega}\mathscr{E}_1, \qquad (x,t)\in Q_L, \quad (3.7)$$

where $c_0 = (n/2)\exp(\kappa T)$. We shall assume that $f(\overline{v}) = \exp\{2c_0\int_1^{\overline{v}}\omega(t)t^{-1}\,dt\}$ for $\overline{v}\geqslant 1$. Now taking the conditions on $pa_p - a$ and $\tilde{\delta}a$ (see (3.2)) and (1.6.19) into account, from (3.7) we derive

$$z\left(-\overline{w}_t + a^{ij}\overline{w}_{ij}\right) + b^k\overline{w}_k \geqslant \left[-z'' + \frac{1}{8}\frac{z'^2}{z} - \mu_1|z'| - (c_1\sigma_1 + c_2\sigma_2)z\right]\mathscr{E}_1\overline{w}$$
$$-|z'|\tilde{\mu}_2\overline{w}, \qquad c_1 = 16c_0ne^{\kappa T},\ c_2 = 8c_0. \qquad (3.8)$$

Let $z = 1 + e^{\alpha m} - e^{\alpha u}$. It is obvious that the expression in square brackets in (3.8) exceeds the quantity

$$\alpha(\alpha - \mu_2)e^{\alpha u} - (c_1\sigma_1 + c_2\sigma_2)(1 + e^{\alpha m}).$$

Let $\alpha = \mu_2 + 1$, and suppose that σ_1 and σ_2 are so small that

$$(\mu_2 + 1)e^{-(\mu_2+1)m} > (c_1\sigma_1 + c_2\sigma_2)[1 + e^{(\mu_2+1)m}]. \qquad (3.9)$$

In addition, taking into account that $|z'| \leqslant \alpha e^{\alpha m}$, in place of (3.8) we then write

$$z\left(-\overline{w}_t + a^{ij}\overline{w}_{ij}\right) + b^k\overline{w}_k > \alpha e^{\alpha m}\tilde{\mu}_2\overline{w}, \qquad (x,t)\in Q_L. \qquad (3.10)$$

Setting $\overline{w} = e^{\gamma t}\overline{\overline{w}}$, $\gamma = \alpha e^{\alpha m}\tilde{\mu}_2$, from (3.10) (taking into account that $z \geqslant 1$) we obtain

$$z\left(-\overline{\overline{w}}_t + a^{ij}\overline{\overline{w}}_{ij}\right) + b^k\overline{\overline{w}}_k > 0, \qquad (x,t)\in Q_L. \qquad (3.11)$$

In view of (3.11) it is obvious that $\overline{\overline{w}}$ cannot have a maximum in Q_L. Therefore,

$$\max_Q \overline{w} \leqslant \max\left\{\max_\Gamma \overline{w},\ \max_{\{v\leqslant L\}} \overline{w}\right\}e^{\gamma T},$$
$$\max_Q w \leqslant \frac{\max z}{\min z}\max\left\{\int_0^{M_1^2} f(t)\,dt,\ \int_0^{L^2} f(t)\,dt\right\}e^{\gamma T}. \qquad (3.12)$$

Taking into account the form of the function z and the fact that f is increasing, from (3.12) it is easy to obtain

$$\int_0^{\overline{M}_1^2} f(t)\,dt \leqslant \max\left\{\int_0^{(aM_1)^2} f(t)\,dt,\ \int_0^{(aL)^2} f(t)\,dt\right\}, \qquad (3.13)$$

where $\overline{M}_1 \equiv \max_Q|\nabla u|$, $a = [(1 + e^{(\mu_2+1)m})e^{(\kappa+\gamma)T}]^{1/2}$. From (3.13) it is then obvious that

$$\overline{M}_1 \leqslant a\max\{M_1, L\}. \qquad (3.14)$$

Theorem 3.1 is proved.

We record separately an important special case of Theorem 3.1 obtained by assuming that $\omega(\rho) \equiv \mu_1 = \text{const} \geqslant 1$.

THEOREM 3.1′. *Suppose that on the set* $\mathfrak{M}_{Q,m,L}$, *for any* $\tau \in \mathbb{R}^n, |\tau| = 1$,

$$|pA_p^\tau| \leqslant \sqrt{\mu_a A^\tau \mathscr{E}_1}\,|p|^{-1}, \quad |\delta A^\tau| \leqslant \sqrt{\sigma_1 A^\tau \mathscr{E}_1}\,,$$
$$|a - pa_p| \leqslant \mu_2 \mathscr{E}_1 + \tilde{\mu}_2, \quad \delta a \geqslant -\sigma_2 \mathscr{E}_1 |p| - \tilde{\mu}_3 |p|, \quad \mathscr{E}_1 > 0, \tag{3.15}$$

where μ_1, μ_2, $\tilde{\mu}_2$, $\tilde{\mu}_3$, σ_1 *and* σ_2 *are nonnegative constants, and a condition of the form* (3.9) *is satisfied. Then for any solution* $u \in C^{3,1,1}(Q) \cap C^{1,0}(\overline{Q})$ *of* (1.1) *such that* $\max_Q |u| \leqslant m$ *and* $\max_\Gamma |\nabla u| \leqslant M_1$ *the estimate* (3.14) *holds.*

Theorem 3.1′ contains as a special case the known result of Ladyzhenskaya and Ural′tseva regarding the estimate of $\max_Q |\nabla u|$ for solutions of quasilinear uniformly parabolic equations [80]. We note that very recently Ladyzhenskaya and Ural′tseva have strengthened their result for uniformly parabolic equations by removing the conditions that the argument u occur in the coefficients of the equation in a weak manner [85].

Theorem 3.1 (and, in particular, Theorem 3.1′) provides an estimate of $\max_{\overline{Q}} |\nabla u|$ for equations with structure described in terms of the majorant $\mathscr{E}_1$. The next theorem gives such an estimate for equations with structure described in terms of $\mathscr{E}_2$.

THEOREM 3.2. *Suppose that condition* (3.1) *is satisfied and that on the set* $\mathfrak{M}_{Q,m,L}$ *for any* τ, $|\tau| = 1$,

$$|A_p^\tau|\,|p| \leqslant \sqrt{\mu_1 A^\tau \operatorname{Tr} A}\,, \quad |\delta A^\tau| \leqslant \sqrt{\sigma_1 A^\tau \operatorname{Tr} A}\,, \quad \operatorname{Tr} A > 0,$$
$$|a_p|\,|p| \leqslant \sigma_2 \mathscr{E}_2 + \tilde{\mu}_2 |p|, \quad \delta a \geqslant -\sigma_3 \mathscr{E}_2 - \tilde{\mu}_3 |p|, \tag{3.16}$$

where μ_1, $\tilde{\mu}_2$ *and* $\tilde{\mu}_3$ *are arbitrary nonnegative constants, and* σ_1, σ_2 *and* σ_3 *are nonnegative constants which are sufficiently small in dependence on* n, T, μ_1, *and the diameter of* Ω. *Then for any solution* $u \in C^{3,1,1}(Q) \cap C^1(\overline{Q})$ *of* (1.1) *such that* $\max_Q |u| = m$ *and* $\max_\Gamma |\nabla u| = M_1$ *the estimate* (3.3) *holds, where* $\overline{M}_1$ *depends only on* n, m, M_1, L, μ_1, T, *and the diameter* d *of* Ω. *Under the additional condition* $\sum_{i,j=1}^n a^{ij} \geqslant \varepsilon_0 \operatorname{Tr} A$ *on* $\mathfrak{M}_{Q,m,L}$, *where* $\varepsilon_0 = \text{const} > 0$, *an arbitrary constant* $\mu_2 > 0$ *can be admitted in conditions* (3.16) *in place of* σ_2.

PROOF. We consider the identity (3.4) on the set $Q_L \equiv \{(x,t) \in Q : |\nabla u| > L\}$. Arguing exactly as in the derivation of (1.7.2), we obtain inequalities of the form (1.7.2) at all points of Q_L. With these inequalities taken into account, we derive from (3.4)

$$-v_t + a^{ij} v_{ij} \geqslant a^{ij} u_{ki} u_{kj} + a_{p_l} v_l + 2\sqrt{v}\,\delta a$$
$$- \mu_1 n \frac{|\nabla v|^2}{v} \operatorname{Tr} A - \sigma_1 n v \operatorname{Tr} A, \qquad (x,t) \in Q_L. \tag{3.17}$$

Let $z = z(x,t)$ be a positive function in $\overline{Q}$ of class $C^2(\overline{Q})$. We set $v = z\bar{v}$. Taking into account that $v_t = z_t \bar{v} + z \bar{v}_t$, $v_i = z_i \bar{v} + z \bar{v}_i$ and $v_{ij} = z_{ij}\bar{v} + z_i \bar{v}_j + z_j \bar{v}_i + z \bar{v}_{ij}$, from (3.17) we deduce that

$$z\left(-\bar{v}_t + a^{ij}\bar{v}_{ij}\right) + b^k \bar{v}_k \geqslant \left(z_t - a^{ij} z_{ij}\right)\bar{v} + \frac{1}{2} a^{ij} u_{ki} u_{kj} + a_{p_l} z_l \bar{v}$$
$$+ 2\frac{\delta}{|p|} z\bar{v} - \mu_1 n \frac{|\nabla z|^2}{z} \operatorname{Tr} A\bar{v} - \sigma_1 n z \operatorname{Tr} A\bar{v}, \qquad (x,t) \in Q_L, \tag{3.18}$$

where the form of b^k is irrelevant for our purposes. Taking condition (3.16) on $|a_p|$ and δa into account, from (3.18) we derive

$$z\left(-\bar v_t + a^{ij}\bar v_{ij}\right) + b^k\bar v_k$$

$$\geqslant\left\{-a^{ij}z_{ij} - \left[\mu_1 n\frac{|\nabla z|^2}{z} + \sigma_2|\nabla z| - (n\sigma_1 + 2\sigma_3)z\right]\operatorname{Tr}A + [z_t - \tilde\mu_2|\nabla z| - 2\tilde\mu_3 z]\right\}\bar v,$$

$$(x, t) \in Q_L. \quad (3.19)$$

We set $z = \alpha + e^{\kappa t} - |x|^2$, $\alpha = (N+1)d^2$ and $d = \max_{\bar\Omega}|x| > 0$, where $\kappa > 0$ and $N > 0$ are constants. Taking into account that $|\nabla z| \leqslant 2d\sqrt{n}$, $z_t = \kappa e^{\kappa t}$, $|\nabla z|^2 z^{-1} \leqslant 4nN^{-1}$, $z \leqslant (1+N)d^2 + e^{\kappa t}$, $z_{ij} = -2\delta_i^j$, where δ_i^j is the Kronecker symbol, and $z_t = \kappa e^{\kappa t}$, in place of (3.19) we write

$$z\left(-\bar v_t + a^{ij}\bar v_{ij}\right) + b^k\bar v_k \geqslant \left\{\left[2 - \frac{4\mu_1 n}{N} - 2\sigma_2 d - (n\sigma_1 + 2\sigma_3)\left((N+1)d^2 + e^{\kappa T}\right)\right]\right.$$

$$\left.\times\operatorname{Tr}A + \left[\kappa e^{\kappa t} - 2\sqrt{n}\,\tilde\mu_2 d - 2\tilde\mu_3(1+N)d^2 - 2\tilde\mu_3 e^{\kappa t}\right]\right\}\bar v. \quad (3.20)$$

We first choose N so that $2 - 4\mu_1 n^2 N^{-1} = 1$, and then set $\kappa = 2\tilde\mu_3 + \tilde\kappa$ and $\tilde\kappa = 1 + 2\sqrt{n}\,\tilde\mu_2 d + 2\tilde\mu_3(1+N)d^2$. We shall now specify the conditions of smallness on the quantities σ_1, σ_2 and σ_3. We assume that σ_1, σ_2 and σ_3 are so small that

$$2\sigma_2 d + (n\sigma_1 + 2\sigma_3)\left((N+1)d^2 + e^{\kappa T}\right) \leqslant 1/2. \quad (3.21)$$

From (3.20) we now obtain

$$z\left(-\bar v_t + a^{ij}\bar v_{ij}\right) + b^k\bar v_k > 0 \quad \text{on } Q_L, \quad (3.22)$$

which shows that $\bar v$ cannot achieve a maximum in Q_L. Therefore,

$$\max_Q v \leqslant \max z \max\left\{\max_\Gamma \bar v,\ \max_{\{v\leqslant L\}} \bar v\right\} \leqslant \frac{\max z}{\min z}\max\left\{\max_\Gamma v,\ L^2\right\}$$

$$\leqslant \frac{(N+1)d^2 + e^{\kappa T}}{Nd^2}\max\left\{\max_\Gamma v,\ L^2\right\}, \quad (3.23)$$

whence we obtain an estimate of the form (3.3) with

$$\overline{M}_1 = \left((N+1)/N + e^{\kappa T}/Nd^2\right)^{1/2}\max(M_2, L).$$

We now suppose that the condition $\sum_{i,j=1}^n a^{ij} \geqslant \varepsilon_0 \operatorname{Tr}A$, $\varepsilon_0 = \text{const} > 0$, is satisfied on the set $\mathfrak{M}_{Q,m,L}$. In this case we suppose that in (3.19) for z there is the function $z = \alpha + e^{\kappa t} + e^{\beta n d} - e^{-\beta\sum_1^n x_k}$, where $\kappa, \alpha = \text{const} > 0$, $\beta = \text{const} > 0$ and $d = \sup_{\bar\Omega}|x|$. Now taking into account that $z \leqslant \alpha + e^{\kappa t} + e^{\beta n d}$, $|\nabla z| \leqslant \sqrt{n}\,\beta e^{-\beta\sum_1^n x_k}$, $|\nabla z|^2 z^{-1} \leqslant \beta^2 e^{2\beta n d}\alpha^{-1}$ and $z_{ij} = -\beta^2 e^{-\beta\sum_1^n x_k}$, we first choose $\beta > 0$ so that $\varepsilon_0\beta^2 - \sigma_2\sqrt{n}\,\beta = 1/2\varepsilon_0\beta^2$. Then

$$-a^{ij}z_{ij} - \sigma_2|\nabla z| \geqslant 1/2\varepsilon_0\beta^2 e^{-\beta n d}.$$

We then choose $\alpha > 1$ so that

$$1/2\varepsilon_0\beta^2 e^{-\beta n d} - \mu_0 n\beta^2 e^{2\beta n d}\alpha^{-1} \geqslant (\varepsilon_0/4)\beta^2 e^{-\beta n d},$$

and we assume that σ_1 and σ_3 are so small that

$$(n\sigma_1 + 2\sigma_3)(\alpha + e^{\kappa T} + e^{\beta n d}) \leqslant \tfrac{1}{8}\varepsilon_0 \beta^2 e^{-\beta n d}, \tag{3.24}$$

where $\kappa = 2\tilde{\mu}_3 + \tilde{\kappa}$ and $\tilde{\kappa} = 1 + \tilde{\mu}_2\sqrt{n}\,\beta e^{\beta n d} + 2\tilde{\mu}_3(\alpha + e^{\beta n d})$. From (3.19) we then obtain (3.22), from which, exactly as in the preceding case, we derive the estimate

$$\max_Q v \leqslant \frac{\alpha + e^{\kappa T} + e^{\beta n d}}{\alpha + e^{\beta n d}} \max\left\{\max_\Gamma v,\, L^2\right\},$$

from which (3.3) follows with

$$\overline{M}_1 = \left(\frac{\alpha + e^{\kappa T} + e^{\beta n d}}{\alpha + e^{\beta n d}}\right)^{1/2} \max(M_1, L).$$

Theorem 3.2 is proved.

As an example of an equation to which Theorem 3.2 is applicable, we consider an equation of the form

$$-u_t + \left(\frac{1 + |\nabla u|^2}{|\nabla u|^2}\delta_i^j - \frac{u_{x_i}u_{x_j}}{|\nabla u|^2}\right)u_{x_i x_j} = a(x, t, u, \nabla u). \tag{3.25}$$

The analysis of (1.7.13) carried out in §1.7 shows that all conditions of the first part of Theorem 3.2 are satisfied for equation (3.25) if its right side $a(x, t, u, p)$ satisfies the condition

$$|a_p| \leqslant \mu_2, \quad \delta a \geqslant -\mu_3 \quad \text{on } \mathfrak{M}_{Q,m,L}. \tag{3.26}$$

As we shall see in §5, conditions (3.26) are best possible in a certain sense.

To conclude this section we note that an a priori estimate of $\max_Q|\nabla u|$ in terms of $\max_Q|u|$ and $\max_\Gamma|\nabla u|$ for equations of specific structure was obtained in [136], where it was assumed that the leading coefficients of the equation have the special form

$$\begin{gathered} a^{ij}(x, t, u, p) = \lambda(x, t, u, p)b^{ij}(p/|p|) + \lambda_1(x, t, u, p)p_i p_j, \\ i, j = 1, \ldots, n, \quad n \geqslant 2, \quad \lambda > 0 \quad \text{on } \mathfrak{M}_{Q,m,L}. \end{gathered} \tag{3.27}$$

Conditions on the growth of the right side of an equation with the principal part prescribed in (3.27) are given in [136] in terms of the majorant $\mathscr{E}_1$. We remark that for $\lambda_1 < 0$ in many cases it is more advantageous to impose conditions on the growth of the right side in terms of $\mathscr{E}_2$ using Theorem 3.2 (see, in particular, (3.25)).

§4. Existence theorems for a classical solution of the first boundary value problem

Theorems 1.3, 2.1, 2.2, 3.1, and 3.2 enable us to obtain the following existence theorems.

THEOREM 4.1. *Suppose that the functions $a^{ij}(x, t, u, p)$, $i, j = 1, \ldots, n$, and $a(x, t, u, p)$ are continuously differentiable and satisfy condition* (1.2) *in $\overline{Q} \times \mathbb{R} \times \mathbb{R}^n$, where $Q = \Omega \times (0, T]$, and Ω is a bounded domain in $\mathbb{R}^n$, $n \geqslant 1$, of class $C^{2+\gamma}$, $0 < \gamma < 1$. Assume that for any solution $v \in C^{2,1}(\overline{Q})$ of problem* (1.9) *for any $\tau \in [0, 1]$ the estimate $\max_Q|v| \leqslant m$ holds, where m does not depend either on v or on $\tau \in [0, 1]$. Suppose that on the set $\{D_0 \times (0, T)\} \times \{|u| \leqslant m\} \times \{|p| > l\}$, where D_0 is defined by condition* (1.3.2) *and $l =$ const $\geqslant 0$, inequalities* (2.2) *and* (2.5) *hold while* *condition* (2.3) *is also satisfied. Let condition* (3.2) *be satisfied on the set* $\overline{Q} \times \{|u| \leqslant m\} \times \{|p| > L\}$, *where* $L = \text{const} \geqslant 0$. *Then for any* $\varphi(x,t) \in C_{2+\gamma}(\overline{Q})$, $0 < \gamma < 1$, *satisfying condition* (1.11) *there exists at least one solution* $u \in C_{2+\alpha}(\overline{Q})$ (*for some* $\alpha \in (0,1)$) *of problem* (1.3).

PROOF. Suppose first that $\Omega \in C^3$, $\varphi \in C^{3,1}(\overline{Q}) \cap C_{2+\gamma}(\overline{Q})$ and the functions $a^{ij}(x,t,u,p)$, $i,j = 1,\dots,n$, and $a(x,t,u,p)$ belong to the class $C^2(\overline{Q} \times \mathbb{R} \times \mathbb{R}^n)$. From results of Friedman [123] pertaining to linear parabolic equations it then follows that any solution $u(x,t) \in C^{2,1}(\overline{Q})$ also belongs to the class $C^{3,1,1}(Q)$, so that it is possible to use Theorems 2.1 and 3.1. From Theorems 2.1, 3.1, and 1.3 it then follows that problem (1.3) has a solution $u \in C_{2+\alpha}(\overline{Q})$ for some $\alpha \in (0,\gamma]$. Suppose now that $\Omega \in C^{2+\gamma}$, $0 < \gamma < 1$, a^{ij}, $a \in C^1$ and $\varphi \in C_{2+\gamma}(\overline{Q})$. We approximate the domain Ω by expanding domains $\tilde{\Omega} \subset \Omega$ of class C^3 which converge in the limit to the original domain Ω in such a way that the $C^{2+\gamma}$-norms of their boundaries are uniformly bounded, and in the cylinders $\tilde{Q} = \tilde{\Omega} \times (0,T]$ we consider problems of the form

$$\begin{gathered} -\tilde{u}_t + \tilde{a}^{ij}(x,t,\tilde{u},\nabla\tilde{u})\tilde{u}_{x_ix_j} - \tilde{a}(x,t,\tilde{u},\nabla\tilde{u}) = 0 \quad \text{in } \tilde{Q}, \\ \tilde{u} = \tilde{\varphi} \quad \text{on } \tilde{\Gamma}, \end{gathered} \tag{4.1}$$

where the functions $\tilde{a}^{ij}(x,t,u,p)$, $i,j = 1,\dots,n$, $\tilde{a}(x,t,u,p)$, and $\tilde{\varphi}(x,t)$ approximate the respective functions $a^{ij}(x,t,u,p)$, $i,j = 1,\dots,n$, $a(x,t,u,p)$, and $\varphi(x,t)$ in such a way that $\tilde{a}^{ij}$, $\tilde{a} \in C^2$, $\tilde{\varphi} \in C^{3,1}(\tilde{Q}) \cap C_{2+\gamma}(\tilde{Q})$, the conditions (2.2), (2.5), (2.3) and (3.2) hold for $\tilde{a}^{ij}$ and $\tilde{a}$ uniformly with respect to the approximation parameters, and conditions (1.11) hold for $\tilde{\varphi}$ and $\tilde{\Gamma}$. This approximation of the equation and its domain is standard (see, for example, [163], pp. 518 and 519). Since, because of the uniform estimate $\max_{\tilde{Q}}(|\tilde{u}| + |\nabla\tilde{u}|) \leqslant \overline{M}_1$ and the theorem of Ladyzhenskaya and Ural′tseva (see Theorem 1.2), the uniform estimate $\|\tilde{u}\|_{1+\alpha,\tilde{Q}} \leqslant c_1$ holds for solutions of problems of the form (4.1), by the results of Friedman [123] we obtain a uniform estimate of the norm, $\|\tilde{u}\|_{2+\alpha,\bar{\tilde{Q}}} \leqslant c_2$. It is obvious that there exists a sequence $\{\tilde{u}_n\}$ converging to a function u in $C^2_{\text{loc}}(Q)$. It is easy to see that the limit function $u \in C_{2+\alpha}(\overline{Q})$, $\alpha \in (0,\gamma]$, satisfies (1.1) in Q and coincides with φ on Γ. Theorem 4.1 is proved.

The following theorems are proved in a similar way on the basis of the results of the foregoing sections.

THEOREM 4.2. *Suppose that the functions* $a^{ij}(x,t,u,p)$, $i,j = 1,\dots,n$, *and* $a(x,t,u,p)$ *are continuously differentiable and satisfy condition* (1.2) *in* $\overline{Q} \times \mathbb{R} \times \mathbb{R}^n$, *where* $Q = \Omega \times (0,T]$, Ω *is a bounded strictly convex domain in* $\mathbb{R}^n$, $n \geqslant 1$, *where* $\Omega \in C^{2+\gamma}$, $0 < \gamma < 1$, *and* $T = \text{const} > 0$. *Assume that for any solution* $v \in C^{2,1}(\overline{Q})$ *of problem* (1.9) *for all* $\tau \in [0,1]$ *the estimate* $\max_Q|v| \leqslant m$ *holds, where* m *does not depend either on* v *or on* $\tau \in [0,1]$. *Suppose that on the set* $\{D_0 \times (0,T]\} \times \{|u| \leqslant m\} \times \{|p| > l\}$, *where* D_0 *is defined by* (1.3.2) *and* $l = \text{const} \geqslant 0$, *inequality* (2.2) *holds and condition* (2.3) *is also satisfied. Suppose that either* (3.2) *or* (3.16) *is satisfied on the set* $\overline{Q} \times \{|u| \leqslant m\} \times \{|p| > L\}$, *where* $L = \text{const} > 0$. *Then for any* $\varphi(x,t) \in C_{2+\gamma}(\overline{Q})$ *satisfying condition* (1.11) *there exists at least one solution* $u \in C_{2+\alpha}(\overline{Q})$ (*for some* $\alpha \in (0,1)$) *of problem* (1.3).

THEOREM 4.3. *Suppose that the functions* $a^{ij}(x, t, u, p)$, $i, j = 1, \dots, n$, *and* $a(x, t, u, p)$ *are continuously differentiable and satisfy condition* (1.2) *in* $\overline{Q} \times \mathbb{R} \times \mathbb{R}^n$, *where* $Q = \Omega \times (0, T]$, Ω *is a bounded domain in* $\mathbb{R}^n$, $n \geqslant 1$, *of class* $C^{2+\gamma}$, $0 < \gamma < 1$, *and* $T = \text{const} > 0$. *Suppose that for any solution* $v \in C^{2,1}(\overline{Q})$ *of problem* (1.9) *for all* $\tau \in [0, 1]$ *the estimate* $\max_Q |v| \leqslant m$ *holds, where* m *does not depend either on* v *or on* $\tau \in [0, 1]$. *Suppose that on the set* $\{D_0 \times (0, T]\} \times \{|u| \leqslant m\} \times \{|p| > l\}$, *where* D_0 *is defined by* (1.3.2) *and* $l = \text{const} > 0$, *the inequalities* (2.15) *and* (2.5) *hold, while on the set* $\overline{Q} \times \{|u| \leqslant m\} \times \{|p| > L\}$, *where* $L = \text{const} > 0$, *condition* (3.2) *is satisfied. Then for any* $\varphi = \varphi(x) \in C_{2+\gamma}(\overline{\Omega})$ *satisfying condition* (1.11) *there exists at least one solution* $u \in C_{2+\alpha}(\overline{Q})$ (*for some* $\alpha \in (0, 1)$) *of problem* (1.3).

THEOREM 4.4. *Suppose that the functions* $a^{ij}(x, t, u, p)$, $i, j = 1, \dots, n$, *and* $a(x, t, u, p)$ *are continuously differentiable and satisfy condition* (1.2) *on* $\overline{Q} \times \mathbb{R} \times \mathbb{R}^n$, *where* $Q = \Omega \times (0, T]$, Ω *is a bounded strictly convex domain in* $\mathbb{R}^n$, $n \geqslant 1$, *and* $\Omega \in C^{2+\gamma}$, $0 < \gamma < 1$. *Suppose that for any solution* $v \in C^{2,1}(\overline{Q})$ *of problem* (1.9) *for all* $\tau \in [0, 1]$ *the estimate* $\max_Q |v| \leqslant m$ *holds, where* m *does not depend either on* v *or on* $\tau \in [0, 1]$. *Suppose that on the set* $\{D_0 \times (0, T]\} \times \{|u| \leqslant m\} \times \{|p| > l\}$, *where* D_0 *is defined by* (1.3.2) *and* $l = \text{const} \geqslant 0$, *inequality* (2.15) *holds, while on the set* $\overline{Q} \times \{|u| \leqslant m\} \times \{|p| > L\}$, *where* $L = \text{const} \geqslant 0$, *either condition* (3.2) *or condition* (3.16) *is satisfied. Then for any* $\varphi = \varphi(x) \in C_{2+\gamma}(\overline{\Omega})$ *satisfying condition* (1.1) *there exists at least one solution* $u \in C_{2+\alpha}(\overline{Q})$ (*for some* $\alpha \in (0, 1)$) *of problem* (1.3).

We remark that although in Theorems 4.1–4.4 the estimate $\max_Q |v| \leqslant m$ for solutions of problem (1.9) is postulated, in essence these theorems represent unconditional results, since many sufficent conditions guaranteeing such an estimate are known. Examples of such estimates can be found in [80] and [136]. In particular (see [80]), the estimate $\max_Q |v| \leqslant m$ is ensured by the following conditions: for any $(x, t) \in Q$ and $u \in \mathbb{R}$

$$a^{ij}(x, t, u, 0)\xi_i\xi_j \geqslant 0, \quad ua(x, t, u, 0) \geqslant -c_1u^2 - c_2, \tag{4.2}$$

where $c_1, c_2 = \text{const} \geqslant 0$.

A theorem for the existence of a classical solution of (1.3) for equations with the special structure (3.27) was obtained in [136].

§5. Nonexistence theorems

In this section results are established which show that the main structural conditions under which existence theorems for a classical solution of problem (1.3) were proved in §4 are caused by the essence of the matter. The proof uses the following proposition which is based on applying the strict maximum principle for parabolic equations.

LEMMA 5.1. *Suppose that* $u \in C^{2,1}(Q) \cap C^{1,0}(\overline{Q})$ *satisfies* (1.1) *in* Q, *and* $\omega \in C^{2,1}(Q) \cap C(\overline{Q})$ *satisfies inequality* (2.1) *in* Q *for any constant* $c \geqslant 0$. *Let* S *be an open subset on* $\partial\Omega$, *and let* $\Gamma_S = S \times (0, T]$. *Suppose that* $u \leqslant \omega$ *on* $\Gamma \setminus \Gamma_S$ *and* $\partial\omega/\Omega\boldsymbol{v} = -\infty$ *on* Γ_S, *where* $\boldsymbol{v}$ *is the unit vector of the inner normal to* Γ_S. *Then* $u \leqslant \omega$ *in* $\overline{Q}$.

Lemma 5.1 is a variant of Lemma 2.1 and can be proved, for example, by means of Lemma 2.1. Propositions of the type of Lemma 5.1 were used in [127], [5], [6],

[163], [136], [38] and others. In particular, a proof of Lemma 5.1 can be found in [136] (see p. 418).

We first present a result obtained in [136].

THEOREM 5.1. *Let Ω be a bounded domain in $\mathbb{R}^n$, $n \geqslant 1$, and suppose that the functions $a^{ij}(x, t, u, p)$, $i, j = 1, \ldots, n$, and $a(x, t, u, p)$ satisfy the conditions*

$$|a(x, t, u, p)| \geqslant \mathscr{E}_1(x, t, u, p)\psi(|p|) \quad \text{in } \overline{Q} \times \{|u| \geqslant m_0\} \times \{|p| \geqslant l_0\}, \tag{5.1}$$

where $\psi(\rho)$ is a positive continuous function such that $\int^{+\infty}(d\rho/\rho\psi(\rho)) < +\infty$, m_0 and l_0 are nonnegative constants, and

$$\begin{gathered} \frac{a(x, t, u, p)}{\mathscr{E}_2(x, t, u, p)} \to +\infty \quad \text{as } p \to \infty \\ (\text{uniformly in } (x, t) \in Q, |u| \geqslant m_0). \end{gathered} \tag{5.2}$$

Then there exists a function $\varphi(x, t) \in C^\infty(\overline{Q})$ with $\partial\varphi/\partial t \not\equiv 0$ in Q for which problem (1.3) *has no classical solution.*

Theorems 4.1, 4.2, and 5.1 show that the majorants $\mathscr{E}_1$ and $\mathscr{E}_2$ (up to the factor $\psi(|p|)$ for the majorant $\mathscr{E}_1$, where $\int^{+\infty}(d\rho/\rho\psi(\rho)) = +\infty$) are the bounds of growth of the right side of (1.1), which growth is admissible in order that problem (1.3) be solvable for any sufficiently smooth function $\varphi(x, t)$ provided that condition (2.3) is satisfied. We now ask whether the condtion $\varphi(x, t) = \varphi(x)$ is natural in the general situation, i.e., when condition (2.3) is not imposed. The next theorem is an answer to this question.

THEOREM 5.2. *Let Ω be a bounded domain in $\mathbb{R}^n$, $n \geqslant 1$, and suppose that the functions $a^{ij}(x, t, u, p)$, $i, j = 1, \ldots, n$, and $a(x, t, u, p)$ satisfy the following conditions*:

$$\mathscr{E}_1 \leqslant 1/\delta(|p|) \quad \text{for } (x, t) \in \overline{Q}, |u| \geqslant m_0, |p| \geqslant l_0, \tag{5.3}$$

where $\delta(\tau)$, $0 \leqslant \tau < +\infty$, is a positive, continuous, increasing function such that $\int^{+\infty}(d\tau/\tau\delta(\tau)) + \infty$, $m_0, l_0 =$ const $\geqslant 0$; *and*

$$\begin{gathered} \mathscr{E}_2(x, t, u, p) \to 0 \quad \text{as } p \to \infty \\ (\text{uniformly in } (x, t) \in \overline{Q}, |u| \geqslant m_0), \end{gathered} \tag{5.4}$$

$$|a(x, t, u, p)| \leqslant \alpha(|p|)\mathscr{E}_1 + \beta\mathscr{E}_2 \quad \text{for } (x, t) \in \overline{Q}, |u| \geqslant m_0, |p| \geqslant l_0, \tag{5.5}$$

where $\alpha(\tau)$, $0 \leqslant \tau < +\infty$, is a positive, continuous, increasing function such that $\int^{+\infty}(d\tau/\tau\alpha(\tau)) = +\infty$, $\beta =$ const > 0. *Then there exists a function $\varphi(x, t) \in C^\infty(\overline{Q})$ with $\partial\varphi/\partial t \not\equiv 0$ in Q for which problem* (1.3) *has no classical solution.*

PROOF. Suppose the point $y_0 \in \partial\Omega$ is such that there exists a sphere $K_{2\rho}(x_0)$ of radius 2ρ, $\rho > 0$, with center at the point $x_0 \in \Omega$ contained in Ω and tangent to $\partial\Omega$ at y_0. Let the diameter of Ω be equal to the number $R > 0$. We denote by $r = r(x)$ the distance from x to y_0, i.e., $r(x) = \operatorname{dist}(x, y_0)$, $x \in \Omega$. In the cylinder $Q^\rho = \Omega^\rho \times (0, T]$, where $\Omega^\rho \equiv \{x \in \Omega: r(x) \geqslant \rho\}$, we consider the function

$$\omega(x, t) = 2t + \hat{m} + h(r), \qquad r = r(x), h(r) \geqslant 0, \tag{5.6}$$

where $\hat{m} = \max(m_0, m_1)$, $m_1 = \max_{\Gamma \setminus \Sigma_\rho} |u|$, $\Sigma_\rho = \{x \in \partial\Omega:\ r < \rho\} \times (0, T]$, and the decreasing function $h(r) \in C^2((\rho, R)) \cap C([\rho, R])$ will be defined below. It is obvious that for all $c = \text{const} \geqslant 0$

$$\begin{aligned}\mathscr{L}(\omega + c) &\equiv -\omega_t + a^{ij}(x, t, \omega + c, \nabla\omega)\omega_{ij} - a(x, t, \omega + c, \nabla\omega)\\ &= -2 + h'' a^{ij} r_i r_j + h' a^{ij} r_{ij} - a, \qquad (x, t) \in Q^\rho, \end{aligned} \tag{5.7}$$

where $r_i \equiv \partial r/\partial x_i$ and $r_{ij} \equiv \partial^2 r/\partial x_i\, \partial x_j$, $i, j = 1, \ldots, n$. Taking into account that

$$\nabla\omega = h'\nabla r, \quad |\nabla\omega| = |h'|,$$

$$\mathscr{E}_1(x, t, \omega + c, \nabla\omega) = a^{ij} r_i r_j h'^2, \quad a^{ij} r_{ij} = (1/r)\left[\operatorname{Tr} A - a^{ij} r_i r_j\right],$$

where $A \equiv \|a^{ij}(x, t, \omega + c, \nabla\omega)\|$, we deduce from (5.7) that

$$\mathscr{L}(\omega + c) = -2 + \mathscr{E}_1 \frac{h''}{h'^2} + \frac{h'}{r}\left[\operatorname{Tr} A - a^{ij} r_i r_j\right] - a(x, t, \omega + c, \nabla\omega). \tag{5.8}$$

We shall assume that $h'' > 0$ and $h'(r) \leqslant -\hat{l}$, where $\hat{l} = \max(l_0, l_1)$ and the number $l_1 > 0$ will be defined below. Taking into account that $0 \leqslant \operatorname{Tr} A - a^{ij} r_i r_j$ and using (5.5),we obtain

$$\mathscr{L}(\omega + c) \leqslant -2 + h''\mathscr{E}_1/h'^2 + \alpha(|h'|)\mathscr{E}_1 + \beta\mathscr{E}_2, \qquad (x, t) \in Q^\rho. \tag{5.9}$$

It may be assumed with no loss of generality that $\alpha(\tau)/\delta(\tau) \to 0$ as $\tau \to +\infty$, so that $\alpha(\tau) \leqslant c\delta(\tau)$ for all sufficiently large τ. Then, taking conditions (5.3) and (5.4) into account, we deduce from (5.9) that

$$\mathscr{L}(\omega + c) \leqslant -1 + (h''/h'^2)(1/\delta(|h'|)), \tag{5.10}$$

if l_1 is taken to be a sufficiently large number. For $h(r)$ it is possible to select a function possessing the following properties:

$$\begin{gathered} h''(h')^{-2} \leqslant \delta(|h'|) \quad \text{for } r \in [\rho, R],\ h'(\rho) = -\infty,\ h(R) = 0,\\ h''(r) \geqslant 0, \quad h'(r) \leqslant -\hat{l} \quad \text{on } [\rho, R]. \end{gathered} \tag{5.11}$$

Indeed, we set

$$\kappa = \max\left(1, (R - \rho)\Big/\int_{\hat{l}}^{+\infty} \frac{d\tau}{\tau^2\delta(\tau)}\right)$$

and let the number $\alpha > 0$ be defined by

$$\int_\alpha^{+\infty} \frac{d\tau}{\tau^2\delta(\tau)} = R - \rho, \quad \text{if } \kappa = 1, \qquad \alpha = \hat{l}, \quad \text{if } \kappa = (R - \rho)\Big/\int_{\hat{l}}^{+\infty} \frac{d\tau}{\tau^2\delta(\tau)}. \tag{5.12}$$

Obviously $\alpha \geqslant \hat{l}$. We define $h(r)$ on $[\rho, R]$ by the parametric equations

$$h = \kappa\int_\alpha^\tau \frac{d\tau}{\tau\delta(\tau)}, \quad r = \rho + \kappa\int_\tau^{+\infty} \frac{d\tau}{\tau^2\delta(\tau)}, \qquad \alpha \leqslant \tau < +\infty. \tag{5.13}$$

It is obvious that $h' = -\tau \leqslant -\hat{l}$, $h'(\rho) = -\infty$, $h(R) = 0$, and $h''/h'^2 = \delta(|h'|)/\kappa \leqslant \delta(|h'|)$. From (5.10) and (5.11) it then follows that

$$\mathscr{L}(\omega + c) \leqslant 0, \quad \forall c = \text{const} \geqslant 0,\ (x, t) \in Q^\rho. \tag{5.14}$$

We consider an arbitrary solution $u(x, t) \in C^{2,1}(Q) \cap C(\overline{Q})$ of (1.1). We shall compare the functions $u(x, t)$ and $\omega(x, t)$ in the cylinder Q^ρ. We denote by $\partial\Omega^\rho$ the boundary of the domain Ω^ρ lying at the base of the cylinder Q^ρ. Let $S^\rho \equiv \partial\Omega^\rho \setminus \partial\Omega$ and $\Gamma_S^\rho \equiv S^\rho \times (0, T]$. From the definition of the function ω it follows that $u \leqslant \omega$ on $\Gamma^\rho \setminus \Gamma_{S^\rho}$, where

$$\Gamma^\rho \equiv (\partial\Omega^\rho \times (0, T]) \cup (\Omega^\rho \times \{t = 0\}),$$

and that $\partial\omega/\partial\nu = -\infty$ on Γ_{S^ρ}, where ν is the unit vector of the inner normal to Γ_{S^ρ}. Therefore, by Lemma 5.1

$$u(x, t) \leqslant \max_{\overline{Q^\rho}} \omega(x, t) \equiv m_*, \qquad (x, t) \in \overline{Q^\rho}, \tag{5.15}$$

where

$$m_* = 2T + \hat{m} + \kappa \int_\alpha^{+\infty} \frac{d\tau}{\tau\delta(\tau)} < +\infty,$$

In the cylinder $\tilde{Q}^{\varepsilon,\rho} \equiv \{K_\rho(y_0) \cap K_{2\rho-\varepsilon}(x_0)\} \times (0, T]$, $0 < \varepsilon < \rho$, we consider the function

$$\tilde{\omega}(x, t) = 3t + m_* + \tilde{h}(\tilde{r}), \qquad \tilde{r} = \tilde{r}(x) = \operatorname{dist}(x, x_0), \tag{5.16}$$

where the function $\tilde{h}(\tilde{r}) \in C^2((\rho, 2\rho - \varepsilon)) \cap C([\rho, 2\rho - \varepsilon])$ satisfies the conditions

$$\begin{gathered} \tilde{h}''(\tilde{h}')^{-2} \leqslant \delta(|\tilde{h}'|) \quad \text{for } \tilde{r} \in [\rho, 2\rho - \varepsilon], \tilde{h}(\rho) = 0, \\ \tilde{h}'(2\rho - \varepsilon) = +\infty, \quad \tilde{h}''(\tilde{r}) \geqslant 0, \quad \tilde{h}'(\tilde{r}) \geqslant \tilde{l} = \text{const} > 0 \quad \text{on } [\rho, 2\rho - \varepsilon], \end{gathered} \tag{5.17}$$

$l = \max(l_0, l_2)$, and the constant l_2 will be specified below. It is obvious that all the conditions (5.17) are satisfied by the function $\tilde{h}(\tilde{r})$ defined on $[\rho, 2\rho - \varepsilon]$ by the parametric equations

$$\tilde{h} = \tilde{\kappa} \int_{\tilde{\alpha}}^{\tau} \frac{d\tau}{\tau\delta(\tau)}, \quad \tilde{r} = 2\rho - \varepsilon - \tilde{\kappa} \int_\tau^{+\infty} \frac{d\tau}{\tau^2\delta(\tau)}, \quad \tilde{\alpha} \leqslant \tau \leqslant +\infty, \tag{5.18}$$

where

$$\tilde{\kappa} = \max\left(1, (\rho - \varepsilon) \Big/ \int_{\tilde{l}}^{+\infty} \frac{d\tau}{\tau^2\delta(\tau)}\right),$$

and the number $\tilde{\alpha} > 0$ is determined by

$$\begin{gathered} \int_{\tilde{\alpha}}^{+\infty} \frac{d\tau}{\tau^2\delta(\tau)} = \rho - \varepsilon, \quad \text{if } \tilde{\kappa} = 1, \\ \tilde{\alpha} = \tilde{l}, \quad \text{if } \tilde{\kappa} = (\rho - \varepsilon) \Big/ \int_{\tilde{l}}^{+\infty} \frac{d\tau}{\tau^2\delta(\tau)}. \end{gathered} \tag{5.19}$$

Taking into account that for $\tilde{\omega}$ the expression $\mathscr{L}(\omega + c)$, $c = \text{const} \geqslant 0$, is obtained by replacing h by $\tilde{h}$ and r by $\tilde{r}$ on the right side of (5.7) and also noting that

$$\begin{gathered} \tilde{h}' a^{ij} \tilde{r}_{ij} = \tilde{h}' \tilde{r}^{-1} \left[\operatorname{Tr} A - a^{ij} \tilde{r}_i \tilde{r}_j\right] \leqslant \tilde{h}' \rho^{-1} \operatorname{Tr} A = \mathscr{E}_2 \rho^{-1}, \\ A \equiv \|a^{ij}(x, t, \tilde{\omega} + c, \nabla\omega)\|, \quad |\nabla\tilde{\omega}| = \tilde{h}', \quad \mathscr{E}_1 = a^{ij} \tilde{r}_i \tilde{r}_j \tilde{h}'^2, \end{gathered}$$

we obtain the inequality

$$\mathscr{L}(\tilde{\omega} + c) \leqslant -3 + \frac{\tilde{h}''}{\tilde{h}'^2} \mathscr{E}_1 + \frac{\mathscr{E}_2}{\rho} - a(x, t, \tilde{\omega} + c, \nabla\tilde{\omega}), \qquad (x, t) \in \tilde{Q}^{\varepsilon,\rho}. \tag{5.20}$$

Taking conditions (5.3)–(5.5) into account, we observe that for sufficiently large l_2, depending only on ρ, β and the way the functions $\alpha(|p|)/\delta(|p|)$ and $\mathscr{E}_2(x, t, u, p)$ tend to 0 as $p \to \infty$, the inequalities

$$\mathscr{E}_2\rho^{-1} \leqslant 1, \quad -a(x, t, \tilde{\omega} + c, \nabla\tilde{\omega}) \leqslant 1, \qquad (x, t) \in \tilde{Q}^{\varepsilon,\rho} \tag{5.21}$$

hold. With (5.3) taken into account, from (5.20) and (5.21) follows

$$\mathscr{L}(\tilde{\omega} + c) \leqslant -1 + (\tilde{h}''/\tilde{h}'^2)(1/\delta(\tilde{h}')), \qquad (x, t) \in \tilde{Q}^{\varepsilon,\rho}. \tag{5.22}$$

In addition, taking into account that $\tilde{h}''(\tilde{h}')^{-2} \leqslant \delta(\tilde{h}')$ (see (5.17)), we get

$$\mathscr{L}(\tilde{\omega} + c) \leqslant 0, \quad \forall c = \text{const} \geqslant 0, \quad (x, t) \in \tilde{Q}^{\varepsilon,\rho}. \tag{5.23}$$

We shall compare the solution $u(x, t) \in C^{2,1}(Q) \cap C(\overline{Q})$ under consideration of (1.1) with the function $\tilde{\omega}(x, t)$ in the cylinder $\tilde{Q}^{\varepsilon,\rho}$. On the lower case of this cylinder $u(x, t) \leqslant \tilde{\omega}(x, t)$, since from the very definition of $\tilde{\omega}$ it follows that $\tilde{\omega} \geqslant m_* > \hat{m} > m_1 = \max_{\Gamma\setminus\Sigma_\rho}|u| \geqslant \max_{\Omega\times\{t=0\}}|u|$. The lateral surface of the cylinder $\tilde{Q}^{\varepsilon,\rho}$ is the union of two surfaces the first of which is a part of the surface $\{(r = \rho) \cap \Omega\} \times (0, T]$ and the second a part of the surface $\{(\tilde{r} = 2\rho - \varepsilon) \cap \Omega\} \times (0, T]$. By what has been proved above, on the first of these surfaces $u(x, t) \leqslant m_* \leqslant \tilde{\omega}(x, t)$ (see (5.15) and (5.16)), while on the second

$$\partial\tilde{\omega}/\partial\nu = -\partial\tilde{\omega}/\partial\tilde{r} = -\tilde{h}'(2\rho - \varepsilon) = -\infty$$

(see (5.17)). Applying Lemma 5.1, we conclude that $u(x, t) \leqslant \tilde{\omega}(x, t) \leqslant m^*$, where

$$m^* = 3T + m_* + \tilde{\kappa}\int_{\tilde{\alpha}}^{\infty} \frac{d\tau}{\tau\delta(\tau)}, \qquad (x, t) \in \overline{\tilde{Q}^{\varepsilon,\rho}}.$$

By the continuity of $u(x, t)$ in $\overline{Q}$ this implies, in particular, the estimate

$$u(y_0, T/2) \leqslant m^*. \tag{5.24}$$

The estimate (5.24) shows that under the conditions of Theorem 5.2 problem (1.3) has no classical solution for some $\varphi(x, t) \in C^\infty(\overline{Q})$, $\partial\varphi/\partial t \not\equiv 0$ in Q. Indeed, taking for $\varphi(x, t)$ a function of the class $C^\infty(\overline{Q})$ such that $\varphi(x, t) = 0$ outside $Q \cap \{(|x - y_0| \leqslant \rho) \times (T/4, 3T/4)\}$ and $\varphi(y_0, T/2) = m^* + 1$, we obtain a contradiction to (5.24). Theorem 5.2 is proved.

CHAPTER 3
LOCAL ESTIMATES OF THE GRADIENTS OF SOLUTIONS OF QUASILINEAR ELLIPTIC EQUATIONS AND THEOREMS OF LIOUVILLE TYPE

§1. Estimates of $|\nabla u(x_0)|$ in terms of $\max_{K_\rho(x_0)}|u|$

Let Ω be an arbitrary domain in $\mathbb{R}^n$, $n \geqslant 2$.

LEMMA 1.1. *Let $u \in C^3(\Omega)$ be a solution of* (1.1.1) *in the domain Ω, where $a(x, u, p) = b(x, u, p) + \hat{b}(x, u, p)$. On the set $K_{\rho,L}(x_0) \equiv \{x \in K_\rho(x_0)\colon |\nabla u| > L\}$, where $K_\rho(x_0)$ is a ball of radius $\rho > 0$ with center at the point $x_0 \in \mathbb{R}^n$ which* *together with its closure is contained in* Ω, *and* $L = \text{const} \geqslant 0$, *suppose that at this solution the following conditions are satisfied*:[1]

$$a^{ij}(x,u,p)\xi_i\xi_j \geqslant 0, \qquad \xi \in \mathbb{R}^n, \tag{1.1}$$

$$\begin{gathered} \mathscr{E}_1 > 0, \quad |pA^\tau_p| \leqslant \sqrt{(\mu_0/n)A^\tau\mathscr{E}_1}\,|p|^{-1}, \quad |\delta A^\tau| \leqslant \sqrt{(\mu_1/n)A^\tau\mathscr{E}_1}\,, \\ |A^\tau_p|\,|p| \leqslant \sqrt{(\mu_2/n^2)A^\tau\mathscr{E}_1}\,|p|^{r/2-1}, \quad \|A\| \leqslant (\mu_3/n)\mathscr{E}_1|p|^{r-2}, \end{gathered} \tag{1.2}$$

$$pb_p - b \geqslant -\mu_4\mathscr{E}_1, \quad \delta b \geqslant -\mu_5\mathscr{E}_1|p|, \quad |b_p| \leqslant \left(\mu_6/\sqrt{n}\right)\mathscr{E}_1|p|^{-2\varepsilon_1}, \tag{1.3}$$

$$\begin{gathered} |p\hat{b}_p| \leqslant \hat{\mu}_4|\hat{b}|, \quad |\delta_-\hat{b}| \leqslant \hat{\mu}_5|\hat{b}|\,|p|, \quad |\hat{b}_p| \leqslant \left(\hat{\mu}_6/\sqrt{n}\right)|\hat{b}|\,|p|^{-2\hat{\varepsilon}_1}, \\ \operatorname{sgn}|\hat{b}|\operatorname{Tr}A|p|^2 \leqslant \hat{\mu}_7\mathscr{E}_1, \end{gathered} \tag{1.4}$$

$$(\operatorname{sgn}|\hat{b}|)|b| \leqslant \hat{\mu}_8\mathscr{E}_1, \tag{1.5}$$

where

$$\mathscr{E}_1 = a^{ij}(x,u,p)p_ip_j, \quad A^\tau = a^{ij}(x,u,p)\tau_i\tau_j, \qquad \tau \in \mathbb{R}^n, |\tau| = 1,$$

$$\delta = \frac{p_k}{|p|}\frac{\partial}{\partial x_k} + |p|\frac{\partial}{\partial u}, \quad \delta_-\hat{b} \equiv \frac{p_k}{|p|}\frac{\partial \hat{b}}{\partial x_k} + |p|\left(\frac{\partial \hat{b}}{\partial u}\right)_-, \quad \left(\frac{\partial \hat{b}}{\partial u}\right)_- = \min\left(\frac{\partial b}{\partial u}, 0\right),$$

$$\|A\| = \left(\sum_{i,j=1}^n (a^{ij})^2\right)^{1/2}, \quad a^{ij} = a^{ij}(x,u,p), \quad b = b(x,u,p), \quad \hat{b} = \hat{b}(x,u,p),$$

$$x \in K_\rho(x_0), \quad u = u(x), \quad p = \nabla u(x), \quad p_k = \frac{\partial u(x)}{\partial x_k}, \qquad k = 1,\dots,n,$$

and the constants μ_i, $\hat{\mu}_i$, r, ε_1, *and* $\hat{\varepsilon}_1$ *satisfy the conditions*

$$\begin{gathered} \mu_i \geqslant 0, \quad i = 0,1,\dots,6; \qquad \hat{\mu}_i \geqslant 0, \quad i = 4,\dots,8; \\ 0 \leqslant r < 2, \quad \varepsilon_1 > 0, \quad \hat{\varepsilon}_1 > 0. \end{gathered} \tag{1.6}$$

Then, for any positive nondecreasing function $z(u) \in C^2(\kappa)$, *where* κ *is the range of* u *in* $K_\rho(x_0)$, *and for any number* $\theta > 0$, *the following alternative holds*: *either at the point* $x_* \in K_\rho(x_0)$ *of the maximum in* $\overline{K_\rho(x_0)}$ *of the function* $\bar{\omega}(x)$ *defined by*

$$\begin{gathered} \bar{\omega} = \omega/z(u(x)), \qquad \omega = w^{\alpha+1}/(\alpha+1), \qquad \alpha = \text{const} > 0, \\ w = v\zeta, \quad v = \sum_{k=1}^n u_k^2, \quad u_k = u_{x_k}(x), \qquad k = 1,\dots,n, \\ \zeta = \zeta(\xi), \quad \zeta \in C^2([0,1]), \quad \zeta(0) = 0, \quad \zeta(\xi) > 0 \quad \text{for } \xi > 0, \xi(1) = 1, \\ \xi = \xi(x) = \begin{cases} 1 - |x-x_0|^2\rho^{-2} & \text{in } K_\rho(x_0), \\ 0 & \text{outside } K_\rho(x_0), \end{cases} \end{gathered} \tag{1.7}$$

the inequality

$$\begin{aligned} -z'' + \frac{\alpha - \gamma - \nu_0}{\alpha+1}\frac{z'^2}{z} - \nu_1 z' - \nu_2 z - \left(\frac{\nu_3}{\rho^2\theta^{2\varepsilon}} + \frac{\nu_4}{\rho^2\theta^{4\hat{\varepsilon}_1}} + \frac{\nu_5}{\rho\theta^{2\varepsilon_1}} + \frac{\nu_6}{\rho\theta^{2\hat{\varepsilon}_1}}\right)z \\ + \frac{\alpha+1}{4}\frac{a^2}{\mathscr{E}_1\operatorname{Tr}A|p|^2}z \leqslant 0 \end{aligned} \tag{1.8}$$

---

[1] As usual, $\operatorname{sgn} c$ ($c \in \mathbb{R}$) is defined to be -1 if $c < 0$, 0 if $c = 0$, and 1 if $c > 0$.

is satisfied, where

$$\begin{gathered}\nu_0 = \mu_0 + \hat{\mu}_4^2\hat{\mu}_7 + \hat{\mu}_7, \qquad \nu_1 = \mu_4 + \hat{\mu}_4\hat{\mu}_8 + \hat{\mu}_8,\\ \nu_2 = 4(\alpha+1)(\mu_1 + \mu_5 + \hat{\mu}_5^2\hat{\mu}_7 + \hat{\mu}_5\hat{\mu}_8),\\ \nu_3 = 4(\alpha+1)\sigma^{-2}[3\mu_2 + \mu_3(\gamma^{-1}+4)],\\ \nu_4 = 4(\alpha+1)\sigma^{-2}\hat{\mu}_6^2\hat{\mu}_7, \qquad \nu_5 = 2(\alpha+1)\sigma^{-1}\mu_6,\\ \nu_6 = 2(\alpha+1)\sigma^{-1}\hat{\mu}_6\hat{\mu}_8, \qquad \gamma > 0, \quad \sigma = \min(\varepsilon/2, \varepsilon_1, \hat{\varepsilon}_1), \end{gathered} \tag{1.9}$$

or

$$|\nabla u(x_0)| \leqslant (z_0/z_*)^{1/2(\alpha+1)}\max(L, \theta), \tag{1.10}$$

where $z_0 \equiv z(u(x_0))$ *and* $z_* \equiv z(u(x_*))$.

PROOF. Applying to (1.1.1) the operator $u_k(\partial/\partial x_k)$, for the function $v \equiv \Sigma_1^n u_k^2$ in $K_\rho(x_0)$ we obtain

$$1/2a^{ij}v_{ij} = a^{ij}u_{ki}u_{kj} + 1/2(a_{p_l} - a^{ij}_{p_l}u_{ij})v_l + \sqrt{v}(\delta a - \delta a^{ij}u_{ij}). \tag{1.11}$$

For the function $w = v\zeta$ (see (1.7)) we then have

$$\begin{aligned} a^{ij}w_{ij} = {} & 2\zeta a^{ij}u_{ki}u_{kj} + (a_{p_l} - a^{ij}_{p_l}u_{ij})\left(w_l - \frac{\zeta_l}{\zeta}w\right) + 2\sqrt{v}\,\zeta(\delta a - \delta a^{ij}u_{ij})\\ & + 2a^{ij}w_i\frac{\zeta_j}{\zeta} + \left(a^{ij}\frac{\zeta_{ij}}{\zeta} - 2a^{ij}\frac{\zeta_i}{\zeta}\frac{\zeta_j}{\zeta}\right)w. \end{aligned} \tag{1.12}$$

Multiplying both sides of (1.12) by $f(w) = w^\alpha$, $\alpha > 0$, and setting $\omega = w^{\alpha+1}/(\alpha+1)$, we derive from (1.12) the following identity on the set $\{x \in K_\rho(x_0): |\nabla u(x)| > 0\}$:

$$\begin{aligned} a^{ij}\omega_{ij} = {} & \frac{f'}{f^2}a^{ij}\omega_i\omega_j + 2\zeta a^{ij}u_{ki}u_{kj} + (a_{p_l} - a^{ij}_{p_l}u_{ij})\left(\omega_l - \frac{\zeta_l}{\zeta}fw\right)\\ & + 2f\sqrt{v}\,\zeta(\delta a - \delta a^{ij}u_{ij}) + 2a^{ij}\omega_i\frac{\zeta_j}{\zeta} + \left(a^{ij}\frac{\zeta_{ij}}{\zeta} - 2a^{ij}\frac{\zeta_i}{\zeta}\frac{\zeta_j}{\zeta}\right)fw. \end{aligned} \tag{1.13}$$

Setting $\omega = z(u)\varpi$, we obtain in place of (1.13)

$$\begin{aligned} za^{ij}\varpi_{ij} + b_k\varpi_k = {} & -z''\mathscr{E}_1\varpi + \frac{\alpha}{\alpha+1}\frac{z'^2}{z}\mathscr{E}_1\varpi + 2\zeta a^{ij}u_{ki}u_{kj}f\\ & + (a_pp - a)z'\varpi - (a^{ij})_p pu_{ij}z'\varpi - (a_{p_l} - a^{ij}_{p_l}u_{ij})\frac{\zeta_l}{\zeta}fw \qquad (1.14)\\ & + 2f\sqrt{v}\,\zeta(\delta a - \delta a^{ij}u_{ij}) + 2a^{ij}u_i\frac{\zeta_j}{\zeta}z'\varpi + \left(a^{ij}\frac{\zeta_{ij}}{\zeta} - 2a^{ij}\frac{\zeta_i}{\zeta}\frac{\zeta_i}{\zeta}\right)fw, \end{aligned}$$

where $p\,\partial/\partial p \equiv p_k\,\partial/\partial p_k$ and the form of b_k is irrelevant for what follows. We shall henceforth consider and transform the identity (1.14) only on the set $K_{\rho,L}(x_0)$. Using conditions (1.2) exactly as in §1.6, we obtain the estimates

$$\begin{aligned}
|a_p^{ij} p z' u_{ij} \overline{\omega}| &\leqslant \frac{f\zeta}{4} a^{ij} u_{ik} u_{kj} + \frac{\mu_0}{\alpha+1}\frac{z'^2}{z}\mathscr{E}_1\overline{\omega},\\
|2f\sqrt{v}\,\zeta\delta a^{ij} u_{ij}| &\leqslant \frac{f\zeta}{4} a^{ij} u_{ki} u_{kj} + 4(\alpha+1)\mu_1 z\mathscr{E}_1\overline{\omega},\\
\left|a_{p_l}^{ij} u_{ij}\frac{\zeta_l}{\zeta} fw\right| &\leqslant \frac{f\zeta}{4} a^{ij} u_{ki} u_{kj} + 4(\alpha+1)\mu_2\frac{\zeta'^2}{\zeta^2}\frac{z}{\rho^2 v^{\varepsilon}}\mathscr{E}_1\overline{\omega},\\
\left|2a^{ij} u_i \frac{\zeta_j}{\zeta} z'\overline{\omega}\right| &\leqslant \frac{\gamma}{\alpha+1}\frac{z'^2}{z}\mathscr{E}_1\overline{\omega} + \frac{4(\alpha+1)}{\gamma}\mu_3\frac{\zeta'^2}{\zeta^2}\frac{z}{\rho^2 v^{\varepsilon}}\mathscr{E}_1\overline{\omega}\\
\left|a^{ij}\frac{\zeta_{ij}}{\zeta} fw\right| &\leqslant 8(\alpha+1)\mu_3\frac{|\zeta''|+|\zeta'|}{\zeta}\frac{z}{\rho^2 v^{\varepsilon}}\mathscr{E}_1\overline{\omega},\\
\left|2a^{ij}\frac{\zeta_i}{\zeta}\frac{\zeta_j}{\zeta} fw\right| &\leqslant 8(\alpha+1)\mu_3\frac{\zeta'^2}{\zeta^2}\frac{z}{\rho^2 v^{\varepsilon}}\mathscr{E}_1\overline{\omega},
\end{aligned} \tag{1.15}$$

where $\varepsilon = 1 - r/2 > 0$. We note that in deriving (1.15) we have used, in particular, the inequalities $|\nabla\zeta| \leqslant 2\sqrt{n}/\rho$ and $\|D^2\zeta\| \leqslant 2n/\rho^2$, where $\|D^2\zeta\| = (\sum_{i,j=1}^n \zeta_{i,j}^2)^{1/2}$.

We now estimate the remaining terms $(a_p p - a)z'\overline{\omega}$, $a_{p_l}(\zeta_l/\zeta)fw$ and $2f\sqrt{v}\,\zeta\delta a$ in (1.14). Taking into account that $a = b + \hat{b}$ and applying (1.3)–(1.5) and (1.7), we obtain

$$\begin{aligned}
(pa_p - a)z'\overline{\omega} &= (pb_p - b)z'\overline{\omega} + (p\hat{b}_p - \hat{b})z'\overline{\omega},\\
(pb_p - b)z'\overline{\omega} &\geqslant -\mu_4 z'\mathscr{E}_1\overline{\omega},\\
p\hat{b}_p z'\overline{\omega} = p\frac{\hat{b}_p}{\hat{b}}\hat{b}z'\overline{\omega} &= p\frac{\hat{b}_p}{\hat{b}}(a^{ij}u_{ij} - b)z'\overline{\omega},\\
\hat{b}z'\overline{\omega} &= (a^{ij}u_{ij} - b)z'\overline{\omega},\\
\left|p\frac{\hat{b}_p}{\hat{b}} a^{ij} u_{ij} z'\overline{\omega}\right| &\leqslant \frac{f\zeta}{4} a^{ij} u_{ki} u_{kj} + \frac{\hat{\mu}_4^2\hat{\mu}_7}{\alpha+1}\frac{z'^2}{z}\mathscr{E}_1\overline{\omega},\\
\left|p\frac{\hat{b}_p}{\hat{b}} bz'\overline{\omega}\right| &\leqslant \hat{\mu}_4\hat{\mu}_8 z'\mathscr{E}_1\overline{\omega},\\
|a^{ij}u_{ij}z'\overline{\omega}| &\leqslant \frac{f\zeta}{4} a^{ij} u_{ki} u_{kj} + \frac{\hat{\mu}_7}{\alpha+1}\frac{z'^2}{z}\mathscr{E}_1\overline{\omega},\\
|bz'\overline{\omega}| &\leqslant \hat{\mu}_8 z'\mathscr{E}_1\overline{\omega};\\
a_{p_l}\frac{\zeta_l}{\zeta} fw &= b_{p_l}\frac{\zeta_l}{\zeta} fw + \hat{b}_{p_l}\frac{\zeta_l}{\zeta} fw,\\
\left|b_{p_l}\frac{\zeta_l}{\zeta} fw\right| &\leqslant 2(\alpha+1)\mu_6\frac{|\zeta'|}{\zeta\rho}\frac{z}{v^{\varepsilon_1}}\mathscr{E}_1\overline{\omega},
\end{aligned} \tag{$*$}$$

(continued)

$$\hat{b}_{p_l}\frac{\zeta_l}{\zeta}fw = \frac{\hat{b}_{p_l}}{\hat{b}}\hat{b}\frac{\zeta_l}{\zeta}fw = \frac{\hat{b}_{p_l}}{\hat{b}}\left(a^{ij}u_{ij} - b\right)\frac{\zeta_l}{\zeta}fw,$$

$$\left|\frac{\hat{b}_{p_l}}{\hat{b}}\frac{\zeta_l}{\zeta}fwa^{ij}u_{ij}\right| \leqslant \frac{f\zeta}{4}a^{ij}u_{ki}u_{kj} + 4(\alpha+1)\hat{\mu}_6^2\hat{\mu}_7\frac{\zeta'^2}{\zeta}\frac{z}{\rho^2 v^{2\varepsilon_1}}\mathscr{E}_1\overline{\omega},$$

$$\left|\frac{\hat{b}_{p_l}}{\hat{b}}b\frac{\zeta_l}{\zeta}fw\right| \leqslant (\alpha+1)\hat{\mu}_6\hat{\mu}_8\frac{|\zeta'|}{\zeta}\frac{2}{\rho}\frac{z}{v^{\hat{\varepsilon}_1}}\mathscr{E}_1\overline{\omega};$$

$$2f\sqrt{v}\,\zeta\delta a \geqslant 2f\sqrt{v}\,\zeta\delta b + 2f\sqrt{v}\,\zeta\delta\_\hat{b},$$

$$2f\sqrt{v}\,\zeta\delta b \geqslant 2fw\frac{\delta b}{|p|} \geqslant -2(\alpha+1)\mu_5 z\mathscr{E}_1\overline{\omega},$$

$$2f\sqrt{v}\,\zeta\delta\_\hat{b} = 2f\sqrt{v}\,\zeta\frac{\delta\hat{b}}{\hat{b}}\hat{b} = 2f\sqrt{v}\,\zeta\frac{\delta\hat{b}}{\hat{b}}\left(a^{ij}u_{ij} - b\right),$$

$$\left|2f\sqrt{v}\,\zeta\frac{\delta\_\hat{b}}{\hat{b}}a^{ij}u_{ij}\right| \leqslant \frac{f\zeta}{4}a^{ij}u_{ki}u_{kj} + 4(\alpha+1)\hat{\mu}_5^2\hat{\mu}_7 z\mathscr{E}_1\overline{\omega},$$

$$\left|2f\sqrt{v}\,\zeta\frac{\delta\_\hat{b}}{\hat{b}}b\right| \leqslant 2(\alpha+1)\hat{\mu}_5\hat{\mu}_6 z\mathscr{E}_1\overline{\omega}.$$

We observe that relations (*) are considered only at those points of $K_{\rho,L}(x_0)$ where $\hat{b}(x, u(x), \nabla u(x)) \neq 0$. At those points of $K_{\rho,L}(x_0)$ where $\hat{b} = 0$ the corresponding expressions estimated in (*) are simply absent. Setting $\zeta(\xi) = \xi^{1/\sigma}$, where $\sigma = \min(\varepsilon/2, \hat{\varepsilon}_1, \varepsilon_1)$, we observe that

$$\frac{\zeta^1}{\zeta} = \frac{1}{\sigma\zeta^{\sigma}},\quad \frac{\zeta'^2}{\zeta^2} = \frac{1}{\sigma^2\zeta^{2\sigma}},\quad \frac{|\zeta''| + |\zeta'|}{\zeta} \leqslant \frac{2}{\sigma^2\zeta^{2\sigma}},$$

so that

$$\frac{\zeta'^2}{\zeta^2 v^{\varepsilon}} \leqslant \frac{1}{\sigma^2 w^{\varepsilon}},\quad \frac{|\zeta''| + |\zeta'|}{\zeta v^{\varepsilon}} \leqslant \frac{2}{\sigma^2 w^{\varepsilon}},\quad \frac{\zeta'^2}{\zeta^2 v^{2\hat{\varepsilon}_1}} \leqslant \frac{1}{\sigma^2 w^{2\hat{\varepsilon}_1}},$$
$$\frac{\zeta'}{\zeta v^{\varepsilon_1}} \leqslant \frac{1}{\sigma w^{\varepsilon_1}},\quad \frac{|\zeta'|}{\zeta v^{\hat{\varepsilon}_1}} \leqslant \frac{1}{\sigma w^{\hat{\varepsilon}_1}}. \tag{1.16}$$

From (1.14)–(1.16) and (*), using the inequality

$$1/4 f\zeta a^{ij}u_{ki}u_{kj} \geqslant \frac{\alpha+1}{4}\frac{a^2}{\mathscr{E}_1 \operatorname{Tr} Av}z\overline{\omega}\mathscr{E}_1, \tag{1.17}$$

which follows from the estimate $a^{ij}u_{ki}u_{kj} \geqslant a^2/\operatorname{Tr} A$ and (1.7), we obtain

$$za^{ij}\overline{\omega}_{ij} + b_k\overline{\omega}_k \geqslant \Bigg[-z'' + \frac{\alpha - \gamma - \nu_0}{\alpha+1}\frac{z'^2}{z} - \nu_1 z' - \nu_2 z$$
$$-\left(\frac{\nu_3}{\rho^2 w^{\varepsilon}} + \frac{\nu^4}{\rho^2 w^{2\hat{\varepsilon}_1}} + \frac{\nu_5}{\rho w^{\varepsilon_1}} + \frac{\nu_6}{\rho w^{\hat{\varepsilon}_1}}\right)z + \frac{\alpha+1}{4}\frac{a^2}{\mathscr{E}_1 \operatorname{Tr} Av}z\Bigg]\mathscr{E}_1\overline{\omega}, \tag{1.18}$$

where the quantities ν_i, $i = 0,\dots,6$, are defined by (1.9).

Suppose that x_* is a point of a maximum of the function $\overline{\omega}$ in $\overline{K_\rho(x_0)}$. It is obvious that $x_* \in K_\rho(x_0)$, since $\overline{\omega}(x) = 0$ on the boundary of the ball $K_\rho(x_0)$. To abbreviate the notation we set $F(w) \equiv w^{\alpha+1}/(\alpha+1)$ and denote by Φ the function inverse to F, so that $\Phi(F(w)) \equiv w$. The following alternative holds: either the inequalities

$$|\nabla u(x_*)| > L, \qquad \Phi(z(x_*)\bar{\omega}(x_*)) \geqslant \theta^2 \tag{1.19}$$

hold simultaneously or at least one of these inequalities fails to hold. In the first case, taking into account that the inequality $|\nabla u(x_*)| > L$ makes it possible to consider (1.18) at the point x_* and observing that the second inequality in (1.19) implies the estimate

$$w_* = \Phi(z_*\bar{\omega}_*) \geqslant \theta^2, \tag{1.20}$$

where $w_* \equiv w(x_*)$, $z_* \equiv z(x_*)$, $\bar{\omega}_* \equiv \bar{\omega}(x_*)$, we conclude (taking the necessary conditions for a maximum into account) that at x_* inequality (1.8) holds. To complete the proof of the lemma it suffices to establish that (1.10) holds if at least one of the inequalities (1.19) is violated. Indeed, if $|\nabla u(x_*)| \leqslant L_*$, then $v(x_*) \leqslant L^2$, and from (1.7) we obtain

$$F(v_0) = F(w_0) = \omega_0 = z_0\bar{\omega}_0 \leqslant z_0\bar{\omega}_* = \frac{z_0}{z_*}\omega_* = \frac{z_0}{z_*}F(v_*) \leqslant \frac{z_0}{z_*}F(L^2), \tag{1.21}$$

whence

$$v_0 \leqslant \Phi\left(\frac{z_0}{z_*}F(L^2)\right) = \left(\frac{z_0}{z_*}\right)^{1/(\alpha+1)} L^2. \tag{1.22}$$

If $\Phi(z_*\bar{\omega}_*) \leqslant \theta^2$, then from (1.7) it follows that

$$F(v_0) = F(w_0) = \omega_0 = z_0\bar{\omega}_0 \leqslant z_0\bar{\omega}_* = \frac{z_0}{z_*}F(\Phi(z_*\bar{\omega}_*)) \leqslant \frac{z_0}{z_*}F(\theta^2), \tag{1.23}$$

whence

$$v_0 \leqslant \Phi\left(\frac{z_0}{z_*}F(\theta^2)\right) = \left(\frac{z_0}{z_*}\right)^{1/(\alpha+1)} \theta^2. \tag{1.24}$$

The estimate (1.10) obviously follows from (1.22), (1.24). Lemma 1.1 is proved.

The role of Lemma 1.1 is in that by imposing various conditions on the constants in conditions (1.1)–(1.6) it is possible to choose the function $z(u)$ and the number θ (depending only on the structure of the equation) so that the first condition of the alternative presented in the formulation of Lemma 1.1 is impossible. This provides an a priori estimate of the form (1.10) for the gradient of the solution at the fixed point x_0.

THEOREM 1.1. *Let $u \in C^3(\Omega)$ be a solution of* (1.1.1) *in Ω such that $m_1 \leqslant u(x) \leqslant m_2$, $x \in K_\rho(x_0)$, m_1, m_2 = const, where $K_\rho(x_0) \subset \Omega$, $\rho > 0$, and suppose that at this soluion on the set $K_{\rho,L}(x_0) \equiv \{x \in K_\rho(x_0)\colon |\nabla u| > L\}$ conditions* (1.1)–(1.6) *hold, where*

$$\nu_2 < \tfrac{1}{2}\nu_1^2 e^{-2\nu_1 m}, \qquad m \equiv m_2 - m_1, \nu_1 > 0, \tag{1.25}$$

and ν_1 and ν_2 are defined in (1.9) *with $\alpha = 2(\mu_0 + \hat{\mu}_4^2\hat{\mu}_7 + \hat{\mu}_7)$. Then*

$$|\nabla u(x_0)| \leqslant 2\max(L, \theta), \tag{1.26}$$

where

$$\theta = \max\left\{\left(\frac{8\nu_3 e^{2\nu_1 m}}{\rho^2\nu_1^2}\right)^{1/2\varepsilon}, \left(\frac{8\nu_4 e^{2\nu_1 m}}{\rho^2\nu_1^2}\right)^{1/4\hat{\varepsilon}_1}, \left(\frac{8\nu_5 e^{2\nu_1 m}}{\rho\nu_1^2}\right)^{1/2\varepsilon_1}, \left(\frac{8\nu_6 e^{2\nu_1 m}}{\rho\nu_1^2}\right)^{1/2\hat{\varepsilon}_1}\right\}, \tag{1.27}$$

and ν_3, ν_4, ν_5 *and* ν_6 *are defined in* (1.9) *with* $\alpha = 2\nu_0$, $\gamma = \alpha/2$ *and* $\nu_0 > 0$. *If instead of condition* (1.25)

$$\nu_1 < 1/2m, \quad \nu_2 < 1/4m^2, \qquad m = m_2 - m_1, \tag{1.28}$$

then (1.26) *holds for* θ *defined by*

$$\theta = \max\left\{\left(\frac{16\nu_3 m^2}{\rho^2}\right)^{1/2\varepsilon}, \left(\frac{16\nu_4 m^2}{\rho^2}\right)^{1/4\hat{\varepsilon}_1}, \left(\frac{16\nu_5 m^2}{\rho}\right)^{1/2\varepsilon_1}, \left(\frac{16\nu_6 m^2}{\rho}\right)^{1/2\hat{\varepsilon}_1}\right\}. \tag{1.29}$$

PROOF. We start by proving the first part of the theorem. We set $z(u) = 2e^{\beta m} - e^{\beta(m_2-u)}$, $m = m_2 - m_1$, $\beta = \text{const} > 0$, and we suppose that the number θ is defined by (1.27). We shall show that in the case of such $z(u)$ and θ the first part of the alternative presented in the formulation of Lemma 1.1 cannot hold. Let $\gamma = \alpha/2$ and $\alpha = 2\nu_0$. Taking into account that $z' = \beta e^{\beta(m_2-u)} > 0$, $z'' = -\beta^2 e^{\beta(m_2-u)}$, $e^{\beta m} \leqslant z \leqslant 2e^{\beta m}$ and $e^{\beta(m_2-u)} \geqslant 1$, we then deduce from (1.8) that

$$(\beta^2 - \beta\nu_1) - 2\nu_2 e^{\beta m} - 2\left(\frac{\nu_3}{\rho^2\theta^{2\varepsilon}} + \frac{\nu_4}{\rho^2\theta^{4\hat{\varepsilon}_1}} + \frac{\nu_5}{\rho\theta^{2\varepsilon_1}} + \frac{\nu_6}{\rho\theta^{2\hat{\varepsilon}_1}}\right)e^{\beta m} \leqslant 0. \tag{1.30}$$

We set $\beta = 2\nu_1$. Then $\beta^2 - \beta\nu_1 = 2\nu_1^2$, and, using (1.25), we find that $\beta^2 - \beta\nu_1 - 2\nu_2 e^{\beta m} > \nu_1^2$. On the other hand, from the definition of θ it follows that

$$2\left(\frac{\nu_3}{\rho^2\theta^{2\varepsilon}} + \frac{\nu_4}{\rho^2\theta^{4\hat{\varepsilon}_1}} + \frac{\nu_5}{\rho\theta^{2\varepsilon_1}} + \frac{\nu_6}{\rho\theta^{2\hat{\varepsilon}_1}}\right)e^{2\nu_1 m} \leqslant \nu_1^2.$$

But then it follows from what has been said that the expression on the left of (1.30) is positive, contrary to (1.30). This implies that for the present choice of $z(u)$ and θ inequality (1.30) is impossible. Hence, the second part of the alternative in the formulation of Lemma 1.1 holds, i.e., inequality (1.10) for the $z(u)$ and θ selected. Taking into account that $z_0/z_* \leqslant 2$, we obtain (1.26) with the θ of (1.27).

Suppose now that condition (1.28) holds instead of (1.25). In this case we choose for $z(u)$ the function $z(u) = 2m^2 - (m_2 - u)^2$, $m = m_2 - m_1$, assuming that $m > 0$ (otherwise $u \equiv \text{const}$ in $K_\rho(x_0)$), and we suppose that θ is defined by (1.29). Suppose also that $\gamma = \alpha/2$, $\alpha = 2\nu_0$ and $\nu_0 > 0$. Taking into account that $z' = 2(m_2 - u) \geqslant 0$, $-z'' = 2$ and $m^2 \leqslant z(u) \leqslant 2m^2$, we deduce from (1.8) that

$$2 - 2\nu_1 m - 2\nu_2 m^2 - 2m^2\left(\frac{\nu_3}{\rho^2\theta^{2\varepsilon}} + \frac{\nu_4}{\rho^2\theta^{4\hat{\varepsilon}_1}} + \frac{\nu_5}{\rho\theta^{2\varepsilon_1}} + \frac{\nu_6}{\rho\theta^{2\hat{\varepsilon}_1}}\right) \leqslant 0. \tag{1.31}$$

Now from (1.28) and the definition of θ by (1.29) it follows that inequality (1.31) is impossible. Therefore, by Lemma 1.1 the estimate (1.10) with the θ of (1.29) holds. Theorem 1.1 is proved.

REMARK 1.1. It is obvious that the conditions of Theorem 1.1. admit arbitrary degeneracy of (1.1) on the set $\{|p| \leqslant L\}$ and degeneracy chacterized by the condition $\mathscr{E}_1 > 0$ on the set $\{|p| > L\}$.

§2. An estimate of $|\nabla u(x_0)|$ in terms of $\max_{K_\rho(x_0)} u$ $\left(\min_{K_\rho(x_0)} u\right)$. Harnack's inequality

In this section we first distinguish a class of equations of the form (1.1.1) for whose solutions $|\nabla u(x_0)|$ depends only on the structure of the equation and not on any bounds on the solution itself. On the basis of this estimate a larger class of equations of the form (1.1.1) is distinguished for whose solutions an estimate of $|\nabla u(x_0)|$ depending on $\max_{K_\rho(x_0)} u$ (or $\min_{K_\rho(x_0)} u$) can be established.

THEOREM 2.1. *Let $u \in C^2(\Omega)$ be a solution of* (1.1.1), *and suppose that $a(x, u, p) = b(x, u, p) + \hat{b}(x, u, p)$. Assume that at this solution the following conditions are satisfied on the set $K_{\rho,L}(x_0) \equiv \{x \in K_\rho(x_0): |\nabla u| > L\}$:*

$$\mathscr{E}_1 > 0, \quad |\delta A^\tau| \leqslant \sqrt{\frac{\mu_1}{n} A^\tau \mathscr{E}_1}, \quad |A^\tau p|\,|p| \leqslant \sqrt{\frac{\mu_2}{n^2} A^\tau \mathscr{E}_1}\,|p|^{r/2-1}, \qquad \|A\| \leqslant \frac{\mu_3}{n}\mathscr{E}_1 |p|^{r-2}, \tag{2.1}$$

$$|b_p| \leqslant \frac{\mu_6}{\sqrt{n}}\mathscr{E}_1 |p|^{-2\varepsilon_1}, \tag{2.2}$$

$$|\delta\_\hat{b}| \leqslant \hat{\mu}_5 |\hat{b}|\,|p|, \quad |\hat{b}_p| \leqslant \frac{\hat{\mu}_6}{\sqrt{n}}|b|\,|p|^{-2\hat{\varepsilon}_1}, \quad \operatorname{sgn}|\hat{b}| \operatorname{Tr} A |p|^2 \leqslant \hat{\mu}_7 \mathscr{E}_1, \tag{2.3}$$

$$(\operatorname{sgn}|\hat{b}|)|b| \leqslant \hat{\mu}_8 \mathscr{E}_1, \tag{2.4}$$

where $\mathscr{E}_1$, A^τ, δ and $\delta\_$ are defined in the formulation of Lemma 1.1, *and the constants $\mu_1, \mu_2, \ldots, r, \varepsilon_1$, and $\hat{\varepsilon}_1$ satisfy* (1.6). *Suppose that with some $\beta \in (0, 1]$ on the set $K_{\rho,L}(x_0)$ the inequality*

$$(1-\beta)\frac{a^2}{\mathscr{E}_1 \operatorname{Tr} A |p|^2} + \frac{\delta b}{\mathscr{E}_1 |p|} - \left(\frac{\mu_1}{\beta} + \frac{\hat{\mu}_5^2 \hat{\mu}_7}{\beta} + \hat{\mu}_5 \hat{\mu}_8\right) \geqslant c_0 \tag{2.5}$$

holds, where $c_0 = \text{const} > 0$. Then

$$|\nabla u(x_0)| \leqslant \max(L, \theta), \tag{2.6}$$

where

$$\theta = \max\left\{\left(\frac{4c_0^{-1}\kappa_1}{\rho^2}\right)^{1/2\varepsilon}, \left(\frac{4c_0^{-1}\kappa_2}{\rho^2}\right)^{1/4\hat{\varepsilon}_1}, \left(\frac{4c_0^{-1}\kappa_3}{\rho}\right)^{1/2\varepsilon_1}, \left(\frac{4c_0^{-1}\kappa_4}{\rho}\right)^{1/2\hat{\varepsilon}_1}\right\}, \tag{2.7}$$

and $\kappa_1 = \kappa(\mu_2 + \mu_3)$, $\kappa_2 = \kappa\hat{u}_6^2\hat{\mu}_7$, $\kappa_3 = \kappa\mu_6$, $\kappa_4 = \kappa\hat{\mu}_6\hat{\mu}_8$, $\kappa = 2\sigma^{-2}\max(1/\beta, 12)$, $\varepsilon = 1 - r/2$.

PROOF. Applying the operator $u_k(\partial/\partial x_k)$ to equation (1.1.1) in the ball $K_\rho(x_0)$, we obtain (1.11). Using conditions (2.1)–(2.4) exactly as in the proof of Lemma 1.1, on the set $K_{\rho,L}(x_0)$ we obtain

$$\begin{gathered}\left|a^{ij}_{p_l}u_{ij}\frac{\zeta_j}{\zeta}w\right| \leqslant \frac{\beta}{2}\zeta a^{ij}u_{ki}u_{kj} + \frac{2\mu_2}{\beta}\frac{\zeta'^2}{\zeta^2\rho^2v^\varepsilon}\mathscr{E}_1 w,\\ \left|2\sqrt{v}\,\zeta\delta a^{ij}u_{ij}\right| \leqslant \frac{\beta}{2}\zeta a^{ij}u_{ki}u_{kj} + \frac{2\mu_1}{\beta}\mathscr{E}_1 w,\\ \left|a^{ij}\frac{\zeta_{ij}}{\zeta}w\right| \leqslant \mu_3\frac{8}{\rho^2}\frac{|\zeta''|+|\zeta'|}{\zeta v^\varepsilon}\mathscr{E}_1 w,\\ 2a^{ij}\frac{\zeta_i}{\zeta}\frac{\zeta_j}{\zeta}w \leqslant 8\mu_3\frac{\zeta'^2}{\zeta^2\rho^2v^\varepsilon}\mathscr{E}_1 w,\end{gathered} \tag{2.8}$$

where $\varepsilon = 1 - r/2 > 0$ and $\beta = \text{const} \in (0,1]$;

$$\begin{aligned}\left|a_{p_l}\frac{\zeta_l}{\zeta}w\right| &\leqslant \left|b_{p_l}\frac{\zeta_l}{\zeta}w\right| + \left|\hat{b}_{p_l}\frac{\zeta_l}{\zeta}w\right|\\ &\leqslant \left(2\mu_6\frac{|\zeta'|}{\zeta\rho v^{\varepsilon_1}} + \frac{2}{\beta}\hat{\mu}_6^2\hat{\mu}_7\frac{\zeta'^2}{\zeta^2\rho^2v^{2\hat{\varepsilon}_1}} + 2\hat{\mu}_6\hat{\mu}_8\frac{|\zeta'|}{\zeta\rho v^{\hat{\varepsilon}_1}}\right)\mathscr{E}_1 w - \frac{\beta}{2}\zeta a^{ij}u_{ki}u_{kj};\end{aligned} \tag{2.9}$$

$$\begin{aligned}2\sqrt{v}\,\zeta\delta a &\geqslant 2\sqrt{v}\,\zeta\delta b + 2\sqrt{v}\,\zeta(\delta b)_-\\ &\geqslant 2\sqrt{v}\,\zeta\delta b - \left(\frac{2\hat{\mu}_5^2\hat{\mu}_7}{\beta} + 2\hat{\mu}_5\hat{\mu}_8\right)\mathscr{E}_1 w - \beta\frac{\zeta}{2}a^{ij}u_{ki}u_{kj}.\end{aligned} \tag{2.10}$$

In addition, taking into account that

$$2(1-\beta)\zeta a^{ij}u_{ki}u_{kj} \geqslant 2(1-\beta)\frac{a^2}{\operatorname{Tr} A} = 2(1-\beta)\frac{a^2}{\operatorname{Tr} A|p|^2\mathscr{E}_1}\mathscr{E}_1 w, \tag{2.11}$$

and choosing $\zeta(\xi) = \xi^{1/\sigma}$, where $\sigma = \min(\varepsilon/2, \varepsilon_1, \hat{\varepsilon}_1)$, we deduce from (1.11) and (2.8)–(2.11) that

$$\begin{aligned}a^{ij}w_{ij} + b^k w_k \geqslant \Bigg\{&\left[2(1-\beta)\frac{a^2}{\operatorname{Tr} A|p|^2\mathscr{E}_1} + \frac{2\delta b}{\mathscr{E}_1|p|} - \left(\frac{2\mu_1 + 2\hat{\mu}_5^2\hat{\mu}_7}{\beta} + 2\hat{\mu}_5\hat{\mu}_8\right)\right]\\ &- \kappa\left[\frac{\mu_2+\mu_3}{\rho^2w^\varepsilon} + \frac{\hat{\mu}_6^2\hat{\mu}_7}{\rho^2w^{2\hat{\varepsilon}_1}} + \frac{\mu_6}{\rho w^{\varepsilon_1}} + \frac{\hat{\mu}_6\hat{\mu}_8}{\rho w^{\hat{\varepsilon}_1}}\right]\Bigg\}\mathscr{E}_1 w,\\ &x \in K_{\rho,L}(x_0),\end{aligned} \tag{2.12}$$

where the form of b^k is irrelevant for what follows and $\kappa = 2\sigma^{-2}\max(1/\beta, 12)$. Let x_* be the point of a maximum of the function w in $\overline{K_\rho(x_0)}$ (it is obvious that $x_* \in K_\rho(x_0)$). The following alternative holds: either the inequalities $|\nabla u(x_*)| > L$ and $w(x_*) \geqslant \theta^2$ are satisfied simultaneously, or at least one of these inequalities fails to hold. In the first case, taking the form of the number θ into account (see (2.7)), we conclude that $x_* \in K_{\rho,L}(x_0)$, and that at this point we have

$$\kappa\left[\frac{\mu_2+\mu_3}{\rho^2w^\varepsilon} + \frac{\hat{\mu}_6\hat{\mu}_7}{\rho^2w^{2\hat{\varepsilon}_1}} + \frac{\mu_6}{\rho w^{\varepsilon_1}} + \frac{\hat{\mu}_6\hat{\mu}_8}{\rho w^{\hat{\varepsilon}_1}}\right] < c_0.$$

From (2.12) and (2.5) it then follows that $a^{ij}w_{ij} > 0$ at x_*. However, this inequality contradicts the fact that $x_* \in K_\rho(x_0)$ is a maximum point for w. Hence, the first assertion of the alternative cannot hold. From the second part of the alternative we easily obtain (2.6). Theorem 2.1 is proved.

We note that several other classes of equations of the form (1.1.1) for whose solutions $u(x)$ an estimate of $|\nabla u(x_0)|$ depending only on the structure of the equation (and not depending, in particular, on any bounds on the solution itself) can be established were distinguished in [157]. It is obvious that Theorem 2.1 admits degeneration of ellipticity of the same sort as Theorem 1.1 does.

We now distinguish some classes of equations for which $|\nabla u(x_0)|$ can be estimated in terms of $u(x_0)$ and $\sup_{K_\rho(x_0)} u$ ($\inf_{K_\rho(x_0)} u$). To this end in (1.1.1) we make the change of unknown function $u = \varphi(\bar{u})$. Equation (1.1.1) then becomes

$$\bar{a}^{ij}\bar{u}_{ij} = \bar{a}, \tag{2.13}$$

where

$$\bar{a}^{ij}(x, \bar{u}, \bar{p}) = a^{ij}(x, \varphi(\bar{u}), \varphi'(\bar{u})\bar{p}),$$

$$\bar{a}(x, \bar{u}, \bar{p}) = \frac{1}{\varphi'(\bar{u})}a(x, \varphi(\bar{u}), \varphi'(\bar{u})\bar{p}) - \frac{\varphi''(\bar{u})}{\varphi'(\bar{u})}\bar{\mathscr{E}}_1,$$

$$\bar{\mathscr{E}}_1 = a^{ij}(x, \varphi(\bar{u}), \varphi'(\bar{u})\bar{p})\bar{p}_i\bar{p}_j = \mathscr{E}_1[\varphi'(\bar{u})]^{-2}.$$

We note also that $p = \varphi'(\bar{u})\bar{p}$ and $\bar{A}^\tau \equiv \bar{a}^{ij}(x, \bar{u}, \bar{p})\tau_i\tau_j = A^\tau$, $\tau \in \mathbb{R}^n$, $|\tau| = 1$. We choose $\varphi(\bar{u})$ so that $\varphi''(\varphi')^{-1} = \text{const}$. More precisely, let

$$u = \varphi(\bar{u}) \equiv m - \delta + e^{\bar{u}}, \qquad m \equiv \inf_{K_\rho(x_0)} u, \quad \delta = \text{const} > 0. \tag{2.14}$$

LEMMA 2.1. *Suppose that on the set $K_{\rho,l}(x_0) \equiv \{x \in K_\rho(x_0): |\nabla u(x)| > l\}$ at a solution $u \in C^3(\Omega)$ of* (1.1.1) *satisfying the condition $u(x) \geqslant m$ in $K_\rho(x_0)$ the following conditions are satisfied*:

$$\begin{gathered}\mathscr{E}_1 > 0, \quad |pA^\tau_p| \leqslant \sqrt{\frac{\mu_0}{n}A^\tau\mathscr{E}_1}\,|p|^{-1}, \quad |\delta A^\tau| \leqslant \sqrt{\frac{\mu_1}{n}A^\tau\mathscr{E}_1}\,|p|^{-1}, \\ |A^\tau_p|\,|p| \leqslant \sqrt{\frac{\mu_2}{n^2}A^\tau\mathscr{E}_1}\,|p|^{-1}, \quad \|A\| \leqslant \mu_3\mathscr{E}_1|p|^{-2},\end{gathered} \tag{2.15}$$

$$\begin{gathered}a = b + \hat{b}, \quad pb_p - b \geqslant -\mu_4\mathscr{E}_1|p|^{-1}, \quad \delta b \geqslant -\mu_5\mathscr{E}_1, \quad |b_p| \leqslant \frac{\mu_6}{\sqrt{n}}\mathscr{E}_1|p|^{-2}, \\ \delta\mathscr{E}_1 \geqslant -\mu_5'\mathscr{E}_1, \quad |(\mathscr{E}_1)_p| \leqslant \frac{\mu_6'}{\sqrt{n}}\mathscr{E}_1|p|^{-1}, \quad a_+ \leqslant \mu_7 \operatorname{Tr} A|p|,\end{gathered} \tag{2.16}$$

$$\begin{gathered}|p\hat{b}_p - \hat{b}| \leqslant \hat{\mu}_4|\hat{b}|, \quad |\delta_-\hat{b}| \leqslant \hat{\mu}_5|\hat{b}|, \quad |\hat{b}_p| \leqslant \frac{\hat{\mu}_6}{\sqrt{n}}|\hat{b}|\,|p|^{-1}, \\ \operatorname{sgn}|\hat{b}|\operatorname{Tr} A|p|^2 \leqslant \hat{\mu}_7\mathscr{E}_1, \quad (\operatorname{sgn}|\hat{b}|)|b| \leqslant \frac{\hat{\mu}_8}{|p|}\mathscr{E}_1,\end{gathered} \tag{2.17}$$

$$2-\frac{p(\mathscr{E}_1)_p}{\mathscr{E}_1}+\frac{(1-\beta)\mathscr{E}_1}{\operatorname{Tr} A|p|^2}-\frac{2\mu_7(1-\beta)}{L}-2\left(\mu_0+\frac{\mu_1}{L^2}\right)\beta^{-1}$$
$$-\frac{\mu_4+\mu_5+\mu_5'}{L}-\left(\frac{\hat\mu_5}{L}+\hat\mu_4\right)^2\hat\mu_7\beta^{-1}-\left(\frac{\hat\mu_5}{L}+\hat\mu_4\right)\left(\frac{\hat\mu_8}{L}+1\right)\geqslant c_0, \tag{2.18}$$

where $L=\text{const}\geqslant l/\delta$, $\beta=\text{const}\in[0,1]$, $l=\text{const}\geqslant 0$, *and the constants* $\mu_0,\dots,\hat\mu_8$ *are nonnegative. Then on the set* $K_{\rho,\bar l}(x_0)$, *where* $\bar l=L$, *at the solution* $\bar u(x)$ *of* (2.13) *corresponding to the choice of the function* φ *by* (2.14), *all the conditions of Theorem* 2.1 *in the case* $r=0$, $\varepsilon_1=1/2$, $\hat\varepsilon_1=1/2$ *are satisfied.*

PROOF. Lemma 2.1 is proved by direct verification that the conditions of Theorem 2.1 are satisfied.

THEOREM 2.2. *Let* $u\in C^2(\Omega)$ *be a solution of* (1.1.1) *in the domain* Ω *such that* $u(x)\geqslant m_1$ *in* $K_\rho(x_0)$. *Suppose that at this solution on the set* $K_{\rho,l}(x_0)\equiv\{x\in K_\rho(x_0)\colon |\nabla u|>l\}$ *conditions* (2.15)–(2.17) *are satisfied, and also that for some numbers* $\beta\in[0,1]$ *and* $L\geqslant l\delta^{-1}$, *where* δ *is a fixed positive number, condition* (2.18) *is satisfied. Then*

$$|\nabla u(x_0)|\leqslant \max(L,\theta)[u(x_0)-m_1+\delta], \tag{2.19}$$

where

$$\theta=\frac{4c_0^{-1}(\kappa_1+\kappa_2+\kappa_3+\kappa_4)}{\rho},$$

c_0 *is the constant in* (2.18), $\kappa_1=\kappa(\mu_2+\mu_3)$, $\kappa_2=\kappa\hat\mu_6^2\hat\mu_7$, $\kappa_3=\kappa(\mu_6/L+\mu_6')$, $\kappa_4=\kappa\hat\mu_6$ *and* $\kappa=2\sigma^{-2}\max(2/\beta,4+2\sqrt{n})$. *If conditions* (2.15)–(2.17) *are satisfied on the set* $K_{\rho,0}(x_0)$ *with*

$$\mu_1=\mu_4=\mu_5=\mu_6=\mu_5'=\mu_7=\hat\mu_8=\hat\mu_5=0 \tag{2.20}$$

and if on the set $K_{\rho,0}(x_0)$

$$2-\frac{P(\mathscr{E}_1)_p}{\mathscr{E}_1}+\frac{(1-\beta)\mathscr{E}_1}{\operatorname{Tr} A|p|^2}-\frac{2\mu_0}{\beta}\geqslant c_0=\text{const}>0,\quad p\hat b_p-\hat b\geqslant 0, \tag{2.21}$$

then the estimate

$$|\nabla u(x_0)|\leqslant c_1\frac{u(x_0)-m_1}{\rho} \tag{2.22}$$

holds, where $c_1=c_0^{-1}(\kappa_1+\kappa_2+\kappa_3+\kappa_4)$.

PROOF. Suppose first that the conditions of the first part of the theorem are satisfied. In view of Lemma 2.1 the function $\bar u$ defined by (2.14) is a solution of (2.13), and at this solution on the set $\overline{K}_{\rho,l}(x_0)\equiv\{x\in K_\rho(x_0)\colon |\nabla\bar u|>\bar l\}$, where $\bar l=L\geqslant l/\delta$, all the conditions of Theorem 2.1 are satisfied. From Theorem 2.1 it then follows that

$$|\nabla\bar u(x_0)|\leqslant \sup(L,\theta).$$

Taking into account that $|\nabla u(x_0)|=e^{\bar u(x_0)}|\nabla\bar u(x_0)|$ and $e^{\bar u(x_0)}=u(x_0)-m_1+\delta$, from this we deduce (2.19). We now prove the second part of the theorem. Under the conditions of the second part of the theorem it is easy to see that for (2.13) a condition of the form (2.5) is satisfied. Since all the remaining conditions of

Theorem 2.1 are also satisfied in the case $L = 0$, for $|\nabla \bar{u}(x_0)|$ we have the estimate $|\nabla \bar{u}(x_0)| \leqslant \theta$, from which (2.22) obviously follows. Theorem 2.2 is proved. An estimate of $|\nabla u(x_0)|$ in terms of $u(x_0)$ and $\sup_{K_\rho(x_0)} u(x)$ can be established in a similar way. It is obvious that Theorem 2.2 admits degeneration of ellipticity of the same sort as Theorem 1.1 does.

A result analogous but not identical to Theorem 2.2 was established earlier in [157]. We note, finally, that from an inequality of the form (2.21) it is possible to derive the following Harnack inequality (see [157], p. 89).

THEOREM 2.3. *Let $u \in C^3(\Omega)$ be a nonnegative solution of* (1.1.1) *in a domain $\Omega \in \mathbb{R}^n$, $n \geqslant 2$. Suppose that at this solution on the set $\{x \in \Omega: |\nabla u| > 0\}$ conditions* (2.15)–(2.17), (2.20) *and* (2.21) *are satisfied. Then for any compact subregion Ω' of Ω there exists a constant $K > 1$, depending only on the structure of the equation and* $\operatorname{dist}(\partial\Omega, \Omega')$, *such that*

$$\max_{\Omega'} u \leqslant K \min_{\Omega'} u. \tag{2.23}$$

PROOF. We set $x - x_0 = \xi|x - x_0| \equiv r\xi$, where $x \in \overline{K_\rho(x_0)} \subset \Omega$, $\rho > 0$, and we consider the function $v = v(r) \equiv u(x_0 + r\xi)$ in $[0, \rho]$. It is obvious that $v'(r) = \xi \cdot \nabla u(x)$. Taking into account that $K_{\rho-r}(x) \subset K_\rho(x_0)$, we deduce from (2.2) (with $m_1 \geqslant 0$) that

$$v'(r) \leqslant |v'(r)| \leqslant c_1 \frac{v(r)}{\rho - r}, \qquad 0 \leqslant r < \rho. \tag{2.24}$$

Integrating (2.24) over $[0, r]$, $r \in (0, \rho/2]$, we find that $v(r) \leqslant v(0)(1 - r/\rho)^{-c_1}$ for $r \in [0, \rho/2]$. This inequality can be rewritten in the form

$$u(x) \leqslant u(x_0)\left(1 - \frac{|x - x_0|}{\rho}\right)^{-c_1}, \qquad x \in K_{\rho/2}(x_0). \tag{2.25}$$

It follows from (2.25) by standard arguments that there exists a constant $K > 0$ such that for any pair of points $x, y \in \Omega'$ the inequality $u(x) \leqslant Ku(y)$ holds, which is equivalent to (2.23). Theorem 2.3 is proved.

Results analogous to Theorem 2.1–2.3 were obtained in [157] under different conditions on the structure of (1.1.1).

§3. Two-sided Liouville theorems

If a function $u \in C^3(\mathbb{R}^n)$ is a solution of (1.1.1) in the entire space $\mathbb{R}^n$ then, following [166], we shall call such a function an *entire solution* of (1.1.1). We set

$$m(\rho) = \sup_{|x| \leqslant \rho} |u(x)|.$$

THEOREM 3.1. *Suppose that the functions $a^{ij}(x, u, p)$, $i, j = 1, \dots, n$, and $a(x, u, p)$ in* (1.1.1) *satisfy conditions* (1.1)–(1.6) *on the sets $\mathfrak{M}_l \equiv \mathbb{R}^n \times \mathbb{R} \times \{|p| > l\}$ with constants $\mu_0, \dots, \hat{\varepsilon}_1$ depending, in general, on l, and suppose that*

$$\mu_1 + \mu_5 + \hat{\mu}_5^2\hat{\mu}_7 + \hat{\mu}_5\hat{\mu}_8 = 0. \tag{3.1}$$

Then any entire solution u of (1.1.1) *satisfying the condition $m(\rho) = o(\ln \rho)$, $\rho \to \infty$, is a constant. If in addition to* (3.1) *the condition*

$$\mu_4 + \hat{\mu}_4\hat{\mu}_8 + \hat{\mu}_8 = 0 \tag{3.2}$$

is satisfied, then any entire solution of (1.1.1) *for which* $m(\rho) = o(\sqrt{\rho})$, $\rho \to \infty$, *is a constant. Finally, if in addition to* (3.1) *and* (3.2) *the condition*

$$\mu_6 + \hat{\mu}_6\hat{\mu}_8 = 0 \tag{3.3}$$

is satisfied, then any entire solution of (1.1.1) *for which* $m(\rho) = o(\rho)$ *as* $\rho \to \infty$ *is a constant.*

PROOF. Let the number $l > 0$ and point $x_0 \in \mathbb{R}^n$ be fixed. We set

$$m_1(\rho) = \min_{K_\rho(x_0)} u(x), \qquad m_2(\rho) = \max_{K_\rho(x_0)} u(x),$$

$$m(\rho) = \operatorname*{osc}_{K_\rho(x_0)} u(x) = m_2(\rho) - m_1(\rho).$$

We first suppose that condition (3.1) is satisfied. Then the quantity ν_2 defined in (1.9) is equal to zero. Therefore, assuming $\mu_4 + \hat{\mu}_4\hat{\mu}_8 + \hat{\mu}_8 > 0$, we note that for the function $u(x)$ in the ball $K_\rho(x_0)$ all the conditions of the first part of Theorem 1.1 are satisfied with $m_1 = m_1(\rho)$, $m_2 = m_2(\rho)$ and $L = l$. (In particular, (1.25) is trivially satisfied.) Then for $|\nabla u(x_0)|$ there is an estimate of the form (1.26) with $L = l$ and θ defined by (1.27) (with m replaced by $m(\rho)$). From the condition $m(\rho) = o(\ln \rho)$, $\rho \to \infty$, and the form of θ it follows that $\theta \to 0$ as $\rho \to \infty$. Therefore, choosing ρ so large that $\theta < l$, we obtain the estimate $|\nabla u(x_0)| \leqslant 2l$. Since $l > 0$ is arbitrary, this implies that $\nabla u(x_0) = 0$. Taking into account that x_0 is an arbitrary point in $\mathbb{R}^n$, we conclude that $u(x) \equiv \text{const}$ in $\mathbb{R}^n$. The second and third parts of Theorem 3.1 are proved in an entirely similar way using the second part of Theorem 1.1. Theorem 3.1 is proved. Theorem 3.1 is a generalization of the two-sided Liouville theorems proved in [166] and [36].

COROLLARY 3.1. *Suppose that for the functions* $a^{ij}(x, u, p)$, $i, j = 1, \ldots, n$, *and* $a(x, u, p)$ *on the sets* $\mathfrak{M}_{m,l} \equiv \mathbb{R}^n \times \{|u| \leqslant m\} \times \{|p| > l\}$ *conditions* (1.1)–(1.6) *are satisfied with constants* $\mu_0, \ldots, \hat{\varepsilon}_1$ *depending on* l *and* m. *Suppose also that* (*for all* $l > 0$ *and* $m > 0$) *condition* (3.1) *is satisfied. Then any entire solution of* (1.1.1) *which is bounded in all of* $\mathbb{R}^n$ *is a constant.*

PROOF. We redefine $a^{ij}(x, u, p)$, $i, j = 1, \ldots, n$, and $a(x, u, p)$ in an appropriate way for $|u| \geqslant m$, and reduce the conditions of Corollary 3.1 to the conditions of the first case of Theorem 3.1, from which the result of Corollary 3.1 then follows.

As examples we present some special cases of Theorem 3.1.

1. Suppose that (1.1.1) has the form

$$a^{ij}(\nabla u)u_{x_ix_j} = b(u, \nabla u), \quad b(u, p) = f(u)h(p), \tag{3.4}$$

where conditions of the form (1.2) for the matrix $A \equiv \|a^{ij}(p)\|$ and the conditions $|f| \leqslant c_1$, $\partial f/\partial u \geqslant 0$, $|ph_p - p| \leqslant c_2\mathscr{E}_1$, $|h_p| \leqslant c_3\mathscr{E}_1|p|^{-2\varepsilon_1}$, $\varepsilon_1 > 0$ and $h \geqslant 0$, where $c_1, c_2 = \text{const} \geqslant 0$, are satisfied on the sets $\mathfrak{M}_l$. Then any entire solution of (3.4) such that $m(\rho) = o(\ln \rho)$, $\rho \to \infty$, is a constant.

2. Suppose that (1.1.1) has the form

$$a^{ij}(\nabla u)u_{x_ix_j} = \hat{b}(u, \nabla u), \tag{3.5}$$

where conditions of the form (1.2) for the matrix $A = \|a^{ij}(p)\|$ and the conditions $\partial\hat{b}/\partial u \geqslant 0$, $|p\hat{b}_p| \leqslant c_1|\hat{b}|$, $|\hat{b}_p| \leqslant c_2|\hat{b}|\,|p|^{-2\hat{\varepsilon}_1}$ and $\hat{\varepsilon}_1 > 0$ are satisfied on the sets $\mathfrak{M}_l$

(we note that for $\hat{\varepsilon}_1 = 1/2$ the last conditions are satisfied, for example, by any function $\hat{b} = f(|p|)$ such that $|f'(t)|\,|t| \leqslant c|f(t)|$, $c = \text{const} > 0$, $0 \leqslant t < +\infty$). Then any entire solution of (3.5) such that $m(\rho) = o(\rho)$, $\rho \to \infty$, is a constnt.

The next result may be called a weak two-sided Liouville theorem.

THEOREM 3.2. *Suppose that on the sets* $\mathfrak{M}_{(l,L)} \equiv \mathbb{R}^n \times \mathbb{R}^n \times \{l < |p| < L\}$, $l > 0$, $L > l$, *the functions* $a^{ij}(x, u, p)$, $i, j = 1, \ldots, n$, *and* $a(x, u, p)$ *satisfy conditions* (1.1)–(1.5) *for any* $r \geqslant 0$ *and* $\varepsilon_1 = \hat{\varepsilon}_1 = 1/2$ *with constants* $\mu_0, \mu_1, \ldots, \hat{\mu}_8$ *depending on* l *and* L. *Let* u *be an entire solution of* (1.1.1) *such that* $\sup_{\mathbb{R}^n}|\nabla u| < +\infty$. *Then the following assertions are true*:

1) *If for all* $l > 0$ *and* $L > l$ *condition* (3.1) *is satisfied and* $m(\rho) = o(\ln \rho)$, $\rho \to \infty$, *then* $u \equiv \text{const}$ *in* $\mathbb{R}^n$.

2) *If for all* $l > 0$ *and* $L > l$ *conditions* (3.1) *and* (3.2) *are satisfied and* $m(\rho) = o(\sqrt{\rho})$, $\rho \to \infty$, *then* $u \equiv \text{const}$ *in* $\mathbb{R}^n$.

3) *If for all* $l > 0$ *and* $L > l$ *conditions* (3.1)–(3.3) *are satisfied and* $m(\rho) = o(\rho)$, $\rho \to \infty$, *then* $u \equiv \text{const}$ *in* $\mathbb{R}^n$.

PROOF. Let $\sup_{\mathbb{R}^n}|\nabla u| = M_1$. Then for any $x_0 \in \mathbb{R}^n$ and $\rho > 0$ conditions (1.1)–(1.6) (with constants $\mu_1, \ldots, \hat{\varepsilon}_1$ depending on l and M) and the condition $m_1 \leqslant u(x) \leqslant m_2$ with $m_1 = \min_{K_\rho(x_0)} u$ and $m_2 = \max_{K_\rho(x_0)} u$ are satisfied for the function $u(x)$ on the set $K_{\rho,l}(x_0) \equiv \{x \in K(x_0)\colon |\nabla u(x)| > l\}$. Redefining the functions $a^{ij}(x, u, p)$, $i, j = ,\ldots, n$, and $a(x, u, p)$ in an appropriate way for $|p| \geqslant M_1$, we reduce the conditions of Theorem 3.2 to the conditions of Theorem 3.1, from which all the results of Theorem 3.2 follow.

The next assertion follows from Corollary 3.1 in a similar way.

COROLLARY 3.1′. *Suppose that on the sets* $\mathfrak{M}_{m(l,L)} \equiv \mathbb{R}^n \times \{|u| \leqslant m\} \times \{l < |p| < L\}$ *where* m, l, *and* L *are any positive numbers, the functions* $a^{ij}(x, u, p)$, $i, j = 1, \ldots, n$, *and* $a(x, u, p)$ *satisfy conditions* (1.1)–(1.5) *for any* $r \geqslant 0$ *and* $\varepsilon_1 = \hat{\varepsilon}_1 = 1/2$ *with constants* $\mu_0, \mu_1, \ldots, \hat{\mu}_8$ *depending on* m, l, *and* L. *Let* u *be an entire solution of* (1.1.1) *such that* $\sup_{\mathbb{R}^n}|u| + \sup_{\mathbb{R}^n}|\nabla u| < +\infty$. *Suppose that condition* (3.1) *is satisfied for all* $m > 0$, $l > 0$ *and* $L > 0$. *Then* $u \equiv \text{const}$ *in* $\mathbb{R}^n$.

REMARK 3.1. It is obvious that in Theorem 3.2 and Corollary 3.1′ the degree of elliptic nonuniformity plays no role.

Theorem 3.2 implies, in particular, the following known result (see [149] and [166]): and entire solution of the equation of a surface of constant mean curvature K, i.e., of an equation of the form (1.1.10) with $\mathscr{H}(x, u, p) \equiv K = \text{const}$, which satisfies the conditions $m(\rho) = o(\ln \rho)$ and $|\nabla u| \leqslant \text{const}$ in $\mathbb{R}^n$, is identically constant in $\mathbb{R}^n$. Indeed, in this case conditions (1.1)–(1.5) are obviously satisfied on the set $\mathfrak{M}_{l,M_1}$ (we assume here that $a = b + \hat{b}$, $b = nK(1 + |p|^2)^{3/2}$ and $\hat{b} \equiv 0$), since the expressions on the right sides of (1.1)–(1.5) are bounded below by a positive number $c_1(l, M_1)$, while the moduli of the expressions on the left sides of these inequalities do not exceed a constant $c_2(l, M_1)$.

Below we shall present a result which asserts, under specific conditions, that any entire solution of (1.1.1) whose oscillation in the entire space does not exceed a certain quantity determined by the structure of the equation is identically constant. It is obvious that results of this sort also belong to the class of Liouville theorems.

THEOREM 3.3. *Suppose that on the sets* $\mathfrak{M}_{m,l} \equiv \mathbb{R}^n \times \{|u| \leqslant m\} \times \{|p| > l\}$ *the functions* $a^{ij}(x, u, p)$, $i, j = 1, \dots, n$, *and* $a(x, u, p)$ *satisfy conditions* (1.1)–(1.6) *with constants* $\mu_0, \dots, \hat{\varepsilon}_1$ *depending on* m *and* l. *Let* u *be an entire solution of* (1.1.1), *and suppose that for any* $m \in (0, \operatorname{osc}_{\mathbb{R}^n} u)$ *and* $l > 0$ *at least one of the conditions*

$$\nu_1^2(2\nu_2)^{-1} > 1, \quad \operatorname*{osc}_{\mathbb{R}^n} u < \left(\ln\left(\nu_1^2/2\nu_2\right)\right)(2\nu_1)^{-1} \leqslant c_1 = \text{const}, \tag{3.6}$$

or

$$\operatorname*{osc}_{\mathbb{R}^n} u < \min\left\{(2\nu_1)^{-1}, \left(2\sqrt{\nu_2}\right)^{-1}\right\} \leqslant c_2 = \text{const} \tag{3.7}$$

is satisfied, where ν_1 *and* ν_2 *are defined in* (1.9) *with* $\alpha = 2(\mu_0 + \hat{\mu}_4^2\hat{\mu}_7 + \hat{\mu}_7)$, *and* c_1 *and* c_2 *do not depend on* l *and* m. *Then* $u \equiv \text{const}$ *in* $\mathbb{R}^n$.

PROOF. We fix $x_0 \in \mathbb{R}^n$ and $\rho > 0$. From the conditions of Theorem 3.3 it follows that on the set $K_{\rho,l}(x_0)$ for the function u all the conditions of the first or second part of Theorem 1.1 with $m_1 = \inf_{K_\rho(x_0)} u(x)$ and $m_2 = \sup_{K_\rho(x_0)} u(x)$ are satisfied. Then for $u(x)$ we have $|\nabla u(x_0)| \leqslant 2\max(l, \theta)$, where θ is defined either by (1.27) or by (1.29). Suppose that for fixed x_0 and $l > 0$ the radius ρ tends to ∞. Taking into account that $\lim_{\rho\to\infty}\theta = 0$ (since $m = m_2(\rho) - m_1(\rho)$ is bounded), we obtain in both cases the estimate $|\nabla u(x_0)| \leqslant 2l$, from which Theorem 3.3 easily follows since x_0 and l are arbitrary. Theorem 3.3 is proved.

We consider, for example, the equation

$$e^{\lambda_i u} u_{x_i x_i} = 0, \qquad \lambda_i = \text{const} > 0, \quad i = 1, \dots, n. \tag{3.8}$$

It is easy to see that conditions (1.2) are satisfied on the sets $\mathfrak{M}_{m,l}$ in the case of this equation with constants $\mu_0 = \mu_2 = 0$, $\mu_1 = \lambda^2 c^{4\lambda m} n$, $\lambda = \max_{i=1,\dots,n} \lambda_i$, $r = 0$ and $\mu_3 = \mu_3(\lambda, m)$, and conditions (1.3)–(1.5) are satisifed with zero constants. In view of (1.9) we then have $\nu_1 = 0$ and $\nu_2 = 4\mu_1$. Applying Theorem 3.3 (with condition (3.7)), we conclude that any entire solution of (3.8) with $\operatorname{osc}_{\mathbb{R}^n} u < \min\{1, 1/4\sqrt{n}\,\lambda e^{2\lambda}\}$ is constant.

§4. One-sided Liouville theorems

THEOREM 4.1. *Suppose that on the sets* $\mathfrak{M}_l = \mathbb{R}^n \times \mathbb{R} \times \{|p| > l\}$, $l = \text{const} > 0$, *the functions* $a^{ij}(x, u, p)$, $i, j = 1, \dots, n$, *and* $a(x, u, p)$ *satisfy conditions* (2.15)–(2.17) *with constants* $\mu_0, \dots, \hat{\mu}_8$ *depending on* l *and satisfying conditions* (2.20). *Suppose also that conditions* (2.21) *are satisfied with a constant* c_0 *not depending on* l. *Then any entire solution of* (1.1.1) *satisfying the condition*

$$\sup_{|x|\leqslant\rho} u = o(\rho), \quad \rho \to +\infty \qquad \left(\inf_{|x|\leqslant\rho} u = o(\rho), \quad \rho \to +\infty\right), \tag{4.1}$$

is identically constant in $\mathbb{R}^n$.

PROOF. We shall prove Theorem 4.1, for example, under the condition $\inf_{|x|\leqslant\rho} u = o(\rho)$, $\rho \to +\infty$. From (2.22) with m_1 replaced by $m_1(\rho) = \inf_{K_\rho(x_0)} u$ it follows that $|\nabla u(x_0)| = 0$ if in (2.22) we pass to the limit as $\rho \to \infty$. Since the point x_0 is arbitrary, this implies that $\nabla u \equiv 0$ in $\mathbb{R}^n$, i.e., $u \equiv \text{const}$ in $\mathbb{R}^n$. Theorem 4.1 is proved. A result similar to Theorem 4.1 was also established in [157].

We shall now prove a one-sided Liouville theorem of less traditional character but which pertains to a larger class of equations than that considered in Theorem 4.1.

THEOREM 4.2. *Suppose that for any* $l > 0$ *the functions* $a^{ij}(x, u, p)$, $i, j = 1, \ldots, n$, *and* $a(x, u, p)$ *satsify conditions* (1.1)–(1.6) *on the set* $\mathscr{P}_l \equiv \mathbb{R}^n \times \mathbb{R} \times \{|p|' > l\}$, *with constants* $\mu_0, \ldots, \hat{\mu}_8$ *depending, in general, on* l, *and constants* $r = 0$, $\varepsilon_1 = 1/2$ *and* $\hat{\varepsilon}_1 = 1/2$. *Suppose also that*

$$\mu_4 + \hat{\mu}_4\hat{\mu}_8 + \hat{\mu}_8 = 0, \qquad \mu_1 + \mu_5 + \hat{\mu}_5^2\hat{\mu}_7 + \mu_5\hat{\mu}_8 = 0, \qquad \mu_6 = 0. \tag{4.2}$$

Then there exists a number $\alpha > 1$, *depending only on the structure of equation* (1.1.1), *such that any entire solution of* (1.1.1) *satisfying for some* $\beta \in [0, 1]$ *the condition*

$$\inf_{|x| \leqslant \rho} u = o(\rho^{\beta}), \qquad |\nabla u| = O(\rho^{(1-\beta)/\alpha}), \qquad \rho \to +\infty, \tag{4.3}$$

is a constant. In the case $\beta = 1$ *in* (4.3) *this result is preserved if conditions* (1.1)–(1.6) *are satisfied on the sets* $\mathfrak{M}_{(l,L)} \equiv \mathbb{R}^n \times \mathbb{R} \times \{l < |p| < L\}$ *with constants* $\mu_0, \ldots, \hat{\mu}_7$ *depending on* l *and* L.

PROOF. We fix a point $x_0 \in \mathbb{R}^n$. Because of Lemma 1.1, either at the point x_* of the maximum in $\overline{K_\rho(x_0)}$ of the function $\overline{\omega}(x)$ defined by (1.7) inequality (1.8) holds, or (1.10) holds with L replaced by l. Suppose that the number $\alpha > 0$ from (1.8) satisfies the following condition: there exists a number $\kappa \in (1/2, 1)$ such that

$$\alpha - \gamma - \nu_0 = \kappa\alpha, \qquad (2\kappa - 1)\alpha > 1, \tag{4.4}$$

where κ is a fixed number in $(0, 1)$ (it is obvious that (4.4) is satisfied if $\alpha > 2(\kappa + \nu_0) + 1$). We set $q = 2(2\alpha\kappa/(\alpha + 1) - 1)$. It is obvious that $q > 0$. Setting in (1.8)

$$z(u) = (u - m_1 + \delta)^2, \qquad \delta = \text{const} > 1, m_1 = m_1(\rho) = \inf_{K_\rho(x_0)} u,$$

in view of (4.4) we then find that

$$-z'' + \frac{\alpha - \gamma - \nu_0}{\alpha + 1}\frac{z'^2}{z} \geqslant q. \tag{4.5}$$

Therefore, in the alternative indicated above inequality (1.8) can be replaced by

$$q - ((\nu_3 + \nu_1)/\rho^2\theta^2)z_* \leqslant 0, \tag{4.6}$$

where ν_3 and ν_4 are defined in (1.9) for the values of α and γ chosen above and where $z_* \equiv z(u(x_*))$; here we have also taken into account that the equalities $\nu_4 = 0$ and $\nu_6 = 0$ follow from our assumptions. We set $\theta = (4\nu_3 z_*/q\rho^2)^{1/2}$. With this choice of θ the left side of (4.6) exceeds the positive number $q/2$, which contradicts (4.6). Thus, for the chosen $z(u)$ and θ we have

$$v_0 \equiv |\nabla u(x_0)|^2 \leqslant \max\left\{\left(\frac{z_0}{z_*}\right)^{1/(\alpha+1)} l, \left(\frac{z_0}{z_*}\right)^{1/(\alpha+1)} \theta^2\right\}, \tag{4.7}$$

where $z_0 \equiv z(u(x_0))$. Since x_* is the point of the maximum of $\overline{\omega} = \omega/z$, we have $\omega_0/z_0 \leqslant \omega_*/z_*$, from which it follows that

$$z_* \leqslant \frac{\omega_*}{\omega_0} z_0 = \frac{w_*^{\alpha+1}}{w_0^{\alpha+1}} \frac{v_*^{\alpha+1}}{v_0^{\alpha+1}} z_0.$$

Then

$$\left(\frac{z_0}{z_*}\right)^{1/(\alpha+1)}\theta^2 \leqslant \frac{2(\nu_3+\nu_4)}{q\rho^2} z_*^{1-1/(\alpha+1)} z_0^{1/(\alpha+1)} \leqslant \frac{2(\nu_3+\nu_4)}{q\rho^2}\frac{v_*^\alpha}{v_0^\alpha} z_0, \tag{4.8}$$

where $v_* = |\nabla u(x_*)|^2$.

From (4.7) and (4.8) we obtain

$$v_0^{\alpha+1} \leqslant \max\left\{\left(\frac{z_0}{z_*}\right)^{1/(\alpha+1)} v_0^\alpha l, \frac{2(\nu_3+\nu_4)}{q}\frac{z_0 v_*^\alpha}{\rho^2}\right\}. \tag{4.9}$$

Suppose that some number $l > 0$ is fixed. Taking into account that

$$\frac{\sqrt{z_0 v_*^\alpha}}{\rho} \leqslant \frac{\sqrt{(u_0+\delta)^2 + m_1^2(\rho)}}{\rho^\beta}\frac{\left(\sup_{K_\rho(x_0)}|\nabla u|\right)^\alpha}{\rho^{1-\beta}}, \qquad \beta \in [0,1], \tag{4.10}$$

because of (4.3) we can choose ρ so large that

$$\frac{2(\nu_3+\nu_4)}{q}\frac{z_0 v_*^\alpha}{\rho^2} < z_0^{1/(\alpha+1)} v_0^\alpha l$$

(since $z_0 \geqslant 1$, and $v_0^\alpha l$ is a constant number for a fixed point $x_0 \in \mathbb{R}^n$). Therefore, taking further into account that $z_* \geqslant 1$, we obtain

$$v_0^{\alpha+1} \leqslant z_0^{1/(\alpha+1)} v_0^\alpha l. \tag{4.11}$$

Since l is arbitrary, it follows that $v_0 = 0$. Taking into account that x_0 is an arbitrary point of $\mathbb{R}^n$, we conclude that $\nabla u \equiv 0$ in $\mathbb{R}^n$, i.e., $u \equiv$ const in $\mathbb{R}^n$. Theorem 4.2 is proved.

A version of Theorem 4.2 corresponding to the condition

$$\sup_{|x|\leqslant\rho} u = o(\rho^\beta), \quad |\nabla u| = o(\rho^{(1-\beta)/\alpha}), \qquad \rho \to \infty$$

(cf. (4.3)), can be established in a similar way. In [157] a somewhat different class of equations of the form (1.1.1) was distinguished for which the result of Theorem 4.2 is established in the case $\beta = 1$ in condition (4.3). The following generalization of Theorem 4.2 is valid.

THEOREM 4.2′. *Suppose that all the conditions of Theorem* 4.2 *are satisfied with the exception of the conditions* $r = 0$ *and* $\hat{\varepsilon}_1 = 1/2$. *Suppose that in place of the latter conditions the relations* $\varepsilon = 1 - r/2 > 0$ *and* $\hat{\varepsilon}_1 = \varepsilon/2$ *hold. Then there exists a number* $\alpha > 1$ *depending only on the structure of* (1.1.1) *such that any entire solution of* (1.1.1) *satisfying for some* $\beta \in [0,1]$ *the condition*

$$\begin{gathered}\inf_{|x|\leqslant\rho} u = o(\rho^\beta) \quad \left(\sup_{|x|\leqslant\rho} u = o(\rho^\beta)\right),\\ \sup_{|x|\leqslant\rho}|\nabla u| = O(\rho^{(1-\beta)/(\alpha-1+\varepsilon)}), \qquad \rho\to\infty,\end{gathered} \tag{4.12}$$

is a constant.

Theorem 4.2′ is proved in the same way as Theorem 4.2.

PART II
QUASILINEAR ($\mathcal{A}$, b)-ELLIPTIC EQUATIONS

Parts II and III of the monograph are devoted to the investigation of questions of solvability of boundary value problems for quasilinear degenerate elliptic and parabolic equations. A large number of books and papers (especially in the case of linear equations) have been devoted to the study of these questions. Many have been devoted to the study of linear elliptic and parabolic equations which degenerate on the boundary. The theory of weighted function spaces arose in connection with the study of such equations by methods of functional analysis. A bibliography of these papers can be found, for example, in the monographs [102] and [113]. In particular, these questions have been studied by M. V. Keldysh [60], S. G. Mikhlin [91], M. I. Vishik [12], L. D. Kudryavcev [73], [74], S. M. Nikol′skiĭ [95], [96], A. V. Bitsadze [9], V. P. Glushko [17], V. A. Kondrat′ev [64], V. G. Maz′ya [89], and many other mathematicians.

G. Fichera [137] formulated boundary value problems for a general linear equation of second order with a nonnegative characteristic form, and proved theorems on the existence of certain generalized solutions of these problems. Existence and uniqueness theorems for nonregular generalized solutions and existence theorems for smooth solutions of the first boundary value problem under broad conditions on the structure of the general linear equation with nonnegative characteristic form were obtained by O. A. Oleĭnik [99], [102]. In the monograph [102] results of other mathematicians in this area and also an extensive bibliography are presented. We note, in particular, A. M. Il′in [56], M. I. Freĭdlin [120], [121], J. J. Kohn and L. Nirenberg [141], R. S. Phillips and L. Sarason [158], E. V. Radkevich [107], and A. L. Treskunov [117]. In these papers linear equations were considered under particular smoothness conditions on their coefficients. In this case reduction of the boundary condition on part of the boundary is characteristic. In the papers of S. N. Kruzhkov [69], M. K. V. Murthy and G. Stampacchia [31], and N. S. Trudinger [55] so-called weakly degenerate linear elliptic equations were studied, while the author [28], [43] studied weakly degenerate linear parabolic equations for which the boundary value problems in formulations traditional for nondegenerate equations are well posed. We remark that the presence of weak degeneracy implies a particular nonregularity of the equation.

A number of papers have been devoted to the study of boundary value problems for certain quasilinear elliptic and parabolic equations admitting implicit degeneracy. Equations arising in boundary-layer theory with gradual acceleration and also in problems of nonstationary filtration have been studied by O. A. Oleĭnik, A. S. Kalashnikov, Jou Yuh-lin, E. S. Sabinina, and others ([97], [100], [101], [59], [60], [108], [109]). Yu. A. Dubinskiĭ [18], [19] considered quasilinear elliptic and parabolic equations of order $2m$ depending linearly on the derivatives of order m such that degeneration occurs at points where the derivatives of order $m - 1$ appearing in the equation, raised to some power, vanish. In the case $m = 2$ Dubinskiĭ's condition on the linearity of the gradient in the equation was recently removed by a doctoral student at the Leningrad Branch of the Steklov Institute of Mathematics, P. Z. Mkrtchyan [93]. The solvability of nonlinear degenerate equations arising in the theory of control processes of diffusion type has been studied by N. V. Krylov [71], [72]. Some one-dimensional quasilinear parabolic equations with implicit degeneracy have been considered by P. A. Raviart [159]. In the above mentioned work (the majority of which is related to very specific equations) the degeneracy has an implicit character exclusively, i.e., points of degeneration of ellipticity or parabolicity depend on the solution under consideration.

The solvability of boundary value problems for some classes of quasilinear elliptic and parabolic equations of second order admitting fixed degeneracy (i.e., degeneracy not depending on the solution under consideration) has been studied by M. I. Freĭdlin [122] and G. M. Fateeva [119]. They, however, assumed that the derivatives of the solution occur in the equation linearly. Moreover, in [122] sufficient smallness of the boundary function and its derivatives was assumed. In [44]–[47] and [49]–[53] the author constructed a theory of boundary value problems for large classes of quasilinear degenerate elliptic and parabolic equations of second order admitting, in particular, fixed degeneration of ellipticity or parabolicity. In particular, the classes of equations he considered include linear second-order equations with a nonnegative characteristic form. Parts II and III of the present monograph are devoted to an exposition of this theory.

In a bounded domain $\Omega \subset \mathbb{R}^n$, $n \geqslant 2$, we consider the quasilinear equation

$$\mathscr{L}u \equiv -\frac{d}{dx_i} l^i(x, u, \nabla u) + l_0(x, u, \nabla u) = f(x), \tag{1}$$

where

$$\frac{d}{dx_i} l^i(x, u, \nabla u) \equiv \sum_{i,j=1}^{n} \frac{\partial l^i}{\partial u_{x_j}} u_{x_i x_j} + \frac{\partial l^i}{\partial u} u_{x_i} + \frac{\partial l^i}{\partial x_i}.$$

We say that equation (1) has *$(A, \mathbf{b})$-structure* in Ω if there exist a matrix $A \equiv \|a^{ij}(x)\|$ or order n, a vector $\mathbf{b} \equiv (b^1(x), \dots, b^n(x))$, and functions $l'^i(x, u, q)$, $i = 1, \dots, n$, and $l'_0(x, u, q)$ such that

$$\mathbf{l}(x, u, p) = A^* \mathbf{l}'(x, u, Ap), \quad l_0(x, u, p) = l'_0(x, u, Ap) + b^i(x) p_i. \tag{2}$$

We call the functions $l'^i(x, u, q)$, $i = 1, \dots, n$, and $l'_0(x, u, q)$ the *reduced coefficients* of this equation. The reduced coefficients of an equation having $(A, \mathbf{b})$-structure are invariants of this equation (with respect to nondegenerate smooth transformations of the independent variables). We call an equation of the form (1) having $(A, \mathbf{b})$-structure in a domain Ω *$(A, \mathbf{b})$-elliptic* (*strictly $(A, \mathbf{b})$-elliptic*) in Ω if

$$\frac{\partial l'^i(x, u, q)}{\partial q_j} \eta_i \eta_j \geqslant 0, \quad \eta = A\xi, \xi \in \mathbb{R}^n, x \in \overline{\Omega}, u \in \mathbb{R}, q = Ap, p \in \mathbb{R}^n \tag{3}$$

$$\left[\frac{\partial l'^i(x, u, q)}{\partial q_j} \eta_i \eta_j > 0, \quad \eta = A\xi \neq 0, \xi \in \mathbb{R}^n, x \in \overline{\Omega}, u \in \mathbb{R}, q = Ap, p \in \mathbb{R}^n \right]. \tag{3'}$$

We observe that the matrix A may admit arbitrary degeneracy on any subset of Ω. If A degenerates in Ω, then even a strictly $(A, \mathbf{b})$-elliptic equation of the form (1) is a degenerate elliptic equation in Ω, since by (2)

$$\frac{\partial l^i(x, u, p)}{\partial p_j}\xi_i\xi_j = \frac{\partial l'^k(x, u, Ap)}{\partial q_s}A_k\xi A_s\xi,$$

where $A_k\xi \equiv a^{kr}\xi_r$, $k = 1,\ldots,n$, so that the form $(\partial l^i/\partial p_j)\xi_i\xi_j$ degenerates at any point $x \in \Omega$ where A is degenerate.

The $(A, \mathbf{b})$-elliptic equations include a large class of quasilinear equations with nonnegative characteristic form. In particular, they contain: 1) nondegenerate elliptic equations ($A \equiv I$, $\mathbf{b} \equiv \mathbf{0}$), I the identity matrix); 2) nondegenerate parabolic equations ($A = \left\|\begin{smallmatrix} 0 & \cdots & 0 \\ \vdots & & I \\ 0 & & \end{smallmatrix}\right\|$, $\mathbf{b} = (1, 0,\ldots,0)$); 3) nondegenerate ultraparabolic equations ($A = \|\begin{smallmatrix} 0 & 0 \\ 0 & I\end{smallmatrix}\|$, $\mathbf{b} = (1,\ldots,1, 0,\ldots,0)$); 4) quasilinear equations of first order ($A = 0$, $\mathbf{b} = (b^1,\ldots,b^n)$); and 5) linear equations with nonnegative characteristic form

$$-\frac{d}{dx_i}\left(\alpha^{ij}(x)u_{x_j}\right) + \beta^i(x)u_{x_i} + c(x)u = f(x), \tag{4}$$

where $\alpha^{ij} = \alpha^{ji}$, $i, j = 1,\ldots,n$, and $\mathfrak{A} \equiv \|\alpha^{ij}(x)\|$ is a nonnegative definite matrix in Ω ($A = \mathfrak{A}^{1/2}$, $\mathbf{b} \equiv \beta = (\beta^1,\ldots,\beta^n)$); nondivergence linear equations with nonnegative characteristic form also reduce to equations of the form (4) provided that their leading coefficients are sufficiently smooth functions.

$(A, \mathbf{0})$-elliptic equations ($\mathbf{b} \equiv \mathbf{0}$) are an important special case of $(A, \mathbf{b})$-elliptic equations. In particular, $(A, \mathbf{0})$-elliptic equations include the Euler equations for variational problems regarding a minimum of integrals of the form

$$\int_\Omega [\mathcal{F}(x, u, A\nabla u) - f(x)u]\, dx, \qquad u|_{\{x\in\partial\Omega: A\boldsymbol{\nu}\neq\mathbf{0}\}} = 0, \tag{5}$$

where $(\partial\mathcal{F}(x, u, q)/\partial q_i\partial q_j)\eta_i\eta_j \geqslant 0$, $\eta = A\xi$, $\xi \in \mathbb{R}$; $A\nabla u \equiv (A_1\nabla u,\ldots,A_n\nabla u)$, $A_i\nabla u \equiv a^{ij}u_{x_j}$, $i = 1,\ldots,n$, and $\boldsymbol{\nu} = (\nu_1,\ldots,\nu_n)$ is the unit vector of the inner normal to $\partial\Omega$. We henceforth call the vector $A\nabla u$ the A-gradient of the function u, and we call its components $A_1\nabla u,\ldots,A_n\nabla u$ the A-derivatves of this function. Indeed, the Euler equation just mentioned has the form

$$-\frac{d}{dx_i}\left[a^{ki}(x)\frac{\partial\mathcal{F}(x, u, A\nabla u)}{\partial q_k}\right] + \frac{\partial\mathcal{F}(x, u, A\nabla u)}{\partial u} = f(x) \tag{6}$$

and it is easily seen to have the structure of an $(A, \mathbf{0})$-elliptic equation with the reduced coefficients

$$l'^i(x, u, q) = \frac{\partial\mathcal{F}(x, u, q)}{\partial q_i}, \quad i = 1,\ldots,n, \qquad l'_0(x, u, q) = \frac{\partial\mathcal{F}(x, u, q)}{\partial u}. \tag{7}$$

In the case

$$\tilde{A} = \left\|\begin{matrix} 0 & \cdots & 0 \\ \vdots & & A \\ 0 & & \end{matrix}\right\|, \quad \tilde{\mathbf{b}} = (1, b^1,\ldots,b^n)$$

we call the corresponding $(\tilde{A}, \tilde{\mathbf{b}})$-elliptic equation $(A, \mathbf{b})$-*parabolic*. We usually consider an $(A, \mathbf{b})$-parabolic equation in a cylinder $Q = \Omega \times (T_1, T_2)$, where $\Omega \subset \mathbb{R}^n$,

$n \geqslant 1$. $(A, \mathbf{0})$-parabolic equations are an important special case of $(A, \mathbf{b})$-parabolic equations. The study of certain questions of the theory of heat conduction, diffusion, etc. leads to such equations.

The $(A, \mathbf{0})$-elliptic and $(A, \mathbf{0})$-parabolic equations are the central objects of our investigations. A detailed study of boundary value problems for such equations is carried out in Part III of the monograph. In Part II we consider general $(A, \mathbf{b})$-elliptic equations. A formulation of the general boundary value problem (in particular, of the first, second, and third problems) for an $(A, \mathbf{b})$-elliptic equation in a bounded domain $\Omega \subset \mathbb{R}^n$, $n \geqslant 2$, is given in Chapter 5. For example, the first boundary value problem for such an equation has the form

$$\mathscr{L}u = f(x) \quad \text{in } \Omega, \qquad u = 0 \quad \text{on } \Gamma \subset \partial\Omega, \tag{8}$$

where Γ is the so-called $(A, \mathbf{b})$-elliptic boundary of the domain Ω defined with the structure of the equation taken into account. It is obvious that the choice of Γ should ensure that problem (8) is well posed. If (1) is a strictly $(A, \mathbf{b})$-elliptic equation and the matrix A is sufficiently smooth, then the $(A, \mathbf{b})$-elliptic boundary Γ can be defined by

$$\Gamma = \Sigma \cup \Sigma'_-, \tag{9}$$

where Σ is the noncharacteristic part of $\partial\Omega$, Σ' is the characteristic part of $\partial\Omega$, $\Sigma'_- \equiv \{x \in \Sigma' \colon b(x) \equiv -b^i \nu_i < 0\}$, and $(\nu_1, \dots, \nu_n) \equiv \boldsymbol{\nu}$ is the unit vector of the inner normal to $\partial\Omega$. Below we also use the following notation: $\Sigma'_+ \equiv \{x \in \Sigma' \colon b(x) > 0\}$, $\Sigma'_0 \equiv \{x \in \Sigma' \colon b(x) = 0\}$, and $\Sigma'_{0,-} \equiv \Sigma'_0 \cup \Sigma'_-$. It is obvious that in this case

$$\Sigma = \{x \in \partial\Omega \colon A\boldsymbol{\nu} \neq \mathbf{0}\}, \quad \Sigma' = \{x \in \partial\Omega \colon A\boldsymbol{\nu} = \mathbf{0}\}. \tag{10}$$

The definition of Γ by formulas (9) and (10) also applies in a number of cases of equations which are not strictly $(A, \mathbf{b})$-elliptic but under the condition that the matrix A be sufficiently smooth. For less smooth matrices A the definition of the $(A, \mathbf{b})$-elliptic boundary Γ by (9) applies, but then the sets Σ and Σ' must be defined somewhat differently than in (10) (see §5.3).

We study questions of the existence and uniqueness of solutions of problem (8) in the following classes of generalized solutions: a) generalized solutions of energy type; b) A-regular generalized solutions; and c) regular generalized solutions. The latter two classes of generalized solutions may have different degrees of regularity. Actually, the general boundary value problem is studied in the first class of generalized solutions, but for brevity we shall here discuss the formulation of only the first boundary value problem.

To define generalized solutions of energy type we introduce the energy space $H \equiv H^{0,\Sigma}_{m,\mathbf{m}}(A, \Omega)$ defined as the completion of the set $\tilde{C}^1_{0,\Sigma}(\Omega)$ in the norm

$$\|u\|_H \equiv \|u\|_{m,\Omega} + \sum_{i=1}^{n} \|A_i \nabla u\|_{m_i,\Omega}. \tag{11}$$

We define a generalized solution of energy type of problem (8) in the case of so-called $(A, \mathbf{b}, m, \mathbf{m})$-elliptic equations, where $m = \text{const} > 1$ and $\mathbf{m} = (m_1, \dots, m_n)$, $m_i > 1$, $i = 1, \dots, n$. An $(A, \mathbf{b})$-elliptic equation of the form (1) is called $(A, \mathbf{b}, m, \mathbf{m})$-*elliptic* in Ω if the reduced coefficients $l'^i(x, u, q)$, $i = 1, \dots, n$, and $l'_0(x, u, q)$ satisfy growth conditions as u, $q \to \infty$ such that for any function

$u \in H^{0,\Sigma}_{m,\mathbf{m}}(A,\Omega)$ the following conditions are satisfied: $l'^i(x,u,A\nabla u) \in L^{m'_i}(\Omega)$, $1/m_i + 1/m'_i = 1$, $i = 1,\dots,n$; $l'_0(x,u,A\nabla u) \in L^{m'}(\Omega)$, $1/m + 1/m' = 1$, and the expression $A\nabla u$ for a function $u \in H^{0,\Sigma}_{m,\mathbf{m}}(A,\Omega)$ denotes the so-called generalized A-gradient of this function (see §§4.1, 5.2 and 5.3). Functions in $H^{0,\Sigma}_{m,\mathbf{m}}(A,\Omega)$ vanish on $\Sigma \subset \partial\Omega$ in a particular sense (see §4.2). A *generalized solution* (of energy type) of problem (8) for an $(A,\mathbf{b},m,\mathbf{m})$-elliptic equation is any function $u \in H^{0,\Sigma}_{m,\mathbf{m}}(A,\Omega)$ satisfying the identity

$$\int_\Omega \left[\mathbf{l}'(x,u,A\nabla u)\cdot A\nabla\eta + l'_0(x,u,A\nabla u)\eta - u\frac{\partial}{\partial x_i}(b^i\eta)\right] dx$$
$$= \int_\Omega f\eta\,dx, \quad \forall \eta \in \tilde{C}^1_{0,\Sigma\cup\Sigma'_+}(\overline{\Omega}). \tag{12}$$

By an *A-regular generalized solution* of problem (8) for an $(A,\mathbf{b})$-elliptic equation in the domain Ω we mean any function $u \in L^\infty(\Omega) \cap H^{0,\Sigma}_m(A,\Omega)$ for all $m > 1$ (where $H^{0,\Sigma}_m(A,\Omega) \equiv H^{0,\Sigma}_{m,\mathbf{m}}(A,\Omega)$ with $\mathbf{m} = (m,\dots,m)$) having a bounded A-gradient in Ω (i.e., $A\nabla u \in \mathbf{L}^\infty(\Omega)$) and satisfying (12).

A *regular generalized solution* of problem (8) for an $(A,\mathbf{b})$-elliptic equation in Ω is any function $u \in L^\infty(\Omega) \cap H^{0,\Sigma}_m(A,\Omega)$ for all $m > 1$ having $\nabla u \in L^\infty(\Omega)$ and such that

$$\int_\Omega [\mathbf{l}(x,u,\nabla u)\cdot\nabla\eta + l_0(x,u,\nabla u)\eta]\,dx = \int_\Omega f\eta\,dx \quad \forall\eta \in \tilde{C}^1_{0,\Sigma\cup\Sigma'_+}(\overline{\Omega}). \tag{13}$$

We also consider regular generalized solutions of problem (8) having bounded second derivatives in Ω.

In Part II we investigate questions of the existence and uniqueness of a generalized solution of energy type of the general boundary value problem for $(A,\mathbf{b},m,\mathbf{m})$-elliptic equations, and existence and uniqueness of regular and strongly regular generalized solutions of the first boundary value problem for $(A,\mathbf{b})$-elliptic equations. In describing results on the solvability of boundary value problems in the class of generalized solutions of energy type, for simplicity we here consider the case of the first boundary value problem (8). We investigate this generalized solvability of problem (8) in the language of operator equations in suitable Banach spaces. This reformulation of the problem requires the introduction of two more function spaces. Let X be the completion of $\tilde{C}^1_{0,\Sigma}(\overline{\Omega})$ in the norm

$$\|u\|_X = \|u\|_H + \|u\|_{L^2(\beta,\Omega_\beta)} + \|u\|_{L^2(b,\Sigma'_+)}, \tag{14}$$

and let Y be the completion of $\tilde{C}^1_{0,\Sigma}(\overline{\Omega})$ in the norm

$$\|u\|_Y = \|u\|_X + \sum_{i=1}^n \|u_{x_i}\|_{L^2(|b^i|,\Omega_{b^i})}, \tag{15}$$

where

$$\beta = \sum_{i=1}^n \left(|b^i| + \left|\frac{\partial b^i}{\partial x_i}\right|\right), \qquad \Omega_\beta \equiv \{x \in \Omega : \beta > 0\},$$
$$\Omega_{b^i} \equiv \{x \in \Omega : b^i \neq 0\}, \qquad i = 1,\dots,n.$$

It turns out that $Y \to X$, $Y \to H$ and $L^{m'}(\Omega) \to H^* \to X^* \to Y^*$. However, there is no imbedding $X \to H$. We consider the operator $\mathscr{L}$: $\tilde{C}^1_{0,\Sigma}(\overline{\Omega}) \subset X \to Y^*$ defined by

$$\mathscr{L} = \mathscr{A} + \mathscr{B}, \quad \mathscr{A}: \tilde{C}^1_{0,\Sigma}(\overline{\Omega}) \subset X \to H^* \subset Y^*, \quad \mathscr{B}: \tilde{C}^1_{0,\Sigma}(\overline{\Omega}) \subset X \to Y^*,$$

$$\langle \mathscr{A}u, \eta \rangle = \int_\Omega (\mathbf{l}' \cdot A\nabla\eta + l'_0\eta)\, dx, \quad \langle \mathscr{B}u, \eta \rangle = -\int_\Omega u(b^i\eta)_{x_i}\, dx + \int_{\Sigma'_1} bu\eta, \tag{16}$$

$$u, \eta \in \tilde{C}^1_{0,\Sigma}(\overline{\Omega}).$$

The operator (16) is bounded and continuous, and it may therefore be considered extended to the entire space X. Henceforth $\mathscr{L}$: $X \to Y^*$ always denotes this extended operator. The problem of finding a generalized solution of (8) is equivalent to solving the operator equation

$$\mathscr{L}u = \mathscr{F}, \tag{17}$$

where $\mathscr{F} \in H^*$ is defined by the formula $\langle \mathscr{F}, \eta \rangle = \int_\Omega f\eta\, dx$, $\eta \in H$. The operator $\mathscr{L}$: $X \to Y^*$ will therefore be called the *operator corresponding to problem* (8). In investigating the solvability of the operator equation (17) we proceed from the theory of equations in a Banach space with operators possessing properties of coercivity and monotonicity (or semibounded-variation) type. This theory has been constructed by F. E. Browder, G. Minty, J. L. Lions, Yu. A. Dubinskiĭ, M. M. Vaĭnberg, and others. The special features of the equations we consider require the construction of a new scheme of operator equations of this type using a triple of Banach spaces H, X, Y such that $Y \to X$, $Y \to H$ and $H^* \to X^* \to Y^*$, while the operators $\mathscr{L}$ under study act from X to Y^* and have the form $\mathscr{L} = \mathscr{A} + \mathscr{B}$, where $\mathscr{A}$: $X \to H^*$ and $\mathscr{B}$: $X \to Y^*$ (see §4.7).

The principal difficulty in the study of equations (17) is due to the *incompatibility* of the operators $\mathscr{A}$ and $\mathscr{B}$ caused by the circumstance that only A-derivatives of the function u participate in the construction of $\mathscr{A}$, while only ordinary derivatives of this function participate in the construction of $\mathscr{B}$. This is manifest, in particular, in that $\mathscr{B}$ is not continuous in the norm of the basic energy space $H \equiv H^{0,\Sigma}_{m,\mathbf{m}}(A, \Omega)$. Therefore, in the formulations of results on the solvability of (17) there are conditions effecting the compatibility of $\mathscr{A}$ and $\mathscr{B}$ in addition to conditions of coercivity and monotonicity (or semiboundedness of the variation) for the operator $\mathscr{L}$. A collection of these conditions leads to the solvability of (17). Replacement of the conditions of coercivity and monotonicity of $\mathscr{L}$ by the condition of strong monotonicity of this operator (see §4.6) leads to unique solvability of (17) and continuous dependence of the solution on its right side.

Simply verifiable algebraic criteria are given in Chapter 5 for the conditions of coercivity, monotonicity, and strong monotonicity for the operator $\mathscr{L}$ corresponding to problem (8) to be satisfied. These criteria are given in terms of the reduced coefficients of (1) and the components of the vector $\mathbf{b}$ for an arbitrary $(A, \mathbf{b}, m, \mathbf{m})$-elliptic equation. However, verification of the conditions responsible for the compatibility of $\mathscr{A}$ and $\mathscr{B}$ requires specification of more concrete classes of $(A, \mathbf{b}, m, \mathbf{m})$-elliptic equations. In any case this verification is realized in Part III for the classes of $(A, \mathbf{0}, m, \mathbf{m})$-elliptic and $(A, \mathbf{0}, m, \mathbf{m})$-parabolic equations; for the first of these classes the verification is trivial. We emphasize that for $(A, \mathbf{0}, m, \mathbf{m})$-elliptic and

$(A, \mathbf{0}, m, \mathbf{m})$-parabolic equations existence theorems are obtained for the general boundary value problem (in particular, also problem (8)) all conditions in which are accompanied by simply verifiable criteria for their validity.

In the case of linear equations with a nonnegative characteristic form of the form (4) (and of a somewhat more general form), which are automatically $(A, \mathbf{b}, 2, \mathbf{2})$-elliptic equations relative to the matrix $A = \mathfrak{A}^{1/2}$ and $\mathbf{b} = \beta$, there is no need for conditions of compatibility of $\mathscr{A}$ and $\mathscr{B}$ (see Theorem 4.6.2). Therefore, the results obtained in Chapter 5 imply definitive results on the existence of generalized solutions in the class $H_{2,\mathbf{2}}^{0,\Sigma}(A, \Omega)$ of the general boundary value problem for linear equations with a nonnegative characteristic form of both divergence and nondivergence type. We remark that in [14], [99], [102], [107] and [120] only the "pure" boundary value problems (the first, second, and third) were investigated with the required completeness.

One of the central results of Part II is the theorem on solvability of problem (8) for $(A, \mathbf{b})$-elliptic equations in the class of regular generalized solutions (see Theorem 6.2.3 and Remark 6.2.4). There are simple examples of $(A, \mathbf{b})$-elliptic and, in particular, linear $(A, \mathbf{0})$-elliptic equations possessing the required smoothness in a sufficiently smooth domain $\Omega \subset \mathbb{R}^n$, $n \geqslant 2$, for which problem (8) has no regular generalized solution, and for these equations it turns out that the following condition is not satisfied:

$$A\boldsymbol{\nu} \neq \mathbf{0} \quad \text{on the entire boundary } \partial\Omega. \tag{18}$$

An example of such an equation is presented in Chapter 6 (see (6.2.11)). The essence of these examples is that such equations have bounded solutions in Ω with derivatives which tend to ∞ on approaching those points of the boundary where $A\boldsymbol{\nu} = \mathbf{0}$. Therefore, in studying the question of existence of regular generalized solutions of (8) it is natural to assume that condition (18) is satisfied. Under this condition problem (8) assumes the form

$$\mathscr{L}u = f(x) \quad \text{in } \Omega, \qquad u = 0 \quad \text{on } \partial\Omega, \tag{19}$$

and the integral identity (13) can be satisfied only for all $\eta \in \tilde{C}_0^1(\overline{\Omega})$.

To prove this solvability we consider problems of the form (19) for regularized equations obtained from the given equation by regularization both of the matrix A and of the reduced coefficients l''^i, $i = 1, \dots, n$. A uniform estimate of the maximum moduli of solutions and their gradients in Ω is established for solutions of the regularized problems; this is the key feature in the proof of solvability of problem (19). In connection with this estimate we go beyond the framework of divergence $(A, \mathbf{b})$-elliptic equations and consider nondivergence $(A, \mathbf{b})$-elliptic equations of the form

$$\hat{\alpha}^{ij}(x, u, \hat{\nabla} u) u_{\hat{j}\hat{i}} - \hat{\alpha}(x, u, \hat{\nabla} u) - b^i(x) u_{x_i} = 0, \tag{20}$$

where $u_{\hat{i}}$ and $u_{\hat{i}\hat{j}}$ are the A-derivatives of u of first and second orders respectively, $\hat{\nabla} u \equiv (u_{\hat{1}}, \dots, u_{\hat{n}})$, the u_{x_i}, $i = 1, \dots, n$, are the ordinary derivatives of u, and $u_{\hat{i}}$ is defined as the derivative of u in the direction of the vector $\mathbf{a}^i$ defined by the ith row of the matrix A (if $\mathbf{a}^i = \mathbf{0}$ at x, then by definition $u_{\hat{i}} = 0$ at this point). We call an equation of the form (20) (A, **b**)*-elliptic* (*strictly* (A, **b**)*-elliptic*) in Ω if

$$\hat{\alpha}^{ij}(x,u,q)\eta_i\eta_j \geqslant 0, \qquad \eta = A\xi,\ \xi \in \mathbb{R}^n,\ x \in \overline{\Omega},\ u \in \mathbb{R},\ q = Ap,\ p \in \mathbb{R}^n \tag{21}$$

$$\left[\hat{\alpha}^{ij}(x,u,q)\eta_i\eta_j > 0, \qquad \eta = A\xi \neq 0,\ \xi \in \mathbb{R}^n,\ x \in \overline{\Omega},\ u \in \mathbb{R},\ q = Ap,\ p \in \mathbb{R}^n\right]. \tag{21'}$$

The divergence (A, **b**)-elliptic equations of the form (1) considered above are a special case of nondivergence (A, **b**)-elliptic equations, since because of condition (2) differentiation of the first term in (1) leads to equation (20) with $\hat{\alpha}^{ij}$ and $\hat{\alpha}$ defined by (6.1.4). In finding the a priori estimate $\max_\Omega(|u| + |\nabla u|) \leqslant M$ with a constant M not depending on the ellipticity constant of the equation for sufficiently smooth solutions of a nondegenerate (A, **b**)-elliptic equation of the form (20), we thereby obtain the uniform estimate indicated above for solutions of the regularized problems approximating (19). In connection with the basic a priori estimate of $\max_\Omega|\nabla u|$ for solutions of (20) we make essential use of the results and methods we applied in studying the Dirichlet problem for nonuniformly elliptic equations in Chapter 1. After an estimate of $\max_\Omega(|u| + |\nabla u|)$ for solutions of regularized equations of the form (19) has been obtained, these equations can be considered uniformly elliptic and boundedly nonlinear. Using known results of Ladyzhenskaya and Ural′tseva, we then establish the existence of solutions of the regularized problems with uniformly bounded $\max_{\overline{\Omega}}(|u| + |\nabla u|)$. From the family of such solutions a sequence is selected which converges to a regular generalized solution of the original problem (19).

One of the main conditions on the structure of (A, **b**)-elliptic equations in Theorem 6.2.3 is condition (6.2.15). It should be mentioned that in the case of linearity of the equation condition (6.2.15) goes over into one of the conditions (see (6.2.19)) distinguished by Oleĭnik in studying the solvability of the first boundary value problem for linear equations with nonnegative characteristic form. Examples show (see (6.2.17) and (6.2.20)) that such conditions are due to the essence of the matter. In Chapter 6 the result is applied to some nonregular variational problems.

A theorem on the existence and uniqueness of a regular generalized solution of (19) possessing bounded second derivatives in Ω is also established in Chapter 6. In this connection it is necessary to obtain an a priori estimate of the maximum moduli of the second derivatives of solutions of (A, **b**)-elliptic equations of the form (20) again using the results of Chapter 1 on nonuniformly elliptic equations. In particular, essential use is here made of local estimates of the gradients of solutions of (1.1.1) on $\partial\Omega$. In establishing the a priori estimate of the second derivatives there arise additional strong restrictions on the structure of the equation which make it almost linear. However, examples show that these restrictions are caused by the essence of the matter.

The results indicated above on the solvability of (19) in classes of regular generalized solutions pertain to the case of a sufficiently smooth domain Ω. However, by applying the results of Chapter 2, it is possible in an altogether analogous way to obtain similar results for (A, **b**)-parabolic equations in a cylinder $Q = \Omega \times (T_1, T_2)$. Bacause of the limited size of this monograph, we shall not present an exposition of such results.

CHAPTER 4
SOME ANALYTIC TOOLS USED IN THE INVESTIGATION OF SOLVABILITY OF BOUNDARY VALUE PROBLEMS FOR $(A, \mathbf{b})$-ELLIPTIC EQUATIONS

§1. Generalized A-derivatives

Suppose that in a domain $\Omega \subset \mathbb{R}^n$, $n \geqslant 2$, there is given a square matrix $A = \|a^{ij}(x)\|$ of order n with elements satisfying the condition

$$a^{ij} \in L^{m_i}(\Omega), \qquad m_i \geqslant 1, \quad i, j = 1, \dots, n. \tag{1.1}$$

The matrix A may degenerate on any subset of Ω. We denote by $\tilde{C}^1_{\mathrm{loc}}(\Omega)$ the set of all functions belonging to $\tilde{C}^1(\Omega')$ (see the basic notation) for any Ω' such that $\overline{\Omega}' \subset \Omega$. The mapping

$$u \to (A_1 \nabla u, \dots, A_n \nabla u), \qquad A_i \nabla u \equiv \sum_{j=1}^{n} a^{ij}(x) u_{x_j}, \quad i = 1, \dots, n, \tag{1.2}$$

can be considered as a linear operator acting from $\tilde{C}^1_{\mathrm{loc}}(\Omega) \subset L^m_{\mathrm{loc}}(\Omega)$ into $\mathbf{L}^{\mathbf{m}}_{\mathrm{loc}}(\Omega)$, where $m \geqslant 1$ and $\mathbf{L}^{\mathbf{m}}(\Omega) = L^{m_1}(\Omega) \times \cdots \times L^{m_n}(\Omega)$. We call this operator the operator of taking the A-gradient. Since in $L^m_{\mathrm{loc}}(\Omega)$ and $\mathbf{L}^{\mathbf{m}}_{\mathrm{loc}}(\Omega)$ it is possible to introduce natural topologies, the question of the existence of the weak closure of the operator of taking the A-gradient is meaningful. We assume that the following condition is satisfied:

$$\text{the operator of taking the } A\text{-gradient admits weak closure.} \tag{1.3}$$

Condition (1.3) obviously means that if $u_n \to 0$ weakly in $L^m_{\mathrm{loc}}(\Omega)$ and $A\nabla u_n \to \mathbf{v}$ weakly in $\mathbf{L}^{\mathbf{m}}_{\mathrm{loc}}(\Omega)$, where $u_n \in \tilde{C}^1_{\mathrm{loc}}(\Omega)$, $n = 1, 2, \dots$, then $\mathbf{v} = \mathbf{0}$ a. e. in Ω.

DEFINITION 1.1. If condition (1.3) is satisfied, we say that a function $u \in L^m_{\mathrm{loc}}(\Omega)$ has a *generalized A-gradient* $A\nabla u \in \mathbf{L}^{\mathbf{m}}_{\mathrm{loc}}(\Omega)$ in Ω if u belongs to the domain of the weak closure of the operator of taking the A-gradient, and the vector $A\nabla u$ is the value of that operator at the function u. We denote the components of the vector $A\nabla u$ by $A_1\nabla u, \dots, A_n \nabla u$ and call them the *generalized A-derivatives* of u in Ω; here obviously $A_i \nabla u \in L^{m_i}_{\mathrm{loc}}(\Omega)$, $i = 1, \dots, n$.

Thus, if condition (1.3) is satisfied a function $u \in L^m_{\mathrm{loc}}(\Omega)$ has a generalized A-gradient $A\nabla u \in \mathbf{L}^{\mathbf{m}}_{\mathrm{loc}}(\Omega)$ if and only if there exist a sequence $\{u_n\}$, $u_n \in \tilde{C}^1_{\mathrm{loc}}(\Omega)$, $n = 1, 2, \dots$, and a vector-valued function $\mathbf{v} \in \mathbf{L}^{\mathbf{m}}_{\mathrm{loc}}(\Omega)$ such that $u_n \to u$ weakly in $L^m_{\mathrm{loc}}(\Omega)$ and $A\nabla u_n \to \mathbf{v}$ weakly in $\mathbf{L}^{\mathbf{m}}_{\mathrm{loc}}(\Omega)$; in this case $A\nabla u = \mathbf{v}$.

LEMMA 1.1. *Suppose that condition* (1.1) *is satisfied, and*

$$a^{ij}, \partial a^{ij}/\partial x_j \in L^{m'}_{\mathrm{loc}}(\Omega), \qquad 1/m + 1/m' = 1, \qquad i, j = 1, \dots, n, \tag{1.4}$$

where the $\partial a^{ij}/\partial x_j$ are ordinary (*Sobolev*) *generalized derivatives of the functions a^{ij} with respect to x_j. Then for any $\mathbf{m} = (m_1, \dots, m_n)$, where $m_i \geqslant 1$, $i = 1, \dots, n$, the operator of taking the A-gradient admits weak closure* (*i.e., condition* (1.3) *is satisfied*).

PROOF. Let $\{u_n\}$, $u_n \in \tilde{C}^1_{\mathrm{loc}}(\Omega)$, $n = 1, 2, \dots$, be a sequence such that $u_n \to 0$ weakly in $L^m_{\mathrm{loc}}(\Omega)$ and $A\nabla u_n \to \mathbf{v}$ weakly in $\mathbf{L}^{\mathbf{m}}_{\mathrm{loc}}(\Omega)$. Then for all $n = 1, 2, \dots$

$$\int_\Omega u_n \left(A_i \nabla \eta + \frac{\partial a^{ij}}{\partial x_j} \eta \right) dx = - \int_\Omega A_i \nabla u_n \eta \, dx, \quad \forall \eta \in \tilde{C}^1_0(\Omega), \tag{1.5}$$

where $\tilde{C}_0^1(\Omega)$ denotes the set of all functions in $\tilde{C}^1(\Omega)$ having compact support in Ω. Passing to the limit as $n \to \infty$ in (1.5), we find that $\int_\Omega v_i \eta \, dx = 0$ for all $\eta \in \tilde{C}_0^1(\Omega)$, $i = 1, \dots, n$, whence it obviously follows that $v_i = 0$ a. e. in Ω, $i = 1, \dots, n$, i.e., $\mathbf{v} = \mathbf{0}$ a. e. in Ω. This proves Lemma 1.1.

Lemma 1.1 shows that fulfillment of condition (1.3) is ensured by imposing a certain regularity condition (condition (1.4)) on the elements of the matrix A. It will be shown below that fulfillment of (1.3) is also ensured by a condition of sufficiently weak degeneracy of A.

LEMMA 1.2. *Suppose that conditions* (1.1) *and* (1.4) *are satisfied, and that the function* $u \in L^m_{\text{loc}}(\Omega)$ *has a generalized A-gradient* $A\nabla u \in \mathbf{L}^m_{\text{loc}}(\Omega)$. *Then*

$$\int_\Omega u\left(A_i \nabla\eta + \frac{\partial a^{ij}}{\partial x_j}\eta\right) dx = -\int_\Omega A_i \nabla u \eta dx, \quad \forall \eta \in \tilde{C}_0^1(\Omega), i = 1, \dots, n, \tag{1.6}$$

where $A_i\nabla u$ is the ith component of the generalized A-gradient of u in Ω.

PROOF. Let $\{u_n\}$, $u_n \in \tilde{C}^1_{\text{loc}}(\Omega)$, $n = 1, 2, \dots$, be a sequence such that $u_n \to u$ weakly in $L^m_{\text{loc}}(\Omega)$ and $A\nabla u_n \to A\nabla u$ weakly in $\mathbf{L}^m_{\text{loc}}(\Omega)$. Passing to the limit as $n \to \infty$ in the identities (1.5) written for terms of this sequence, we obtain (1.6). Lemma 1.2 is proved.

The following converse of Lemma 1.2 holds under conditions on the matrix A which are somewhat stronger than (1.1) and (1.4).

LEMMA 1.3. *Suppose the elements of the matrix A satisfy the condition*

$$a^{ij} \in \text{Lip}(\Omega), \qquad i, j = 1, \dots, n, \tag{1.7}$$

and that for some functions $u \in L^m_{\text{loc}}(\Omega)$ *and* $\mathbf{v} \in \mathbf{L}^m_{\text{loc}}(\Omega)$, $m > 1$,

$$\int_\Omega u\left(A_i \nabla\eta + \frac{\partial a^{ij}}{\partial x_j}\eta\right) dx = -\int_\Omega v_i \eta \, dx, \quad \forall \eta \in \tilde{C}_0^1(\Omega), i = 1, \dots, n, \tag{1.8}$$

where the $v_i \in L^m_{\text{loc}}(\Omega)$ *are the components of the vector-valued function* $\mathbf{v}$. *Then the function* $u \in L^m_{\text{loc}}(\Omega)$ *has a generalized A-gradient* $A\nabla u \in \mathbf{L}^m(\Omega)$, *and* $A\nabla u = \mathbf{v}$.

PROOF. Since it follows from (1.7) that conditions (1.1) and (1.4) are satisfied, condition (1.3) is satisfied by Lemma 1.1. We shall prove that then there exists a sequence $\{u_n\}$, $u_n \in \tilde{C}_0^1(\Omega)$, $n = 1, 2, \dots$, for which the following conditions hold:

$$u_n \to u \quad \text{weakly in } L^m_{\text{loc}}(\Omega), \qquad A\nabla u_n \to \mathbf{v} \quad \text{weakly in } \mathbf{L}^m_{\text{loc}}(\Omega). \tag{1.9}$$

We consider the averaging

$$u_h(x) = \int_{\mathbf{R}^n} \omega_h(x - y) u(y) \, dy, \qquad h > 0,$$
$$x \in \Omega_h \equiv \{s \in \Omega: \text{dist}(x, \partial\Omega) > h\}, \tag{1.10}$$

of the function u with the infinitely differentiable, normalized kernel

$$\omega_h(z) = \frac{1}{\kappa_n h^n}\omega\left(\frac{|z|}{h}\right), \qquad \omega(t) = \begin{cases} \lambda(t) & \text{for } t \in [0, 1], \\ 0 & \text{for } t > 1, \end{cases} \tag{1.11}$$

where $\kappa_n = \sigma_n \int_0^1 \lambda(t)t^{n-1}\,dt$, σ_n is the surface area of the unit ball in $\mathbb{R}^n$, and the function $\lambda(t)$ defined on $[0,1]$ satisfies the conditions

$$\lambda \in C^\infty([0,1]), \quad \int_0^1 \lambda(t)t^{n-1}\,dt > 0, \quad \lambda(t) \geqslant 0,$$

$$\lambda^{(k)}(1) = 0, \qquad k = 0,1,\dots. \tag{1.12}$$

It follows from (1.11) and (1.12) that $\omega_h(z) \in \tilde{C}_0^1(\mathbb{R}^n)$, $\omega_h(z) = 0$ for $|z| > h$, $\omega_h(z) \geqslant 0$ for $|z| \leqslant h$, and $\int_{\mathbb{R}^n} \omega_h(z)\,dz = 1$.

It is obvious that at any point $x \in \Omega_h$, for any $j = 1,\dots,n$,

$$\frac{\partial u_h}{\partial x_j} = \int_{\mathbb{R}^n} \frac{\partial \omega_h(x-y)}{\partial x_j} u(y)\,dy = \int_\Omega \frac{\partial \omega_h(x-y)}{\partial x_j} u(y)\,dy. \tag{1.13}$$

For fixed $i \in \{1,\dots,n\}$ at the point $x \in \Omega_h$ we compute the expression

$$\begin{aligned} A_i \nabla u_h &= \int_\Omega a^{ij}(x) \frac{\partial \omega_h(x-y)}{\partial x_j} u(y)\,dy = -\int_\Omega a^{ij}(x) \frac{\partial \omega_h(x-y)}{\partial y_j} u(y)\,dy \\ &= -\int_\Omega a^{ij}(y) \frac{\partial \omega_h(x-y)}{\partial y_j} u(y)\,dy - \int_\Omega [a^{ij}(x) - a^{ij}(y)] \frac{\partial \omega_h(x-y)}{\partial y_j}\,dy. \end{aligned} \tag{1.14}$$

By (1.8) with $\eta = \omega_h$, from (1.14) we obtain

$$\begin{aligned} A_i \nabla u_h &= \int_\Omega v_i(y)\omega_h(x-y)\,dy + \int_\Omega \frac{\partial a^{ij}(y)}{\partial y_j} \omega_h(x-y)u(y)\,dy \\ &\quad - \int_\Omega [a^{ij}(x) - a^{ij}(y)] \frac{\partial \omega_h(x-y)}{\partial y_j} u(y)\,dy \\ &\equiv v_{ih} + J_h, \end{aligned} \tag{1.15}$$

where J_h can also be written in the form

$$J_h = \left(\frac{\partial a^{ij}}{\partial x_j} u \right)_h + a^{ij} u_{hx_j} - (a^{ij}u)_{hx_j}. \tag{1.16}$$

We first prove that $A_i \nabla u_h \to v_i$ weakly in $L^m_{\text{loc}}(\Omega)$. In view of (1.15) and the known fact of strong convergence of v_{ih} to v_i in $L^m_{\text{loc}}(\Omega)$, for this it suffices to prove that $J_h \to 0$ weakly in $L^m_{\text{loc}}(\Omega)$. Suppose that some compact subregion Ω' ($\overline{\Omega}' \subset \Omega$) in Ω is fixed. We first prove that

$$\lim_{h \to 0} \int_{\Omega'} J_h \eta\,dx = 0$$

for all $\eta \in \tilde{C}_0^1(\Omega')$. Let $h < \operatorname{dist}(\partial\Omega', \partial\Omega)$. Multiplying both sides of (1.16) by $\eta \in \tilde{C}_0'(\Omega')$ and integrating over Ω', we obtain

$$\int_{\Omega'} J_h \eta\,dx = \int_{\Omega'} \left[\left(\frac{\partial a^{ij}}{\partial x_j} u \right)_h \eta - \frac{\partial a^{ij}}{\partial x_j} u_h \eta - a^{ij} u_h \eta_{x_j} + (a^{ij}u)_h \eta_{x_j} \right] dx. \tag{1.17}$$

Letting h tend to zero in (1.17) and using the properties of locally integrable functions, we find that

$$\lim_{h \to 0} \int_{\Omega'} J_h \eta\,dx = 0.$$

Since Ω' is an arbitrary strictly interior subregion of Ω, to complete the proof of weak convergence $J_h \to 0$ as $h \to 0$ in $L^m_{\text{loc}}(\Omega)$ it suffices to verify that the norms $\|A_i \nabla u_h\|_{m,\Omega'}$ are bounded uniformly in $h \in \mathbb{R}_+$. Using (1.7), we write

$$\begin{aligned}
2^{1-m}\|J_h\|^m_{m,\Omega'} &\leqslant \int_{\Omega'} dx \left| \int_{K_h(x)} \frac{\partial a^{ij}(y)}{\partial y_j} \omega_h(x-y) u(y)\, dy \right|^m \\
&\quad + \int_{\Omega'} dx \left| \int_{K_h(x)} [a^{ij}(x) - a^{ij}(y)] \frac{\partial \omega_h(x-y)}{\partial y_j} u(y)\, dy \right|^m \\
&\leqslant c\left\{ \||u|_h\|^m_{m,\Omega'} \sum_{j=1}^{n} \left[\int_{\Omega'} dx \int_{K_h(x)} h \left| \frac{\partial \omega_h(x-y)}{\partial x_j} \right| |u(y)| dy \right]^m \right\},
\end{aligned} \tag{1.18}$$

where $K_h(x)$ is the ball of radius h with center at the point $x \in \Omega'$ and $\overline{K_h(x)} \subset \Omega$. To be specific, we suppose that $\lambda(t)$ in (1.11) is chosen in the traditional manner, i.e.,

$$\lambda(t) = \begin{cases} \exp\left(t^2(t^2-1)^{-1}\right) & \text{for } t \in [0,1), \\ 0 & \text{for } t = 1. \end{cases} \tag{1.19}$$

In view of (1.11) and (1.19) we have

$$\left| \frac{\partial \omega_h(z)}{\partial z_j} \right| \leqslant \frac{1}{\kappa_n h^{n+1}} \tilde{\omega}\left(\frac{|z|}{h} \right), \tag{1.20}$$

where

$$\tilde{\omega}(t) = \begin{cases} \tilde{\lambda}(t) & \text{for } t \in [0,1], \\ 0 & \text{for } t > 1, \end{cases} \qquad \tilde{\lambda}(t) = \frac{2t}{1-t^2} e^{t^2/(t^2-1)}. \tag{1.21}$$

It is obvious that the function $\tilde{\lambda}(t)$ together with the function (1.19) satisfies all the conditions (1.12). Setting

$$\tilde{\omega}_h(z) = \frac{1}{\tilde{\kappa}_n h^n} \tilde{\omega}\left(\frac{|z|}{h} \right), \qquad \tilde{\kappa}_n = \sigma_n \int_0^1 \tilde{\lambda}(t) t^{n-1}\, dt, \tag{1.22}$$

by (1.11) and (1.19)–(1.21) we obtain

$$h\left| \frac{\partial \omega_h(z)}{\partial z_j} \right| \leqslant \frac{\tilde{\kappa}_n}{\kappa_n} \tilde{\omega}_h(z), \qquad j = 1, \dots, n. \tag{1.23}$$

From (1.18) and (1.23) we then obtain

$$\|J_h\|_{m,\Omega'} \leqslant c\left[\||u|_h\|_{m,\Omega'} + \||\tilde{u}|_h\|_{m,\Omega'} \right], \tag{1.24}$$

where $|\tilde{u}|_h$ is the average of the function $|u|$ with respect to the new kernel $\tilde{\omega}_h$. Since $u \in L^m(\Omega')$ and $\||u|_h\|_{m,\Omega'} \leqslant c\|u\|_{m,\Omega'}$, $\||\tilde{u}|_h\|_{m,\Omega'} \leqslant c\|u\|_{m,\Omega'}$, from (1.24) we obtain the estimate $\|J_h\|_{m,\Omega'} \leqslant c$ with a constant c not depending on $h \in \mathbb{R}_+$. Thus, $J_h \to 0$ weakly in $L^m_{\text{loc}}(\Omega)$, and hence $A_i \nabla u_h \to v_i$ weakly in $L^m_{\text{loc}}(\Omega)$. Taking into account that $u_h \in C^\infty_{\text{loc}}(\Omega)$ and $u_h \to u$ in $L^m_{\text{loc}}(\Omega)$ as $h \to 0$, we conclude that for any sequence $\{u_{h_n}\}$, where $\lim_{n\to\infty} h_n = 0$, the following conditions are satisfied: $u_{h_n} \to u$ weakly in $L^m_{\text{loc}}(\Omega)$, and $A\nabla u_{h_n} \to \mathbf{v}$ weakly in $\mathbf{L}^m_{\text{loc}}(\Omega)$; from this it follows that $\mathbf{v} \in \mathbf{L}^m_{\text{loc}}(\Omega)$ is the generalized A-gradient of the function $u \in L^m_{\text{loc}}(\Omega)$. Lemma 1.3 is proved.

The next proposition follows from the proof of Lemma 1.3.

COROLLARY 1.1. *Suppose that condition* (1.7) *is satisfied, and that the function* $u \in L^m_{\mathrm{loc}}(\Omega)$ *has a generalized A-gradient* $A\nabla u \in \mathbf{L}^m_{\mathrm{loc}}(\Omega)$. *Let* u_h *be the average of* u *defined by* (1.10)–(1.12) *and* (1.19). *Then* $A\nabla u_h \to A\nabla u$ *weakly in* $\mathbf{L}^m_{\mathrm{loc}}(\Omega)$.

In the special case where the matrix $A \equiv \|a^{ij}\|$ is constant in Ω this result can be strengthened by replacing weak convergence $A\nabla u_h \to A\nabla u$ in $\mathbf{L}^m_{\mathrm{loc}}(\Omega)$ by strong convergence in the assertion of Corollary 1.1. Indeed, if the matrix A is constant, then $J_h \equiv 0$ in Ω. From (1.15) and the strong convergence of v_{ih} to v_i in $L^m_{\mathrm{loc}}(\Omega)$, $i = 1, \dots, n$, we obtain the assertion.

To conclude this section we observe that it is no trouble to define generalized A-derivatives of any order $k > 1$ relative to a square matrix A of order n^k defined in a domain $\Omega \subset \mathbb{R}^n$.

§2. Generalized limit values of a function on the boundary of a domain

Let Ω be a bounded domain in $\mathbb{R}^n$, $n \geqslant 2$. We suppose that in Ω there is given a square matrix $A \equiv \|a^{ij}(x)\|$ of order n satisfying conditions (1.1) and (1.3) for fixed $m \geqslant 1$ and $\mathbf{m} = (m_1, \dots, m_n)$, $m_i \geqslant 1$, $i = 1, \dots, n$. Let Γ be an arbitrary part of the boundary $\partial\Omega$ (in particular, we admit the cases $\Gamma = \partial\Omega$ or $\Gamma = \varnothing$, the empty set).

We denote by $H \equiv H^{0,\Gamma}_{m,\mathbf{m}}(A, \Omega)$ the completion of the set $\tilde{C}^1_{0,\Gamma}(\Omega)$ in the norm

$$\|u\|_H \equiv \|u\|_{m,\Omega} + \|A\nabla u\|_{\mathbf{m},\Omega}, \tag{2.1}$$

where $\|A\nabla u\|_{\mathbf{m},\Omega} = \Sigma_1^n \|A_i \nabla u\|_{m_i,\Omega}$. It is obvious that H is a Banach space. In the case $\Gamma = \partial\Omega$ the space H is denoted by $H \equiv H^0_{m,\mathbf{m}}(A, \Omega)$, and in the case $\Gamma = \varnothing$ by $H \equiv H_{m,\mathbf{m}}(A, \Omega)$.

Let $\Pi \subset \partial\Omega$ be an arbitrary set. We assign to each function $u \in \tilde{C}^1(\overline{\Omega})$ its value $u|_\Pi$ on Π, i.e., we consider the mapping

$$u \to u|_\Pi. \tag{2.2}$$

The mapping (2.2) can be considered a linear operator with domain contained in $H_{m,\mathbf{m}}(A, \Omega)$ (with $m, \mathbf{m}$, and A fixed above) and range in $L^1_{\mathrm{loc}}(\Pi)$. We call this operator the *operator of taking the limit value on* Π.

We suppose that for the set Π the following condition is satisfied:

$$\text{the operator of taking the limit value on } \Pi \text{ admits closure.} \tag{2.3}$$

Condition (2.3) obviously means that if $u_n \to 0$ in H and $u_n|_\Pi \to \varphi$ in $L^1_{\mathrm{loc}}(\Pi)$, where $u_n \in \tilde{C}^1(\overline{\Omega})$, $n = 1, 2, \dots$, then $\varphi = 0$ a.e. on Π.

DEFINITION 2.1. If condition (2.3) is satisfied we say that the function $u \in H_{m,\mathbf{m}}(A, \Omega)$ has a *generalized limit value* $u|_\Pi$ on the set Π if the function u belongs to the domain of the closure of the operator of taking the limit value on the set Π, and $u|_\Pi$ is the value of this operator at the function u.

Thus, if condition (2.3) is satisfied a function $u \in H_{m,\mathbf{m}}(A, \Omega)$ has a generalized limit value on the set Π if and only if there exist a sequence $\{u_n\}$, $u_n \in \tilde{C}^1(\overline{\Omega})$, $n = 1, 2, \dots$, and a function $\varphi \in L^1_{\mathrm{loc}}(\Pi)$ such that $u_n \to u$ in H and $u_n|_\Pi \to \varphi$ in $L^1_{\mathrm{loc}}(\Pi)$; in this case $u|_\Pi = \varphi$.

We henceforth say that a function $u \in H_{m,\mathbf{m}}(A, \Omega)$ has a generalized limit value $u|_\Pi \in L^r_{\mathrm{loc}}(\Pi)$ on Π if there exist a sequence $\{u_n\}$, $u_n \in \tilde{C}^1(\overline{\Omega})$, $n = 1, 2, \dots$, and a function $\varphi \in L^r_{\mathrm{loc}}(\Pi)$ such that $u_n \to u$ in $H_{m,\mathbf{m}}(A, \Omega)$ and $u_n|_\Pi \to \varphi$ in $L^r_{\mathrm{loc}}(\Pi)$; in this case $u|_\Pi = \varphi$.

The next lemma illustrates the role which generalized limit values on $\Pi \subset \partial\Omega$ can play for functions possessing these values.

LEMMA 2.1. *Let Ω be a bounded domain in $\mathbb{R}^n$, $n \geqslant 2$, of class C^1, let $a^{ij} \in W_2^1(\Omega)$, $i, j = 1, \dots, n$, and suppose that for some subset $\Pi \subset \partial\Omega$ condition* (2.3) *is satisfied with $m = 2$ and $\mathbf{m} = \mathbf{2}$. Then for any function $u \in H_{2,\mathbf{2}}(A, \Omega)$ having a generalized limit value $u|_\Pi \in L^2_{\mathrm{loc}}(\Pi)$ the identities*

$$\int_\Omega u\left(A_i\nabla\eta + \frac{\partial a^{ij}}{\partial x_j}\eta\right)dx = -\int_\Omega A_i\nabla u\eta\,dx - \int_\Pi A_i\boldsymbol{\nu} u|_\Pi\,\eta\,ds, \qquad \forall\eta \in \tilde{C}^1_{0,\partial\Omega\setminus\Pi}(\overline{\Omega}),\quad i = 1, \dots, n, \tag{2.4}$$

hold, where $\boldsymbol{\nu}$ is the unit vector of the inner normal to $\partial\Omega$ and $\tilde{C}^1_{0,\partial\Omega\setminus\Pi}(\overline{\Omega})$ denotes the set of all functions in $\tilde{C}^1(\overline{\Omega})$ equal to 0 in some neighborhood of $\partial\Omega\setminus\Pi$ (*the neighborhood depends on the function η*).

PROOF. Let u be a function possessing the properties indicated in the formulation of the lemma. Then there exists a sequence $\{u_n\}$, $u_n \in \tilde{C}^1(\overline{\Omega})$, $n = 1, 2, \dots$, such that $u_n \to u$ in $H_{2,\mathbf{2}}(A, \Omega)$ and $u_n|_\Pi \to u|_\Pi$ in $L^1_{\mathrm{loc}}(\Omega)$ as $n \to \infty$. In view of the condition $a^{ij} \in W_2^1(\Omega)$, $i, j = 1, \dots, n$, and Sobolev's imbedding theorem we have $A_i\boldsymbol{\nu} \equiv (A\boldsymbol{\nu})_i \in L^2(\Pi)$, $i = 1, \dots, n$, so that for each function u_n, $n = 1, 2, \dots$,

$$\int_\Omega u_n\left(A_i\nabla\eta + \frac{\partial a^{ij}}{\partial x_j}\eta\right)dx = -\int_\Omega A_i\nabla u_n\eta\,dx - \int_\Pi A_i\boldsymbol{\nu} u_n\eta\,ds, \qquad \forall\eta \in \tilde{C}^1_{0,\partial\Omega\setminus\Pi}(\overline{\Omega}),\quad i = 1, \dots, n.$$

Passing to the limit as $n \to \infty$ in these identities, we obtain (2.4). Lemma 2.1 is proved.

We now present sufficient conditions on a matrix A and a set $\Pi \subset \partial\Omega$ guaranteeing that (2.3) is satisfied.

LEMMA 2.2. *Suppose that conditions* (1.1) *and* (1.3) *are satisfied, and suppose that for the matrix A and set $\Pi \subset \partial\Omega$ the following condition holds*:

> *for each point $x_0 \in \operatorname{int}\Pi$ there exist surfaces P_{x_0} and $\hat{P}_{x_0}$ of class C^2, $x_0 \in \hat{P}_{x_0} \subset P_{x_0} \subset \Pi$, and numbers d_0 and $\hat{d}_0$, $0 < \hat{d}_0 \leqslant d_0$, such that*:
>
> 1) *the set $\omega_{x_0} \equiv \{x \in \mathbb{R}^n: x = y + t\boldsymbol{\nu}(y),\ y \in P_{x_0},\ t \in [0, d_0]\}$, where $\boldsymbol{\nu}(y)$ is the unit vector of the inner normal to Π at y, is contained in $\overline{\Omega}$*;
>
> 2) *the elements of A are continuous, while the elements of A^*A, where A^* is the adjoint matrix of A, are continuously differentiable on the set ω_{x_0}*;
>
> 3) *the vector $A^*(y)A(y)\boldsymbol{\nu}(y)$ does not lie in the tangent plane to $\partial\Omega$ at the point y for all $y \in \hat{P}_{x_0}$*;

4) *the set* $\hat{\omega}_{x_0} \equiv \{x \in \mathbb{R}^n\colon x = y + tA^*(y)A(y)\nu(y),\ y \in \hat{P}_{x_0},\ t \in [0, \hat{d}_0]\}$ *is contained in* Ω_{x_0};(¹) *and*

5) *there exists a constant* $c_0 > 0$ *such that for any* $y \in \hat{P}_{x_0}$, $t \in [0, \hat{d}_0]$ *and* $\xi \in \mathbb{R}^n$, *the inequality* $|A(y)\xi| \leqslant c_0|A(y + tA^*(y)A(y)\nu(y))\xi|$ *holds.* (2.5)

Then the operator of taking the limit value on the set Π *admits closure, i.e., condition* (2.3) *is satisfied.*

PROOF. Suppose that the function $u \in H_{m,\mathbf{m}}(A, \Omega)$ and the sequence $\{u_n\}$, $u_n \in \tilde{C}^1(\bar{\Omega})$, $n = 1, 2, \ldots$, satisfy the conditions $u_n \to u$ in H and $u_n|_\Pi \to \varphi$ in $L^1_{\mathrm{loc}}(\Pi)$ as $n \to \infty$. This implies, in particular, that

$$u_n \to u \quad \text{in } H_{m_*}(A, \Omega), \qquad u_n|_\Pi \to \varphi \quad \text{in } L^1_{\mathrm{loc}}(\Pi), \tag{2.6}$$

where $m_* = \min(m, m_1, \ldots, m_n)$, and $H_{m_*}(A, \Omega) \equiv H_{m,\mathbf{m}}(A, \Omega)$ with $m = m_*$ and $\mathbf{m} = (m_*, \ldots, m_*)$. We note that condition (2.6) is invariant under orthogonal transformations of the variables $x_1, \ldots, x_n$. To prove the lemma it suffices to establish that (2.6) implies the equality $\varphi = 0$ a.e. on int Π. Let $x_0 \in$ int Π. Taking into account that conditions (2.5) and (2.6) are invariant under orthogonal transformations of the variables $x_1, \ldots, x_n$,(²) we shall assume that the coordinate system $Ox_1, \ldots, x_n$ is local with respect to the point x_0, i.e., the axis Ox_n is directed along the inner normal ν to the surface $\hat{P}_{x_0} \subset P_{x_0} \subset \Pi$, while the remaining axes are situated in the plane tangent to P_{x_0} at x_0. It may be assumed that the surface $\hat{P}_{x_0}$ is given in this coordinate system by the equation

$$y_n = \Phi(y'), \qquad y' \equiv (y_1, \ldots, y_{n-1}), \quad |y'| \leqslant \delta_0 = \text{const} > 0. \tag{2.7}$$

It follows from (2.5) that Φ belongs to the class $C^2\{|y'| \leqslant \delta_0\}$.

We consider the mapping defined by

$$\begin{aligned} &(y', t) \to x = (x_1, \ldots, x_n), \qquad |y'| \leqslant \delta_0,\ t \in [0, \hat{d}_0],\\ &x_i = y_i + t[A^*(y)A(y)\nu(y)]_i, \qquad i = 1, \ldots, n-1,\\ &x_n = \Phi(y') + t[A^*(y)A(y)\nu(y)]_n, \end{aligned} \tag{2.8}$$

where $d_0 = \text{const} > 0$, $\nu(y)$ is the unit vector of the inner normal to $\hat{P}_{x_0}$ at the point $y \equiv (y', \Phi(y'))$, and $(A^*(y)A(y)\nu(y))_i$ is the ith component of the vector $A^*(y)A(y)\nu(y)$, $i = 1, \ldots, n$. It is obvious that the Jacobian J of the transformation (2.8) is not equal to 0 at the point $(0, \ldots, 0)$, since for $(y', t) = (0, 0)$

$$\begin{aligned} \partial x_i/\partial y_j = \delta_i^j,\quad j = 1, \ldots, n; \qquad &\partial x_n/\partial t = [A^*(0)A(0)\nu(0)]_n;\\ \partial x_n/\partial y_j = 0, \qquad &j = 1, \ldots, n-1, \end{aligned} \tag{2.9}$$

(¹) Since $A^*A\nu = A\nu \cdot A\nu \geqslant 0$, by condition 3) the vector $A^*A\nu$ always makes an acute angle with the inner normal to $\partial\Omega$.

(²) We remark that by making an orthogonal transformation $x = C(x - x_0)$ we obtain the validity of conditions of the form (2.5) and (2.6) for the matrix $\bar{A} = AC^{-1}$ and the vector $\bar{\nu} = C\nu$, where in verifying these conditions we in particular take account of the equalities $A^*A\nu \cdot \nu = \bar{A}^*\bar{A}\bar{\nu} \cdot \bar{\nu}$ and $A\nabla_x u = \bar{A}\nabla_{\bar{x}} u$, $u \in \tilde{C}^1(\Omega)$.

and $[A^*(0)A(0)\nu(0)]_n = |A(0)\nu(0)|^2 > 0$. Taking into account that (2.5) implies the continuity of all the first derivatives of the functions $x_i(y', t)$, $i = 1, \dots, n$, defining the transformation (2.8), we may assume that the numbers $\delta_0 > 0$ and $\hat{d}_0 > 0$ are so small that $J > 0$ on $(|y'| \leqslant \delta_0) \times (0 \leqslant t \leqslant \hat{d}_0)$. The transformation (2.8) is then a diffeomorphism between $(|y'| \leqslant \delta_0) \times (0 \leqslant t \leqslant \hat{d}_0)$ and the set $\hat{\omega}_{x_0} \subset (\bar{\Omega})$.

Let $f \in \tilde{C}^1(\hat{\omega}_{x_0})$, and let y be a fixed point on $\hat{P}_{x_0} = \{ y = (y', \varphi(y')) \colon |y'| \leqslant \delta_0 \}$. Taking into account that $W^1_{m_*}([0, \hat{d}_0]) \to C([0, \hat{d}_0])$, we have

$$\begin{aligned} |f(y)|^{m_*} &\leqslant \max_{t \in [0, \hat{d}_0]} |f(y) + tA^*(y)A(y)\nu(y)|^{m_*} \\ &\leqslant c\left[\int_0^{\hat{d}_0} |f(y + \tau(A^*A\nu)(y))|^{m_*}\, d\tau \right. \\ &\qquad \left. + \int_0^{\hat{d}_0} |(A^*A\nu)(y) \cdot \nabla f(y + \tau(A^*A\nu)(y))|^{m_*}\, d\tau \right], \end{aligned} \tag{2.10}$$

where $c = 2^{m_*-1}\max(1, \hat{d}_0^{-1})$. Writing $A^*A\nu \cdot \nabla f$ in the form $A\nu \cdot A\nabla f$ and using conditions 2) and 5) of (2.5), we have

$$\begin{aligned} |f(y)|^{m_*} \leqslant c_1 &\left[\int_0^{\hat{d}_0} |f(y + \tau A^*(y)A(y)\nu(y))|^{m_*}\, d\tau \right. \\ &\left. + \int_0^{\hat{d}_0} |A(y + \tau A^*(y)A(y)\nu(y))\nabla f(y + \tau A^*(y)A(y)\nu(y))|^{m_*}\, d\tau \right], \end{aligned} \tag{2.11}$$

where c_1 depends on c, $\max_{\hat{P}_{x_0}} |a^{ij}|$ and the constant c_0 from condition 5) in (2.5). Integrating (2.11) over $\hat{P}_{x_0}$, we obtain

$$\int_{\hat{P}_{x_0}} |f|^{m_*}\, ds \leqslant c_1 \int_0^{\hat{d}_0} \int_{\hat{P}_{x_0}} \left(|f|^{m_*} + |A\nabla f|^{m_*} \right) ds\, d\tau. \tag{2.12}$$

Taking into account the smoothness of the surface $\hat{P}_{x_0}$ and the positivity of the Jacobian of the transformation (2.8), from (2.12) it is easy to derive the estimate

$$\int_{\hat{P}_{x_0}} |f|^{m_*}\, ds \leqslant c_2 \int_{\hat{\omega}_{x_0}} \left(|f|^{m_*} + |A\nabla f|^{m_*} \right) dx, \tag{2.13}$$

where c_2 depends on c_1, $\max_{|y'| \leqslant \delta_0} |\nabla \Phi|$ and $\min_{(|y'| \leqslant \delta_0) \times [0, \hat{d}_0]} J$. Setting $f = u_n$, $n = 1, 2, \dots$, in (2.13), where $\{u_n\}$ is the sequence of (2.6), and passing to the limit as $n \to \infty$, we find that $\varphi \in L^{m_*}(\hat{P}_{x_0})$ and

$$\int_{\hat{P}_{x_0}} |\varphi|^{m_*}\, ds \leqslant c_2 \lim_{n \to \infty} \int_{\omega_{x_0}} \left(|u_n|^{m_*} + |A\nabla u_n|^{m_*} \right) dx = 0. \tag{2.14}$$

Thus, in view of (2.14) and the fact that the point $x_0 \in \operatorname{int} \Pi$ is arbitrary, we conclude that $\varphi = 0$ a.e. on $\operatorname{int} \Pi$, i.e., condition (2.5) is satisfied. Lemma 2.2 is proved.

LEMMA 2.3. *Suppose that conditions* (1.1) *and* (1.3) *are satisfied, and suppose that condition* (2.5) *holds for the matrix* A *and the set* $\Pi \subset \partial\Omega$. *Then any function* $u \in H_{m,\mathbf{m}}(A, \Omega)$ *has on* Π *a generalized limit value* $u|_\Pi \in L^{m_*}_{\mathrm{loc}}(\Pi)$, *where* $m_* = \min(m, m_1, \dots, m_n)$. *Moreover, for each point* $x_0 \in \operatorname{int} \Pi$ *there exists a neighborhood*

$\hat{P}_{x_0} \subset \Pi$ *such that*

$$\int_{\hat{P}_{x_0}} \left| u|_{\Pi} \right|^{m_*} ds \leqslant c \int_{\hat{\omega}_{x_0}} \left(|u|^{m_*} + |A \nabla u|^{m_*} \right) dx, \tag{2.15}$$

where $\hat{\omega}_{x_0} \equiv \{x \in \mathbb{R}^n : x = y + tA^*(y)A(y)\boldsymbol{\nu}(y), y \in \hat{P}_{x_0}, t \in [0, \hat{d}_0]\}$ *(see condition* 4) *in* (2.5)) *and the constant c does not depend on the function* $u \in H_{m,\mathbf{m}}(A, \Omega)$. *The value* $u|_{\Pi}$ *is assumed by the function* $u \in H_{m,\mathbf{m}}(A, \Omega)$ *in the following sense*:

$$\lim_{\tau \to 0} \int_{\hat{P}_{x_0}} \left| u(y + \tau A^*(y)A(y)\boldsymbol{\nu}(y)) - u|_{\Pi}(y) \right|^{m_*} ds = 0. \tag{2.16}$$

PROOF. Let $u \in H_{m,\mathbf{m}}(A, \Omega)$. Then there exists a sequence $\{u_n\}$, $u_n \in \tilde{C}^1(\overline{\Omega})$, $n = 1, 2, \ldots$, such that $u_n \to u$ in H. From the proof of Lemma 2.2 it follows that for all $x_0 \in \operatorname{int} \Pi$ we have (see (2.13) with $f = u_n - u_m$, $n, m = 1, 2, \ldots$)

$$\int_{\hat{P}_{x_0}} |u_n - u_m|^{m_*} ds \leqslant c_2 \int_{\hat{\omega}_{x_0}} \left(|u_n - u_m|^{m_*} + |A \nabla (u_n - u_m)|^{m_*} \right) dx. \tag{2.17}$$

Since the right side of (2.17) tends to 0 as $n, m \to \infty$, and the point $x_0 \in \operatorname{int} \Pi$ is arbitrary, we conclude that there exists a function $\varphi \in L^{m_*}_{\text{loc}}(\Pi)$ such that $u_n \to \varphi$ in $L^{m_*}_{\text{loc}}(\Pi)$. It then follows from Lemma 2.2 that on Π the function $u \in H$ has the generalized limit value $u|_{\Pi} = \varphi \in L^{m_*}_{\text{loc}}(\Pi)$. Setting now in (2.13) (for any fixed point $x_0 \in \operatorname{int} \Pi$) $f = u_n$, $n = 1, 2, \ldots$, and passing to the limit as $n \to \infty$, we obtain (2.15). We now prove (2.16). Suppose the numbers $\delta_0 > 0$ and $\hat{d}_0 > 0$ are so small that (2.8) is a diffeomorphism between $(|y'| \leqslant \delta_0) \times (0 \leqslant t \leqslant \hat{d}_0)$ and $\hat{\omega}_{x_0} \subset \overline{\Omega}$ (see the proof of Lemma 2.2). It is obvious that for any $f \in \tilde{C}^1(\hat{\omega}_{x_0})$ and any $y \in \hat{P}_{x_0} \equiv \{y \equiv (y', \Phi(y')) : |y'| \leqslant \delta_0\}$ we have

$$\begin{aligned} |f(y + tA^*(y)A(y)\boldsymbol{\nu}(y)) - f(y)| &\leqslant \int_0^t \left| \frac{d}{d\tau} [f(y + \tau A^*(y)A(y)\boldsymbol{\nu}(y))] \right| d\tau \\ &= \int_0^t |A^*(y)A(y)\boldsymbol{\nu}(y) \cdot \nabla f(y + \tau A^*(y)A(y)\boldsymbol{\nu}(y))| d\tau \\ &= \int_0^t |A(y)\boldsymbol{\nu}(y) \cdot A(y) \nabla f(y + \tau A^*(y)A(y)\boldsymbol{\nu}(y))| d\tau \\ &\leqslant c \int_0^t |A(y) \nabla f(y + \tau A^*(y)A(y)\boldsymbol{\nu}(y))| d\tau, \qquad t \in (0, \hat{d}_0], \end{aligned} \tag{2.18}$$

where the constant c depends only on $\max_{\hat{P}_{x_0}} |a^{ij}|$, $i, j = 1, \ldots, n$. Raising (2.18) to the power m_* and integrating over $\hat{P}_{x_0}$, we obtain

$$\begin{aligned} &\int_{\hat{P}_{x_0}} |f(y + tA^*(y)A(y)\boldsymbol{\nu}(y)) - f(y)|^{m_*} ds \\ &\qquad \leqslant c_1 \int_0^t \int_{\hat{P}_{x_0}} |A(y) \nabla f(y + \tau A^*(y)A(y)\boldsymbol{\nu}(y))|^{m_*} ds\, d\tau, \qquad t \in [0, \hat{d}_0]. \end{aligned} \tag{2.19}$$

Taking account of the smoothness of the surface $\hat{P}_{x_0}$, the positivity of the Jacobian J of the transformation (2.8), and condition 5) of (2.5), we deduce from (2.19) that

$$\int_{\hat{P}_{x_0}} |f(y + tA^*(y)A(y)\boldsymbol{\nu}(y)) - f(y)|^{m_*} ds \leqslant c_2 \int_{(\hat{\omega}_{x_0})_t} |A \nabla f|^{m_*} dx, \tag{2.20}$$

where $(\hat{\omega}_{x_0})_t \equiv \{x \in \mathbb{R}^n: x = y + \tau A^*(y)A(y)\nu(y), y \in \hat{P}_{x_0}, \tau \in [0, t]\} \subset \hat{\omega}_{x_0} \subset \overline{\Omega}$, $t \in (0, \hat{d}_0]$, and c_2 depends on c_1, $\max_{|y'| \leqslant \delta_0} |\nabla\Phi|$, $\min J$, and the constant c_0 of condition 5) in (2.5).

Substituting $f = u_n$, $n = 1, 2, \ldots$, into (2.20), we obtain

$$\int_{\hat{P}_{x_0}} |u_n(y + tA^*(y)A(y)\nu(y)) - u_n(y)|^{m_*} ds$$

$$\leqslant c_2 \int_{(\hat{\omega}_{x_0})_t} |A\nabla u_n|^{m_*} dx, \qquad n = 1, 2, \ldots. \tag{2.21}$$

From the condition $u_n \to u$ in H it follows, in particular, that for some subsequence (which we denote again by $\{u_n\}$)

$$\lim_{n\to\infty} \int_{\hat{P}_{x_0}} |(u_n - u)(y + tA^*(y)A(y)\nu(y))|^{m_*} ds = 0 \quad \text{for a.e. } t \in [0, \hat{d}_0], \tag{2.22}$$

$$\lim_{n\to\infty} \int_{(\hat{\omega}_{x_0})_t} |A\nabla u_n - A\nabla u|^{m_*} dx = 0. \tag{2.23}$$

Moreover, by what has been proved above

$$\lim_{n\to\infty} \int_{\hat{P}_{x_0}} |u_n - u|_\Pi|^{m_*} ds = 0. \tag{2.24}$$

The equality (2.16) obviously follows from (2.21)–(2.24). Lemma 2.3 is proved. The following version of Lemmas 2.2 and 2.3 is proved in an entirely similar manner.

LEMMA 2.4. *Suppose conditions* (1.1) *and* (1.3) *are satisfied, and suppose that for some set* $\Pi \in \partial\Omega$ *and matrix* A *the following conditions hold*:

1) *the elements of* A *are continuous in the closure* $\overline{\omega} \subset \overline{\Omega}$ *of some n-dimensional neighborhood* ω *of the set* $\Pi \subset \partial\Omega$, *and the elements of* A^*A *are continuously differentiable there*;

2) *for some number* $d_0 > 0$ *the set* $\omega_\Pi \equiv \{x \in \mathbb{R}^n: x = y + tA^*(y)A(y)\nu(y),\ y \in \Pi,\ t \in [0, d_0]\}$ *is contained in* $\overline{\omega} \subset \overline{\Omega}$;

3) *the mapping* $x = y + tA^*(y)A(y)\nu(y)$, $y \in \Pi$, $t \in [0, d_0]$, *is a homeomorphism between* $\Pi \times [0, d_0]$ *and* ω_Π; (2.25)

4) *the set* Π *admits a parametric prescription* $y = \Phi(s')$, $s' \in S \subset \mathbb{R}^{n-1}$, *where* $\Phi \in C^2(S)$ *and* $\sup_S |\nabla\Phi| \leqslant \text{const}$;

5) *the Jacobian* J *of the transformation* $(s', t) \to x$, $s' \in S$, $t \in [0, d_0]$ *defined by the formula* $x = \Phi(s') + tA^*(\Phi(s'))A(\Phi(s'))\nu(\Phi(s'))$, $s' \in S$, $t \in [0, d_0]$, *is positive, and*

6) *there exists a constant* $c_0 > 0$ *such that, for any* $y \in \Pi$, $t \in [0, d_0]$, *and* $\xi \in \mathbb{R}^n$,

$$|A(y)\xi| \leqslant c_0 |A(y + tA^*(y)A(y)\nu(y))\xi|.$$

Then the operator of taking the limit value on Π *admits closure, any function* $u \in H_{m,\mathbf{m}}(A, \Omega)$ *has on* Π *a generalized limit value* $u|_{\Pi} \in L^{m_*}(\Pi)$, *where* $m_* = \min(m, m_1, \ldots, m_n)$, *and, moreover,*

$$\int_{\Pi} |u|_{\Pi}|^{m_*}\, ds \leqslant c \int_{\omega_\Pi} \left(|u|^{m_*} + |A\nabla u|^{m_*}\right) dx, \tag{2.26}$$

where c *does not depend on* u, *and*

$$\lim_{\tau \to 0} \int_{\Pi} |u(y + \tau A^*(y)A(y)\nu(y)) - u|_{\Pi}(y)|^{m_*}\, \partial s = 0. \tag{2.27}$$

§3. The regular and singular parts of the boundary $\partial\Omega$

Suppose that a matrix $A \equiv \|a^{ij}(x)\|$ satisfies conditions (1.1) and (1.3) in a domain $\Omega \subset \mathbb{R}^n$, $n \geqslant 2$, for some index $\mathbf{m} = (m_1, \ldots, m_n)$, $m_i \geqslant 1$, $i = 1, \ldots, n$. Let $m \geqslant 1$.

DEFINITION 3.1. A set $\Pi \subset \partial\Omega$ is called a *regular part* of the boundary (relative to the matrix A and the indices m and $\mathbf{m}$) if the operator of taking the limit value on Π defined in §2 admits closure (i.e., condition (2.3) holds for Π). A set $\mathscr{P} \subset \partial\Omega$ is called a *singular part* of the boundary $\partial\Omega$ (relative to the matrix A and the indices m and $\mathbf{m}$) if for $\mathscr{P}$ the following condition is satisfied:

$$\text{the set } \tilde{C}^1_{0,\mathscr{P}}(\overline{\Omega}) \text{ is dense in } H_{m,\mathbf{m}}(A, \Omega). \tag{3.1}$$

If condition (2.3) is satisfied for a set $\Pi \subset \partial\Omega$, i.e., if Π is a regular part of $\partial\Omega$, then any function in $H^{0,\Pi}_{m,\mathbf{m}}(A, \Omega)$ vanishes on Π in the following generalized sense: if $u \in H^{0,\Pi}_{m,\mathbf{m}}(A, \Omega)$ and the sequence $\{u_n\}$, $u_n \in \tilde{C}^1(\overline{\Omega})$, $n = 1, 2, \ldots$, converges to u in the norm (2.1) and, in addition, $u_{n|\Pi} \to \varphi$ in $L^1_{\text{loc}}(\Pi)$, then necessarily $\varphi = 0$ a.e. on Π. In particular, if a sequence $\{u_n\}$, $u_n \in \tilde{C}^1_{0,\Pi}(\overline{\Omega})$, $n = 1, 2, \ldots$, converges to a function $u \in C^1(\overline{\Omega})$ in the norm (2.1), then necessarily $u|_{\Pi} = 0$.Thus, among smooth functions in $\overline{\Omega}$ which belong to $H^{0,\Pi}_{m,\mathbf{m}}(A, \Omega)$ there are none which do not vanish identically on Π. As was shown above (see (2.5)), the regularity of a set $\Pi \subset \partial\Omega$ is ensured by the assumption of sufficient smoothness of Π and of the matrix A near Π and the nondegeneracy on Π of the vector $A\boldsymbol{\nu}$, where $\boldsymbol{\nu}$ is the unit vector of the inner normal to Π (i.e., $A\boldsymbol{\nu} \neq \mathbf{0}$ on Π). In this case any function $u \in H_{m,\mathbf{m}}(A, \Omega)$ has on Π a generalized limit value $u|_{\Pi} \in L^{m_*}_{\text{loc}}(\Pi)$. Other criteria for the regularity of Π will be given in what follows. It is obvious that the fulfillment of condition (2.3) for a set $\Pi \subset \partial\Omega$ implies the validity of the negation of condition (3.1) for Π (i.e., a regular part of $\partial\Omega$ is certainly not singular).

If condition (3.1) holds for a set $\mathscr{P} \subset \partial\Omega$, then the negation of condition (2.3) holds for $\mathscr{P}$. Indeed, suppose (3.1) is satisfied for $\mathscr{P}$, and let the sequence $\{u_n\}$, $u_n \in \tilde{C}^1_{0,\mathscr{P}}(\overline{\Omega})$, $n = 1, 2, \ldots$, converge in H to a function $u \in C^1(\overline{\Omega})$, $u \neq 0$, on $\mathscr{P}$ (the existence of such a sequence follows from (3.1)). Let $\{\tilde{u}_n\}$ be the stationary sequence with $\tilde{u}_n \equiv u$, $n = 1, 2, \ldots$ We set $v_n = u_n - \tilde{u}_n$, $n = 1, 2, \ldots$. It is obvious that $v_n \to 0$ in H, but $v_n|_{\mathscr{P}} \equiv -u|_{\mathscr{P}}$ does not tend to zero in $L^1_{\text{loc}}(\mathscr{P})$. Thus, condition (2.3) does not hold for $\mathscr{P}$. Sufficient conditions for condition (3.1) to hold for a set $\mathscr{P} \subset \partial\Omega$ are presented below.

LEMMA 3.1. *Suppose that a set* $\mathscr{P} \subset \partial\Omega$ *satisfies the following condition:*

$\mathscr{P}$ is the union of a finite or countable set of surfaces $\mathscr{P}_k$, $k = 1, 2, \ldots$, of class C^2 such that $\overline{\mathscr{P}}_k \cap \overline{\mathscr{P}}_l = \varnothing$ for $k \neq l$, and there exists a number $\delta_0 > 0$ such that for each surface $\mathscr{P}_k$ the following conditions are satisfied:

1) for each $\delta \in (0, \delta_0)$ there exists a surface with boundary $\mathscr{P}_{k,\delta}$ of class C^2 containing $\mathscr{P}_k$, a tubular half neighborhood of which $\omega_{k,\delta} \equiv \{x \in \mathbb{R}^n\colon x = y + d\boldsymbol{\nu}(y), 0 \leqslant d \leqslant \delta, y \in \mathscr{P}_{k,\delta}\}$, where $\boldsymbol{\nu}(y)$ is the unit vector of the normal to $\mathscr{P}_{k,\delta}$ varying continuously along $\mathscr{P}_{k,\delta}$ and coinciding on $\mathscr{P}_k$ with the unit vector of the unit normal to $\partial\Omega$, decomposes the domain Ω into parts $\Omega \cap \omega_{k,\delta}$ and $\Omega \setminus \omega_{k,\delta}$ such that $\overline{\Omega \cap \omega_{k,\delta}} \cap \overline{\Omega \setminus \omega_{k,\delta}} \subset \{x \in \mathbb{R}^n\colon x = y + \delta\boldsymbol{\nu}(y), y \in \mathscr{P}_{k,\delta}\}$;

2) there exists constants $c_1 > 0$ and $c_2 > 0$ (not depending on k or δ) such that $\omega_{k,\delta}$ is contained in a $c_1\delta$-neighborhood of the set $\mathscr{P}_k$ in $\mathbb{R}^n$ (i.e., for any $x \in \omega_{k,\delta}$, $\operatorname{dist}(x, \mathscr{P}_k) < c_1\delta$), and $\operatorname{meas}_n \omega_{k,\delta} \leqslant c_2\delta$; (3.2)

3) for any point $x \in \omega_{k,\delta}$ there exists a unique point $y \equiv y(x) \in \mathscr{P}_{k,\delta}$ for which $\operatorname{dist}(x, \mathscr{P}_{k,\delta}) = \operatorname{dist}(x, y)$;

4) $\omega_{k,\delta} \cap \omega_{l,\delta} = \varnothing$ for any $k \neq l$ and all $\delta \in (0, \delta_0)$;

5) in $\omega_{k,\delta} \cap \overline{\Omega}$ the functions $x \to A_i(x)\boldsymbol{\nu}(y(x))$, $i = 1, \ldots, n$, satisfy the Hölder conditions

$$|A_i(x)\boldsymbol{\nu}(y(x)) - A_i(x')\boldsymbol{\nu}(y(x'))| \leqslant c_3|x - x'|^{\alpha_i},$$

where $\alpha_i > 1/m_i'$, $1/m_i + 1/m_i' = 1$, $i = 1, \ldots, n$, and the constants c_3 and α_i do not depend on k or δ;

6) $A_i(y)\boldsymbol{\nu}(y) = 0$ on $\mathscr{P}_k$, $i = 1, \ldots, n$.

Then $\mathscr{P}$ is a singular part of $\partial\Omega$.

PROOF. To prove the density of $\tilde{C}^1_{0,\mathscr{P}}(\overline{\Omega})$ in $H_{m,\mathbf{m}}(A, \Omega)$ (see (3.1)) it suffices to show that for any $u \in \tilde{C}^1(\overline{\Omega})$ there is a sequence $\{u_n\}$, $u_n \in \tilde{C}^1_{0,\mathscr{P}}(\overline{\Omega})$, $n = 1, 2, \ldots$, converging to u in H, since $\tilde{C}^1(\overline{\Omega})$ is dense in H. We shall first assume that $k = 1$ in (3.2). In this case in place of $\mathscr{P}_1$ and $\mathscr{P}_{1\delta}$ we use the notation $\mathscr{P} \equiv \mathscr{P}_1$ and $\mathscr{P}_\delta \equiv \mathscr{P}_{1,\delta}$. We also set $\omega_\delta \equiv \{x \in \Omega\colon x = y + d\boldsymbol{\nu}(y),\ y \in \mathscr{P}_\delta,\ 0 \leqslant d \leqslant \delta\}$ and $\omega_{\delta/2,\delta} = \omega_\delta \setminus \omega_{\delta/2}$, $0 < \delta < \delta_0$. We put

$$\zeta_\delta(x) = \begin{cases} 0, & x \in \omega_{\delta/2}, \\ 2[d(x) - \delta/2]/\delta, & x \in \omega_{\delta/2,\delta}, \\ 1, & x \in \Omega \setminus \omega_\delta, \end{cases} \tag{3.3}$$

where $\delta \in (0, \delta_0)$, and $d(x) \equiv \operatorname{dist}(x, \mathscr{P}_\delta)$ is a function defined on ω_δ. From (3.2) and the equality $\nabla d(x) = \boldsymbol{\nu}(y(x))$, $x \in \omega_\delta$, it follows that $\zeta_\delta \in \tilde{C}^1_{0,\mathscr{P}}(\overline{\Omega})$, and

$$\nabla\zeta_\delta(x) = \begin{cases} 0, & x \in \omega_{\delta/2} \cup (\Omega \setminus \omega_\delta), \\ (2/\delta)\boldsymbol{\nu}(y(x)), & x \in \omega_{\delta/2,\delta}, \end{cases} \tag{3.4}$$

where $y(x)$ is the point on $\mathscr{P}_\delta$ closest to x. For the chosen function $u \in \tilde{C}^1(\overline{\Omega})$ we set $u_\delta(x) = u(x)\zeta_\delta(x)$. It is obvious that $u_\delta \in \tilde{C}^1_{0,\mathscr{P}}(\overline{\Omega})$. We compute

$$\begin{aligned}\|u - u_\delta\|_H &= \|u(1-\zeta_\delta)\|_{m,\Omega} + \sum_{i=1}^n \|A_i[\nabla u(1-\zeta_\delta) - u\nabla\zeta_\delta]\|_{m_i,\Omega} \\ &\leqslant \|u\|_{m,\omega_\delta} + \sum_{i=1}^n \|A_i\nabla u\|_{m_i,\omega_\delta} + \sum_{i=1}^n \|uA_i\nabla\zeta_\delta\|_{m_i,\omega_{\delta/2,\delta}},\end{aligned} \tag{3.5}$$

where we have noted that $\zeta_\delta = 1$ outside ω_δ. In view of (3.4), for each $i \in \{1,\dots,n\}$ we have

$$\|uA_i\nabla\zeta_\delta\|^{m_i}_{m_i,\omega_{\delta/2,\delta}} = \left(\frac{2}{\delta}\right)^{m_i}\int_{\omega_{\delta/2,\delta}} |A_i(x)\nu(y(x))|^{m_i}|u(x)|^{m_i}\,dx. \tag{3.6}$$

Because of condition 2) in (3.2), for any $x \in \omega_{\delta/2,\delta}$ there is a point $z \in \mathscr{P}$ such that $\operatorname{dist}(x, z) \leqslant C_1\delta$. From conditions 2), 5), and 6) of (3.2) it then follows that

$$|A_i(x)\nu(y(x))| \leqslant c_3(c_1\delta)^{\alpha_i}, \qquad i = 1,\dots,n. \tag{3.7}$$

Taking further into account that $\operatorname{meas}_n \omega_{\delta/2,\delta} \leqslant c_2\delta$ (see condition 2) of (3.2)), we obtain

$$\left(\frac{2}{\delta}\right)^{m_i}\int_{\omega_{\delta/2,\delta}} |A_i(x)\nu(y(x))|^{m_i}|u(x)|^{m_i}\,dx \leqslant c\delta^{m_i(-1+\alpha_i+1/m_i)} = c\delta^{m_i\varepsilon_i}, \tag{3.8}$$

where $\varepsilon_i = \alpha_i - 1/m_i' > 0$ and the constant c does not depend on δ. From (3.5)–(3.8) it obviously follows that $u_\delta \to u$ in H as $\delta \to 0$. Choosing an arbitrary sequence $\{\delta_n\}$, $\delta_n \in (0, \delta_n)$, $\delta_n \to 0$, we conclude that $u_{\delta_n} \to u$ in H, where $u_{\delta_n} \in \tilde{C}^1_{0,\mathscr{P}}(\overline{\Omega})$. Thus, $\tilde{C}^1_{0,\mathscr{P}}(\overline{\Omega})$ is dense in H. Suppose now that k in (3.2) is arbitrary. Taking condition 4) in (3.2) into account and denoting by $\zeta^{(k)}_\delta$ the function defined by a formula of the form (3.3) for the component $\mathscr{P}_k$ of $\mathscr{P}$, we set

$$\zeta_\delta(x) = \prod_{k=1,2,\dots} \zeta^{(k)}_\delta(x). \tag{3.9}$$

We observe that at each fixed point $x \in \Omega$ only one of the factors in (3.9) can be different from 1. It is obvious that $\zeta_\delta \in \tilde{C}^1_{0,\mathscr{P}}(\overline{\Omega})$, and $\zeta_\delta = 1$ in $\Omega \setminus \bigcup_{k=1,2,\dots}\omega_{k,\delta}$ where the sets $\omega_{k,\delta}$ are defined in a manner analogous to the sets ω_δ (see (3.2)). For a fixed function $u \in \tilde{C}^1_{0,\mathscr{P}}(\overline{\Omega})$ we set

$$u_\delta(x) = u(x)\zeta_\delta(x). \tag{3.10}$$

It is obvious that $u_\delta \in \tilde{C}^1_{0,\mathscr{P}}(\overline{\Omega})$. Arguing exactly as in the case $k = 1$, we establish that

$$u_\delta \to u \quad \text{in } H, \tag{3.11}$$

whence it follows that $\tilde{C}^1_{0,\mathscr{P}}(\overline{\Omega})$ is dense in H. Lemma 3.1 is proved.

COROLLARY 3.1. *If the conditions of Lemma 3.1 are satisfied, the spaces $H^{0,\mathscr{P}}_{m,\mathbf{m}}(A,\Omega)$ and $H_{m,\mathbf{m}}(A,\Omega)$ coincide. In particular, the set $\tilde{C}^1(\overline{\Omega})$ is (densely) contained in $H^{0,\mathscr{P}}_{m,\mathbf{m}}(A,\Omega)$.*

PROOF. Assertions of Corollary 3.1 follow in an obvious way from the density of $\tilde{C}^1_{0,\mathscr{P}}(\overline{\Omega})$ in $H^{0,\mathscr{P}}_{m,\mathbf{m}}(A,\Omega)$ and the definition of the spaces $H^{0,\Gamma}_{m,\mathbf{m}}(A,\Omega)$.

§4. Some imbedding theorems

In this section we present anisotropic analogues of some imbedding theorems of S. L. Sobolev and also some multiplicative inequalities first established by V. P. Il'in [57]. For convenience of the reader a brief presentation of the proofs of these results is given; the results are established within a framework sufficient for our needs.

The following result, related to the question of equivalent norms in the space $W_p^1(\Omega)$, is well known (see, for example, [8]).

LEMMA 4.1. *Let Ω be a bounded strongly Lipschitz domain in $\mathbb{R}^n$, $n \geq 2$. Then for any $p \geq 1$ and any function $u \in W_p^1(\Omega)$*

$$\|u\|_{p,\Omega} \leqslant c_1\left(\sum_{i=1}^{n} \|u_{x_i}\|_{p,\Omega} + \|u\|_{1,\Omega}\right), \tag{4.1}$$

where c_1 depends only on n, p, and Ω. Suppose further that Γ is a subset of positive $(n-1)$-dimensional measure on $\partial\Omega$. Then for any $p \geqslant 1$ and any function $u \in W_p^1(\Omega)$

$$\|u\|_{p,\Omega} \leqslant c_2\left(\sum_{i=1}^{n} \|u_{x_i}\|_{p,\Omega} + \int_\Gamma |u|\,ds\right), \tag{4.2}$$

where c_2 depends only on n, p, Ω, and Γ.

We denote by $W_{p,\mathbf{q}}^1(\Omega)$, $p \geqslant 1$, $\mathbf{q} = (q_1, \dots, q_n)$, $q_i \geqslant 1$, the Banach space of functions $u \in L^p(\Omega)$ having generalized derivatives $u_{x_i} \in L^{q_i}(\Omega)$, $i = 1, \dots, n$, with the norm

$$\|u\|_{W_{p,\mathbf{q}}^1(\Omega)} = \|u\|_{p,\Omega} + \|\nabla u\|_{\mathbf{q},\Omega} \equiv \|u\|_{p,\Omega} + \sum_{i=1}^{n} \|u_{x_i}\|_{q_i,\Omega}. \tag{4.3}$$

For the same indices p and $\mathbf{q}$ as in (4.3) we denote by $H_{p,\mathbf{q}}(\Omega)$ the closure of $\tilde{C}^1(\overline{\Omega})$ in the norm (4.3) (see the basic notation). If the domain Ω is strongly Lipschitz, then the spaces $H_{p,\mathbf{q}}(\Omega)$ and $W_{p,\mathbf{q}}^1(\Omega)$ are known to coincide.

We denote by $\tilde{H}_{\mathbf{q}}(\Omega)$, $\mathbf{q} = (q_1, \dots, q_n)$, $q_i \geqslant 1$, $i = 1, \dots, n$, the closure of $\tilde{C}^1(\Omega)$ in the norm

$$\|u\|_{\tilde{H}_{\mathbf{q}}(\Omega)} = \|u\|_{1,\Omega} + \|\nabla u\|_{\mathbf{q},\Omega}. \tag{4.4}$$

From what has been said above it follows that for a strongly Lipschitz domain Ω the space $\tilde{H}_{\mathbf{q}}(\Omega)$ coincides with $W_{1,\mathbf{q}}^1(\Omega)$. From the imbedding theorems established below it follows that $\tilde{H}_{\mathbf{q}}(\Omega)$ coincides with $W_{p,\mathbf{q}}^1(\Omega)$ for all $p \in [1, \hat{l}]$, where $\hat{l}$ is a certain limit index defined as a function of $\mathbf{q}$ and n.

LEMMA 4.2. *Suppose numbers $q_i \geqslant 1$, $i = 1, \dots, n$, and $s \geqslant 1$ are fixed, and that for some $\alpha \in [0, 1]$ the index l satisfies the conditions*

$$\frac{1}{l} = \frac{\alpha}{\hat{l}} + \frac{1-\alpha}{s},$$

$$\hat{l} = \frac{n}{\sum_{i=1}^n 1/q_i - 1} \quad \textit{for} \ \sum_{i=1}^{n} \frac{1}{q_i} > 1,\ n \geqslant 2,$$

$$\hat{l} \in (2, +\infty) \quad \textit{for} \ \sum_{i=1}^{n} \frac{1}{q_i} = 1,\ n \geqslant 2,$$

$$\hat{l} \in (2, +\infty] \quad \textit{for} \ \sum_{i=1}^{n} \frac{1}{q_i} < 1,\ n \geqslant 2 \ \textit{and for}\ n = 1. \tag{4.5}$$

Then for any function $u \in \tilde{C}_0^1(\Omega)$, $\Omega \subset \mathbb{R}^n$, $n \geqslant 1$,

$$\|u\|_{l,\Omega} \leqslant c\|\nabla u\|_{\mathbf{q},\Omega}^{\alpha}\|u\|_{s,\Omega}^{1-\alpha}, \tag{4.6}$$

where $\|\nabla u\|_{\mathbf{q},\Omega} \equiv \sum_{i=1}^n \|u_{x_i}\|_{q_i,\Omega}$ *and the constant c in* (4.6) *depends only on* n, $\mathbf{q}$, s, *and* α *for* $\sum_1^n 1/q_i > 1$, *and on* n, l, s, α *and* Ω *for* $\sum_1^n 1/q_i \leqslant 1$.

PROOF. For any function $u \in \tilde{C}_0^1(\Omega)$

$$\|u\|_{l,\Omega} \leqslant \|u\|_{\hat{l},\Omega}^{\alpha}\|u\|_{s,\Omega}^{1-\alpha}, \qquad 1/l = \alpha/\hat{l} + (1-\alpha)/s. \tag{4.7}$$

We now use the estimate

$$\|u\|_{\hat{l},\Omega} \leqslant \left(\prod_{i=1}^n h_i\right)^{1/n} \|\nabla u\|_{\mathbf{q},\Omega}, \tag{4.8}$$

where $h_i = 1 + \hat{l}/q_i'$, $1/q_i + 1/q_i' = 1$, $i = 1,\dots,n$. (4.8) follows, for example, from the arguments of [87]. It is a consequence of the inequality

$$\left(\int_{\mathbb{R}^n}\prod_{i=1}^n |u_i|^{1/(n-1)}\,dx\right)^{n-1} \leqslant \prod_{i=1}^n\left(\int_{\mathbb{R}^n}\left|\frac{\partial u_i}{\partial x_i}\right|dx\right), \tag{4.9}$$

which is valid for any functions $u_i \in \tilde{C}_0^1(\mathbb{R}^n)$, $i = 1,\dots,n$ (see [87], p. 25). Indeed, if, for example, $n \geqslant 2$ and $\sum_1^n 1/q_i > 1$, then, considering the function u to be extended by zero to all of $\mathbb{R}^n$, setting $u_i = |u|^{h_i}$, $i = 1,\dots,n$, and taking into account that $\sum_1^n h_i = \hat{l}(n-1)$, from (4.9) we deduce that

$$\left(\int_{\mathbb{R}^n}|u|^{\hat{l}}\,dx\right)^{n-1} \leqslant \prod_{i=1}^n h_i\int_{\mathbb{R}^n}|u|^{\hat{l}/q_i'}\left|\frac{\partial u}{\partial x_i}\right|dx. \tag{4.10}$$

Applying the Hölder inequality and then raising both sides of the inequality so obtained to the power $1/\hat{l}(n-1)$, we obtain

$$\|u\|_{\hat{l},\Omega} \leqslant \left(\prod_{i=1}^n h_i\right)^{1/\hat{l}(n-1)} \|u\|_{\hat{l},\Omega}^{(\sum_{i=1}^n 1/q_i')/(n-1)}\|\nabla u\|_{\mathbf{q},\Omega}^{n/\hat{l}(n-1)}. \tag{4.11}$$

Taking into account that $1 - (\sum_1^n 1/q_i')/(n-1) = n/\hat{l}(n-1)$, from (4.11) we deduce (4.8). In the case $\sum_1^n 1/q_i \leqslant 1$, $n \geqslant 2$, (4.8) is established in a similar way. Substituting into (4.7) in place of $\|u\|_{\hat{l},\Omega}$ the right side of (4.8), we obtain (4.6) with a constant $c = (\prod_1^n h_i)^{\alpha/n}$. This proves Lemma 4.2 for $n \geqslant 2$. In the case $n = 1$ inequality (4.6) obviously follows from (4.7) and the estimate

$$\max_{\mathbb{R}}|u| \leqslant \int_{\mathbb{R}}\left|\frac{du}{dx}\right|dx, \quad \forall u \in C_0(\mathbb{R}).$$

Thus, Lemma 4.2 may be considered proved.

REMARK 4.1. Under the conditions of Lemma 4.2,

$$\|u\|_{l,\Omega} \leqslant c\left(\prod_{i=1}^n \|u_{x_i}\|_{q_i,\Omega}^{\alpha/n}\right)\|u\|_{s,\Omega}^{1-\alpha}. \tag{4.6$'$}$$

Indeed, together with (4.8) we obviously also have

$$\|u\|_{\hat{l},\Omega} \leqslant \prod_{i=1}^n h_i^{1/n}\|u_{x_i}\|_{q_i,\Omega}^{1/n}. \tag{4.8$'$}$$

Then (4.6′) follows from (4.7) and (4.8′).

REMARK 4.2. We denote by $\overset{\circ}{\tilde{H}}_{\mathbf{q}}(\Omega)$ the closure of $\tilde{C}_0^1(\Omega)$ in the norm $\|\cdot\|_{\tilde{H}_{\mathbf{q}}(\Omega)}$. From Lemma 4.2 it then follows easily that if condition (4.5) is satisfied any function $u \in \overset{\circ}{\tilde{H}}_{\mathbf{q}}(\Omega)$ belongs to $L^l(\Omega)$ and satisfies (4.6).

LEMMA 4.3. *Suppose the bounded, strongly Lipschitz domain $\Omega \subset \mathbb{R}^n$, $n \geqslant 2$, satisfies the weak $1/\lambda$-horn condition* ([8], §8) *for $\lambda = (\lambda_1, \ldots, \lambda_n)$, $\lambda_i = 1 - \sum_{j=1}^n 1/q_j + n/q_i > 0$, $q_i > 1$, $i = 1, \ldots, n$. Let*

$$\hat{l} = \frac{n}{\sum 1/q_i - 1} \quad \text{for } \sum_{i=1}^n \frac{1}{q_i} > 1, \qquad \hat{l} \in (2, \infty) \quad \text{for } \sum_{i=1}^n \frac{1}{q_i} = 1,$$
$$\hat{l} \in (2, \infty] \quad \text{for } \sum_{i=1}^n \frac{1}{q_i} < 1. \tag{4.12}$$

Then there is the imbedding $\tilde{H}_{\mathbf{q}}(\Omega) \to L^{\hat{l}}(\Omega)$. In particular, for any function $u \in \tilde{H}_{\mathbf{q}}(\Omega)$

$$\|u\|_{\hat{l},\Omega} \leqslant c\|u\|_{\tilde{H}_{\mathbf{q}}(\Omega)}, \tag{4.13}$$

where the constant c depends only on n, $\hat{l}$, $\mathbf{q}$, and Ω. If the indices l and $\mathbf{q}$ satisfy condition (4.5) *for some $\alpha \in (0, 1)$ and $s \in [1, \hat{l})$, then there is a compact imbedding of $\tilde{H}_{\mathbf{q}}(\Omega)$ in $L^l(\Omega)$, and the inequality*

$$\|u\|_{l,\Omega} \leqslant c\|u\|^{\alpha}_{\tilde{H}_{\mathbf{q}}(\Omega)}\|u\|^{1-\alpha}_{s,\Omega}, \quad \forall u \in \tilde{H}_{\mathbf{q}}(\Omega), \tag{4.14}$$

holds with the constant c depending only on n, $\mathbf{q}$, α, s, and Ω.

PROOF. Applying a familiar method (see [8], [57] and [87]), we extend the given function $u \in \tilde{H}_{\mathbf{q}}(\Omega)$ to a larger domain $\hat{\Omega} \supset \Omega$ in such a way that the extended function $\hat{u}$ belongs to $\overset{\circ}{\tilde{H}}_{\mathbf{q}}(\hat{\Omega})$ and satisfies the inequalities

$$\|\hat{u}\|_{\tilde{H}_{\mathbf{q}}(\hat{\Omega})} \leqslant c_1\|u\|_{\tilde{H}_{\mathbf{q}}(\Omega)}, \qquad \|\hat{u}\|_{s,\hat{\Omega}} \leqslant c_2\|u\|_{s,\Omega} \tag{4.15}$$

with constants c_1 and c_2 not depending on u. Using Remark 4.2 and the compactness of the imbedding of $H_1(\Omega)$ in $L^1(\Omega)$, it is easy to establish Lemma 4.3.

Let Γ be a fixed subset of positive $(n\text{-}1)$-dimensional surface measure on $\partial\Omega$. We denote by $\widetilde{H_{\mathbf{q}}^{0,\Gamma}}(\Omega)$ the closure of the set $\tilde{C}^1_{0,\Gamma}(\Omega)$ (see the basic notation) in the norm

$$\|u\|_{\widetilde{H_{\mathbf{q}}^{0,\Gamma}}(\Omega)} \equiv \|\nabla u\|_{\mathbf{q},\Omega}. \tag{4.16}$$

From Lemma 4.1 (see (4.2)) it follows that (4.16) actually defines a norm and that this norm is equivalent to $\|u\|_{W^1_{q_*,\mathbf{q}}(\Omega)}$, where $q_* = \min(q_1, \ldots, q_n)$. It is obvious that $\widetilde{H_{\mathbf{q}}^{0,\Gamma}}(\Omega) \subset \tilde{H}_{\mathbf{q}}(\Omega)$. We remark that in the case $\Gamma = \partial\Omega$ the space $\widetilde{H_{\mathbf{q}}^{0,\partial\Omega}}(\Omega)$ coincides with the space $\overset{\circ}{\tilde{H}}_{\mathbf{q}}(\Omega)$ defined in Remark 4.2. We further denote by $H_{p,\mathbf{q}}^{0,\Gamma}(\Omega)$ the closure of $\tilde{C}^1_{0,\Gamma}(\Omega)$ in the norm (4.3). It follows from the next lemma that in the case of a strongly Lipschitz domain Ω the space $\widetilde{H_{\mathbf{q}}^{0,\Gamma}}(\Omega)$ is isomorphic to $H_{p,\mathbf{q}}^{0,\Gamma}(\Omega)$ for all $p \leqslant \hat{l}$, where $\hat{l}$ is the same limit index as in Lemma 4.3.

LEMMA 4.4. *Suppose Ω satisfies the same conditions as in Lemma 4.3, and let the index $\hat{l}$ be defined by* (4.12). *Then there is the imbedding $\widetilde{H_{\mathbf{q}}^{0,\Gamma}}(\Omega) \to L^{\hat{l}}(\Omega)$. In particular, for any function $u \in \widetilde{H_{\mathbf{q}}^{0,\Gamma}}(\Omega)$*

$$\|u\|_{\hat{l},\Omega} \leqslant c\|\nabla u\|_{\mathbf{q},\Omega}, \tag{4.17}$$

where the constant c depends on n, $\hat{l}$, $\mathbf{q}$, and Ω. If the indices l and $\mathbf{q}$ satisfy (4.5) *for some $\alpha \in (0, 1)$ and $s \in [1, \hat{l})$, then there is a compact imbedding of $\widetilde{H_{\mathbf{q}}^{0,\Gamma}}(\Omega)$ in $L^l(\Omega)$,* *and*

$$\|u\|_{l,\Omega} \leqslant c\left(\|\nabla u\|_{\mathbf{q},\Omega}\right)^{\alpha}\|u\|_{s,\Omega}^{1-\alpha}, \tag{4.18}$$

where the constant c depends on n, l, $\mathbf{q}$, *s, and* Ω. *In the case* $\Gamma \in \partial\Omega$ *the constants in* (4.17) *and* (4.18) *do not depend on* Ω.

PROOF. Lemma 4.4 follows in an obvious way from Lemmas 4.3 and 4.1.

REMARK 4.3. Whenever imbeddings of $\widetilde{H_{\mathbf{q}}^{0,\Gamma}}(\Omega)$ (in particular $\widetilde{H_{\mathbf{q}}^{0,\phi}}(\Omega) \equiv \tilde{H}_{\mathbf{q}}(\Omega)$) in $L^{\hat{l}}(\Omega)$ or $L^{l}(\Omega)$ are used in the sequel, it is implicitly assumed that the domain Ω satisfies the weak $1/\lambda$-horn condition ([8], §8), where

$$\lambda = (\lambda_1,\dots,\lambda_n), \lambda_i = 1 - \sum_{j=1}^{n} 1/q_j + n/q_i > 0, q_i > 1, i = 1,\dots,n. \tag{4.19}$$

LEMMA 4.5. *Let* Ω *be a bounded, strongly Lipschitz domain in* $\mathbb{R}^n$, $n \geqslant 2$. *Then*

$$\tilde{H}_{\mathbf{q}}(\Omega) \to L^r(\partial\Omega), \tag{4.20}$$

where the index $r \geqslant 1$ *is determined from the relations*

$$\begin{gathered}(n-1)/r = n/q_* - 1, \quad \text{if } q_* = \min(q_1,\dots,q_n) < n,\\ r \in [1, +\infty) \quad \text{for } q_* \geqslant n.\end{gathered} \tag{4.21}$$

PROOF. Lemma 4.5 follows from the imbedding $\tilde{H}_{\mathbf{q}}(\Omega) \to W^1_{\mathbf{q}_*}(\Omega)$ and Sobolev's imbedding theorem.

LEMMA 4.6. *Let* $\Omega \subset \mathbb{R}^n$, $n \geqslant 1$, *be a bounded, strongly Lipschitz domain, and let* $\gamma \subset \partial\Omega$ *be a Lebesgue-measurable set on* $\partial\Omega$ *with either* $\operatorname{meas}_{n-1}\gamma > 0$ *or* $\gamma = \varnothing$. *Let the functions* $\psi_k(x)$, $k = 1,2,\dots$, *form an orthonormal basis in* $L^2(\Omega)$, *and suppose that the indices* $q_1,\dots,q_n$, $q_i \geqslant 1$, $i = 1,\dots,n$, *satisfy the condition*

$$\sum_{i=1}^{n} \frac{1}{q_i} < \frac{n+2}{2} \quad \text{for } n \geqslant 2,$$

ensuring compactness of the imbedding of $\widetilde{H_{\mathbf{q}}^{0,\gamma}}(\Omega)$ *in* $L^2(\Omega)$. *Then for any* $\varepsilon > 0$ *there is a number* N_ε *such that for any function* $u \in \widetilde{H_{\mathbf{q}}^{0,\gamma}}(\Omega)$

$$\|u\|_{2,\Omega} \leqslant \left(\sum_{k=1}^{N_\varepsilon} (u, \psi_k)^2\right)^{1/2} + \varepsilon\|u\|_{\widetilde{H_{\mathbf{q}}^{0,\gamma}}(\Omega)}, \tag{4.22}$$

where the number N_ε *does not depend on u.*

PROOF. Condition (4.21) can be rewritten in the form

$$\hat{l} > 2 \qquad \text{for } \sum_{i=1}^{n} \frac{1}{q_i} > 1, \quad n \geqslant 2, \tag{4.23}$$

where $\hat{l}$ is defined as a function of $\mathbf{q}$ and n in (4.5). From Lemma 4.3 with $s = 1$, $\alpha = 1/2(1 - 1/\hat{l})^{-1} \in (0,1)$ and $l = 2$ it then actually follows that the imbedding $\widetilde{H_{\mathbf{q}}^{0,\gamma}}(\Omega) \to L^2(\Omega)$ is compact. The remainder of the proof of Lemma 4.6 is the same as that of Lemma 6.1 of [80] (Chapter V, §6), where the case $\mathbf{q} = (q,\dots,q)$ is considered.

§5. Some imbedding theorems for functions depending on time

In this section we consider functions depending on the independent variables x and t, where $x = (x_1, \ldots, x_n)$ and $t \in \mathbb{R}$. The variables $x_1, \ldots, x_n$ are called the *spatial variables*, and the variable t is called the *time*. We denote by $L^{p,p_0}(Q)$, where $p \geqslant 1$, $p_0 \geqslant 1$, $Q = \Omega \times [T_1, T_2]$, $\Omega \subset \mathbb{R}^n$, $n \geqslant 1$ and $T_1, T_2 \in \mathbb{R}$, the Banach space of all measurable functions in Q with norm

$$\|u\|_{p,p_0,Q} \equiv \left(\int_{T_1}^{T_2}\left(\int_\Omega |u|^p\,dx\right)^{p_0/p} dt\right)^{1/p_0}. \tag{5.1}$$

In particular, for $p_0 = p$ the space $L^{p,p_0}(Q)$ coincides with $L^p(Q)$.

LEMMA 5.1. *If a function* $u \in L^{q,q_0}(Q) \cap L^{s,s_0}(Q)$, *then* $u \in L^{p,p_0}(Q)$ *and*

$$\|u\|_{p,p_0,Q} \leqslant \|u\|_{q,q_0,Q}^{\kappa}\|u\|_{s,s_0,Q}^{1-\kappa} \tag{5.2}$$

provided that

$$1/p = \kappa/q + (1-\kappa)/s, \qquad 1/p_0 = \kappa/q_0 + (1-\kappa)/s_0, \qquad \kappa \in [0,1]. \tag{5.3}$$

PROOF. Lemma 5.1 is a well-known fact which is established by twofold application of Hölder's inequality.

We denote by $W^{0;1}_{p,p_0;\mathbf{q},\mathbf{q}_0}(Q)$, where $p \geqslant 1$, $p_0 \geqslant 1$, $\mathbf{q} = (q_1, \ldots, q_n)$, $\mathbf{q}_0 = (q_{01}, \ldots, q_{0n})$, $q_i \geqslant 1$ and $q_{0i} \geqslant 1$, $i = 1, \ldots, n$, the Banach space with elements which are functions $u \in L^{p,p_0}(Q)$ having generalized spatial derivatives $u_{x_i} \in L^{q_i,q_{0i}}(Q)$ (in the sense of Sobolev), $i = 1, \ldots, n$, and the norm of the form

$$\|u\|_{W^{0;1}_{p,p_0;\mathbf{q},\mathbf{q}_0}(Q)} = \|u\|_{p,p_0,Q} + \|\nabla u\|_{\mathbf{q},\mathbf{q}_0,Q}, \tag{5.4}$$

where

$$\|\nabla u\|_{\mathbf{q},\mathbf{q}_0,Q} \equiv \sum_{i=1}^{n} \|u_{x_i}\|_{q_i,q_{0i},Q}.$$

In particular, for $p = p_0$ and $\mathbf{q} = \mathbf{q}_0$ we denote the corresponding space by $W^{0;1}_{p,\mathbf{q}}(Q)$. If the base Ω of the cylinder Q is a bounded, strongly Lipschitz domain in $\mathbb{R}^n$, then $\overset{\circ}{W}_{p,p_0;\mathbf{q},\mathbf{q}_0}(Q)$ coincides with the closure of $\tilde{C}^1(Q)$ in the norm (5.4). We denote the latter space by $H_{p,p_0;\mathbf{q},\mathbf{q}_0}(Q)$.

Let $\Gamma = \gamma \times (T_1, T_2)$, $\gamma \subset \partial\Omega$, where we henceforth always assume that either $\operatorname{meas}_{n-1}\gamma > 0$ or $\gamma = \varnothing$. We denote by $\widetilde{H^{0,\Gamma}_{\mathbf{q},\mathbf{q}_0}}(Q)$ the closure of the set $\tilde{C}^1_{0,\Gamma}(Q)$ of all functions in $\tilde{C}^1(Q)$ which are equal to 0 outside some n-dimensional neighborhood of Γ in the norm

$$\|u\|_{\tilde{H}_{\mathbf{q},\mathbf{q}_0}(Q)} = \|u\|_{2,\infty,Q} + \|\nabla u\|_{\mathbf{q},\mathbf{q}_0,Q}, \tag{5.5}$$

where $\|u\|_{2,\infty,Q} = \operatorname{ess\,sup}_{t\in[T_1,T_2]}\|u\|_{2,\Omega}$. In the case $\gamma = \partial\Omega$ we denote this Banach space by $\overset{\circ}{H}_{\mathbf{q},\mathbf{q}_0}(Q)$, while in the case $\gamma = \varnothing$ we denote it by $\tilde{H}_{\mathbf{q},\mathbf{q}_0}(Q)$. It is obvious that $\tilde{H}_{\mathbf{q},\mathbf{q}_0}(Q)$ coincides with $H_{2,\infty;\mathbf{q},\mathbf{q}_0}(Q)$. From the imbedding theorems presented below it will follow that $\widetilde{H^{0,\Gamma}_{\mathbf{q},\mathbf{q}_0}}(Q)$ is imbedded in $H_{p,p_0;\mathbf{q},\mathbf{q}_0}(Q)$ for particular pairs of indices p, p_0.

In the case $\mathbf{q} = \mathbf{q}_0$ we denote $\widetilde{H^{0,\Gamma}_{\mathbf{q},\mathbf{q}_0}}(Q)$ by $\widetilde{H^{0,\Gamma}_{\mathbf{q}}}(Q)$. Analogous simplifications of the notation for spaces connected with "double norms" of the type (5.1) will be used below without special mention. For example, for $p = p_0$ the space $H_{p,p_0;\mathbf{q},\mathbf{q}_0}(Q)$ is denoted by $H_{p;\mathbf{q},\mathbf{q}_0}(Q)$, etc.

LEMMA 5.2. *Suppose that numbers $q_i \geqslant 1$, $q_{0i} \geqslant 1$, $i = 1,\dots,n$, $p \in [1, +\infty]$, $p_0 \in [1, +\infty]$, $r \in [1, +\infty]$ and $r_0 \in [1, +\infty]$ are fixed, and suppose that for any $\alpha \in [0,1]$ and $\beta \in [0,1]$ the indices l and l_0 satisfy the conditions*

$$\frac{1}{l} = \frac{\alpha}{\hat{l}} + \frac{(1-\alpha)\beta}{p} + \frac{(1-\alpha)(1-\beta)}{r},$$

$$\frac{1}{l_0} = \frac{\alpha}{\hat{l}_0} + \frac{(1-\alpha)\beta}{p_0} + \frac{(1-\alpha)(1-\beta)}{r_0},$$

$$\hat{l} = \frac{n}{\sum_{i=1}^n 1/q_i - 1} \quad \text{for } \sum_{i=1}^n \frac{1}{q_i} > 1,\ n \geqslant 2,$$

$$\hat{l} \in (2, +\infty) \quad \text{for } \sum_{i=1}^n \frac{1}{q_i} = 1,\ n \geqslant 2, \tag{5.6}$$

$$\hat{l} \in (2, +\infty] \quad \text{for } \sum_{i=1}^n \frac{1}{q_i} < 1,\ n \geqslant 2 \quad \text{and} \quad \hat{l}_0 = n\left(\sum_{i=1}^n \frac{1}{q_{0i}}\right)^{-1} \quad \text{for } n = 1.$$

Then for any function $u \in \tilde{C}^1_{0,\partial\Omega\times(T_1,T_2)}(Q)$

$$\|u\|_{l,l_0,Q} \leqslant c\|\nabla u\|^{\alpha}_{\mathbf{q},\mathbf{q}_0,Q}\|u\|^{(1-\alpha)\beta}_{p,p_0,Q}\|u\|^{(1-\alpha)(1-\beta)}_{r,r_0,Q}, \tag{5.7}$$

where c depends only on n, $\mathbf{q}$, $\mathbf{q}_0$, α, and β for $\sum_1^n 1/q_i > 1$ and on n, $\hat{l}$, $\hat{l}_0$, α, β, and Ω for $\sum_1^n 1/q_i \leqslant 1$.

PROOF. Let $u \in \tilde{C}^1_{0,\partial\Omega\times(T_1,T_2)}(Q)$. In view of Remark 4.1, for all $t \in [T_1, T_2]$ inequality (4.6) holds for some $s \geqslant 1$. We raise both sides of this inequality to the power l_0 and integrate with respect to t from T_1 to T_2. Applying also the inequality

$$\|u\|_{s,\Omega} \leqslant \|u\|^{\beta}_{p,\Omega}\|u\|^{1-\beta}_{r,\Omega}, \qquad \frac{1}{s} = \frac{\beta}{p} + \frac{1-\beta}{r}, \qquad \beta \in [0,1], \tag{5.8}$$

we then obtain

$$\begin{aligned}\int_{T_1}^{T_2}\|u\|^{l_0}_{l,\Omega}\,dt &\leqslant c\int_{T_1}^{T_2}\left(\prod_{i=1}^n \|u_{x_i}\|^{\alpha l_0/n}_{q_i,\Omega}\right)\|u\|^{(1-\alpha)l_0}_{s,\Omega}\,dt \\ &\leqslant c\int_{T_1}^{T_2}\left(\prod_{i=1}^n \|u_{x_i}\|^{\alpha l_0/n}_{q_i,\Omega}\right)\|u\|^{(1-\alpha)\beta l_0}_{\beta,\Omega}\|u\|^{(1-\alpha)(1-\beta)l_0}_{r,\Omega}\,dt.\end{aligned} \tag{5.9}$$

Applying the Hölder inequality to the integral on the right side of (5.9), we obtain an upper bound for it in terms of

$$\prod_{i=1}^n\left(\int_{T_1}^{T_2}\|u_{x_i}\|^{\alpha l_0/n\mu_i}_{q_i,\Omega}\,dt\right)^{\mu_i}\left(\int_{T_1}^{T_2}\|u\|^{(1-\alpha)\beta l_0/\nu}\,dt\right)^{\nu}\left(\int_{T_1}^{T_2}\|u\|^{(1-\alpha)(1-\beta)l_0/\sigma}\,dt\right)^{\sigma} \tag{5.10}$$

for any $\mu_i \geqslant 0$, $i = 1,\dots,n$, $\nu \geqslant 0$ and $\sigma \geqslant 0$, $\sum_1^n \mu_i + \nu + \sigma = 1$. We choose μ_i, ν, and σ so that $\alpha l_0/n\mu_i = q_{0i}$, $i = 1,\dots,n$, $(1-\alpha)\beta l_0/\nu = p_0$ and $(1-\alpha)(1-\beta)/\sigma = r_0$. From this we find that $\mu_i = \alpha l_0/nq_{0i}$, $i = 1,\dots,n$, $\nu = (1-\alpha)\beta l_0/p_0$, and $\sigma = (1-\alpha)(1-\beta)/r_0$. It is obvious that the values of μ_i, ν, and σ thus found are nonnegative, and from (5.6) it follows that $\Sigma_1^n \mu_i + \nu + \sigma = 1$. Thus, it was legitimate to apply the Hölder inequality. The estimate (5.7) obviously follows from (5.9) and (5.10) with Young's inequality taken into account. Lemma 5.2 is proved.

COROLLARY 5.1. *Let the numbers* $q_i \geqslant 1$ *and* $q_{0i} \geqslant 1$, $i = 1,\dots,n$, *be fixed, and suppose that for some* $\alpha \in [0,1]$ *the indices* $\bar{l}$ *and* $\bar{l}_0$ *satisfy the conditions*

$$1/\bar{l} = \alpha/\hat{l} + (1-\alpha)/2, \quad 1/\bar{l}_0 = \alpha/\hat{l}_0, \tag{5.11}$$

where $\hat{l}$ *and* $\hat{l}_0$ *are defined in* (5.6). *Then for any function* $u \in \tilde{C}^1_{0,\partial\Omega\times(T_1,T_2)}(\overline{Q})$

$$\|u\|_{\bar{l},\bar{l}_0,Q} \leqslant c\|\nabla u\|^{\alpha}_{\mathbf{q},\mathbf{q}_0,Q}\|u\|^{1-\alpha}_{2,\infty,Q} \leqslant c\|u\|_{\mathring{V}_{\mathbf{q},\mathbf{q}_0}}(Q), \tag{5.12}$$

where c depends only on n, $\mathbf{q}$, $\mathbf{q}_0$, *and* α *in the case* $\Sigma_1^n 1/q_i > 1$ *and on n,* $\hat{l}$, $\hat{l}_0$, α, *and* Ω *in the case* $\Sigma_1^n 1/q_i \leqslant 1$.

PROOF. Corollary 5.1 is a special case of Lemma 5.2 with $\beta = 0$, $r = 2$ and $r_0 = +\infty$.

COROLLARY 5.2. 1) *Suppose that for fixed* $q_i \geqslant 1$ *and* $q_{0i} \geqslant 1$, $i = 1,\dots,n$, *for some* $\alpha \in [0,1]$ *and* $\beta \in [0,1]$ *the indices l and* l_0 *satisfy the conditions*

$$1/l = \alpha/\hat{l} + (1-\alpha)/2, \quad 1/l_0 = \alpha/\hat{l}_0 + (1-\alpha)\beta, \tag{5.13}$$

where $\hat{l}$ *and* $\hat{l}_0$ *are the same as in* (5.6). *Then*

$$\|u\|_{l,l_0,Q} \leqslant c\|\nabla u\|^{\alpha}_{\mathbf{q},\mathbf{q}_0,Q}\|u\|^{(1-\alpha)\beta}_{2,1,Q}\|u\|^{(1-\alpha)(1-\beta)}_{2,\infty,Q}, \quad \forall u \in \tilde{C}^1_{0,\partial\Omega\times(T_1,T_2)}(\overline{Q}). \tag{5.14}$$

2) *If*

$$\begin{gathered}1/l = \alpha/\hat{l} + (1-\alpha)/2, \quad 1/l_0 = \alpha/\hat{l}_0 + (1-\alpha)\beta/2,\\ \alpha \in [0,1], \quad \beta \in [0,1],\end{gathered} \tag{5.15}$$

where $\hat{l}$ *and* $\hat{l}_0$ *are the same as in* (5.6), *then*

$$\|u\|_{l,l_0,Q} \leqslant c\|\nabla u\|^{\alpha}_{\mathbf{q},\mathbf{q}_0,Q}\|u\|^{(1-\alpha)\beta}_{2,Q}\|u\|^{(1-\alpha)(1-\beta)}_{2,\infty,Q}, \quad \forall u \in \tilde{C}^1_{0,\partial\Omega\times(T_1,T_2)}(\overline{Q}). \tag{5.16}$$

3) *If*

$$1/l = \alpha/\hat{l} + (1-\alpha)/2, \quad 1/l_0 = \alpha/\hat{l}_0 + (1-\alpha)/2, \qquad \alpha \in [0,1], \tag{5.17}$$

where $\hat{l}$ *and* $\hat{l}_0$ *are the same as in* (5.6), *then*

$$\|\nabla u\|_{l,l_0,Q} \leqslant c\|\nabla u\|^{1-\alpha}_{\mathbf{q},\mathbf{q}_0,Q}\|u\|^{1-\alpha}_{2,\infty,\Omega}, \quad \forall u \in \tilde{C}^1_{0,\partial\Omega\times(T_1,T_2)}(\overline{Q}). \tag{5.18}$$

For $\Sigma_1^n 1/q_i > 1$ *the constants in* (5.14), (5.16), *and* (5.18) *depend on n,* $\hat{l}$, $\hat{l}_0$, α, *and* β, *while for* $\Sigma_1^n 1/q_i \leqslant 1$ *they further depend on* Ω *as well.*

PROOF. Corollary 5.2 distinguishes a number of special cases of Lemma 5.2: part 1) corresponds to the case $p = 2$, $p_0 = 1$, $r = 2$, $r_0 = +\infty$; part 3) corresponds to the case $\beta = 1$, $p = 2$, $p_0 = 2$, $r = 2$, $r_0 = +\infty$.

LEMMA 5.3. *Let* $Q = \Omega \times (T_1, T_2)$ *be a cylinder with a bounded, strongly Lipschitz domain* Ω *as its base, and let* $\Gamma = \gamma \times (T_1, T_2)$, $\gamma \subset \partial\Omega$, *where either* $\operatorname{meas}_{n-1}\gamma > 0$ *or* $\gamma = \varnothing$. *Let the numbers* $q_i \geqslant 1$ *and* $q_{0i} \geqslant 1$, $i = 1,\dots,n$, *be fixed, and suppose that the indices* $\bar{l}$ *and* $\bar{l}_0$ *satisfy conditions* (5.11) *for some* $\alpha \in [0,1]$. *Suppose* Ω *satisfies the strong* $1/\boldsymbol{\lambda}$*-horn condition* ([8], §8) *for* $\boldsymbol{\lambda}$ *of the form* (4.19). *Then there is the imbedding*

$\widetilde{H^{0,\Gamma}_{\mathbf{q},\mathbf{q}_0}}(Q) \to L^{l,l_0}(Q)$, *and for any function* $u \in \widetilde{H^{0,\Gamma}_{\mathbf{q},\mathbf{q}_0}}(Q)$ *inequalities of the form* (5.12) *hold with a constant c depending on* n, $\bar{l}$, $\bar{l}_0$, $\mathbf{q}$, $\mathbf{q}_0$, *and* Ω. *If instead of the indices* $\bar{l}$ *and* $\bar{l}_0$ *indices* l *and* l_0 *are considered which satisfy conditions* (5.13) *for some* $\alpha \in (0,1)$ *and* $\beta \in (0,1)$, *then for any* $\varepsilon > 0$

$$\|u\|_{l,l_0,Q} \leqslant \varepsilon\|u\|_{\tilde{H}_{\mathbf{q},\mathbf{q}_0}(Q)} + c_1\varepsilon^{-\lambda}\|u\|_{2,1,Q}. \tag{5.19}$$

If instead of (5.13) *conditions* (5.15) *hold for the indices* l *and* l_0 *(for some* $\alpha \in (0,1)$ *and* $\beta \in (0,1)$*), then for any* $\varepsilon > 0$

$$\|u\|_{l,l_0,Q} \leqslant \varepsilon\|u\|_{\tilde{H}_{\mathbf{q},\mathbf{q}_0}(Q)} + c_2\varepsilon^{-\lambda}\|u\|_{2,Q}. \tag{5.20}$$

The constants c_1 *and* c_2 *in* (5.19) *and* (5.20) *depend only on* n, $\hat{l}$, $\hat{l}_0$, α, β *and* Ω,(3) *while* $\lambda > 0$ *depends only on* α *and* β. *In the case* $\Gamma = \partial\Omega \times (T_1, T_2)$ *and* $\sum_1^n 1/q_i > 1$ *the constants* c, c_1, *and* c_2 *in the above inequalities do not depend on* Ω.

PROOF. Let $u \in \tilde{C}^1_{0,\Gamma}(\overline{\Omega})$. We extend this function to the set $\hat{Q} = \hat{\Omega} \times (T_1, T_2)$, $\hat{\Omega} \supset \Omega$, in such a way that the extended function $\hat{u}$ belongs to $\tilde{C}^1_{0,\partial\Omega\times(T_1,T_2)}(\overline{Q})$ and satisfies the inequalities

$$\|\nabla\hat{u}\|_{\mathbf{q},\hat{\Omega}} \leqslant c_0\big(\|\nabla u\|_{\mathbf{q},\Omega} + \|u\|_{1,\Omega}\big), \quad \|\hat{u}\|_{2,\hat{\Omega}} \leqslant c_0\|u\|_{2,\Omega}, \qquad t \in [T_1, T_2], \tag{5.21}$$

where c_0 depends only on Ω and $\mathbf{q}$. This extension can be realized by using the methods of [8], [57] and [87], for example. From Corollary 5.1 with (5.21) taken into account we then obtain

$$\|u\|_{\bar{l},\bar{l}_0,Q} \leqslant c\|u\|_{\tilde{H}_{\mathbf{q},\mathbf{q}_0}(Q)}, \quad \forall u \in \tilde{C}^1_{0,\Gamma}(\overline{Q}). \tag{5.22}$$

The first part of Lemma 5.3 follows easily from (5.22). If now l and l_0 satisfy conditions (5.13), then by case 1) of Corollary 5.2 for the function $\hat{u}$ we have

$$\|\hat{u}\|_{l,l_0,\hat{Q}} \leqslant c\|\nabla\hat{u}\|^{\alpha}_{\mathbf{q},\mathbf{q}_0,\hat{Q}}\|\hat{u}\|^{(1-\alpha)(1-\beta)}_{2,\infty,\hat{Q}}\|\hat{u}\|^{(1-\alpha)\beta}_{2,1,Q}, \tag{5.23}$$

where $\alpha, \beta \in (0,1)$. Applying Young's inequality, we find that for any $\varepsilon > 0$

$$\|\hat{u}\|_{l,l_0,\hat{Q}} \leqslant \varepsilon\big[\|\nabla\hat{u}\|_{\mathbf{q},\mathbf{q}_0,\hat{Q}} + \|\hat{u}\|_{2,\infty,\hat{Q}}\big] + c\varepsilon^{-\lambda}\|\hat{u}\|_{2,1,\hat{Q}}, \tag{5.24}$$

where $\lambda = [1-\beta(1-\alpha)]/(1-\alpha)\beta$. Using (5.21), we easily obtain (5.19). Assuming that the indices l and l_0 satisfy (5.15), the estimate (5.20) is established in a similar way. Lemma 5.3 is proved.

REMARK 5.1. It is obvious that conditions (5.11) on the indices $\bar{l}$ and $\bar{l}_0$ are obtained from (5.15) for $\beta = 0$. In order that the indices l and l_0 in (5.15) (in particular, the indices $\bar{l}$ and $\bar{l}_0$ in (5.11)) satisfy the inequalities $l > 2$ and $l_0 > 2$ [$\bar{l} > 2$ and $\bar{l}_0 > 2$] it is necessary and sufficient that the following conditions hold:

$$\hat{l} > 2 \quad \text{for} \sum_{i=1}^{n}\frac{1}{q_i} > 1, n \geqslant 2, \qquad \frac{\alpha}{\bar{l}_0} + \frac{(1-\alpha)\beta}{2} < \frac{1}{2}, \quad \alpha, \beta \in [0,1]$$
$$\left[\hat{l} > 2 \quad \text{for} \sum_{i=1}^{n}\frac{1}{q_i} > 1, n \geqslant 2, \qquad \frac{\alpha}{\bar{l}_0} < \frac{1}{2}, \quad \alpha \in [0,1]\right]. \tag{5.25}$$

The assertion follows in an obvious way from (5.15) and (5.11).

(3) It is obvious that an inequality of the form (5.20) follows from (5.19), but in this case the constant c_2 in (5.20) also depends on $T_2 - T_1$.

REMARK 5.2. Lemma 5.3 implies, in particular, the following results.

1. There is the imbedding

$$\widetilde{H^{0,\Gamma}_{\mathbf{q},\mathbf{q}_0}}(Q) \to L^{\bar{l}}(Q), \qquad \bar{l} = 2 + \hat{l}_0 - 2\frac{\hat{l}_0}{\hat{l}}, \tag{5.26}$$

where $\hat{l}$ and $\hat{l}_0$ are defined in (5.6), and

$$\|u\|_{\bar{l},Q} \leqslant c(\|u\|_{2,\infty,Q} + \|\nabla u\|_{\mathbf{q},\mathbf{q}_0,Q}), \quad \forall u \in \widetilde{H^{0,\Gamma}_{\mathbf{q},\mathbf{q}_0}}(Q), \tag{5.27}$$

where $c = c(n, \mathbf{q}, \Omega)$ for $\Sigma_1^n 1/q_i > 1$ and $c = c(n, \mathbf{q}, \hat{l}, \hat{l}_0, \Omega)$ for $\Sigma_1^n 1/q_i \leqslant 1$. In particular, for $\mathbf{q} = \mathbf{q}_0$ the index $\bar{l}$ in (5.26) is determined by the formula $\bar{l} = (n+2)/\Sigma_1^n 1/q_i$.

2. For any index l that satisfies for some $\beta \in (0, 1)$ the condition

$$\frac{1}{l} = \frac{(1/2 - \beta)(1/\hat{l} - 1/2)}{1/2 - \beta - 1/\hat{l} + 1/\hat{l}_0} + \frac{1}{2}, \tag{5.28}$$

and for every $\varepsilon > 0$,

$$\|u\|_{l,Q} \leqslant \varepsilon\|u\|_{\tilde{H}_{\mathbf{q},\mathbf{q}_0}(Q)} + c_1\varepsilon^{-\lambda}\|u\|_{2,1,Q}, \quad \forall u \in \tilde{H}_{\mathbf{q},\mathbf{q}_0}(Q). \tag{5.29}$$

If, however, the index l satisfies the condition

$$\frac{1}{l} = \frac{(1 - \beta)(1/\hat{l} - 1/2)}{1 - \beta - 2(1/\hat{l} - 1/\hat{l}_0)} + \frac{1}{2}, \qquad \beta \in (0, 1), \tag{5.30}$$

then for every $\varepsilon > 0$

$$\|u\|_{l,Q} \leqslant \varepsilon\|u\|_{\tilde{H}_{\mathbf{q},\mathbf{q}_0}(Q)} + c_2\varepsilon^{-\lambda}\|u\|_{2,Q}, \quad \forall u \in \tilde{H}_{\mathbf{q},\mathbf{q}_0}(Q). \tag{5.31}$$

The constants c_1 and c_2 in (5.29) and (5.31) depend only on n, $\hat{l}$, $\hat{l}_0$, β, and Ω, while $\lambda > 0$ depends only on β. In the case $\Gamma = \partial\Omega \times (T_1, T_2)$ and $\Sigma_1^n 1/q_i > 1$ the constants c_1 and c_2 do not depend on Ω. We note further that $\bar{l} \geqslant 2$ ($l \geqslant 2$) if and only if $\hat{l} \geqslant 2$ for $\Sigma 1/q_i > 1$, $n \geqslant 2$, and $\bar{l} > 2$ ($l > 2$) if and only if $\hat{l} > 2$ for $\Sigma 1/q_i > 1$, $n \geqslant 2$.

LEMMA 5.4. *Let $Q = \Omega \times [T_1, T_2]$ be a cylinder with a bounded, strongly Lipschitz domain $\Omega \subset \mathbb{R}^n$ as base of the cylinder Q. Then for any p, p_0 and $\mathbf{q}$, $\mathbf{q}_0$, where $p \geqslant 1$, $p_0 \geqslant 1$, $\mathbf{q} = (q_1, \dots, q_n)$, $\mathbf{q}_0 = (q_{01}, \dots, q_{0n})$, $q_i \geqslant 1$ and $q_{0i} \geqslant 1$, $i = 1, \dots, n$, there is the imbedding*

$$H_{p,p_0;\mathbf{q},\mathbf{q}_0}(Q) \to L^1(\partial\Omega \times (T_1, T_2)). \tag{5.32}$$

PROOF. (5.32) is obtained from the known imbeddings

$$H_{p,p_0;\mathbf{q},\mathbf{q}_0}(Q) \to W_1^{1,0}(Q) \to L^1(\partial\Omega \times (T_1, T_2)).$$

REMARK 5.3. Whenever imbeddings of $\widetilde{H^{0,\Gamma}_{\mathbf{q},\mathbf{q}_0}}(Q)$ in $L^{\hat{l},\hat{l}_0}(Q)$ or $L^{l,l_0}(Q)$ are used in the sequel, it is implicitly assumed that the domain Ω lying in the base of the cylinder Q satisfies the strong $1/\lambda$-horn condition for λ of the form (4.19).

§6. General operator equations in a Banach space

In this section X and Y denote real, separable Banach spaces. We assume that Y and X are reflexive and $Y \to X$, where this imbedding is not only continuous but also dense. Let X^* and Y^* be the dual spaces of X and Y respectively. From the preceding conditions it follows that $X^* \to Y^*$, and this imbedding is also not only continuous but dense. The inner product (duality) between X and X^* and also between Y and Y^* we denote by $\langle \cdot , \cdot \rangle$.

We consider an operator $\mathscr{L}$: $X \to Y^*$ (nonlinear, in general). We call this operator *locally coercive* if there exists a number $\rho > 0$ such that $\langle \mathscr{L}u, u \rangle \geqslant 0$ for all $u \in Y$, $\|u\|_X = \rho$.

We call an operator $\mathscr{L}$: $X \to Y^*$ *coercive* if

$$\lim_{u \in Y, \|u\|_X \to +\infty} (\langle \mathscr{L}u, u \rangle / \|u\|_X) = +\infty. \tag{6.1}$$

We say that an operator $\mathscr{L}$: $X \to Y^*$ has *semibounded variation* if for each $\rho > 0$, for any $u, v \in Y$ such that $\|u\|_X \leqslant \rho$ and $\|v\|_X \leqslant \rho$,

$$\langle \mathscr{L}u - \mathscr{L}v, u - v \rangle \geqslant -\gamma(\rho, \|u - v\|'_X), \tag{6.2}$$

where $\gamma(\rho, \tau)$ is a continuous, nonnegative function satisfying the condition

$$\lim_{\varepsilon \to +0} (\gamma(\rho, \varepsilon\tau)/\varepsilon) = 0, \quad \forall \rho \geqslant 0, \forall \tau \geqslant 0, \tag{6.3}$$

while the norm $\|\cdot\|'_X$ is compact relative to $\|\cdot\|_X$, i.e., such that from a sequence $\{u_\nu\}$ bounded in the norm $\|\cdot\|_X$ it is possible to select a subsequence which converges to u in the norm $\|\cdot\|'_X$.

An operator $\mathscr{L}$: $X \to Y^*$ is called *monotone* (*strictly monotone*) if

$$\begin{gathered} \langle \mathscr{L}u - \mathscr{L}v, u - v \rangle \geqslant 0, \quad \forall u, v \in Y \\ [\langle \mathscr{L}u - \mathscr{L}v, u - v \rangle > 0, \quad \forall u, v \in Y, u \neq v]. \end{gathered} \tag{6.4}$$

An operator $\mathscr{L}$: $X \to Y^*$ is called *uniformly monotone* (*strongly monotone*) if for any $u, v \in Y$

$$\begin{gathered} \langle \mathscr{L}u - \mathscr{L}v, u - v \rangle \geqslant \gamma(\|u - v\|_X) \\ [\langle \mathscr{L}u - \mathscr{L}v, u - v \rangle \geqslant \delta(\|u - v\|_X)\|u - v\|_X], \end{gathered} \tag{6.5}$$

where $\gamma(\rho)$ is a continuous, strictly increasing function on $[0, +\infty)$ which is equal to 0 for $\rho = 0$, and $\delta(\rho)$ is a continuous, nondecreasing function equal to 0 only at $\rho = 0$ and such that $\delta(\rho) \to +\infty$ as $\rho \to +\infty$.

It is obvious that for an operator $\mathscr{L}$: $X \to Y^*$ strong monotonicity implies uniform monotonicity, uniform monotonicity implies strict monotonicity, and strict monotonicity implies monotonicity; monotonicity implies that the operator has semibounded variation. The definitions given above are well known in the special case where $X = Y$. Below we study operator equations of the form $\mathscr{L}u = \mathscr{F}$, where $\mathscr{F} \in Y^*$, and, in particular $\mathscr{L}u = 0$. We present first of all the well-known lemma on the acute angle (see, for example [145]) which will be used in the proof of solvability of equations $\mathscr{L}u = \mathscr{F}$.

Let $\mathscr{P}_n$ be an n-dimensional vector space. In $\mathscr{P}_n$ we introduce a norm $\|v\|$ and an inner product $v \cdot w$, where $v, w \in \mathscr{P}_n$. Let $|v|$ denote the norm in $\mathscr{P}_n$ generated by this inner product in $\mathscr{P}_n$. Then there exist positive constants c_1 and c_2 such that $c_1\|v\| \leqslant |v| \leqslant c_2\|v\|$ for all $v \in \mathscr{P}_n$. Let T be a continuous transformation in $\mathscr{P}_n$.

LEMMA 6.1 (on the acute angle). *Suppose that there exists a number $\rho > 0$ such that $Tv \cdot v \geqslant 0$ for all $v \in \mathscr{P}_n$, $\|v\| = \rho$. Then there exists at least one element $v \in \mathscr{P}_n$ such that $\|v\| \leqslant \rho$ and $Tv = 0$.*

In order to formulate existence theorems for the equations indicated above, we make some further assumptions. We suppose that in addition to the spaces X and Y there is given a third real, separable, reflexive space H and that the following condition is satisfied:

$$\text{there exists a set } \mathfrak{N}\text{, which is contained and dense in each of the spaces } Y, X, \text{ and } H, \text{ such that } \|u\|_H \leqslant c_1\|u\|_X \leqslant c_2\|u\|_Y \text{ for all } u \in \mathfrak{N}, \text{ where the constants } c_1 \text{ and } c_2 \text{ do not depend on } u \in \mathfrak{N}. \tag{6.6}$$

Condition (6.6) implies the imbeddings $H^* \to X^* \to Y^*$. The duality between H and H^* we denote in the same way as that between X and X^*, i.e., by $\langle\,\cdot\,,\cdot\rangle$. We further suppose that Y can be identified with some (algebraic) subspace in H, and $Y \to H$. Suppose there are given operators $\mathscr{A}\colon X \to H^*$ and $\mathscr{B}\colon X \to Y^*$. We assume that the operator $\mathscr{L}\colon X \to Y^*$ satisfies the condition

$$\mathscr{L} = \mathscr{A} + \mathscr{B}, \text{ where } \mathscr{A}\colon X \to H^* \text{ is bounded and demicontinuous,}^{(4)} \text{ and } \mathscr{B}\colon X \to Y^* \text{ is linear and continuous.} \tag{6.7}$$

We denote by V the following subspace of X:

$$V = \{u \in X\colon \mathscr{B}u \in H^*\}. \tag{6.8}$$

Suppose that the following condition is satisfied:

$$\text{the restriction of } \mathscr{B}\colon X \to Y^* \text{ to the set } V \cap Y \subset X \text{ is a bounded linear operator from } (V \cap Y) \subset Y \text{ into } X^*. \tag{6.9}$$

We suppose finally that the following conditions are satisfied:

$$\text{the set } V \cap Y \text{ is dense in } X, \tag{6.10}$$

and

$$\text{the function } v \to \langle \mathscr{B}v, v\rangle,\ v \in V, \text{ is continuous in the norm } \|\cdot\|_X. \tag{6.11}$$

THEOREM 6.1. *Suppose that conditions* (6.6), (6.7) *and* (6.9)–(6.11) *are satisfied, and that the operator* $\mathscr{L}\colon X \to Y^*$ *is locally coercive and has semibounded variation. Then the equation* $\mathscr{L}u = 0$ *has at least one solution* $u \in V$.

PROOF. It is obvious that there exists an expanding sequence of finite-dimensional subspaces $\mathscr{P}_N$ of dimension $N = 1, 2, \ldots$, contained in Y and such that $\bigcup_1^\infty \mathscr{P}_N$ is dense in Y and $\bigcup_1^\infty (\mathscr{P}_N \cap V)$ is dense in $Y \cap V$ in the norm $\|\cdot\|_Y$. Then $\bigcup_1^\infty \mathscr{P}_N$ is dense in X and in H. In each $\mathscr{P}_N$ we introduce the norm $\|u\| \equiv \|u\|_X$ and an inner product $u \cdot v$, where $u, v \in \mathscr{P}_N$. Let $|\cdot|$ be the norm in $\mathscr{P}_N$ generated by this inner product. In $\mathscr{P}_N$ we consider the transformation $\mathscr{L}_N$ defined by $\mathscr{L}_N u = \mathscr{L}u|_{\mathscr{P}_N}$, $u \in \mathscr{P}_N$, where $\mathscr{L}u|_{\mathscr{P}_N}$ is the restriction of $\mathscr{L}u$ to $\mathscr{P}_N$. It is obvious that $\mathscr{L}_N u \cdot v = \langle \mathscr{L}u, v\rangle$ for all $u, v \in \mathscr{P}_N$. It follows from (6.7) that the transformation $\mathscr{L}_N$ is continuous. From the condition of local coerciveness of the operator $\mathscr{L}\colon X \to Y^*$ and Lemma 6.1 it follows that for some $\rho > 0$ there exists at least one element

(4) We recall that an operator $T\colon B_1 \to B_2$ (where B_1 and B_2 are Banach spaces) is called *bounded* if any set bounded in B_1 is mapped by this operator into a set bounded in B_2. An operator $T\colon B_1 \to B_2$ is called *demicontinuous* if it is continuous from B_1 with the strong topology to B_2 with the weak topology.

$u_N \in \mathscr{P}_N \subset Y \subset X$ such that

$$\|u_N\|_X \leqslant \rho, \qquad N = 1, 2, \ldots, \tag{6.12}$$

and

$$\langle \mathscr{L} u_N, \xi \rangle = 0, \quad \forall \xi \in \mathscr{P}_N, \quad N = 1, 2, \ldots. \tag{6.13}$$

We consider a sequence $\{u_N\}$, $u_N \in \mathscr{P}_N \subset Y \subset X$, $N = 1, 2, \ldots$, satisfying (6.12) and (6.13). Since X is reflexive, it is possible to extract from $\{u_N\}$ a subsequence which converges weakly in X to an element $u \in X$. Taking into account the reflexivity of the space H^*, the boundedness of the operator $\mathscr{A}$: $X \to H^*$ and the continuity of $\mathscr{B}$: $X \to Y^*$, we conclude that from this subsequence it is possible to extract a new subsequence (for which we preserve the notation of the original sequence) such that

$$\mathscr{A} u_N \to f \quad \text{weakly in } H^*, \qquad \mathscr{B} u_N \to \mathscr{B} u \quad \text{weakly in } Y^*, \tag{6.14}$$

where f is an element of H^*. It obviously follows from (6.13) and (6.14) that

$$\langle f, \xi \rangle + \langle \mathscr{B} u, \xi \rangle = 0, \quad \forall \xi \in \bigcup_{N=1}^{\infty} \mathscr{P}_N. \tag{6.15}$$

Since $\cup_1^\infty \mathscr{P}_N$ is dense in Y and in H, (6.15) makes it possible to identify $\mathscr{B} u$ with the element $-f \in H^*$. Therefore, $u \in V$ and

$$\langle f, \xi \rangle + \langle Bu, \xi \rangle = 0, \quad \forall \xi \in H, \forall \xi \in X. \tag{6.16}$$

We now use the condition that the operator $\mathscr{L}$: $X \to Y^*$ has semibounded variation. We fix an index N, and we suppose that $\xi \in \mathscr{P}_N \cap V$. Then (see (6.2))

$$\langle \mathscr{L} u_N - \mathscr{L} \xi, u_N - \xi \rangle \geqslant -\gamma(\rho, \|u_N - \xi\|'_X), \tag{6.17}$$

where $\rho = \sup_{N=1,2,\ldots} \|u_N\|_X + \|\xi\|_X$. Subtracting from (6.17) the equality $\langle \mathscr{L} u_N, u_N - \xi \rangle = 0$, which follows from (6.13), we obtain

$$-\langle \mathscr{A} \xi, u_N - \xi \rangle - \langle \mathscr{B} \xi, u_N - \xi \rangle \geqslant -\gamma(\rho, \|u_N - \xi\|'_X), \quad \forall \xi \in V \cap \mathscr{P}_N. \tag{6.18}$$

Taking into account that by (6.9) $\mathscr{B}\xi \in X^*$ for all $\xi \in \mathscr{P}_N \cap V$, and also taking account of the weak convergence $u_N \to u$ in X, the properties of the function γ, and the compactness of the norm $\|\cdot\|'_X$ relative to $\|\cdot\|_X$, we deduce from (6.18) that for all $\xi \in \cup_1^\infty (\mathscr{P}_n \cap V)$

$$-\langle \mathscr{A} \xi, u - \xi \rangle - \langle \mathscr{B} \xi, u - \xi \rangle \geqslant -\gamma(\rho, \|u - \xi\|'_X). \tag{6.19}$$

Taking into account the density of $\cup_1^\infty (\mathscr{P}_N \cap V)$ in $(Y \cap V) \subset Y$, the imbeddings $Y \to X$ and $H^* \to X^*$, condition (6.9), and the properties of the function γ, it is easy to prove that (6.19) also holds for all $\xi \in Y \cap V$. Adding (6.19) to an equality of the form (6.16) with ξ replaced by $u - \xi$, where $\xi \in Y \cap V$, we find that for all $\xi \in Y \cap V$

$$\langle f - \mathscr{A} \xi, u - \xi \rangle + \langle \mathscr{B}(u - \xi), u - \xi \rangle \geqslant -\gamma(\rho, \|u - \xi\|'_X). \tag{6.20}$$

Because of conditions (6.10) and (6.11) for the element $u \in V$ we have found and a fixed element $\xi \in V$ there exists a sequence $\{\xi_n\}$, $\xi_n \in Y \cap V$, $n = 1, 2, \ldots$, such that $\xi_n \to \xi$ in X and

$$\langle \mathscr{B}(u - \xi_n), u - \xi_n \rangle \to \langle \mathscr{B}(u - \xi), u - \xi \rangle.$$

From this and (6.20), taking the properties of the operator $\mathscr{A}: X \to H^*$, the function γ, and the norm $\|\cdot\|'_X$ into account, it follows easily that (6.20) holds also for all $\xi \in V$, while for ρ on the right side of (6.20) it is possible to take

$$\rho = \sup_{N=1,2,\dots} \|u_N\|_X + \sup_{n=1,2,\dots} \|\xi_n\|_X,$$

where $\{\xi_n\}$ converges to ξ in X. Thus, (6.20) holds for all $\xi \in V \cup Y$. In (6.20) we then set $\xi = u - \varepsilon\eta$, where $\varepsilon \in (0,1)$ and $\eta \in Y \cap V$. It is obvious that the number ρ in (6.20) does not depend on ε. Dividing the inequality thus obtained by ε, we find

$$\langle f - \mathscr{A}(u - \varepsilon\eta), \eta \rangle + \varepsilon\langle \mathscr{B}\eta, \eta\rangle \geqslant -\gamma(\rho, \varepsilon\|\eta\|'_X)/\varepsilon. \tag{6.21}$$

Letting ε tend to 0 in (6.21), with the demicontinuity of $\mathscr{A}$ and equality (6.3) taken into account, we obtain $\langle f - \mathscr{A}u, \eta\rangle \geqslant 0$ for all $\eta \in Y \cap V$. Replacing η by $-\eta$, we obtain the opposite inequality $\langle f - \mathscr{A}u, \eta \rangle \leqslant 0$ for all $\eta \in Y \cap V$. Thus, the equality $\langle f - \mathscr{A}u, \eta\rangle = 0$ holds for all $\eta \in Y \cap V$; since $Y \cap V$ is dense in X, this implies that $\mathscr{A}u = f$. It then follows from (6.16) that $\mathscr{L}u = 0$. Theorem 6.1 is proved.

COROLLARY 6.1. *Suppose that in Theorem* 6.1 *the condition of local coerciveness of the operator* $\mathscr{L}: X \to Y^*$ *is replaced by the condition of local coerciveness of the operator* $\mathscr{L}_{\mathscr{F}}: X \to Y^*$, *where* $\mathscr{L}_{\mathscr{F}}u = \mathscr{L}u - \mathscr{F}$, $\mathscr{F} \in H^*$, *while all the other conditions of Theorem* 6.1 *are satisfied. Then for the selected* $\mathscr{F} \in H^*$ *the equation* $\mathscr{L}u = \mathscr{F}$ *has at least one solution* $u \in V$.

PROOF. The equation $\mathscr{L}u = \mathscr{F}$ is equivalent to the equation $\mathscr{L}_{\mathscr{F}}u = 0$, where $\mathscr{L}_{\mathscr{F}} = \mathscr{A}_{\mathscr{F}} + \mathscr{B}$, $\mathscr{A}_{\mathscr{F}}u = \mathscr{A}u - \mathscr{F}$, $\mathscr{F} \in H^*$. Since for the operator $\mathscr{L}_{\mathscr{F}}: X \to Y^*$ all the conditions of Theorem 6.1 are satisfied, the latter implies Corollary 6.1.

COROLLARY 6.2. *Suppose that the condition of local coerciveness of the operator* $\mathscr{L}$: $X \to Y^*$ *is replaced by the condition of coerciveness of this operator while all the other conditions of Theorem* 6.1 *are satisfied. Then for all* $\mathscr{F} \in H^*$ *the equation* $\mathscr{L}u = \mathscr{F}$ *has at least one solution.*

PROOF. We fix an arbitrary element $\mathscr{F} \in H^*$. In view of (6.1) for all $u \in Y$ with sufficiently large norm $\|u\|_X$ the inequality $\langle \mathscr{L}_{\mathscr{F}}u, u\rangle \equiv \langle \mathscr{L}u, u\rangle - \langle \mathscr{F}, u\rangle \geqslant 0$ holds, i.e., the operator $\mathscr{L}_{\mathscr{F}}: X \to Y^*$ is locally coercive. The result of Corollary 6.2 then follows from Corollary 6.1.

THEOREM 6.2. *Let* Y, X, *and* H *be the same spaces as in Theorem* 6.1. *Suppose that the operator* $\mathscr{L}: X \to Y^*$ *satisfies the following conditions*: $\mathscr{L} = \mathscr{A} + \mathscr{B}$, $\mathscr{A}: X \to H^*$, $\mathscr{B}: X \to Y^*$, *where* $\mathscr{A}: X \to H^*$ *is demicontinuous and weakly compact*,([5]) *and* $\mathscr{B}$: $X \to Y^*$ *is linear and continuous. Suppose also that the operator* $\mathscr{L}: X \to Y^*$ *is locally coercive. Then the equation* $\mathscr{L}u = \mathscr{F}$ *has at least one solution.*

PROOF. To prove Theorem 6.2 we use the same scheme as in the proof of Theorem 6.1. The existence of approximate solutions u^N of the equation $\mathscr{L}u = 0$ defined by (6.13) can be established in exactly the same way as in Theorem 6.1, since only the demicontinuity and local coerciveness of the operator $\mathscr{L}: X \to Y^*$ are used in this

([5]) Weak compactness of the operator $\mathscr{A}: X \to H^*$ means that weak convergence of $\{u_n\}$ to u in X implies the existence of a subsequence $\{u_\nu\}$ such that $\mathscr{A}u_\nu \to \mathscr{A}u$ weakly in H^*.

part of the theorem. The estimate (6.12) is thus established, which implies the existence of a subsequence $\{u^N\}$ which converges weakly to some element $u \in V$. By the weak compactness of the operator $\mathscr{A}$: $X \to H^*$ it is possible to extract from this sequence a subsequence $\{u^\nu\}$ for which

$$\mathscr{A}u^\nu \to \mathscr{A}u \quad \text{weakly in } H^*, \qquad \mathscr{B}u^\nu \to \mathscr{B}u \quad \text{weakly in } Y^*. \tag{6.22}$$

From (6.13) and (6.22) it then follows that for all $\xi \in \bigcup_1^\infty \mathscr{P}_N$ for the limit element u the equality $\langle \mathscr{A}u, \xi\rangle + \langle \mathscr{B}u, \xi\rangle = 0$ holds. Since $\bigcup_1^\infty \mathscr{P}_N$ is dense in Y and H, the last equality holds also for all $\xi \in H$, i.e., $\mathscr{L}u = 0$. Theorem 6.2 is proved.

REMARK 6.1. Regarding the solvability of the equation $\mathscr{L}u = \mathscr{F}$ we remark that Theorem 6.2 has corollaries altogether analogous to Corollaries 6.1 and 6.2.

THEOREM 6.3. *Let Y, X, and H be the same spaces as in Theorem* 6.1, *and suppose that the operator* $\mathscr{L} = \mathscr{A} + \mathscr{B}$, $\mathscr{A}$: $X \to H^* \subset Y^*$, $\mathscr{B}$: $X \to Y^*$, *satisfies the condition*

$$\langle \mathscr{L}u - \mathscr{L}v, u - v\rangle > 0 \quad \textit{for any } u, v \in V, u \neq v. \tag{6.23}$$

Then for every $\mathscr{F} \in H^*$ the equation $\mathscr{L}u = \mathscr{F}$ has no more than one solution in X. Condition (6.23) *is clearly satisfied if the following condition holds*:

the operator $\mathscr{A}$: $X \to (H^* \subset X^*)$ is strictly monotone, and the operator $\mathscr{B}$: $X \to Y^*$ is linear and such that $\langle \mathscr{B}v, v\rangle \geqslant 0$ for all $v \in V$. (6.24)

PROOF. We first observe that if $u \in X$ is a solution of the equation $\mathscr{L}u = \mathscr{F}$ for some $\mathscr{F} \in H^*$, then $\mathscr{B}u \in H^*$ (and hence $\mathscr{L}u \in H^*$), since in this case $\mathscr{B}u = \mathscr{F} - \mathscr{A}u$, where $\mathscr{A}u \in H^*$. Let u^1 and u^2 be any solutions of the equation $\mathscr{L}u = \mathscr{F}$ for fixed $\mathscr{F} \in H^*$ with $u^1 \neq u^2$. It then follows from (6.23) that

$$\langle \mathscr{L}u^1 - \mathscr{L}u^2, u^1 - u^2\rangle > 0.$$

However, this is impossible, since $\mathscr{L}u^1 = \mathscr{L}u^2 = \mathscr{F}$, and hence

$$\langle \mathscr{L}u^1 - \mathscr{L}u^2, u^1 - u^2\rangle = 0.$$

We note finally that the validity of (6.23) follows from (6.24) in an obvious way. Theorem 6.3 is proved.

THEOREM 6.4. *Let Y, X, and H be the same spaces as in Theorem* 6.1. *Suppose that the operator $\mathscr{L}$: $X \to Y^*$ has the form $\mathscr{L} = \mathscr{A} + \mathscr{B}$, where $\mathscr{A}$: $X \to H^*$ is demicontinuous, and $\mathscr{B}$: $X \to Y^*$ is linear. Suppose also that conditions* (6.10) *and* (6.11) *are satisfied. Suppose finally that the operator $\mathscr{L}$: $X \to Y^*$ is uniformly monotone. Then for all $\mathscr{F} \in H^*$ the equation $\mathscr{L}u = \mathscr{F}$ has no more than one solution in X.*

PROOF. From the uniform monotonicity of $\mathscr{L}$: $X \to Y^*$ and the linearity of $\mathscr{B}$: $X \to Y^*$ it follows that for any $u, v \in Y$

$$\langle \mathscr{A}u - \mathscr{A}v, u - v\rangle + \langle \mathscr{B}(u - v), u - v\rangle \geqslant \tilde{\gamma}(\|u - v\|_X), \tag{6.25}$$

where $\tilde{\gamma}(\rho)$ is the same sort of function as $\gamma(\rho)$ in (6.5). Let u and v be arbitrary elements of the set V. Because of (6.10) there exist sequences $\{u_n\}$ and $\{v_n\}$, $u_n, v_n \in V \cap Y$, $n = 1, 2, \dots$, converging to u and v respectively in X. Taking (6.25) into account, we conclude that for any $n = 1, 2, \dots$

$$\langle \mathscr{A} u_n - \mathscr{A} v_n, u_n - v_n \rangle + \langle \mathscr{B}(u_n - v_n), u_n - v_n \rangle \geqslant \tilde{\gamma}(\|u_n - v_n\|_X). \quad (6.26)$$

Using the demicontinuity of the operator $\mathscr{A}$: $X \to H^*$, the imbedding $H^* \to X^*$, condition (6.11), and the properties of the function $\tilde{\gamma}$, on passing to the limit in (6.25) we find that (6.25) also holds for any $u, v \in V$. Since this condition implies (6.23), Theorem 6.4 follows from Theorem 6.3. Theorem 6.4 is proved.

We now note that the following obvious assertion holds.

LEMMA 6.2. *If the operator $\mathscr{L}$: $X \to Y^*$ is strongly monotone, then it is coercive.*

THEOREM 6.5. *Let Y, X, and H be the same spaces as in Theorem* 6.1. *Suppose that conditions* (6.6), (6.7) *and* (6.9)–(6.11) *are satisfied, and assume that the operator $\mathscr{L}$: $X \to Y^*$ is strongly monotone. Then the restriction $\hat{\mathscr{L}}$: $(V \subset X) \to H^*$ of $\mathscr{L}$ to the set V is a bijection, and the inverse operator $\hat{\mathscr{L}}^{-1}$: $(H^* \subset X^*) \to (V \subset X)$ is continuous.*

PROOF. Taking Lemma 6.2 into account, we conclude that the conditions of Theorem 6.5 obviously imply that all the conditions of Corollary 6.2 and Theorem 6.4 are satisfied. Therefore, the mapping $\mathscr{L}$: $(V \subset X) \to H^*$ is a bijection. We shall prove that the operator $\hat{\mathscr{L}}^{-1}$: $(H^* \subset X^*) \to (V \subset X)$ is continuous. From the conditions of strong monotonicity of the operator $\mathscr{L}$: $X \to Y^*$, the demicontinuity of the operator $\mathscr{A}$: $X \to H^*$, and (6.10) and (6.11) we get

$$\langle \mathscr{L} u - \mathscr{L} v, u - v \rangle \geqslant \delta(\|u - v\|)\|u - v\|_X, \qquad u, v \in V. \qquad (6.27)$$

Indeed, for given elements $u, v \in V$ in view of (6.10) there are sequences $\{u_n\}$ and $\{v_n\}$, $u_n, v_n \in V \cap Y$, converging to u and v respectively in X. Because of (6.11) and the linearity of the operator $\mathscr{B}$: $X \to Y^*$ we have

$$\lim_{n \to \infty} \langle \mathscr{B}(u_n - v_n), u_n - v_n \rangle = \langle \mathscr{B}(u - v), u - v \rangle. \qquad (6.28)$$

Since the strong monotonicity of $\mathscr{L}$: $X \to Y^*$ for any $n = 1, 2, \dots$ implies

$$\begin{aligned} \langle \mathscr{L} u_n - \mathscr{L} v_n, u_n - v_n \rangle &= \langle \mathscr{A} u_n - \mathscr{A} v_n, u_n - v_n \rangle + \langle \mathscr{B}(u_n - v_n), u_n - v_n \rangle \\ &\geqslant \delta(\|u_n - v_n\|_X)\|u_n - v_n\|_X, \end{aligned} \qquad (6.29)$$

on passing to the limit as $n \to \infty$ in (6.29) and taking account of the demicontinuity of the operator $\mathscr{A}$: $X \to H^*$, the imbedding $H^* \to X^*$, equality (6.28), and the properties of the function $\delta(\rho)$, we find that (6.27) holds. Taking into account that $H^* \to X^*$, from (6.27) we easily obtain

$$\delta(\|u - v\|_X) \leqslant \|\mathscr{L} u - \mathscr{L} v\|_{X^*}, \qquad u, v \in V,$$

from which the continuity of $\hat{\mathscr{L}}^{-1}$: $(H^* \subset X^*) \to (V \subset X)$ obviously follows. Theorem 6.5 is proved.

§7. A special space of functions of scalar argument with values in a Banach space

Let $I \equiv [a, b]$ be a compact interval of the real axis $\mathbb{R}$, and let Z be a Banach space with norm $\|\cdot\|_Z$. The set of functions defined on I with values in Z we denote by $\{I \to Z\}$. Below we shall use the spaces $C^m(I; Z)$, $m = 0, 1, \dots$, and $L^p(I; Z)$, $p > 1$, understood in the usual sense (see, for example, [16]).

Let $u \in \{I \to Z\}$. We denote by $T_h u$ the function in $\{I \to Z\}$ defined by any of the formulas

$$\begin{aligned} &\text{(i)} \quad T_h u(t) = \bar{u}(t+h), \\ &\text{(ii)} \quad T_h u(t) = \tilde{u}(t+h), \\ &\text{(iii)} \quad T_h u(t) = \tilde{\tilde{u}}(t+h), \end{aligned} \tag{7.1}$$

where $h \in \mathbb{R}$, $t \in I$, and the functions $\bar{u}$, $\tilde{u}$ and $\tilde{\tilde{u}}$ are defined by

$$\text{(i)} \quad \bar{u}(t) = \begin{cases} u(t), & t \in I, \\ 0, & t \in \mathbb{R} \setminus I; \end{cases}$$

$$\text{(ii)} \quad \tilde{u}(t) = \begin{cases} u(a+\tau), & t = a-\tau, 0 \leqslant \tau \leqslant d \equiv (b-a), \\ u(t), & t \in [a,b], \\ u(b-\tau), & t = b+\tau, 0 \leqslant \tau \leqslant d, \\ 0, & t \in \mathbb{R} \setminus [a-d, b+d]; \end{cases} \tag{7.2}$$

$$\text{(iii)} \quad \tilde{\tilde{u}}(t) = \begin{cases} -u(a+\tau), & t = a-\tau, 0 \leqslant \tau \leqslant d, \\ u(t), & t \in [a,b], \\ u(b-\tau), & t = b+\tau, 0 \leqslant \tau \leqslant d, \\ 0, & t \in \mathbb{R} \setminus [a-d, b+d]. \end{cases}$$

LEMMA 7.1. *If* $u \in L^p(I; Z)$, $1 \leqslant p < +\infty$, *then* $T_h u \in L^p(I; Z)$ *and* $T_h u \to u$ *in* $L^p(I; Z)$.

We now consider the average $S_h u \in \{I \to Z\}$ of the function $u \in L^1(I; Z)$ defined by any of the formulas

$$\begin{aligned} &\text{(i)} \quad (\bar{S}_h u)(t) = \int_{\mathbb{R}} \omega_h(t-\tau)\bar{u}(\tau)\, d\tau, \\ &\text{(ii)} \quad (\tilde{S}_h u)(t) = \int_{\mathbb{R}} \omega_h(t-\tau)\tilde{u}(\tau)\, d\tau, \\ &\text{(iii)} \quad (\tilde{\tilde{S}}_h u)(t) = \int_{\mathbb{R}} \omega_h(t-\tau)\tilde{\tilde{u}}(\tau)\, d\tau, \end{aligned} \tag{7.3}$$

where $\bar{u}, \tilde{u}$ and $\tilde{\tilde{u}}$ are defined by (7.2) and $\omega_h(\eta)$ is an infinitely differentiable normalized averaging kernel. Suppose to be specific that

$$\omega_h(\eta) = \frac{1}{\kappa h}\omega\left(\frac{|\eta|}{h}\right), \qquad \omega(\rho) = \begin{cases} e^{\rho^2 / (\rho^2 - 1)}, & \rho \in [0,1), \\ 0, & \rho \geqslant 1, \end{cases}$$

$$\kappa = \int_{-1}^{1} e^{\rho^2 / (\rho^2 - 1)}\, d\rho. \tag{7.4}$$

It is obvious that $S_h u \in C^\infty(I; Z) \subset L^p(I; Z)$, $p \in [1, +\infty]$. The following known results hold.

LEMMA 7.2. *If* $u \in L^p(I; Z)$, $1 \leqslant p < +\infty$, *then*

$$\lim_{h > 0} \|S_h u - u\|_{L^p(I;Z)} = 0.$$

LEMMA 7.3. *If* $u \in C(I; Z)$, *then* $\lim_{h \to 0} \|\tilde{S}_h u - u\|_{C(I;Z)} = 0$.

We now recall some facts related to the concept of a distribution on $I \equiv [a, b]$ with values in a Banach space Z. This concept was introduced by L. Schwartz ([161]; see also [16], Chapter IV, §1.4). We denote by $\mathscr{D}^*(I; Z)$ the space of all continuous linear mappings of the space $\mathscr{D}(I)$ (i.e., the set $C_0^\infty(I)$ equipped with the Schwartz topology) into the space Z considered with the weak topology. Elements f of the space $\mathscr{D}^*(I; Z)$ are called distributions on I with values in Z. On the set $\mathscr{D}^*(I; Z)$ we introduce a topology converting $\mathscr{D}^*(I; Z)$ into a locally convex space by means of the family of seminorms

$$p_{\varphi,\mathscr{F}}(f) = |(\mathscr{F}, f(\varphi))|, \qquad \varphi \in \mathscr{D}(I), \mathscr{F} \in Z^*, f \in \mathscr{D}^*(I; Z), \tag{7.5}$$

where $(\cdot,\cdot)$ is the inner product of an element in Z and an element in Z^*, which will be called for brevity the *inner product between* Z and Z^*. For convergence of a sequence $\{f_n\}$ to f in $\mathscr{D}^*(I; Z)$ it is then necessary and sufficient that $\lim_{n\to\infty}(F, f_n(\varphi)) = (F, f(\varphi))$ for all $\varphi \in \mathscr{D}(I)$ and all $F \in Z^*$.

If a function $u \in \{I \to Z\}$ is locally Bochner integrable on I, then it can be assigned a distribution f_u on I with values in Z by the rule $u \to f_u$, where

$$f_u(\varphi) = \int_I u(t)\varphi(t)\,dt, \qquad \varphi \in \mathscr{D}(I), \tag{7.6}$$

and the integral in (7.6) is understood as the Bochner integral. We note that the correspondence $u \to f_u$ is one-to-one (with the usual convention that equivalent locally Bochner integrable functions are identified). Therefore, $L^1_{\text{loc}}(I; Z)$ can be identified with a subspace of $\mathscr{D}^*(I; Z)$.

We recall that for any distribution $f \in \mathscr{D}^*(I; Z)$ the derivative $f' \in \mathscr{D}^*(I; Z)$ is defined by

$$f'(\varphi) = -f(\varphi'), \quad \forall\varphi \in \mathscr{D}(I). \tag{7.7}$$

The mapping $f \to f'$ here is linear and continuous in $\mathscr{D}^*(I; Z)$. Recalling that $I = [a, b]$, we denote by $\tilde{I}$ the interval $\tilde{I} \equiv [\tilde{a}, \tilde{b}]$ where $\tilde{a} = a - d$ and $\tilde{b} = b + d$, $d = b - a$.

LEMMA 7.4. *If a function $u \in L^1(I; Z)$ considered as an element of $\mathscr{D}^*(I; Z)$ has a derivative $u' \in L^1(I; Z)$, then the function $\tilde{u}$ defined on $\tilde{I}$ by formula* (ii) *of* (7.2) *belongs to $L^1(\tilde{I}; Z)$, and its derivative $(\tilde{u})'$ when $\tilde{u}$ is considered as an element of $\mathscr{D}^*(\tilde{I}; Z)$ belongs to $L^1(\tilde{I}; Z)$. Here $(\tilde{u})'$ coincides (in the sense of equality of elements in $\mathscr{D}^*(\tilde{I}; Z)$) with the extension $\tilde{u}'$ to $\tilde{I}$ of the derivative $u' \in L^1(I; Z)$ defined by formula* (iii) *in* (7.2).

We note that the extensions $(\tilde{u})'$ and $\tilde{u}'$ considered in Lemma 7.4 also coincide as elements of $L^1(\tilde{I}; Z)$.

Let B_0 be a reflexive Banach space, and let the l_k: $G \subset B_0 \to B_k$ be linear operators defined on an (algebraic) subspace G dense in B_0 with values in the reflexive Banach spaces B_k, $k = 1,\ldots,N$. We suppose that the operators l_k admit closure. On G we define the norm

$$\|u\|_B = \sum_{k=0}^{N} \|l_k u\|_{B_k}, \tag{7.8}$$

where $l_0 u \equiv u$ for all $u \in G$. We denote the closure in the norm (7.8) by B. Obviously B is a Banach space. From the conditions imposed on the operators l_k, $k = 1,\dots,N$, it follows that B can be identified with a subspace of B_0, since the convergence $u_n \to u$, $\tilde{u}_n \to u$ in B_0 and $l_k u_n \to v_k$, $l_k \tilde{u}_n \to \tilde{v}_k$ in B_k, $k = 1,\dots,N$, where $u_n, \tilde{u}_n \in G$, $n = 1,2,\dots$, implies that $v_k = \tilde{v}_k$, $k = 1,\dots,N$. The operators l_k, $k = 1,\dots,N$, can be extended to all of B. The extended operators $l_k\colon B \to B_k$, considered as operators acting from the Banach space B to the Banach spaces B_k, are obviously continuous.

Let $\mathbb{H}$ be a Hilbert space. We suppose further that there exists a Banach space $\hat{B}$ such that $\hat{B} \to B \to B_0 \to \mathbb{H} \to B_0^* \to B^* \to \hat{B}^*$ and $\hat{B} \to B_k \to \mathbb{H} \to B_k^* \to B^k$, $k = 1,\dots,N$, and that these imbeddings are not only continuous but dense. The inner products between $\hat{B}$ and $\hat{B}^*$ and also between B_0 and B_0^*, B and B^*, and B_k and B_k^*, $k = 1,\dots,N$, we denote in the same way by $(\cdot,\cdot)$. It is easy to see that the space B is reflexive and that each linear functional F in B can be given by

$$(F,\eta) = (F_0,\eta) + \sum_{k=1}^{N} (F_k, l_k\eta), \quad \forall \eta \in B, \tag{7.9}$$

where $F_0 \in B_0^*$, $F_k \in B_k^*$, $k = 1,\dots,N$, and F_0 and F_k can be chosen so that

$$\|F\|_{B^*} = \sup\big(\|F_0\|_{B_0^*}, \|F_1\|_{B_1^*},\dots,\|F_N\|_{B_N^*}\big). \tag{7.10}$$

On the other hand, for any $F_0 \in B_0^*,\dots,F_k \in B_k^*$ the right side of (7.9) defines a linear functional in B with norm not exceeding the quantity on the right side of (7.10).([6]) Equality (7.9) is equivalent to the equality

$$F = F_0 + \sum_{k=1}^{N} l_k^* F_k, \tag{7.11}$$

understood in the sense of equality of elements in B^*, and the $l_k^* F_k, k = 1,\dots,N$, are defined by

$$(l_k^* F_k, \eta) \equiv (F_k, l_k \eta), \qquad \eta \in B, k = 1,\dots,N. \tag{7.12}$$

It is obvious that $F_0 \in B^*$ and $l_k^* F_k \in B^*$, $k = 1,\dots,N$, since $B_0^* \to B^*$ and

$$\sup_{\|\eta\|B=1} |(F_k, l_k\eta)| \leqslant \|F_k\|_{B_k^*}\|l_k\eta\|_{B_k} \leqslant \|F_k\|_{B_k^*}, \qquad k = 1,\dots,N.$$

We denote by U the set of all functions in $\{I \to B\}$ having finite norm

$$\|u\|_U = \sum_{k=0}^{N} \|l_k u\|_{L^{p_k}(I;\, B_k)}, \tag{7.13}$$

where $p_k \in (1,+\infty)$, $k = 0,1,\dots,N$. It is easy to verify the completeness of the space U. The inequality

$$\int_I \|u\|_B\,dt = \int_I \sum_{k=0}^{N} \|l_k u\|_{B_k}\,dt \leqslant c \sum_{k=0}^{N} \|l_k u\|_{L^{p_k}(I;\, B_k)} = c\|u\|_U \tag{7.14}$$

implies the imbedding $U \to L^1(I; B)$ and hence also the imbedding $U \to L^1(I; B^*)$.

Considering the mapping $\pi\colon U \to L^{p_0}(I; B_0) \times \cdots \times L^{p_N}(I; B_N)$ defined by the formula $\pi(u) = (l_0 u, l_1 u,\dots,l_N u)$, $u \in U$, and using the form of a linear functional

---

([6]) Analogous facts are proved below in connection with the assertions of Lemma 5.2.1.

in $L^p(I; Z)$, we establish that each linear functional $\mathscr{F}$ in U can be represented in the form

$$\langle \mathscr{F}, \eta \rangle = \int_I \left[(\mathscr{F}_0, \eta) + \sum_{k=1}^{N} (\mathscr{F}_k, l_k \eta) \right] dt, \qquad \eta \in U, \tag{7.15}$$

where $\mathscr{F}_0 \in L^{p'_0}(I; B_0^*)$, $\mathscr{F}_k \in L^{p'_k}(I; B_k^*)$, $k = 1, \ldots, N$, and $\mathscr{F}_0$ and the $\mathscr{F}_k$ can be chosen so that

$$\|\mathscr{F}\|_{U^*} = \sup\left\{ \|\mathscr{F}_0\|_{L^{p'_0}(I;\, B_0^*)}, \ldots, \|\mathscr{F}_N\|_{L^{p'_N}(I;\, B_N^*)} \right\}. \tag{7.16}$$

It is also obvious that for any $\mathscr{F}_0 \in L^{p'_0}(I;\ B_0^*), \ldots, \mathscr{F}_N \in L^{p'_N}(I;\ B_N^*)$ the integral on the right side of (7.15) is a linear functional in U with norm not exceeding the quantity on the right side of (7.16). From what has been said above it follows that $\langle \mathscr{F}, \eta \rangle$ can also be written in the form

$$\langle \mathscr{F}, \eta \rangle = \int_I (\mathscr{F}(\eta), \eta(t))\, dt, \qquad \eta \in U, \tag{7.17}$$

where $\mathscr{F}(t) = \mathscr{F}_0(t) + \sum_1^N l_k^* \mathscr{F}_k(t)$, and $\mathscr{F}_0(t)$ and $l_k^* \mathscr{F}_k(t)$, $k = 1, \ldots, N$, belong to B^* for almost all $t \in I$ (see (7.11) and (7.12)). We denote by $l_k^* \mathscr{F}_k$, $k = 1, \ldots, N$, the linear functionals in U defined by

$$\langle l_k^* \mathscr{F}_k, \eta \rangle \equiv \int_I (l_k^* \mathscr{F}_k(t), \eta(t))\, dt = \int_I (\mathscr{F}_k(t), (l_k \eta)(t))\, dt, \qquad \eta \in U. \tag{7.18}$$

We note that the inequality

$$|\langle l_k^* \mathscr{F}_k, \eta \rangle| \leqslant \|\mathscr{F}_k\|_{L^{p'_k}(I;\, B_k^*)} \|l_k \eta\|_{L^{p_k}(I;\, B_k)}, \qquad k = 1, \ldots, N, \tag{7.19}$$

implies that (7.18) actually defines linear functionals in U. All $\mathscr{F} \in U^*$ can then be written in the form

$$\mathscr{F} = \mathscr{F}_0 + \sum_{k=1}^{N} l_k^* \mathscr{F}_k. \tag{7.20}$$

Because of the obvious estimate

$$\int_I \|\mathscr{F}(t)\|_{B^*}\, dt \leqslant c \|\mathscr{F}\|_{U^*}, \qquad \mathscr{F} \in U^*, \tag{7.21}$$

we have the imbedding $U^* \to L^1(I; B^*)$.

Using Lemma 7.2 and the form of the norm in U and U^*, we establish the following proposition.

LEMMA 7.4. *For any $u \in U$ the average $\tilde{S}_h u$ (see* (ii) *in* (7.3)) *belongs to U and tends to u in U as $h \to 0$. For any $\mathscr{F} \in U^*$ the average $\tilde{\tilde{S}}_h \mathscr{F}$ (see* (iii) *in* (7.3)) *belongs to U^* and tends to $\mathscr{F}$ in U^* as $h \to 0$.*

We now distinguish an important subspace of U. We denote by $\mathscr{W}$ the (algebraic) subspace

$$\mathscr{W} = \{ u \in U \colon u' \in U^* \}, \tag{7.22}$$

where u' denotes the derivative of u in the sense of distributions on I with values in the Banach space B^* (in view of the imbedding $U \to L^1(I; B^*)$) it is obvious that each element $u \in U$ can be identified with an element of $\mathscr{D}^*(I;\, B^*)$). It is easy to see that with respect to the norm

$$\|u\|_{\mathscr{W}} \equiv \|u\|_U + \|u'\|_{U^*} \tag{7.23}$$

$\mathscr{W}$ is a Banach space. In view of Lemma 7.4 the following assertion holds.

LEMMA 7.5. *For any $u \in \mathscr{W}$ the average $\tilde{S}_h u$ (see* (ii) *in* (7.3)) *belongs to $\mathscr{W}$ and tends to u in $\mathscr{W}$ as $h \to 0$.*

To simplify notation we henceforth write

$$u_h \equiv \tilde{S}_h u. \tag{7.24}$$

COROLLARY 7.1. *The set $C^\infty(I;\, B)$ is dense in $\mathscr{W}$.*

PROOF. Since $u_h \in C^\infty(I;\, B)$, the density of $C^\infty(I;\, B)$ in $\mathscr{W}$ follows from Lemma 7.5.

LEMMA 7.6. $\mathscr{W} \to C(I; \mathbb{H})$.

PROOF. We first prove that $\mathscr{W} \to C(I;\, B^*)$. Let $u \in \mathscr{W}$. Then $u' \in U^* \to L^1(I;\, B^*)$. We set $v = \int_a^t u'(\tau)\,d\tau$. It is obvious that the function v belongs to $C(I;\, B^*)$, and considered as an element of the space $\mathscr{D}^*(I;\, B^*)$ it has derivative v' equal to u'. It is known that u and v then differ on I by a constant quantity $\Phi_0 \in B^*$, i.e., $u(t) = v(t) + \Phi_0$ for almost all $t \in I$. Since $v \in C(I;\, B^*)$, also $u \in C(I;\, B^*)$. Therefore, $w \in C(I;\, B^*)$. We shall prove that this imbedding is continuous. From the estimate $\|v\|_{C(I;\, B^*)} \leqslant \|u'\|_{L^1(I;\, B^*)}$ and an inequality of the form (7.21) it follows that

$$\|v\|_{C(I;\, B^*)} \leqslant c\|u'\|_{U^*}. \tag{7.25}$$

Taking into account that $\Phi_0 = u(t) - v(t)$, we write

$$d\|\Phi_0\|_{B^*} = \int_I \|\Phi_0\|_{B^*}\,dt = \int_I \|u - v\|_{B^*}\,dt \leqslant \int_I \|u\|_{B^*}\,dt + c\|v\|_{C(I;\, B^*)}, \tag{7.26}$$

where $d = b - a$. From (7.25), (7.26), (7.21), (7.14) and the imbedding $\mathscr{W} \to C(I; \mathbb{H})$ we then obtain

$$\|\Phi_0\|_{B^*} \leqslant c(\|u\|_U + \|u'\|_{U^*}) = c\|u\|_{\mathscr{W}}, \tag{7.27}$$

while from (7.25) and (7.27) we obtain

$$\|u\|_{C(I;\, B^*)} \leqslant c\|u\|_{\mathscr{W}}. \tag{7.28}$$

Thus, the imbedding $\mathscr{W} \to C(I;\, B^*)$ is proved.

We now prove the imbedding $\mathscr{W} \to C(I; \mathbb{H})$. We note first of all that for any $v, w \in C^1(I;\, B)$ and all $t_1, t_2 \in I$

$$(v, w)\Big|_{t=t_1}^{t=t_2} = \int_{t_1}^{t_2} \{(v', w) + (v, w')\}\,dt. \tag{7.29}$$

Let $u \in C^1(I;\, B)$. Setting $v = ((t - a)/d)u$ and $w = ((b - t)/d)u$, $d = b - a$, writing (7.29) first for functions v and u with $t_1 = a$ and $t_2 = t \in (a, b]$, and then for the functions w and u with $t_1 = t \in [a, b)$ and $t_2 = b$, and subtracting the second equality from the first, we obtain

$$(u(t), u(t)) = \frac{1}{d}\int_a^b (u, u)\, dt + 2\int_a^t \frac{\tau - a}{d}(u', u)\, d\tau$$
$$-2\int_t^b \frac{b-\tau}{d}(u', u)\, d\tau. \tag{7.30}$$

From (7.30) we obtain

$$\|u(t)\|_{\mathbb{H}}^2 \leqslant d^{-1}\|u\|_{C(I;\,B^*)}\|u\|_{L^1(I;\,B)} + 4\|u'\|_{U^*}\|u\|_U, \qquad t \in I. \tag{7.31}$$

From (7.31) and (7.28) we then easily obtain

$$\|u(t)\|_{C(I;\,\mathbb{H})} \leqslant c\|u\|_{\mathscr{W}}. \tag{7.32}$$

Let u_h and $u_{h'}$ be averages of an arbitrary function $u \in \mathscr{W}$. Applying (7.32) to the functions $u_h - u_{h'}$, we obtain

$$\|u_h - u_{h'}\|_{C(I;\,\mathbb{H})} \leqslant c\|u_h - u_{h'}\|_{\mathscr{W}}. \tag{7.33}$$

In view of Lemma 7.5 and the completeness of $C(I; \mathbb{H})$, from (7.33) it follows easily that $u \in C(I; \mathbb{H})$ (as always, we identify equivalent functions) and also that (7.32) holds for any $u \in \mathscr{W}$. Lemma 7.6 is proved.

In the proof of Lemma 7.6 we have also established the following fact.

COROLLARY 7.2. *For any function $u \in \mathscr{W}$ its averages u_h (see* (7.24) *and* (7.3)) *converge to it in $C(I; \mathbb{H})$, i.e.,*

$$\lim_{h\to 0} \|u_h - u\|_{C(I;\,\mathbb{H})} = 0. \tag{7.34}$$

LEMMA 7.7. (7.29) *is valid for any $v, w \in \mathscr{W}$. For any $u \in \mathscr{W}$*

$$\tfrac{1}{2}(u, u)\Big|_{t_1}^{t_2} = \int_{t_1}^{t_2}(u', u)\, dt, \qquad t_1, t_2 \in I. \tag{7.35}$$

PROOF. The validity of (7.29) for any $v, w \in C^1(I; B)$ was already noted in the proof of Lemma 7.6. Let $u, v \in \mathscr{W}$, and let v_h and w_h be their averages. Passing to the limit as $h \to 0$ in (7.29) written for v_h and w_h, and taking Lemma 7.5 and Corollary 7.2 into account, we find that (7.29) holds for v and w. Equality (7.35) is a special case of (7.29) ($v = w = u$). Lemma 7.7 is proved.

CHAPTER 5
THE GENERAL BOUNDARY VALUE PROBLEM FOR (A, b, m, m)-ELLIPTIC EQUATIONS

§1. The structure of the equations and the classical formulation of the general boundary value problem

In considering differential equations of the form

$$-(d/dx_i)l^i(x, u, \nabla u) + l_0(x, u, \nabla u) = 0, \tag{1.1}$$

where $x = (x_1, \ldots, x_n)$, $n \geqslant 2$, $\nabla u = (u_{x_1}, \ldots, u_{x_n})$, d/dx_i is the symbol of the total derivative with respect to the variable x_i, $i = 1, \ldots, n$, and $l^i(x, u, p)$ and $l_0(x, u, p)$ are given functions in $\Omega \times \mathbb{R} \times \mathbb{R}^n$, we shall usually be dealing with generalized solutions of these equations; this makes it possible to consider functions $l^i(x, u, p)$ and $l_0(x, u, p)$ under very weak assumptions regarding their differential properties with respect to the independent variables. We shall always assume that the functions $l^i(x, u, p)$, $i = 1,\dots,n$, and $l_0(x, u, p)$ satisfy the Carathéodory condition $\Omega \times \mathbb{R} \times \mathbb{R}^n$ (i.e., these functions are measurable with respect to x in Ω for all $u, p \in \mathbb{R} \times \mathbb{R}^n$ and are continuous with respect to u, p in $\mathbb{R} \times \mathbb{R}^n$ for almost all $x \in \Omega$). Additional regularity conditions will be imposed on the $l^i(x, u, p)$ and $l_0(x, u, p)$ when necessary. As always in this monograph in (1.1) summation over twice repeated indices is assumed.

DEFINITION 1.1. We say that an equation of the form (1.1) has $(A, \mathbf{b})$-*structure* in a domain $\Omega \subset \mathbb{R}^n$, $n \geqslant 2$, if there exist a square matrix $A \equiv \|a^{ij}(x)\|$ of order n, a vector $\mathbf{b} \equiv (b^1(x),\dots,b^n(x))$, and functions $l'^i(x, u, p)$, $i = 1,\dots,n$, and $l'_0(x, u, p)$ such that for almost all $x \in \Omega$ and any $u \in \mathbb{R}$ and $p \in \mathbb{R}^n$

$$\begin{aligned} \mathbf{l}(x, u, p) &= A^*(x)\mathbf{l}'(x, u, A(x)p), \\ l_0(x, u, p) &= l'_0(x, u, A(x)p) + b^i(x)p_i, \end{aligned} \tag{1.2}$$

where A^* is the matrix adjoint to A, and

$$\begin{aligned} \mathbf{l}(x, u, p) &= (l^1(x, u, p),\dots,l^n(x, u, p)), \\ \mathbf{l}'(x, u, q) &= (l'^1(x, u, q),\dots,l'^n(x, u, q)). \end{aligned}$$

We call $l'^i(x, u, q)$, $i = 1,\dots,n$, and $l'_0(x, u, q)$ the *reduced coefficients* of the equation.

DEFINITION 1.2. We say that an equation of the form (1.1) has $(A, \mathbf{b}, m, \mathbf{m})$-*structure* in a domain Ω if it has $(A, \mathbf{b})$-structure in this domain relative to a matrix A satisfying conditions (4.1.1), (4.1.3) with $m_i > 1$, $i = 1,\dots,n$, a vector $\mathbf{b}(x)$ such that $b^i \in C(\overline{\Omega})$, $\partial b^i/\partial x_i \in C(\overline{\Omega})$, $i = 1,\dots,n$, and reduced coefficients $l'^i(x, u, q)$, $i = 1,\dots,n$, and $l'_0(x, u, q)$ satisfying the Carathéodory condition in $\Omega \times \mathbb{R} \times \mathbb{R}^n$ and such that for almost all $x \in \Omega$ and any $u \in \mathbb{R}$ and $q \in \mathbb{R}^n$

$$\begin{aligned} |l'^i(x, u, q)| &\leqslant \mu_1\left(\sum_{k=1}^n |q_k|^{m_k/m'_i} + |u|^{m/m'_i} + \varphi_i(x)\right), \qquad i = 1,\dots,n, \\ |l'_0(x, u, q)| &\leqslant \mu_2\left(\sum_{k=1}^n |q_k|^{m_k/m'} + |u|^{m/m'} + \varphi_0(x)\right), \end{aligned} \tag{1.3}$$

where $\mu_1, \mu_2 = \text{const} \geqslant 0$, $\varphi_i \in L^{m'_i}(\Omega)$, $1/m_i + 1/m'_i = 1$, $i = 1,\dots,n$, and $\varphi \in L^{m'}(\Omega)$, $1/m + 1/m' = 1$, $m > 1$.

In the isotropic case ($m_1 = \cdots = m_n = \overline{m}$) inequalities (1.3) are equivalent to

$$\begin{aligned} |\mathbf{l}'(x, u, q)| &\leqslant \mu_1\left(|q|^{\overline{m}/\overline{m}'} + |u|^{m/\overline{m}'} + \overline{\varphi}(x)\right), \\ |l_0(x, u, q)| &\leqslant \mu_2\left(|q|^{\overline{m}/m'} + |u|^{m/m'} + \varphi_0(x)\right), \end{aligned} \tag{1.4}$$

where $\mu_1, \mu_2 = \text{const} \geqslant 0$, $\overline{\varphi} \in L^{\overline{m}'}(\Omega)$, $1/\overline{m} + 1/\overline{m}' = 1$, $\overline{m} > 1$, and $\varphi_0 \in L^{m'}(\Omega)$, $1/m + 1/m' = 1$, $m > 1$.

PROPOSITION 1.1. *Suppose that for almost all $x \in \Omega$ and any $u \in \mathbb{R}$ and $p \in \mathbb{R}^n$*

$$\begin{aligned} |(A^*)_i^{-1}\mathbf{l}(x, u, p)| &\leqslant \mu_1\left(\sum_{k=1}^n |A_k p|^{m_k/m'_i} + |u|^{m/m'_i} + \varphi_i(x)\right), \qquad i = 1,\dots,n, \\ |l_0(x, u, p) - b^i(x)p_i| &\leqslant \mu_2\left(\sum_{k=1}^n |A_k p|^{m_k/m'} + |u|^{m/m'} + \varphi_0(x)\right), \end{aligned} \tag{1.5}$$

where the matrix A is nondegenerate for almost all $x \in \Omega$, $(A^*)_i^{-1}\mathbf{l} \equiv ((A^*)^{-1}\mathbf{l})_i$ is the ith component of the vector $(A^*)^{-1}\mathbf{l}$, while the vector $\mathbf{b}$, *the indices m and* $\mathbf{m} = (m_1,\dots,m_n)$, *and the functions $\varphi_1,\dots,\varphi_n$, φ_0 are the same as in Definition* 1.1. *Then equation* (1.1) *has* $(A, \mathbf{b}, m, \mathbf{m})$*-structure in* Ω.

PROOF. Indeed, for almost all $x \in \Omega$ and any $u \in \mathbb{R}$ and $p \in \mathbb{R}^n$

$$\mathbf{l}(x, u, p) = A^*(A^*)^{-1}\mathbf{l}(x, u, A^{-1}(Ap)),$$
$$l_0(x, u, p) - b^i(x)p_i = l_0(x, u, A^{-1}(Ap)) - \mathbf{b}(x)A^{-1}(Ap).$$

Setting

$$\mathbf{l}'(x, u, q) = (A^*)^{-1}\mathbf{l}(x, u, A^{-1}q),$$
$$l_0'(x, u, q) = l_0(x, u, A^{-1}q) - \mathbf{b}(x)A^{-1}q, \tag{1.6}$$

we then note that in view of (1.5) and (1.6) conditions (1.2) and (1.3) hold. Proposition 1.1 is proved.

DEFINITION 1.3. We call an equation of the form (1.1) having $(A, \mathbf{b})$-structure in Ω $(A, \mathbf{b})$-*elliptic* (*strictly* $(A, \mathbf{b})$-*elliptic*) in Ω if for almost all $x \in \Omega$ and any $u \in \mathbb{R}$, $q = Ap$, $\eta = A\xi$, $p \in \mathbb{R}^n$ and $\xi \in \mathbb{R}^n$ the following condition of A-ellipticity (strict A-ellipticity) is satisfied:

$$\frac{\partial l'^i(x, u, q)}{\partial q_j}\eta_i\eta_j \geqslant 0 \qquad \left[\frac{\partial l'^i(x, u, q)}{\partial q_j}\eta_i\eta_j > 0, \quad \forall \eta \neq 0\right]. \tag{1.7}$$

We call an equation of the form (1.1) having $(A, \mathbf{b}, m, \mathbf{m})$-structure in Ω $(A, \mathbf{b}, m, \mathbf{m})$-*elliptic* in Ω if for almost all $x \in \Omega$ and any $u \in \mathbb{R}$, $q = Ap$, $\eta = A\xi$, $p \in \mathbb{R}^n$ and $\xi \in \mathbb{R}^n$ the following qualified condition of A-ellipticity is satisfied:

$$\frac{\partial l'^i(x, u, q)}{\partial q_j}\eta_i\eta_j \geqslant \nu \sum_{i=1}^{n} |q_i|^{m_i - 2}\eta_i^2, \qquad \nu = \text{const} > 0. \tag{1.8}$$

If equation (1.1) is $(A, \mathbf{b}, m, \mathbf{m})$-elliptic in Ω, then taking into account that

$$\frac{\partial l^i}{\partial p_j}\xi_i\xi_j = a^{si}\frac{\partial l'^s}{\partial q_k}a^{kj}\xi_i\xi_j = \frac{\partial l'^s}{\partial q_k}(A_s\xi)(A_k\xi) = \frac{\partial l'^s}{\partial q_k}\eta_s\eta_k, \tag{1.9}$$

where $\eta = A\xi$, $A_s\xi \equiv (A\xi)_s \equiv a^{si}\xi_i$ and $A_k\xi \equiv a^{kj}\xi_j$, we conclude that (1.1) admits fixed degeneration of ellipticity at all those points $x \in \Omega$ where the matrix $A(x)$ is degenerate. Moreover, it follows from (1.9) and (1.8) that (1.1) admits implicit degeneration of ellipticity on the sets $\{A_i p = 0, i = 1,\dots,j-1, j+1,\dots,n\}$, $i = 1,\dots,n$. Thus, an $(A, \mathbf{b}, m, \mathbf{m})$-elliptic equation, being, like any $(A, \mathbf{b})$-elliptic equation, an equation with nonnegative characteristic form $(\partial l^i/\partial p_j)\xi_i\xi_j$, is not, in general, a strictly $(A, \mathbf{b})$-elliptic equation.

We shall present examples of $(A, \mathbf{b}, m, \mathbf{m})$-elliptic equations.

1. A linear equation of the form

$$-\frac{d}{dx_i}\left(\alpha^{ij}(x)\frac{\partial u}{\partial x_j}\right) + \beta^i(x)\frac{\partial u}{\partial x_i} + c(x)u - f(x) = 0 \tag{1.10}$$

with a nonnegative symmetric matrix $\mathfrak{A} \equiv \|\alpha^{ij}(x)\|$ in a domain $\Omega \subset \mathbb{R}^n$, $n \geqslant 2$, such that $A \equiv \mathfrak{A}^{1/2}$ satisfies conditions (4.1.1) and (4.1.3) with $m = 2$ and $\mathbf{m} = \mathbf{2}$, when this equation is considered under the conditions $\beta^i \in C(\overline{\Omega})$, $\partial\beta^i/\partial x_i \in C(\overline{\Omega})$, $i = 1,\dots,n$, $c \in L^\infty(\Omega)$ and $f \in L^2(\Omega)$, is $(A, \mathbf{b}, 2, \mathbf{2})$-elliptic with respect to the matrix $A = \mathfrak{A}^{1/2}$ and the vector $\mathbf{b} = (\beta^1,\dots,\beta^n)$.

2. An equation of the form (1.1) satisfying the conditions

$$\frac{\partial l^i(x, u, p)}{\partial p_j}\xi_i\xi_j \geqslant \nu \sum_{i=1}^n \left(1 + |p_i|\right)^{m_i - 2}\xi_i^2, \qquad \nu = \text{const} > 0,$$

$$|l^i(x, u, p)| \leqslant \mu_1\left(\sum_{k=1}^n |p_k|^{m_k/m_i'} + |u|^{m/m_i'} + \varphi_i(x)\right), \qquad i = 1,\dots,n,$$

$$|l_0(x, u, p)| \leqslant \mu_2\left(\sum_{k=1}^n |p_k|^{m_k/m'} + |u|^{m/m'} + \varphi_0(x)\right), \tag{1.11}$$

where $\varphi_i \in L^{m_i'}(\Omega)$, $i = 1,\dots,n$, and $\varphi_0 \in L^{m'}(\Omega)$, is obviously an $(A, \mathbf{b}, m, \mathbf{m})$-elliptic equation relative to the identity matrix $A = I$ and the vector $\mathbf{b} \equiv \mathbf{0}$.

3. An equation of the form

$$\partial u/\partial t - (d/dx_i)l^i(t, x, u, \nabla u) + l_0(t, x, u, \nabla u) = 0, \tag{1.12}$$

where $\nabla u = (u_{x_1},\dots,u_{x_n})$, considered in the cylinder $Q = \Omega \times (T_1, T_2) \subset \mathbb{R}^{n+1}$, $n \geqslant 1$, under the assumption that for almost all $(t, x) \in Q$ and any $u \in \mathbb{R}$, $p \in \mathbb{R}^n$, and $\xi \in \mathbb{R}^n$ the conditions

$$\frac{\partial l^i(t, x, u, p)}{\partial p_j}\xi_i\xi_j \geqslant \nu \sum_{i=1}^n \left(1 + |p_i|\right)^{m_i - 2}\xi_i^2, \qquad \nu = \text{const} > 0,$$

$$|l^i(t, x, u, p)| \leqslant \mu_1\left(\sum_{k=1}^n |p_k|^{m_k/m_i'} + |u|^{m/m_i'} + \varphi_i(t, x)\right), \qquad i = 1,\dots,n,$$

$$|l_0(t, x, u, p)| \leqslant \mu_2\left(\sum_{k=1}^n |p_k|^{m_k/m'} + |u|^{m/m'} + \varphi_0(t, x)\right) \tag{1.13}$$

are satisfied, where $\varphi_i \in L^{m_i'}(Q)$, $i = 1,\dots,n$, and $\varphi_0 \in L^{m'}(Q)$, is easily seen to be an $(\tilde{A}, \tilde{\mathbf{b}}, m, \tilde{\mathbf{m}})$-elliptic equation in Q relative to the matrix $\tilde{A}$ of the form

$$\tilde{A} = \begin{Vmatrix} 0 & \cdots & 0 \\ \vdots & \multicolumn{2}{c}{I} \\ 0 & & \end{Vmatrix}$$

of order $n + 1$, the $(n + 1)$-dimensional vector $\tilde{\mathbf{b}} = (1, 0,\dots,0)$, and the indices m and $\tilde{\mathbf{m}} = (2, m_1,\dots,m_n)$, where a condition of the form (1.2) is satisfied with

$$\mathbf{l}(x, u, \tilde{p}) = \left(0, l^1(x, u, p),\dots,l^n(x, u, p)\right),$$
$$\mathbf{l}'(x, u, \tilde{q}) = \left(q_0, l^1(x, u, q),\dots,l^n(x, u, q)\right),$$

$A = \tilde{A}$ and $\mathbf{b} = (1, 0,\dots,0)$.

4. Consider an equation of the form

$$\sum_{k=1}^s \frac{\partial u}{\partial t_k} - \frac{d}{dx_i}l^i(t, x, u, \nabla u) + l_0(t, x, u, \nabla u) = 0, \tag{1.14}$$

where $t = (t_1, \ldots, t_s)$, $s \geqslant 1$, $x = (x_1, \ldots, x_n)$, $n \geqslant 1$, $s + n \geqslant 2$, and $\nabla u = (u_{x_1}, \ldots, u_{x_n})$, taken in a domain $D \subset \mathbb{R}^{n+s}$ under the assumption that for almost all $(t, x) \in D$ and any $u \in \mathbb{R}$, $p \in \mathbb{R}^n$, $\xi \in \mathbb{R}^n$ the inequality

$$\frac{\partial l^i(t, x, u, p)}{\partial p_j} \xi_i \xi_j \geqslant \nu \sum_{i=1}^{n} (1 + |p_i|)^{m_i - 2} \xi_i^2,$$

$$\nu = \text{const} > 0, m_i > 1, i = 1, \ldots, n, \tag{1.15}$$

holds as well as the second and third inequalities in (1.13) with $t = (t_1, \ldots, t_s)$ and $\varphi_i \in L^{m_i'}(D)$, $i = 1, \ldots, n$, $\varphi_0 \in L^{m'}(D)$. It is easy to see that (1.14) is an $(A, \mathbf{b}, m, \mathbf{m})$-elliptic equation in D relative to the matrix

$$A = \left\| \begin{array}{c|c} 0 & 0 \\ \hline 0 & I \end{array} \right\|$$

of order $n + s$, the $(n + s)$-dimensional vector $\mathbf{b} = (1, \ldots, 1, 0, \ldots, 0)$, and the indices m and $\tilde{\mathbf{m}} = (2, \ldots, 2, m_1, \ldots, m_n)$.

5. The equation of first order

$$l_0(x, u) + \beta^i(x) u_{x_i} = f(x), \tag{1.16}$$

considered in a domain $\Omega \subset \mathbb{R}^n$ under the conditions $|l_0(x, u)| \leqslant \mu |u|^{m/m'} + \varphi(x)$, where $\mu = \text{const} \geqslant 0$, $\varphi \in L^{m'}(\Omega)$, and $f \in L^{m'}(\Omega)$ is a $(0, \mathbf{b}, m, \mathbf{0})$-elliptic equation relative to the matrix 0 and the vector $\mathbf{b} = (\beta^1, \ldots, \beta^n)$.

6. An equation of the form

$$-(d/dx_i)\left(a_i(x)|u_{x_i}|^{m_i - 2} u_{x_i}\right) - f(x) = 0, \qquad m_i > 1, i = 1, \ldots, n, \tag{1.17}$$

where $a_i(x) \geqslant 0$ in a domain $\Omega \subset \mathbb{R}^n$, $n \geqslant 2$, is $(A, \mathbf{0}, m, \mathbf{m})$-elliptic in Ω relative to the diagonal matrix A with elements $[a_i(x)]^{1/m_i}$, $i = 1, \ldots, n$, on the main diagonal, $\mathbf{m} = (m_1, \ldots, m_n)$, and any $m > 1$, provided only that the diagonal matrix A satisfies conditions (4.1.1) and (4.1.3).

7. An equation of the form

$$u_t - \frac{d}{dx_i}\left(a_i(t, x)|u_{x_i}|^{m_i - 2} u_{x_i}\right) - f(t, x) = 0, \qquad m_i > 1, i = 1, \ldots, n, \tag{1.18}$$

where $a_i(t, x) \geqslant 0$ in the cylinder $Q = \Omega \times (T_1, T_2) \subset \mathbb{R}^{n+1}$, $n \geqslant 1$, is $(A, \mathbf{b}, m, \mathbf{m})$-elliptic in Q relative to the matrix A of order $n + 1$ of the form

$$A = \left\| \begin{matrix} 0 & \cdot & \cdot & 0 \\ \cdot & \alpha_1 & & 0 \\ \cdot & & \cdot & \cdot \\ \cdot & & 0 & \cdot \\ 0 & & & \alpha_n \end{matrix} \right\|, \qquad \alpha_1 = (a_1)^{1/m_1}, \ldots, \alpha_n = (a_n)^{1/m_n}, \tag{1.19}$$

the $(n + 1)$-dimensional vector $\mathbf{b} = (1, 0, \ldots, 0)$, $\mathbf{m} = (2, m_1, \ldots, m_n)$, and any $m > 1$.

In a bounded domain $\Omega \subset \mathbb{R}^n$, $n \geqslant 2$, we consider an equation of the form (1.1), assuming now that the functions $l^i(x, u, p)$, $i = 1, \ldots, n$, and $l_0(x, u, p)$ belong to the classes $C^1(\overline{\Omega} \times \mathbb{R} \times \mathbb{R}^n)$ and $C(\overline{\Omega} \times \mathbb{R} \times \mathbb{R}^n)$ respectively, while the domain Ω belongs to the class C^2. Suppose that (1.1) is strictly $(A, \mathbf{b})$-elliptic in Ω. We denote by $\nu = (\nu_1, \ldots, \nu_n)$ the unit vector of the inner normal to $\partial\Omega$. Under these assumptions we call the set

$$\Sigma \equiv \left\{ x \in \partial\Omega \colon \frac{\partial l^i(x, u, p)}{\partial p_j} \nu_i \nu_j > 0 \text{ for all } u \in \mathbb{R}, p \in \mathbb{R}^n \right\}$$

the *noncharacteristic part* of the boundary $\partial\Omega$, and the set

$$\Sigma \equiv \left\{ x \in \partial\Omega \colon \frac{\partial l^i(x, u, p)}{\partial p_j} \nu_i \nu_j = 0 \text{ for all } u \in \mathbb{R}, p \in \mathbb{R}^n \right\}$$

the *characteristic part* of the boundary $\partial\Omega$. In view of the equality

$$\left(\partial l^i / \partial p_j\right) \nu_i \nu_j = \left(\partial l'^s / \partial q_k\right) (A_k \nu)(A_s \nu) \tag{1.20}$$

and the condition of strict $(A, \mathbf{b})$-ellipticity of (1.1) we have

$$\Sigma = \{ x \in \partial\Omega \colon A\nu \neq \mathbf{0} \}, \qquad \Sigma' = \{ x \in \partial\Omega \colon A\nu = \mathbf{0} \}. \tag{1.21}$$

Suppose that the set Σ is decomposed in arbitrary fashion into parts Σ_1, Σ_2, Σ_3 such that $\Sigma_1 \cup \Sigma_2 \cup \Sigma_3 = \Sigma$ and $\Sigma_i \cap \Sigma_j = \varnothing$ for $i \neq j$, $i, j = 1, 2, 3$. On $\partial\Omega$ we consider the function

$$b = b(x) = -b^i(x) \nu_i(x), \quad x \in \partial\Omega. \tag{1.22}$$

We denote by $(\Sigma_i)_0$, $(\Sigma_i)_+$ and $(\Sigma_i)_-$ those parts of the set Σ_i $(i = 1, 2, 3)$ on which $b = 0$, $b > 0$ and $b < 0$ respectively. We decompose Σ' into parts Σ'_0, Σ'_+ and Σ'_- in a similar way. Suppose that a piecewise continuous, bounded, positive function λ is defined on the set Σ_3.

Under the above assumptions we consider the following general boundary value problem: find a function $u \in C^2(\Omega) \cap C^1(\Omega \cup \Sigma_2 \cup \Sigma_3) \cap C(\overline{\Omega})$ such that

$$\begin{gathered} -(d/dx_i) l^i(x, u, \nabla u) + l_0(x, u, \nabla uu) = 0 \quad \text{in } \Omega, \\ u = \varphi \quad \text{on } \Sigma_1 \cup \Sigma'_-, \quad \mathbf{l} \cdot \nu + cu = \psi \quad \text{on } \Sigma_2, \quad \mathbf{l} \cdot \nu + (c - \lambda) u + \chi \quad \text{on } \Sigma_3, \end{gathered} \tag{1.23}$$

where

$$c = \begin{cases} 0 & \text{on } (\Sigma_{2,3})_{0,+}, \\ b(x) & \text{on } (\Sigma_{2,3})_-, \end{cases} \qquad (\Sigma_{2,3})_{0,+} = (\Sigma_2)_0 \cup (\Sigma_2)_+ \cup (\Sigma_3)_0 \cup (\Sigma_3)_+,$$

$$(\Sigma_{2,3})_- = (\Sigma_2)_- \cup (\Sigma_3)_-,$$

and φ, ψ, and χ are piecewise continuous functions defined on the sets $\Sigma_1 \cup \Sigma'_-$, Σ_2, and Σ_3 respectively. There are no boundary conditions on the part $\Sigma'_0 \cup \Sigma'_+$ of $\partial\Omega$.

We note that the conditions on Σ_2 and Σ_3 can be rewritten in the more compact form

$$\mathbf{l}(x, u, \nabla u) \cdot \nu - s(x) u = g(x) \quad \text{on } \Sigma_{2,3} \equiv \Sigma_2 \cup \Sigma_3, \tag{1.24}$$

where $g(x)$ is a piecewise continuous function defined on $\Sigma_{2,3}$ and $s(x)$ is a piecewise continuous function defined on $\Sigma_{2,3}$ such that

$$s(x) \geqslant \max(0, -b(x)) \quad \text{on } \Sigma_{2,3}.$$

It is more convenient, however, for us to use the previous form of writing the boundary conditions on $\Sigma_{2,3}$, even though it is more cumbersome.

We consider the main special cases of Problem (1.2).

1. For the choice $\Sigma_1 = \Sigma$, $\Sigma_2 = \Sigma_3 = \varnothing$ we obtain the first boundary value problem: find a function $u \in C^2(\Omega) \cap C(\overline{\Omega})$ such that

$$-(d/dx_i)l^i(x, u, \nabla u) + l_0(x, u, \nabla u) = 0 \quad \text{in } \Omega, \quad u = \varphi \quad \text{on } \Sigma \cup \Sigma'_-. \tag{1.25}$$

2. For the choice $\Sigma_1 = \varnothing$, $\Sigma_2 = \Sigma$, $\Sigma_3 = \varnothing$ we obtain the second boundary value problem: find a function $u \in C^2(\Omega) \cap C^1(\Omega \cup \Sigma) \cap C(\overline{\Omega})$ such that

$$-dl^i/dx_i + l_0 = 0 \quad \text{in } \Omega, \qquad u = \varphi \quad \text{on } \Sigma'_-, \qquad \mathbf{l} \cdot \boldsymbol{\nu} + cu = \psi \quad \text{on } \Sigma, \tag{1.26}$$

where $c = 0$ on $\Sigma_{0,+}$ and $c = b(x)$ on Σ_-.

3. For the choice $\Sigma_1 = \Sigma_2 = \varnothing$, $\Sigma_3 = \Sigma$ we obtain the third boundary value problem: find a function $u \in C^2(\Omega) \cap C^1(\Omega \cup \Sigma) \cap C(\overline{\Omega})$ such that

$$-dl^i/dx_i + l_0 = 0 \quad \text{in } \Omega, \qquad u = \varphi \quad \text{on } \Sigma'_-, \qquad \mathbf{l} \cdot \boldsymbol{\nu} + (\lambda - c)u + \chi \quad \text{on } \Sigma, \tag{1.27}$$

where $c = 0$ on $\Sigma_{0,+}$ and $c = b(x)$ on Σ_-.

In the case of linear dependence of $l(x, u, p)$ on u and p, i.e., in the case $l^i(x, u, p) = \alpha^{ij}(x)p_j + \alpha^i(x)u + g^i(x)$, $i = 1, \ldots, n$, the conditions on Σ_2 and Σ_3 take the respective forms

$$\begin{aligned} \partial u/\partial N + \boldsymbol{\alpha} \cdot \boldsymbol{\nu} u + \mathbf{g} \cdot \boldsymbol{\nu} + cu = \psi \quad &\text{on } \Sigma_2, \\ \partial u/\partial N + \boldsymbol{\alpha} \cdot \boldsymbol{\nu} u + \mathbf{g} \cdot \boldsymbol{\nu} + (c - \lambda)u = \chi \quad &\text{on } \Sigma_3, \end{aligned} \tag{1.28}$$

where $\partial u/\partial N \equiv \alpha^{ij}u_{x_j}\nu_i$ is the derivative of u with respect to the conormal to $\partial\Omega$.

Applying the standard procedure of suitable replacement of the unknown function, one can reduce the boundary conditions in (1.23) to homogeneous form. To abbreviate subsequent formulations we shall henceforth, as a rule, consider the general boundary value problem in the case of homogeneous boundary conditions, i.e., we shall assume that problem (1.23) is already reduced to the form

$$\begin{gathered} -dl^i/dx_i + l_0 = 0 \quad \text{in } \Omega, \qquad u = 0 \quad \text{on } \Sigma \cup \Sigma'_-, \qquad \mathbf{l} \cdot \boldsymbol{\nu} + cu = 0 \quad \text{on } \Sigma_2, \\ \mathbf{l} \cdot \boldsymbol{\nu} + (c - \lambda)u = 0 \quad \text{on } \Sigma_3. \end{gathered} \tag{1.29}$$

Here we note that the equation $-dl^i/dx_i + l_0 = 0$ itself is not, generally speaking, homogeneous due to the arbitrariness of the function $l_0(x, u, p)$. It is natural to consider the formulation of the general boundary value problem for equation (1.1) presented above as classical.

It is easy to see that the formulation of the general boundary value problem of the form (1.23) given above for an $(A, \mathbf{b})$-elliptic equation is invariant under smooth nondegenerate transformations of the independent variables. Indeed, suppose there is given a smooth coordinate transformation

$$\hat{x} = \hat{x}(x), \qquad \det\left\| \frac{\partial \hat{x}_i}{\partial x_j} \right\| \neq 0 \quad \text{in } \Omega. \tag{1.30}$$

Making the substitution (1.30) in an equation of the form (1.1), we obtain the new equation

$$-(d/d\hat{x}_k)\hat{l}^k(\hat{x}, u, \hat{\nabla} u) + \hat{l}_0(\hat{x}, u, \hat{\nabla} u) = 0, \tag{1.31}$$

defined in the corresponding domain $\hat{\Omega} \subset \mathbb{R}^n$. We denote by $\hat{A} \equiv \|\hat{a}^{ik}(\hat{x})\|$ and $\hat{\mathbf{b}} = \hat{\mathbf{b}}(\hat{x})$ the matrix and vector defined by the formulas

$$\hat{a}^{ik}(\hat{x}) = a^{ij}(x)\frac{\partial \hat{x}_k}{\partial x_j}, \quad \hat{b}^k(\hat{x}) = b^i(x)\frac{\partial \hat{x}_k}{\partial x_i}, \qquad i, k = 1,\dots,n. \tag{1.32}$$

Using condition (1.2), it is not hard to verify that (1.31) has $(\hat{A}, \hat{\mathbf{b}})$-structure in $\hat{\Omega}$ relative to the matrix $\hat{A}$ and the vector $\hat{\mathbf{b}}$ defined by (1.32). Here the roles of functions $\hat{l}'^i(\hat{x}, u, \hat{q})$, $i = 1,\dots,n$, and $\hat{l}'_0(\hat{x}, u, \hat{q})$ from identities of the form

$$\hat{\mathbf{l}}(\hat{x}, u, \hat{p}) = \hat{A}^*\hat{\mathbf{l}}'(\hat{x}, u, \hat{A}\hat{p}), \quad \hat{l}_0(\hat{x}, u, \hat{p}) = \hat{l}'_0(\hat{x}, u, \hat{A}\hat{p}) + \hat{b}^k\hat{p}_k \tag{1.33}$$

are played by the functions

$$\hat{l}'^i(\hat{x}, u, \hat{q}) = l'^i(x(\hat{x}), u, \hat{q}),$$
$$\hat{l}'_0(\hat{x}, u, \hat{q}) = l'_0(x(\hat{x}), u, \hat{q}) + a^{ki}l'^k\frac{\partial^2 \hat{x}_m}{\partial x_i \partial \hat{x}_m}. \tag{1.34}$$

The equalities (1.34) make it possible to formulate the following assertion.

PROPOSITION 1.2. *The reduced coefficients* $l'^i(x, u, q)$, $i = 1,\dots,n$ ($l'_0(x, u, q)$) *of an equation of the form* (1.1) *having* $(A, \mathbf{b})$*-structure in a domain* Ω *are invariant under smooth (linear) transformations of the independent variables, i.e., for any smooth (linear) transformation* (1.30) *an equation of the form* (1.1) *having* $(A, \mathbf{b})$*-structure in* Ω *goes over into an equation of the form* (1.1) *having* $(\hat{A}, \hat{\mathbf{b}})$*-structure in* $\hat{\Omega}$, *where* $\hat{\Omega}$ *is the image of* Ω *under the mapping* (1.30), *relative to the matrix* $\hat{A}$ *and the vector* $\hat{\mathbf{b}}$ *defined by formula* (1.32), *while the old and new equations have the same reduced coefficients* l'^i, $i = 1,\dots,n$ (l'_0) (*see* (1.34)).

It is further easy to verify that the vectors $A\nu$ and $\hat{A}\hat{\nu}$ computed at corresponding points of $\partial\Omega$ and $\partial\hat{\Omega}$ differ only by a nonzero scalar factor. From this it follows that as a result of the transformation (1.30) the sets Σ and $\hat{\Sigma}$ as well as Σ' and $\hat{\Sigma}'$ defined in accordance with (1.21) go over into one another. It is also easy to verify that the functions $b(x)$ and $\hat{b}(\hat{x})$ defined according to a formula of the form (1.22) are related at the corresponding points $x \in \partial\Omega$ and $\hat{x} \in \partial\hat{\Omega}$ by the equality $b(x) = c(x)\hat{b}(\hat{x})$, where $c(x) > 0$. This means that as a result of the transformation (1.30) the sets $(\Sigma_i)_+$ and $(\hat{\Sigma}_i)_+$, $(\Sigma_i)_{0,-}$ and $(\hat{\Sigma}_i)_{0,-}$, Σ'_0 and $\hat{\Sigma}'_0$, Σ'_+ and $\hat{\Sigma}'_+$ and Σ'_- and $\hat{\Sigma}'_-$ go over into one another. Taking into account that the left sides of the boundary conditions in (1.23) are invariant under the substitution (1.30), we conclude from what has been said that the boundary conditions on the sets $\Sigma_1 \cup \Sigma'_-$, Σ_2 and Σ_3 go over into completely analogous conditions on the sets $\hat{\Sigma}_1 \cup \hat{\Sigma}'_-$, $\hat{\Sigma}_2$, and $\hat{\Sigma}_3$. In view of the invariance of the reduced coefficients $l'^i(x, u, q)$ the new equation (1.31) will be $(\hat{A}, \hat{\mathbf{b}})$- (strictly $(\hat{A}, \hat{\mathbf{b}})$-) elliptic in $\hat{\Omega}$ if the original equation was $(A, \mathbf{b})$- (strictly $(A, \mathbf{b})$-) elliptic in Ω. The following assertion now follows from what has been proved.

PROPOSITION 1.3. *The formulation of the general boundary value problem for an* $(A, \mathbf{b})$*-elliptic equation is invariant under any smooth nondegenerate transformation of the independent variables.*

An important special case of $(A, \mathbf{b})$-elliptic equations are the so-called $(A, \mathbf{b})$-parabolic equations considered in the cylinder $Q = \Omega \times (T_1, T_2)$, where $\Omega \subset \mathbb{R}^n$, $n \geqslant 1$, $T_1 < T_2$, $T_1, T_2 = \text{const}$, and defined as $(\tilde{A}, \tilde{\mathbf{b}})$-elliptic equations in $Q \subset \mathbb{R}^{n+1}$ of the form

$$-\sum_{i=1}^{n} \frac{d}{dx_i} l^i(t, x, u, \nabla u) + l_0(t, x, u, \nabla u) + u_t = f(t, x) \tag{1.1'}$$

relative to the matrix

$$\tilde{A} = \begin{Vmatrix} 0 & \cdots & 0 \\ \vdots & & A(t, x) \\ 0 & & \end{Vmatrix}$$

of order $n + 1$ and the $(n + 1)$-dimensional vector $\tilde{\mathbf{b}} \equiv (0, b^1(t, x), \ldots, b^n(t, x))$. It is somewhat more convenient, however, to give an independent definition of an $(A, \mathbf{b})$-parabolic equation in the cylinder Q.

DEFINITION 1.1′. We say that an equation of the form (1.1) has *spatial* $(A, \mathbf{b})$-*structure* in the cylinder $Q = \Omega \times (T_1, T_2)$, $\Omega \subset \mathbb{R}^n$, $n \geqslant 1$, if there exist a square matrix $A = \|a^{ij}(t, x)\|$ of order n, an n-dimensional vector $\mathbf{b} = (b^1(t, x), \ldots, b^n(t, x))$, and functions $l'^i(t, x, u, q)$, $i = 1, \ldots, n$, and $l'_0(t, x, u, q)$ such that for almost all $(t, x) \in Q$ and any $u \in \mathbb{R}$ and $p \in \mathbb{R}^n$

$$\mathbf{l}(t, x, u, p) = A^* \mathbf{l}'(t, x, u, Ap),$$
$$l_0(t, x, u, p) = l'_0(t, x, u, Ap) + b^i(t, x) p_i, \tag{1.2'}$$

where A^* is the matrix adjont to A, $\mathbf{l} = (l^1, \ldots, l^n)$ and $\mathbf{l}' = (l'_1, \ldots, l'_n)$. We call the functions $l'^i(t, x, u, q)$, $i = 1, \ldots, n$, and $l'_0(t, x, u, q)$ the *reduced coefficients* of the equation. An equation of the form (1.1) having spatial $(A, \mathbf{b})$-structure in the cylinder Q is called $(A, \mathbf{b})$-*parabolic* (*strictly* $(A, \mathbf{b})$-*parabolic*) in Q if the following condition of A-parabolicity (strict A-parabolicity) is satisfied: for almost all $(t, x) \in Q$ and any $u \in \mathbb{R}$, $q = Ap$, $\eta = A\xi$, $p \in \mathbb{R}^n$ and $\xi \in \mathbb{R}^n$

$$\frac{\partial l'^i(t, x, q)}{\partial q_j} \eta_i \eta_j \geqslant 0 \qquad \left[\frac{\partial l'^i(t, x, u, q)}{\partial q_j} \eta_i \eta_j > 0, \quad \forall \eta \neq 0 \right]. \tag{1.7'}$$

It is easy to see that an equation of the form (1.1) which has spatial $(A, \mathbf{b})$-structure in the cylinder Q [is $(A, \mathbf{b})$-parabolic (strictly $(A, \mathbf{b})$-parabolic) in Q] also has $(\tilde{A}, \tilde{\mathbf{b}})$-structure) in Q [is $(\tilde{A}, \tilde{\mathbf{b}})$-elliptic (strictly $(\tilde{A}, \tilde{\mathbf{b}})$-elliptic) in Q] relative to the matrix

$$\tilde{A} = \begin{Vmatrix} 0 & \cdots & 0 \\ \vdots & & A \\ 0 & & \end{Vmatrix},$$

the vector $\tilde{\mathbf{b}} = (1, 0, \ldots, 0)$, and the reduced coefficients $\tilde{\mathbf{l}}'(\tilde{x}, u, \tilde{q})$ and $\tilde{l}'_0(\tilde{x}, u, \tilde{q})$, where $\tilde{x} = (t, x)$, $\tilde{q} = (q_0, q_1, \ldots, q_n)$, $\tilde{\mathbf{l}}' = (q_0, l'^1(x, t, u, q), \ldots, l'^n(x, t, u, q))$, $\tilde{l}'_0 = l'_0(x, t, u, q)$ and $q = (q_1, \ldots, q_n)$. The next assertions thus follow from Propositions 1.2 and 1.3.

PROPOSITION 1.2′. *The reduced coefficients* $l'^i(t, x, u, q)$, $i = 1, \ldots, n$ $(l'_0(t, x, u, q))$ *of an equation of the form* (1.1′) *having spatial* $(A, \mathbf{b})$-*structure in a cylinder* Q *are invariant under any smooth* (*linear*) *transformation of the spatial variables, i.e., under* *any smooth (linear) transformation of the spatial variables* $\hat{x} = \hat{x}(x)$, $x \in \overline{\Omega}$, *an equation of the form* (1.1) *having spatial* $(A, \mathbf{b})$*-structure in the cylinder* $Q = \Omega \times (T_1, T_2)$ *goes over into an equation of the form* (1.1) *having spatial* $(\hat{A}, \hat{\mathbf{b}})$*-structure in the cylinder* $\hat{Q} = \hat{\Omega} \times (T_1, T_2)$, *where* $\hat{\Omega}$ *is the image of* Ω *under the mapping* $\hat{x} = \hat{x}(x)$, $\hat{A} = AP^*$, $\hat{\mathbf{b}} = P\mathbf{b}$, *and* P *is the Jacobi matrix of the mapping in question, while the old and new equations have the same reduced coefficients; more precisely*

$$\hat{l}'^i(t, \hat{x}, u, \hat{q}) = l'^i(t, \hat{x}(x), u, \hat{q}), \quad i = 1, \dots, n$$
$$\left[\hat{l}'_0(t, \hat{x}, u, \hat{q}) = l'_0(t, x(\hat{x}), u, \hat{q})\right]. \tag{1.34'}$$

PROPOSITION 1.3′. *For any smooth, nondegenerate change of the spatial variables an* $(A, \mathbf{b})$*-parabolic (strictly* $(A, \mathbf{b})$*-parabolic) equation in the cylinder* $Q = \Omega \times (T_1, T_2)$ *goes over into an* $(\hat{A}, \hat{\mathbf{b}})$*-parabolic (strictly* $(\hat{A}, \hat{\mathbf{b}})$*-parabolic) equation in the cylinder* $\hat{Q} = \hat{\Omega} \times (T_1, T_2)$, *while the form of the general boundary value problem is unchanged.*

We now return to general $(A, \mathbf{b})$-elliptic equations. It will henceforth be more convenient for us to deal with an equation written in the form

$$-(d/dx_i)l^i(x, u, \nabla u) + l_0(x, u, \nabla u) = f(x), \tag{1.35}$$

i.e., an equation with the term $f(x)$, depending only on the independent variables, explicitly distinguished. We say that an equation of the form (1.35) is $(A, \mathbf{b}, m, \mathbf{m})$*-elliptic* (has $(A, \mathbf{b}, m, \mathbf{m})$*-structure*) in a domain Ω if the equation

$$-(d/dx_i)l^i(x, u, \nabla u) + \tilde{l}_0(x, u, \nabla u) = 0$$

has this property, where $\tilde{l}_0(x, u, p) = l_0(x, u, p) - f(x)$.

A smooth solution of an equation of the form (1.35) is any function $u \in C^2(\Omega) \cap C^1(\overline{\Omega})$ for which $l^i(x, u, \nabla u) \in C^1(\overline{\Omega})$, $i = 1, \dots, n$, $l_0(x, u, \nabla u) - f(x) \in C(\overline{\Omega})$ and at all points of Ω

$$-(d/dx_i)l^i(x, u, \nabla u) + l_0(x, u, \nabla u) = f(x).$$

PROPOSITION 1.4. *If the equation* (1.35) *has* $(A, \mathbf{b})$*-structure in a domain* Ω *and* u *is a smooth solution of the equation in* Ω, *then*

$$\int_\Omega \left[\mathbf{l}'(x, u, A\nabla u) \cdot A\nabla\eta + l'_0(x, u, A\nabla u)\eta - u(b^i\eta)_{x_i}\right] dx$$
$$+ \int_{\partial\Omega} \left[\mathbf{l}'(x, u, A\nabla u) \cdot A\boldsymbol{\nu} + bu\right] \eta\, ds$$
$$= \int_\Omega f\eta\, dx, \quad \forall \eta \in \tilde{C}^1(\Omega), \tag{1.36}$$

where $\boldsymbol{\nu}$ *is the unit vector of the inner normal to* $\partial\Omega$.

PROOF. Multiplying an identity of the form (1.35) by an arbitrary function $\eta \in \tilde{C}^1(\overline{\Omega})$, integrating over Ω, and applying the formula for integration by parts with condition (1.2) taken into account, we obtain (1.36). This proves Proposition 1.4.

PROPOSITION 1.5. *If equation* (1.35) *has* $(A, \mathbf{b})$*-structure in the domain* Ω *and* u *is a smooth solution of this equation in* Ω *satisfying the same boundary conditions as in* (1.29), *then*

$$\int_\Omega \left[\mathbf{l}'(x, u, A\nabla u)\cdot A\nabla\eta + l_0'(x, u, A\nabla u)\eta - u(b^i\eta)_{x_i}\right] dx$$

$$+\int_{\Sigma_3}\lambda u\eta\, ds + \int_{(\Sigma_{2,3})_+} bu\eta\, ds$$

$$= \int_\Omega f\eta\, ds, \quad \forall\eta \in \tilde{C}^1_{0,\Sigma_1\cup\Sigma'_+}(\Omega). \tag{1.37}$$

PROOF. Taking into account that $\partial\Omega = \Sigma_1 \cup \Sigma_2 \cup \Sigma_3 \cup \Sigma'$, we rewrite the boundary integral in (1.36) as the sum of integrals

$$\int_{\Sigma_1}(\mathbf{l}'\cdot A\nu + bu)\eta\, ds + \int_{(\Sigma_2)_{0,+}} \mathbf{l}'\cdot A\nu\eta\, ds + \int_{(\Sigma_2)_-}(\mathbf{l}'\cdot A\nu + bu)\eta\, ds$$

$$+\int_{(\Sigma_3)_{0,+}}(\mathbf{l}'\cdot A\nu - \lambda u)\eta\, ds + \int_{(\Sigma_3)_-}\left[\mathbf{l}'\cdot A\nu + (b-\lambda)u\right]\eta\, ds$$

$$+\int_{\Sigma'_{0,-}} bu\eta\, ds + \int_{\Sigma_3}\lambda u\eta\, ds + \int_{(\Sigma_{2,3})_+\cup\Sigma'_+} bu\eta\, ds. \tag{1.38}$$

Taking into account the fact that $\eta = 0$ on $\Sigma_1 \cup \Sigma'_+$ and $b = 0$ on Σ'_0, and the boundary conditions $\mathbf{l}'\cdot A\nu = 0$ on $(\Sigma_2)_+$, $\mathbf{l}'\cdot A\nu + bu = 0$ on $(\Sigma_2)_{0,-}$, $\mathbf{l}'\cdot A\nu - \lambda u = 0$ on $(\Sigma_3)_+$, $\mathbf{l}'\cdot A\nu + (b-\lambda)u = 0$ on $(\Sigma_3)_{0,-}$, and $u = 0$ on Σ'_-, we conclude that this sum of integrals is equal to the sum $\int_{\Sigma_3}\lambda u\eta\, ds + \int_{(\Sigma_{2,3})_+} bu\eta\, ds$, whence (1.37) follows. This proves Proposition 1.5.

Below an identity of the form (1.37) will form the basis for the definition of a generalized solution of the general boundary value problem of the form (1.29) with considerably broader assumptions regarding the structure of the equation and the domain Ω. First, however, we must consider some function spaces and operators connected with this problem.

§2. The basic function spaces and the operators connected with the general boundary value problem for an $(A, \mathbf{b}, m, \mathbf{m})$-elliptic equation

In a bounded domain $\Omega \subset \mathbb{R}^n$, $n \geqslant 2$, of class $\tilde{C}^{(1)}$ we consider an equation of the form (1.35) having $(A, \mathbf{b}, m, \mathbf{m})$-structure in Ω. We denote by $\widetilde{\partial\Omega}$ the set of all interior points of the smooth pieces constituting $\partial\Omega$ assuming that

$$\operatorname*{meas}_{n-1}\left(\partial\Omega\setminus\widetilde{\partial\Omega}\right) = 0,$$

and we assume that

$\widetilde{\partial\Omega} = \Sigma\cup\Sigma'$, where Σ is the regular and Σ' the singular part of $\partial\Omega$ (relative to the matrix A and the indices $m, \mathbf{m}$) in the sense of Definition 4.3.1. (2.1)

We decompose the regular part of Σ in an arbitrary manner into parts Σ_1, Σ_2 and Σ_3 such that $\Sigma_1 \cup \Sigma_2 \cup \Sigma_3 = \Sigma$ and $\Sigma_i \cap \Sigma_j = \varnothing$ for $i \neq j$, $i, j = 1, 2, 3$, and we assume that $\text{meas}_{n-1}(\partial\Sigma_i) = 0$, $i = 1, 2, 3$, where $\partial\Sigma_i$ denotes the boundary of Σ_i on $\partial\Omega$. Suppose that a piecewise continuous, bounded, positive function λ is defined on Σ_3.

The completion of the set $\tilde{C}_{0,\Sigma}(\overline{\Omega})$ in the norm

$$\|u\|_{H_\lambda} = \|u\|_H + \|u\|_{L^2(\lambda,\Sigma_3)} \equiv \|u\|_{m,\Omega} + \|A\nabla u\|_{\mathbf{m},\Omega} + \|u\|_{L^2(\lambda,\Sigma_3)} \tag{2.2}$$

we denote by $H_\lambda \equiv H_{m,\mathbf{m}}(A; \Omega; \Sigma_3, \lambda)$. In the case $\Sigma_3 = \varnothing$ the space H_λ coincides with the space $H \equiv H^{0,\Gamma}_{m,\mathbf{m}}(A, \Omega)$ with $\Gamma = \Sigma_1$ introduced in Chapter 4, §2.

LEMMA 2.1. *The space H_λ is separable and reflexive. Any linear functional F in H_λ can be defined by the equality*

$$\langle F, \eta\rangle = \int_\Omega (f_0\eta + \mathbf{f}\cdot A\nabla\eta)\,dx + \int_{\Sigma_3} \lambda\psi\eta\,ds, \qquad \eta \in \tilde{C}^1_{0,\Sigma_1}(\overline{\Omega}), \tag{2.3}$$

where $f_0 \in L^{m'}(\Omega)$, $1/m + 1/m' = 1$, $\mathbf{f} = (f^1,\dots,f^n)$, $f^i \in L^{m_i'}(\Omega)$, $1/m_i + 1/m_i' = 1$, $i = 1,\dots,n$, $\psi \in L^2(\Omega)$, and f_0, $\mathbf{f}$, and ψ can be chosen so that

$$\|F\|_{H_\lambda^*} = \sup\left(\|f_0\|_{m',\Omega}, \|f^1\|_{m_1',\Omega}, \dots, \|f^n\|_{m_n',\Omega}, \|\psi\|_{L^2(\lambda,\Sigma_3)}\right), \tag{2.4}$$

where $\|F\|_{H_\lambda^*}$ is the norm of F. Conversely, any expression of the form (2.3) *with f_0, $\mathbf{f}$, and ψ satisfying the conditions indicated above defines a linear functional in H_λ with norm not exceeding the quantity on the right side of* (2.4).

PROOF. The separability of the space H_λ is obvious. To prove that H_λ is reflexive we consider the mapping

$$\pi\colon \tilde{C}^1_{0,\Sigma_1}(\Omega) \subset H_\lambda \to L^m(\Omega) \times \mathbf{L}^{\mathbf{m}}(\Omega) \times L^2(\lambda, \Sigma_3),\text{[1]}$$

defined by

$$\pi(u) = \left(u, A_1\nabla u, \dots, A_n\nabla u, u|_{\Sigma_3}\right), \qquad u \in \tilde{C}^1_{0,\Sigma_1}(\overline{\Omega}).$$

Obviously the mapping π is linear and isometric. We extend it to the entire space H_λ. The extended mapping $\pi\colon H_\lambda \to L^m(\Omega) \times \mathbf{L}^{\mathbf{m}}(\Omega) \times L^2(\lambda, \Sigma_3)$ is also linear and isometric. Therefore, the image $\pi(H_\lambda)$ of H_λ is closed in $L^m(\Omega) \times \mathbf{L}^{\mathbf{m}}(\Omega) \times L^2(\lambda, \Sigma_3)$. Taking account of the fact that the latter space is reflexive, we conclude that H_λ is also reflexive. We now prove that any linear functional F in H_λ can be defined by (2.3). We note first of all that the prescription of the functional F on the set $\tilde{C}^1_{0,\Sigma_1}(\overline{\Omega})$ determines it completely, since $\tilde{C}^1_{0,\Sigma_1}(\overline{\Omega})$ is dense in H_λ. The condition $F \in H_\lambda^*$ implies that $F \circ \pi^{-1}$ is a linear functional on the subspace $\pi(H_\lambda)$ of the space $L^m(\Omega) \times \mathbf{L}^{\mathbf{m}}(\Omega) \times L^2(\lambda, \Sigma)$. By the Hahn-Banach theorem this functional can be extended with preservation of norm to the entire space $L^m(\Omega) \times \mathbf{L}^{\mathbf{m}}(\Omega) \times L^2(\lambda, \Sigma_3)$. Now any linear functional Φ in the latter space can be represented in the form

$$\langle \Phi, (v_0, \mathbf{v}, \varphi)\rangle = \int_\Omega (f_0 v_0 + \mathbf{f}\cdot\mathbf{v})\,dx + \int_{\Sigma_3} \lambda\psi\varphi\,ds, \tag{2.5}$$

[1] The space $L^m(\Omega) \times \mathbf{L}^{\mathbf{m}}(\Omega) \times L^2(\lambda, \Sigma_3)$ is equipped with the norm

$$\|(v_0, \mathbf{v}, \varphi)\| = \|v_0\|_{m,\Omega} + \|\mathbf{v}\|_{\mathbf{m},\Omega} + \|\varphi\|_{L^2(\lambda,\Sigma_3)},$$

where $\|\mathbf{v}\|_{\mathbf{m},\Omega} \equiv \|\mathbf{v}\|_{\mathbf{L}^{\mathbf{m}}(\Omega)} \equiv \sum_{i=1}^n \|v^i\|_{m_i,\Omega}$, $(v_0, \mathbf{v}, \varphi) \in L^m(\Omega) \times \mathbf{L}^{\mathbf{m}}(\Omega) \times L^2(\lambda, \Sigma_3)$ and $\mathbf{v} = (v^1, \dots, v^n)$.

where f_0, $\mathbf{f}$, and ψ are as in the formulation of the lemma. Therefore, for all $\eta \in \tilde{C}^1_{0,\Sigma_1}(\Omega)$ the value $\langle F, \eta\rangle$ of the functional F at the function η is determined by the formula

$$\langle F, \eta\rangle = \langle (F \circ \pi^{-1}), (\eta, A\nabla\eta, \eta|_{\Sigma_3})\rangle$$
$$= \int_\Omega (f_0\eta + \mathbf{f}\cdot A\nabla\eta)\,dx + \int_{\Sigma_3} \lambda\psi\eta\,ds, \tag{2.6}$$

and the norm $\|F\|_{H^*_\lambda}$ coincides with the norm of the functional Φ obtained in extending the functional $F \circ \pi^{-1}$, i.e., the norm of F in H_λ is given by (2.4). The last part of the lemma follows easily from the estimate

$$\left|\int_\Omega (f_0\eta + \mathbf{f}\cdot A\nabla\eta)\,dx + \int_{\Sigma_3}\lambda\psi\eta\,ds\right|$$
$$\leqslant \sup\left(\|f_0\|_{m',\Omega}, \|f^1\|_{m'_1,\Omega}, \ldots, \|f^n\|_{m'_n,\Omega}, \|\psi\|_{L^2(\lambda,\Sigma_3)}\right)\|\eta\|_{H_\lambda}, \tag{2.7}$$

which is valid for any $\eta \in \tilde{C}^1_{0,\Sigma_1}(\overline{\Omega})$. The lemma is proved.

LEMMA 2.2. *There is the dense imbedding $H_\lambda \to H$. In the case $m \geqslant 2$, $m_i \geqslant 2$, $i = 1,\ldots,n$, if condition* (4.2.25) *is satisfied for Σ_3 the spaces H_λ and H are isomorphic.*

PROOF. Since any subset of the regular part of $\partial\Omega$ is also regular, it follows that Σ_3 is a regular part of $\partial\Omega$. Then, obviously, any sequence convergent in itself in H_λ can be identified with some element of H, whence the first part of the lemma easily follows. The second part of the lemma follows from (4.2.26) (with $m_* = 2$) and the condition of boundedness of λ on Σ_3. Lemma 2.2 is proved.

Thus, elements of the space $H_{m,\mathbf{m}}(A; \Omega; \Sigma_3, \lambda)$ are functions in $L^m(\Omega)$ having a generalized A-gradient $A\nabla u \in \mathbf{L}^{\mathbf{m}}(\Omega)$ and generalized limit values $u|_{\Sigma_3}$ on the set Σ_3, where $u|_{\Sigma_3} \in L^2(\lambda, \Sigma_3)$.

LEMMA 2.3. *Suppose that some set $\mathscr{P} \subset \Sigma'$ satisfies condition* (4.3.2). *Then the set $\tilde{C}^1_{0,\Sigma_1\cup\mathscr{P}}(\overline{\Omega})$ is dense in $H_\lambda \equiv H^{0,\Sigma_1}_{m,\mathbf{m}}(A; \Omega; \Sigma_3, \lambda)$.*

PROOF. Lemma 2.3 is proved in exactly the same way as Lemma 3.1 of Chapter 1. We note only that on the right side of (4.3.5) appears the additional term $(\int_{\Sigma_3\cap\omega_\delta}\lambda u^2\,ds)^{1/2}$. Because $\Sigma_3 \cap \mathscr{P} = \varnothing$ and $\text{meas}_{n-1}(\partial\Sigma_3) = 0$, this term tends to 0 as $\delta \to 0$. The remainder of the argument is a literal repetition of the proof of Lemma 2.2. This proves Lemma 2.3.

We introduce two more function spaces. We denote by $(\Sigma_i)_0$, $(\Sigma_i)_+$ and $(\Sigma_i)_-$ those parts of the set Σ_i $(i = 2, 3)$ on which the function b defined by (1.22) is respectively equal to 0, greater than 0, and less than 0. We decompose Σ' into parts Σ'_0, Σ'_+, and Σ'_- in a similar way. We assume that

$$\text{meas}_{n-1}\partial(\Sigma_i)_+ = \text{meas}_{n-1}\partial(\Sigma_i)' = 0, \qquad i = 2, 3;$$
$$\text{meas}_{n-1}\partial\Sigma'_+ = \text{meas}_{n-1}\partial\Sigma'_- = 0. \tag{2.8}$$

We set

$$\beta(x) = \sum_{i=1}^n \left(|b^i(x)| + \left|\frac{\partial b^i(x)}{\partial x_i}\right|\right), \quad \Omega_\beta \equiv \{x \in \Omega: \beta(x) > 0\},$$
$$\Omega_{b^i} \equiv \{x \in \Omega: b^i(x) \neq 0\}, \qquad i = 1,\ldots,n.$$

We introduce the following norms:

$$\|u\|_{L^2(\beta,\Omega)} = \left(\int_{\Omega_\beta} \beta u^2\,dx\right)^{1/2}, \quad \|u\|_{L^2(|b|,(\Sigma_{2,3})_\pm \cup \Sigma'_+)} = \left(\int_{(\Sigma_{2,3})_\pm \cup \Sigma'_+} |b| u^2\,ds\right)^{1/2},$$

$$\|u\|_{L^2(|b^i|,\Omega_{b^i})} = \left(\int_{\Omega^{b^i}} |b^i| u^2\,dx\right)^{1/2}, \tag{2.9}$$

where $(\Sigma_{2,3})_\pm \equiv (\Sigma_2)_\pm \cup (\Sigma_3)_\pm$ and $(\Sigma_i)_\pm = (\Sigma_i)_+ \cup (\Sigma_i)_-$, $i = 2, 3$.

The completion of the set $\tilde{C}^1_{0,\Sigma_1}(\overline{\Omega})$ in the norms

$$\|u\|_X = \|u\|_{H_\lambda} + \|u\|_{L^2(\beta,\Omega_\beta)} + \|u\|_{L^2(|b|,(\Sigma_{2,3})_\pm \cup \Sigma'_+)} \tag{2.10}$$

and

$$\|u\|_Y = \|u\|_X + \sum_{i=1}^n \|u_{x_i}\|_{L^2(|b^i|,\Omega_{b^i})} \tag{2.11}$$

we denote respectively by

$$X \equiv X^{0,\Sigma_1}_{m,\mathbf{m}}(A; \mathbf{b}; \Omega; \Sigma_2; \Sigma_3, \lambda) \quad \text{and} \quad Y \equiv Y^{0,\Sigma_1}_{m,\mathbf{m}}(A; \mathbf{b}; \Omega; \Sigma_2; \Sigma_3, \lambda)$$

The following lemma is established in complete analogy to Lemma 2.1.

LEMMA 2.4. *The space X (Y) is separable and reflexive. Any linear functional F in X (Y) can be defined by*

$$\langle F, \eta\rangle = \int_\Omega (f_0\eta + \mathbf{f}\cdot A\nabla\eta)\,dx + \int_\Omega \beta g_0 \eta\,dx$$
$$+ \int_{(\Sigma_{2,3})_\pm \cup \Sigma'_+} |b| h\eta\,ds \left[+ \sum_{i=1}^n |b^i| g_i \eta_{x_i}\,dx\right], \qquad \eta \in \tilde{C}^1_{0,\Sigma_1}(\overline{\Omega}), \tag{2.12}$$

where f_0 and $\mathbf{f}$ are the same functions as in (2.3),

$$g_0 \in L^2(\beta, \Omega_\beta), \quad q \in L^2(\lambda, \Sigma_3), \quad h \in L^2(|b|, (\Sigma_{2,3})_\pm \cup \Sigma'_+),$$
$$g^i \in L^2(|b^i|, \Omega_{b^i}), \qquad i = 1, \dots, n,$$

and f_0, $\mathbf{f}$, g_0, q, h, and the g^i, $i = 1, \dots, n$, can be chosen so that the norm $\|F\|_{X^*}$ ($\|F\|_{Y^*}$) is equal to the supremum of the norms

$$\|f_0\|_{m',\Omega}, \|f^1\|_{m'_1,\Omega}, \dots, \|f^n\|_{m'_n,\Omega}, \qquad \|g_0\|_{L^2(\beta,\Omega_\beta)}, \qquad \|q\|_{L^2(\lambda,\Sigma_3)},$$
$$\|h\|_{L^2(|b|(\Sigma_{2,3})_\pm \cup \Sigma'_+)} \qquad \left[\|g^1\|_{L^2(|b^1|,\Omega_{b^1})}, \dots, \|g^n\|_{L^2(|b^n|,\Omega_{b^n})}\right].$$

Conversely, any expression of the form (2.12) *for which the above conditions for f_0, $\mathbf{f}$, g_0, q, h (and the g^i, $i = 1, \dots, n$) are satisfied defines a linear functional F in X (in Y) with norm not exceeding the indicated supremum.*

We denote by $\hat{H}_\lambda$ the completion of $\tilde{C}^1_{0,\Sigma_1}(\overline{\Omega})$ in the norm $\|u\|_{H_\lambda} + \|u\|_{L^2(\beta,\Omega_\beta)}$. It is obvious that $\hat{H}_\lambda \to H_\lambda$.

LEMMA 2.5. *The space X can be identified with a subspace of $H_\lambda \times L^2(b, \Sigma'_+)$.*

PROOF. The process of proving Lemma 2.4 (altogether analogous to the proof of Lemma 2.1) shows that X can be identified with a subspace of

$$L^m(\Omega) \times \mathbf{L}^{\mathbf{m}}(\Omega) \times L^2(\beta, \Omega_\beta) \times L^2(\lambda, \Sigma_3) \times L^2(|b|, (\Sigma_{2,3})_\pm \cup \Sigma'_+)$$

by writing elements $\mathbf{u}$ of X in the form $\mathbf{u} = (u, \mathbf{v}, \tilde{u}, \varphi, \psi)$, where $u \in L^m(\Omega)$, $\mathbf{v} \in \mathbf{L^m}(\Omega)$, $\tilde{u} \in L^2(\beta, \Omega_\beta)$, $\varphi \in L^2(\lambda, \Sigma_3)$ and $\psi \in L^2(|b|, (\Sigma_{2,3})_\pm \cup \Sigma'_+)$. However, taking into account that the components $\mathbf{v}$, $\tilde{u}$, φ, and the restriction of ψ to $(\Sigma_{2,3})_\pm$ are uniquely determined by the first component u of the element $\mathbf{u} \in X$, we can identify X with a subspace of $\hat{H}_\lambda \times L^2(b, \Sigma'_+)$ by writing elements $\mathbf{u} \in X$ as pairs $\mathbf{u} = (u, \varphi)$, where $u \in \hat{H}_\lambda$ and $\varphi \in L^2(b, \Sigma'_+)$. This proves Lemma 2.5.

REMARK 2.1. Let $\operatorname{meas}_{n-1} \Sigma'_+ > 0$, and suppose that for the set Σ'_+ the following condition is satisfied:(²)

$$\text{the set } \tilde{C}^1_{0,\Sigma_1 \cup \Sigma'_+}(\overline{\Omega}) \text{ is dense in } \hat{H}_\lambda. \tag{2.13}$$

Then X cannot be identified with a subspace of $\hat{H}_\lambda$.

Indeed, in view of (2.13) there exists a sequence $\{u_n\}$, $u_n \in \tilde{C}^1_{0,\Sigma_1\cup\Sigma'_+}(\overline{\Omega})$, $n = 1, 2, \ldots$, converging in H_λ to a given function $u \in \tilde{C}^1_{0,\Sigma_1}(\overline{\Omega})$ which is not equal to 0 on Σ'_+. It is obvious that for the stationary sequence $\{\tilde{u}_n\}$, where $\tilde{u}_n = u$, $n = 1, 2, \ldots$, there is also convergence to u in H_λ. Hence, $\lim_{n\to\infty} \|u_n - \tilde{u}_n\|_{\hat{H}_\lambda} = 0$. It is obvious, however, that

$$\lim_{n\to\infty} \|u_n - \tilde{u}_n\|_{L^2(b, \Sigma'_+)} \neq 0,$$

i.e., $\|u_n - \tilde{u}_n\|_X \nrightarrow 0$ as $n \to \infty$.

LEMMA 2.6. *The space Y can be identified both with a subspace of X and with a subspace of $\hat{H}_\lambda$.*

PROOF. The process of proving Lemma 2.3 (see the proof of Lemma 2.1) shows that Y can be identified with a subspace of

$$L^m(\Omega) \times \mathbf{L^m}(\Omega) \times L^2(\beta, \Omega_\beta) \times L^2\big(|b|, (\Sigma_{2,3})_\pm \cup \Sigma'_+\big) \times \prod_{i=1}^{n} L^2\big(|b^i|, \Omega_{b^i}\big)$$

by writing an element $\mathbf{u} \in Y$ in the form $\mathbf{u} = (u, \mathbf{v}, \tilde{u}, \varphi, \psi, \mathbf{z})$, where $u \in L^m(\Omega)$, $\mathbf{v} \in \mathbf{L^m}(\Omega)$, $\tilde{u} \in L^2(\beta, \Omega_\beta)$, $\varphi \in L^2(\lambda, \Sigma_3)$, $\psi \in L^2(|b|, (\Sigma_{2,3})_\pm \cup \Sigma'_+)$ and $\mathbf{z} \in \prod_1^n L^2(|b^i|, \Omega_{b^i})$. However, as we noted in the proof of Lemma 2.5, the components $\mathbf{v}$, u, φ and the restriction of ψ to the set $(\Sigma_{2,3})_\pm$ are uniquely determined by the first component u of this element, and $\mathbf{v} = A\nabla u$, $\tilde{u} = u$, $\varphi = u|_{\Sigma_3}$ and $\psi|_{(\Sigma_{2,3})_\pm} = u|_{(\Sigma_{2,3})_\pm}$. We shall now prove that the components z_i, $i = 1, \ldots, n$, of the element $\mathbf{u} \in Y$ are also uniquely determined by the first component u of this element. Suppose that the sequence $\{u_k\}$, $u_k \in \tilde{C}^1_{0,\Sigma_1}(\overline{\Omega})$, $k = 1, 2, \ldots$, converges to $\mathbf{u}$ in Y. This implies, in particular, that $u_k \to u$ in $L^m(\Omega)$, $u_k \to u$ in $L^2(\beta, \Omega_\beta)$, and

$$\lim_{k\to\infty} \int_{\Omega^{b^i}} |b^i| (u_{kx_i} - z_i)^2\, dx = 0, \qquad i = 1, \ldots, n.$$

Passing to the limit as $k \to \infty$ in the identities

$$\int_\Omega b^i u_{kx_i} \eta\, dx = -\int_\Omega \left(\frac{\partial b^i}{\partial x_i} u_k \eta + b^i u_k \eta_{x_i} \right) dx, \qquad \eta \in \tilde{C}^1_0(\overline{\Omega}),\ i = 1, \ldots, n, \tag{2.14}$$

(²) Condition (4.3.2) for the set Σ'_+ is a sufficient condition that (2.13) be satisfied; this assertion is proved in exactly the same way as Lemma 2.3.

we obtain

$$\int_\Omega b^i u\eta_{x_i}\,dx = -\int_\Omega\left(\frac{\partial b^i}{\partial x_i}u + b^i\tilde{z}_i\right)\eta\,dx \quad \forall\eta\in\tilde{C}_0^1(\Omega),\ i=1,\dots,n, \tag{2.15}$$

where we have defined $\tilde{z}_i = z_i$ on Ω_{b^i} and $\tilde{z}_i = 0$ on $\Omega\setminus\Omega_{b^i}$, $i=1,\dots,n$.

Since $b^i\tilde{z}_i\in L^1(\Omega)$, $i=1,\dots,n$, this implies that the functions b^iu have generalized derivatives $\partial(b^iu)/\partial x_i$, $i=1,\dots,n$, in Ω, and

$$\partial(b^iu)/\partial x_i = \partial b^iu/\partial x_i + b^i\tilde{z}_i \quad \text{for almost all } x\in\Omega,\ i=1,\dots,n. \tag{2.16}$$

From (2.16) we obtain

$$z_i = b_i^{-1}\left[\partial b^iu/\partial x_i - \partial(b^iu)/\partial x_i\right] \quad \text{for almost all } x\in\Omega_{b^i},\ i=1,\dots,n. \tag{2.17}$$

Using known facts from the theory of generalized derivatives ([97], pp. 43–45), from (2.16) we can also deduce that u has generalized derivatives $\partial u/\partial x_i$ in the domains Ω_{b^i} and $\partial u/\partial x_i = z_i\in L^1_{\text{loc}}(\Omega_{b^i})\cap L^2(|b^i|,\Omega_{b^i})$, $i=1,\dots,n$. By $\partial u/\partial x_i$ we henceforth understand z_i extended by zero by $\Omega\setminus\Omega_{b^i}$, $i=1,\dots,n$. Then in place of (2.16) we can write the equalities

$$\partial(b^iu)/\partial x_i = \partial b^iu/\partial x_i + b^i(\partial u/\partial x_i) \quad \text{for almost all } x\in\Omega,\ i=1,\dots,n. \tag{2.18}$$

In particular, it follows from (2.18) that $b^i\partial u/\partial x_i\in L^1(\Omega)$, $i=1,\dots,n$.

The assertions presented above not only prove that the component $\mathbf{z}=(z_1,\dots,z_n)$ is uniquely determined by the component u, but also establish specific properties of u. From what has been proved it follows that Y can be identified with a subspace of X, and the elements $\mathbf{u}\in Y$ can be written in the form $\mathbf{u}=(u,\varphi)$, where $u\in\hat{H}_\lambda$ and $\varphi\in L^2(b,\Sigma'_+)$, i.e., in the same way as elements $\mathbf{u}$ of X. We shall prove, however, that for an element $\mathbf{u}\equiv(u,\varphi)\in Y$ the second component φ is uniquely determined by the first component u. Indeed, suppose that a sequence $\{u_n\}$, $u_n\in\tilde{C}^1_{0,\Sigma_1}(\overline{\Omega})$, $n=1,2,\dots$, converges to $\mathbf{u}=(u,\varphi)$ in Y. Then, passing to the limit in the identity

$$\int_\Omega b^iu_{nx_i}\eta\,dx = -\int_\Omega\left(\frac{\partial b^i}{\partial x_i}u_n\eta + b^iu_n\eta_{x_i}\right)dx - \int_{\Sigma'_+} b^i\nu_iu_n\eta\,ds, \qquad \eta\in\tilde{C}^1_{0(\Sigma_{2,3})_\pm\cup\Sigma'_-}(\overline{\Omega}), \tag{2.19}$$

we find that

$$\int_\Omega b^iu_{x_i}\eta\,dx = -\int_\Omega\left(\frac{\partial b^i}{\partial x_i}u\eta + b^iu\eta_{x_i}\right)dx - \int_{\Sigma'_+} b^i\nu_i\varphi\eta\,ds, \qquad \eta\in\tilde{C}^1_{0(\Sigma_{2,3})_\pm\cup\Sigma'_-}(\overline{\Omega}), \tag{2.20}$$

where we have used the fact that convergence of $\{u_n\}$ to $\mathbf{u}$ in Y implies, in particular, convergence of $\{u_n\}$ to φ in $L^2(b,\Sigma'_+)$. It follows from (2.20) that the component φ of the element $\mathbf{u}\in Y$ is uniquely determined by u, since, as is evident from (2.20), the values of the integral

$$\int_{\Sigma'_+} b^i\nu_i\varphi\eta\,ds, \qquad \eta\in\tilde{C}^1_{0(\Sigma_{2,3})_\pm\cup\Sigma'_-}(\overline{\Omega})$$

are completely determined by u; $b^i\nu_i \equiv -b \neq 0$ on Σ'_+, and the restrictions of functions $\eta \in \tilde{C}^1_{0,(\Sigma_{2,3})_+\cup\Sigma'_-}(\overline{\Omega})$ to Σ'_+ are dense in $L^2(b, \Sigma'_+)$. We therefore agree henceforth to denote the component φ of an element $\mathbf{u} = (u, \varphi) \in Y$ by $\varphi \equiv [u]_{\Sigma'_+}$. It follows from what has been proved that Y can be identified with a subspace of $\hat{H}_\lambda$, and its elements $\mathbf{u}$ can be written either in the form $\mathbf{u} = (u, [u]_{\Sigma'_+})$ or simply in the form of functions $\mathbf{u} = u \in \hat{H}_\lambda$. Lemma 2.6 is proved.

It is obvious that the following imbeddings ensue from Lemmas 2.1–2.4 and 2.6:

$$X \to \hat{H}_\lambda \times L^2(b, \Sigma'_+), \quad Y \to X, \quad Y \to \hat{H}_\lambda, \quad L^{m'}(\Omega) \to H^*_\lambda \to X^* \to Y^*. \tag{2.21}$$

Thanks to these imbeddings and the existence of a common dense set $\tilde{C}^1_{0,\Sigma_1}(\overline{\Omega})$ in H_λ, X, and Y it is possible to use the same notation $\langle\,\cdot\,,\cdot\rangle$ for the duality between H_λ and H^*_λ, X and X^*, and Y and Y^*.

REMARK 2.2. In connection with the fact that elements of X are realized as pairs $\mathbf{u} = (u, \varphi)$, the same notation should be applied to elements of the original set $\tilde{C}^1_{0,\Sigma_1}(\overline{\Omega})$ whose closure in the norm (2.10) gives X. We agree, however, when considering an element $u \in \tilde{C}^1_{0,\Sigma_1}(\overline{\Omega})$ as an element of X to write it in the usual way, identifying a function $u \in \tilde{C}^1_{0,\Sigma_1}(\overline{\Omega})$ with the pair $(u, u|_{\Sigma'_+})$, where $u|_{\Sigma'_+}$ is the restriction of $u \in \tilde{C}^1_{0,\Sigma_1}(\overline{\Omega})$ to Σ'_+.

We consider the linear operator $\mathscr{B}$: $\tilde{C}^1_{0,\Sigma_1}(\overline{\Omega}) \subset X \to Y^*$ defined by

$$\langle \mathscr{B}u, \eta\rangle = -\int_\Omega u(b^i\eta)_{x_i}\,dx + \int_{(\Sigma_{2,3})_+\cup\Sigma'_+} bu\eta\,ds, \qquad u, \eta \in \tilde{C}^1_{0,\Sigma_1}(\overline{\Omega}). \tag{2.22}$$

It is obvious that for any $u, \eta \in \tilde{C}^1_{0,\Sigma_1}(\overline{\Omega})$

$$|\langle \mathscr{B}u, \eta\rangle| \leqslant \|u\|_X\|\eta\|_Y. \tag{2.23}$$

Because of the linearity of $\mathscr{B}$ and (2.23), the operator $\mathscr{B}$ is bounded (and hence also continuous). It can therefore be extended to the entire space X. Henceforth $\mathscr{B}$: $X \to Y^*$ always denotes this extended operator.

We denote by V an (algebraic) subspace of X which is important for subsequent considerations; it is defined by the condition

$$V = \{\mathbf{u} \in X: \mathscr{B}\mathbf{u} \in H^*_\lambda\}. \tag{2.24}$$

LEMMA 2.7. *Suppose that the following condition is satisfied*:

$$\text{the set } \tilde{C}^1_{0,\Sigma_1\cup\Sigma'_+}(\overline{\Omega}) \text{ is dense in } H_\lambda. \tag{2.25}$$

Then V can be identified with a subspace of $\hat{H}_\lambda$.

PROOF. We note first that the restriction of the operator $\mathscr{B}$: $X \to Y^*$ to V can be completely defined by giving the values $\langle \mathscr{B}\mathbf{u}, \eta\rangle$, $\mathbf{u} \in V$, $\eta \in \tilde{C}^1_{0,\Sigma_1\cup\Sigma'_+}(\overline{\Omega})$, where

$$\langle \mathscr{B}\mathbf{u}, \eta\rangle = -\int_\Omega u(b^i\eta)_{x_i}\,dx + \int_{(\Sigma_{2,3})_+} bu|_{(\Sigma_{2,3})_+}\eta\,ds. \tag{2.26}$$

Indeed, passing to the limit as $n \to \infty$ in an equality of the form (2.22) written for functions $u_n \in \tilde{C}^1_{0,\Sigma_1}(\overline{\Omega})$, $n = 1, 2, \ldots$, converging to a given element $\mathbf{u} = (u, \varphi) \in V$ in X, we obtain

$$\langle \mathscr{B}\mathbf{u}, \eta\rangle = -\int_\Omega u(b^i\eta)_{x_i}\,dx + \int_{(\Sigma_{2,3})_+} bu|_{(\Sigma_{2,3})_+}\eta\,ds$$
$$+ \int_{\Sigma'_+} b\varphi\eta\,ds, \qquad \eta \in \tilde{C}^1_{0,\Sigma_1}(\overline{\Omega}). \tag{2.27}$$

Taking account of the density of $\tilde{C}^1_{0,\Sigma_1\cup\Sigma'_+}(\overline{\Omega})$ in H_λ, we prove the above assertion and, in particular, obtain (2.26). The definition of the subspace V implies the existence of an element $F \in H^*_\lambda$ such that

$$-\int_\Omega u(b^i\eta)_{x_i}\,dx + \int_{(\Sigma_{2,3})_+} bu|_{(\Sigma_{2,3})_+}\eta\,ds + \int_{\Sigma'_+} b\varphi\eta\,ds = \langle F, \eta\rangle, \qquad \eta \in \tilde{C}^1_{0,\Sigma_1}(\overline{\Omega}). \tag{2.28}$$

Let $\{\eta_n\}$, $\eta_n \in \tilde{C}^1_{0,(\Sigma_{2,3})_+\cap\Sigma'_+}(\overline{\Omega})$, $n = 1, 2, \ldots$, be a sequence converging to a fixed function $\eta \in \tilde{C}^1_{0,\Sigma_1}(\overline{\Omega})$ in H_λ. We denote by $\hat{\mathscr{B}}(u, \eta)$ the expression

$$\hat{\mathscr{B}}(u, \eta) \equiv -\int_\Omega u\frac{\partial}{\partial x_i}(b^i\eta)\,dx + \int_{(\Sigma_{2,3})_+} bu|_{(\Sigma_{2,3})_+}\eta\,ds. \tag{2.29}$$

The expression (2.29) is meaningful for any $u \in H_\lambda$ and $\eta \in \tilde{C}^1_{0,\Sigma_1}(\overline{\Omega})$. In view of (2.28) and (2.29),

$$\langle F, \eta\rangle = \lim_{\eta_n\to\eta \text{ in } H_\lambda} \hat{\mathscr{B}}(u, \eta_n), \tag{2.30}$$

where u is the first component of the element $(u, \varphi) \in V$ under consideration. From (2.28)–(2.30) we then obtain

$$\int_{\Sigma'_+} b\varphi\eta\,ds = \lim_{\eta_n\to\eta \text{ in } H_\lambda} \hat{\mathscr{B}}(u, \eta_n - \eta) \tag{2.31}$$

for any sequence $\{\eta_n\}$, $\eta_n \in \tilde{C}^1_{0,\Sigma_1\cup\Sigma'_+}(\overline{\Omega})$, $n = 1, 2, \ldots$, converging to η in H_λ. Taking account of the fact that $b \neq 0$ on Σ'_+ and the arbitrariness of the restrictions of functions η to Σ'_+, we conclude in view of (2.31) that the component φ of the element $\mathbf{u} = (u, \varphi) \in V$ is completely determined by the first component $u \in \hat{H}_\lambda$. Lemma 2.7 is proved.

We agree to denote the component $\varphi \in L^2(b, \Sigma'_+)$ of an element $(u, \varphi) \in V$ by $(u)_{\Sigma'_+}$. Thus,

$$\int_{\Sigma'_+} b(u)_{\Sigma'_+}\eta\,ds = \lim_{\eta_n\to\eta \text{ in } H_\lambda} \hat{\mathscr{B}}(u, \eta_n - \eta), \tag{2.32}$$

where η and η_n are the same as in (2.31). We henceforth write elements $\mathbf{u} = (u, (u)_{\Sigma'_+})$ of V simply as functions $u \in H_\lambda$. In particular, (2.26) can be rewritten in the form

$$\langle \mathscr{B}u, \eta\rangle = -\int_\Omega u(b^i\eta)_{x_i}\,dx + \int_{(\Sigma_{2,3})_+} bu|_{(\Sigma_{2,3})_+}\eta\,ds, \qquad u \in V, \eta \in \tilde{C}^1_{0,\Sigma_1\cup\Sigma'_+}(\overline{\Omega}). \tag{2.33}$$

REMARK 2.3. In view of Lemma 2.3, condition (2.25) is certainly satisfied if

$$\text{condition (4.3.2) is satisfied for the set } \Sigma'_+. \tag{2.34}$$

If condition (2.25) is satisifed, V can be identified with a subspace of $\hat{H}_\lambda$. We now pose the following question: under what condition is a function $u \in \hat{H}_\lambda$ contained in V?

LEMMA 2.8. *Suppose condition* (2.25) *is satisfied. A function u belonging to* $\hat{H}_\lambda$ *is an element of the subspace V if and only if the following conditions hold*:

1) *There exists a sequence* $\{u_k\}$, $u_k \in \tilde{C}^1_{0,\Sigma_1}(\overline{\Omega})$, $k = 1, 2, \ldots$, *such that*

$$\lim_{k\to\infty} \|u_k - u\|_{H_\lambda} = 0, \qquad \lim_{k,s\to\infty} \|u_k - u_s\|_X = 0. \tag{2.35}$$

2) *For any function* $\eta \in \tilde{C}^1_{0,\Sigma_1\cup\Sigma'_+}(\overline{\Omega})$ *and any sequence* $\{\eta_k\}$, $\eta_k \in \tilde{C}^1_{0,\Sigma_1\cup\Sigma'_+}(\overline{\Omega})$, $k = 1, 2, \ldots$, *converging in* H_λ *to* η,

$$\lim_{k\to\infty} \hat{\mathscr{B}}(u, \eta_k - \eta) = 0.$$

PROOF. If $u \in V$, then the validity of (2.35) follows from the definition of V and Lemma 2.7. In particular, condition 2) of (2.35) follows from (2.32), since for all $\eta \in \tilde{C}^1_{0,\Sigma_1\cup\Sigma'_+}(\overline{\Omega})$ the left side of (2.32) is equal to 0. Suppose now that conditions (2.35) are satisfied for some function $u \in \hat{H}_\lambda$. From condition 1) in (2.35) it then follows that u is the first component of some element $\mathbf{u} = (u, \varphi) \in X$. Noting (see (2.27) and (2.29)) that $\langle \mathscr{B}(u,\varphi), \eta\rangle = \hat{\mathscr{B}}(u, \eta)$ for all $\eta \in \tilde{C}^1_{0,\Sigma_1\cup\Sigma'_+}(\overline{\Omega})$, we deduce from condition 2) of (2.35) that for the element $\mathbf{u} = (u, \varphi)$ the linear mapping $\eta \to \langle \mathscr{B}(u, \varphi), \eta\rangle$ is continuous on $\tilde{C}^1_{0,\Sigma_1\cup\Sigma'_+}(\overline{\Omega})$ in the norm of H_λ. Since $\tilde{C}^1_{0,\Sigma_1\cup\Sigma'_+}(\overline{\Omega})$ is dense in H_λ, this implies that $\mathscr{B}(u, \varphi) \in H^*_\lambda$, i.e., $(u, \varphi) \in V$. Thus, if (2.35) is satisfied the function u is the first component of some element of V with which this element can be identified. This proves Lemma 2.8.

We denote by $\mathscr{A}$: $\tilde{C}^1_{0,\Sigma_1}(\overline{\Omega}) \subset X^* \to H^*_\lambda \subset X^* \subset Y^*$ the operator defined by

$$\langle \mathscr{A}u, \eta\rangle = \int_\Omega [\mathbf{l}'(x, u, A\nabla u)\cdot A\nabla\eta + l'_0(x, u, A\nabla u)\eta]\,dx$$

$$+ \int_{\Sigma_3} \lambda u|_{\Sigma_3}\eta\,ds, \qquad u, \eta \in \tilde{C}^1_{0,\Sigma_1}(\overline{\Omega}), \tag{2.36}$$

where $\mathbf{l}'(x, u, q)$ and $l'_0(x, u, q)$ are the functions in (1.2). It follows from Lemma 2.1 and (1.3) that for all $u \in \tilde{C}^1_{0,\Sigma_1}(\overline{\Omega})$ formula (2.36) defines a linear functional $\mathscr{A}u \in H^*_\lambda$ in the space H_λ, since, using Hölder's inequality, we easily obtain

$$|\langle \mathscr{A}u, \eta\rangle| \leqslant c\|\eta\|_{H_\lambda}, \qquad u, \eta \in \tilde{C}^1_{0,\Sigma_1}(\overline{\Omega}), \tag{2.37}$$

where the constant c depends on $\|\mathscr{A}u\|_{H_\lambda}$.

LEMMA 2.9. *The operator* $\mathscr{A}$: $\tilde{C}^1_{0,\Sigma_1}(\overline{\Omega}) \subset X \to H^*_\lambda$ *is bounded and continuous.*

PROOF. We consider an operator $\mathscr{N}$: $L^p(\Omega)\times \mathbf{L}^{\mathbf{p}}(\Omega) \to L^q(\Omega)$, where $p \geqslant 1$, $\mathbf{p} = (p_1, \ldots, p_n)$, $p_i \geqslant 1$, $i = 1, \ldots, n$, and $q \geqslant 1$, defined by

$$\mathscr{N}(z, \mathbf{z}) = \Phi(x, z(x), \mathbf{z}(x)), \qquad (z, \mathbf{z}) \in L^p(\Omega)\times\mathbf{L}^{\mathbf{p}}(\Omega), \tag{2.38}$$

where $\Phi(x, z, \mathbf{z})$ satisfies the Carathéodory condition in $\Omega\times\mathbb{R}\times\mathbb{R}^n$ and the condition $\Phi(x, z(x), \mathbf{z}(x)) \in L^q(\Omega)$ for any $z \in L^p(\Omega)$ and $\mathbf{z} \in \mathbf{L}^{\mathbf{p}}(\Omega)$. It is well known (see [66], pp. 31–41) that such an operator is bounded and continuous. Using the conditions imposed on $\mathbf{l}(x, u, q)$ and $l'_0(x, u, q)$ in Definition 1.1 (see, in particular, (1.3)), we can easily reduce the proof of Lemma 2.9 to an application of the theorem of Krasnosel'skiĭ indicated above. Lemma 2.9 is proved.

COROLLARY 2.1. *The operator* $\mathscr{A}$: $\tilde{C}^1_{0,\Sigma_1}(\overline{\Omega}) \subset X \to H^*_\lambda$ *admits extension to the entire space* X *with preservation of boundedness and continuity.*

PROOF. Corollary 2.1 follows from Lemma 2.9 in an obvious way. We shall write out the explicit form of the values $\langle \mathscr{A}u, \eta\rangle$, $\mathbf{u} = (u, \varphi) \in X$, $\eta \in H_\lambda$, where $\mathscr{A}$ denotes the extended operator (2.36). Suppose that a sequence $\{u_k\}$, $u_k \in \tilde{C}^1_{0,\Sigma_1}(\overline{\Omega})$, $k = 1, 2, \ldots$, converges to $\mathbf{u} = (u, \varphi) \in X$ in X. Then $u_k \to u$ in $L^m(\Omega)$, $A\nabla u_k \to A\nabla u$ in $\mathbf{L}^{\mathbf{m}}(\Omega)$, and $u_k|_{\Sigma_3} \to u_k|_{\Sigma_3}$ in $L^2(\lambda, \Sigma_3)$, where $A\nabla u$ is the generalized A-gradient, and $u|_{\Sigma_3}$ is the generalized limit value u on Σ_3 of the function $u \in H_\lambda$. Applying Krasnosel'skiĭ's theorem, we establish that the functions $l'^i(x, u_k, A\nabla u_k)$ converge to $l'^i(x, u, A\nabla u)$ in $L^{m_i'}(\Omega)$, $i = 1, \ldots, n$, and the functions $l_0'(x, u_k, A\nabla u_k)$ converge to $l_0'(x, u, A\nabla u)$ in $L^{m'}(\Omega)$. From what has been proved it follows that

$$\langle \mathscr{A}\mathbf{u}, \eta\rangle = \int_\Omega [\mathbf{l}'(x, u, A\nabla u)\cdot A\nabla\eta + l_0'(x, u, A\nabla u)\eta]\, dx + \int_{\Sigma_3} \lambda u|_{\Sigma_3}\eta|_{\Sigma_3}\, ds, \tag{2.39}$$

where $\mathbf{u} = (u, \varphi) \in X$ and $\eta \in H_\lambda$; here $A\nabla u$ and $A\nabla\eta$ denote the generalized A-gradients of u and η, while $u|_{\Sigma_3}$ and $\eta|_{\Sigma_3}$ denote their generalized limit values on Σ_3. Corollary 2.1 is proved.

Henceforth $\mathscr{A}$: $X \to H^*_\lambda$ always denotes the extended operator $\mathscr{A}$ indicated above. We denote by $\mathscr{L}$: $X \to Y^*$ the operator defined as follows:

$$\mathscr{L} = \mathscr{A} + \mathscr{B}, \text{ where } \mathscr{A}\colon X \to H^*_\lambda \subset X^* \subset Y^* \text{ is defined by (2.39) and } \mathscr{B}\colon X \to Y^* \text{ is defined by (2.22).} \tag{2.40}$$

The next assertion follows from the properties of $\mathscr{A}$ and $\mathscr{B}$ established above.

LEMMA 2.10. *The operator* $\mathscr{L}$: $X \to Y^*$ *defined by* (2.40) *is bounded and continuous.*

It is obvious that the operator $\mathscr{L}$: $X \to Y^*$ defined by (2.40) is uniquely determined by its values $\langle \mathscr{L}u, \eta\rangle$, $u, \eta \in \tilde{C}^1_{0,\Sigma_1}(\overline{\Omega})$ which have the form

$$\langle \mathscr{L}u, \eta\rangle = \int_\Omega \left[\mathbf{l}'\cdot A\nabla\eta + l_0'\eta - u\frac{\partial}{\partial x_i}(b^i\eta)\right] dx$$

$$+ \int_{(\Sigma_{2,3})_+\cup\Sigma'_+} bu\eta\, ds + \int_{\Sigma_3} \lambda u\eta\, ds, \qquad u, \eta \in \tilde{C}^1_{0,\Sigma_1}(\overline{\Omega}). \tag{2.41}$$

Comparing (1.37) and (2.41), we conclude that $\mathscr{L}$: $X \to Y^*$ can naturally be called the operator corresponding to the general boundary value problem for equation (1.35) having $(A, \mathbf{b}, m, \mathbf{m})$-structure in the domain Ω.

§3. A generalized formulation of the general boundary value problem for $(A, \mathbf{b}, m, \mathbf{m})$-elliptic equations

In a bounded domain $\Omega \subset \mathbb{R}^n$, $n \geqslant 2$, of class $\tilde{C}^{(1)}$ we consider an equation of the form (1.35) having $(A, \mathbf{b}, m, \mathbf{m})$-structure in this domain under the assumption that condition (2.1) is satisfied and the regular part $\Sigma \subset \partial\Omega$ has been decomposed into sets Σ_1, Σ_2, Σ_3, while Σ_i and Σ' have been decomposed into subsets $(\Sigma_i)_0$, $(\Sigma_i)_+$ and $(\Sigma_i)_-$, $i = 1, 2, 3$, and Σ'_0, Σ'_+ and Σ'_-, as described at the beginning of §2. In particular, it is assumed that condition (2.8) is satisfied. We suppose that on Σ_3 there is given a piecewise continuous, bounded, positive function λ. We suppose also that

(2.25) is satisfied. We assume the validity of the conditions listed above throughout this section. Under such conditions, in §2 we introduced and studied the spaces H_λ, X and Y (see (2.2), (2.10) and (2.11)), the subspace V (see (2.24) and (2.22)), and the operator $\mathscr{L}\colon X \to Y^*$ (see (2.40)) corresponding to a general boundary value problem of the form

$$-dl^i/dx_i + l_0 = f \quad \text{in } \Omega, \qquad u = 0 \quad \text{on } \Sigma_1 \cup \Sigma'_-,$$
$$\mathbf{l}' \cdot Av + cu = 0 \quad \text{on } \Sigma_2, \qquad \mathbf{l}' \cdot Av + (c - \lambda)u = 0 \quad \text{on } \Sigma_3, \tag{3.1}$$

where $c = 0$ on $(\Sigma_{2,3})_{0,+}$, $c = b(x)$ on $(\Sigma_{2,3})_-$ and $b(x)$ is defined by (2.22).

By a *generalized solution* (*of energy type*)([3]) *of problem* (3.1) we mean any function $u \in V$ satisfying the operator equation

$$\mathscr{L}u = F, \tag{3.2}$$

where $\mathscr{L}\colon X \to Y^*$ is the operator corresponding to problem (3.1) (see (2.40)), and the element $F \in H_\lambda^*$ is defined by $\langle F, \eta \rangle = \int_\Omega f(x)\eta(x)\,dx$, $\eta \in \tilde{C}^1_{0,\Sigma_1}(\overline{\Omega})$.

In view of condition (2.25) and Lemma 2.7 any element $u \in V$ is a function $u(x)$ belonging to the space $\hat{H}_\lambda$. Since $V \subset X$, from the definition of the space X and the regularity of the part $\Sigma \subset \partial\Omega$ it follows that any function $u \in V$ has a generalized limit value $u|_{\tilde{\Sigma}}$ on the set $\tilde{\Sigma} \equiv \Sigma_1 \cup (\Sigma_{2,3})_+ \cup \Sigma_3$, where the function u assumes its limit value $u|_{\tilde{\Sigma}}$ in the following stronger sense (as compared with the general definition of the generalized limit value of a function $u \in H_{m,\mathfrak{m}}(A, \Omega)$ given in §4.2): there exists a sequence $\{u_k\}$, $u_k \in \tilde{C}^1_{0,\Sigma_1}(\overline{\Omega})$, $k = 1, 2, \dots$, such that $u_k \to u$ in $\hat{H}_\lambda$ and $u_k|_{\tilde{\Sigma}} \to u|_{\tilde{\Sigma}}$ in $L^1_{\text{loc}}(\tilde{\Sigma}) \cap L^2(|b|, (\Sigma_{2,3})_+) \cap L^2(\lambda, \Sigma_3)$. Indeed, the convergence of the sequence $\{u_k\}$ in X implies, in particular, the convergence of $\{u_k|_{\tilde{\Sigma}}\}$ in $L^1_{\text{loc}}(\tilde{\Sigma})$ to some element $\psi \in L^1_{\text{loc}}(\tilde{\Sigma})$, where

$$\psi|_{\Sigma_1} = 0, \quad \psi|_{(\Sigma_{2,3})_+} \in L^2(|b|, (\Sigma_{2,3})_+), \quad \psi|_{\Sigma_3} \in L^2(\lambda, \Sigma_3);$$

this element ψ is the generalized limit value of u on $\tilde{\Sigma}$. Henceforth in this chapter we agree to understand by the words "a function $u \in \hat{H}_\lambda$ has generalized limit value $u|_{\tilde{\Sigma}}$ on the set $\tilde{\Sigma}$" that this property is satisfied in the stronger sense indicated above.

Taking into account the properties of elements of V formulated above and also the density of $\tilde{C}^1_{0,\Sigma_1 \cup \Sigma'_-}(\overline{\Omega})$ in H_λ, which follows from (2.25) (see Remark 2.3), we can say that a generalized solution of problem (3.1) is any function $u \in V \subset \hat{H}_\lambda$ satisfying the integral identity

$$\begin{aligned}\langle \mathscr{L}u, \eta \rangle \equiv \int_\Omega \left[\mathbf{l}'(x, u, A\nabla u) \cdot A\nabla\eta + l'_0(x, u, A\nabla u)\eta - u\frac{\partial}{\partial x_i}(b^i\eta) \right] dx \\ + \int_{\Sigma_3} \lambda u|_{\tilde{\Sigma}}\eta\, ds + \int_{(\Sigma_{2,3})_+} bu|_{\tilde{\Sigma}}\eta\, ds \\ = \langle F, \eta \rangle \quad \forall \eta \in \tilde{C}^1_{0,\Sigma_1 \cup \Sigma'_-}(\overline{\Omega}),\end{aligned} \tag{3.3}$$

where $A\nabla u$ is the generalized A-gradient, and $u|_{\tilde{\Sigma}}$ is the generalized limit value of u on $\tilde{\Sigma}$.

([3]) Since throughout this chapter we shall not consider generalized solutions of other types for problem (3.1), in this chapter we henceforth call the solutions of problem (3.1) defined here simply *generalized solutions*.

PROPOSITION 3.1. *Any function* $u \in \hat{H}_\lambda$ *having generalized limit value*

$$u|_{\tilde{\Sigma}} \in L^1_{\text{loc}}(\tilde{\Sigma}) \cap L^2(|b|, (\Sigma_{2,3})_\pm) \cap L^2(\lambda, \Sigma_3)$$

and satisfying (3.3) *belongs to the subspace* V *and is hence a generalized solution of problem* (3.1).

PROOF. We use Lemma 2.8. From the condition $u \in \hat{H}_\lambda$ and the existence of a generalized limit value for u it follows that there is a sequence $\{u_k\}$, $u_k \in \tilde{C}^1_{0,\Sigma_1}(\overline{\Omega})$, $k = 1, 2, \dots$, for which

$$\lim_{k\to\infty} \|u_k - u\|_{H_\lambda} = 0 \quad \text{and} \quad \lim_{k,s\to\infty} \|u_k - u_s\|_X = 0.$$

On the other hand, for all $\eta \in \tilde{C}^1_{0,\Sigma_1\cup\Sigma'_+}(\overline{\Omega})$ and for all $\{\eta_k\}$, $\eta_k \in \tilde{C}^1_{0,\Sigma_1\cup\Sigma'_+}(\overline{\Omega})$, $k = 1, 2, \dots$, converging to η in H_λ it follows from (3.3) that

$$\lim_{k\to\infty} \hat{\mathscr{B}}(u, \eta_k - \eta) = 0.$$

Thus, both conditions of (2.35) are satisfied, and so $u \in V$. This proves Proposition 3.1.

REMARK 3.1. Below in order to impart a closed character to further results on the solvability of problem (3.1), it is convenient for us in (3.2) (and hence also in (3.3)) to understand by F an arbitrary element of H^*_λ. We therefore agree to assume in (3.1) that $f \equiv F \in H^*_\lambda$. It should be born in mind, however, that replacement of the function $f \in L^{m'}(\Omega)$ in (3.1) by an arbitrary element $F \in H^*_\lambda$ actually distorts the form of the problem. In view of Lemma 2.1 the problem (3.1) with $f \equiv F \in H^*_\lambda$, realized in the form (2.3), can be rewritten in the form

$$-d\tilde{l}^i/dx_i + \tilde{l}_0 = 0 \quad \text{in } \Omega, \qquad u = 0 \quad \text{on } \Sigma_1 \cup \Sigma'_-, \qquad \tilde{\mathbf{l}}' \cdot A\nu + cu = \mathbf{f} \cdot A\nu \quad \text{on } \Sigma_2,$$
$$\tilde{\mathbf{l}}' \cdot A\nu + (c - \lambda)u = \mathbf{f} \cdot A\nu + \lambda\psi \quad \text{on } \Sigma_3, \tag{3.1'}$$

where $c = 0$ on $(\Sigma_{2,3})_{0,+}$ and $c = b(x)$ on $(\Sigma_{2,3})_-$.

REMARK 3.2. It follows from Proposition 1.5 that any solution of problem (3.1) smooth in $\overline{\Omega}$ is also a generalized solution of this problem. If u is a generalized solution of (3.1), then its vanishing on Σ_1 is ensured by its membership in the space $H_\lambda \equiv H^{0,\Sigma_1}_{m,\mathbf{m}}(A; \Omega; \Sigma_3, \lambda)$ and the condition of regularity of the set $\Sigma_1 \subset \Sigma$.

In distinguishing concrete conditions guaranteeing the regularity of Σ, it becomes clear in what sense the vanishing of a function u on Σ_1 follows from the fact that it belongs to H_λ. We note that if, for example, condition (4.2.5) is satisfied for the set Σ_1, a function $u \in H_\lambda$ vanishes on Σ_1 in the sense that for all interior points of Σ_1 equalities of the form (4.2.16) are satisfied with $u|_\pi$ replaced by 0. That a generalized solution satisfies the remaining boundary conditions in (3.1) is determined by the integral identity (3.3) itself. It is obvious that if a generalized solution of problem (3.1) and the functions forming (1.35) are sufficiently smooth while $\Omega \in C^2$, then such a generalized solution is also a classical solution of this problem.

§4. Conditions for existence and uniqueness of a generalized solution of the general boundary value problem

In a bounded domain $\Omega \subset \mathbb{R}^n$, $n \geqslant 2$, of class $\tilde{C}^{(1)}$ we consider the general boundary value problem of the form (3.1) for an equation (1.35) having $(A, \mathbf{b}, m, \mathbf{m})$-structure in Ω under the assumption that conditions (2.1), (2.8), and

(2.34) are satisfied. Under such conditions a definition of a generalized solution (of energy type) of problem (3.1) was given in §3. To prove the existence and uniqueness of such a solution of problem (3.1) we use the results of §4.6. It follows from the results of §2 that for the spaces $H \equiv H_\lambda$, X and Y considered there, which arise in connection with the generalized formulation of problem (3.1), all the conditions imposed on H, X, and Y in §4.6 are satisfied. In particular, condition (4.6.6) holds, where the role of the set $\mathfrak{N}$ in (4.6.6) is played here by the set $\tilde{C}^1_{0,\Sigma_1}(\overline{\Omega})$. Condition (4.6.7) is obviously satisfied for the operator $\mathscr{L}$: $X \to Y^*$ corresponding to problem (3.1). We remark that in this section the operator $\mathscr{L}$: $X \to Y^*$ is always understood to be the operator corresponding to problem (3.1) (see (2.40)), and the operators $\mathscr{A}$: $X \to H^*$ and $\mathscr{B}$: $X \to Y^*$ its components (see (2.39) and (2.22)), where $H \equiv H_\lambda$, X and Y are the spaces defined in §2.

The following results on generalized solvability of problem (3.1) ensue directly from the results of §4.6.

THEOREM 4.1. *Assume that the following conditions hold*:

1) *For the equation* (1.35) *having* $(A, \mathbf{b}, m, \mathbf{m})$*-structure in a bounded domain* $\Omega \subset \mathbb{R}^n$, $n \geqslant 2$, *of class* $\tilde{C}^{(1)}$ *conditions* (2.1), (2.8), *and* (2.34) *are satisfied.*

2) *Condition* (4.6.9) *is satisfied for the operator* $\mathscr{B}$: $X \to Y^*$.

3) *For the subspace* V *defined by* (2.24) *a condition of the form* (4.6.10) *is satisfied, i.e.,* $V \cap Y$ *is dense in* X.

4) *The function* $v \to \langle \mathscr{B}v, v\rangle$, $v \in V$, *is continuous in the norm* $\|\cdot\|_X$.

Suppose also that the operator $\mathscr{L}$: $X \to Y^*$ *is locally coercive* (*coercive*) *and has semibounded variation. Then problem* (3.1) *with* $F = 0$ (*problem* (3.1) *for every* $F \in H^*_\lambda$) *has at least one generalized solution.*

Theorem 4.1 follows from Theorem 4.6.1 and Corollary 4.6.2. The following results on uniqueness of a generalized solution of problem (3.1) ensue from Theorems 4.6.3 and 4.6.4.

THEOREM 4.2. *Suppose that condition* 1) *of Theorem* 4.1 *and a condition of the form* (4.6.23) *are satisfied* (*in particular, condition* (4.6.23) *is satisfied if the operator* $\mathscr{A}$: $X \to H^*$ *is strictly monotone and* $\langle \mathscr{B}v, v\rangle \geqslant 0$ *for all* $v \in V$). *Then for every* $F \in H^*_\lambda$ *problem* (3.1) *has no more than one generalized solution.*

THEOREM 4.3. *Suppose that conditions* 1), 3), *and* 4) *of Theorem* 4.1 *are satisfied, and that the operator* $\mathscr{L}$: $X \to Y^*$ *is uniformly monotone. Then for every* $F \in H^*_\lambda$ *problem* (3.1) *has no more than one generalized solution.*

The next theorem follows from Theorem 4.6.5.

THEOREM 4.4. *Suppose that conditions* 1)–4) *of Theorem* 4.1 *are satisfied, and that the operator* $\mathscr{L}$: $X \to Y^*$ *is strongly monotone. Then for every* $F \in H^*_\lambda$ *problem* (3.1) *has exactly one generalized solution, and the restriction* $\hat{\mathscr{L}}$: $(V \subset X) \to (H^*_\lambda \subset Y^*)$ *of the operator* $\mathscr{L}$: $X \to Y^*$ *to the set* V *is a homeomorphism.*

Finally, from Theorem 4.6.2 we obtain the following result.

THEOREM 4.5. *Suppose that condition* 1) *of Theorem* 4.1 *is satisfied, and that the operator* $\mathscr{A}$: $X \to H_\lambda^*$ *is weakly compact, while* $\mathscr{L}$: $X \to Y^*$ *is locally coercive* (*coercive*). *Then problem* (3.1) *with* $F = 0$ (*problem* (3.1) *for every* $F \in H_\lambda^*$) *has at least one generalized solution.*

Below we shall present sufficient conditions for the validity of conditions 2) and 3) of the above theorems, and also algebraic criteria for the conditions of local coercivity, coercivity, monotonicity, and strong monotonicity for the operator $\mathscr{L}$: $X \to Y^*$ to be satisfied.

LEMMA 4.1. *Suppose that*

$$\|u\|_{L^2(|b|,\Sigma'_-)} \leqslant c\|u\|_Y, \quad \forall u \in \tilde{C}^1_{0,\Sigma_1}(\overline{\Omega}), \tag{4.1}$$

where the constant c *does not depend on the function* $u \in \tilde{C}^1_{0,\Sigma_1}(\overline{\Omega})$. *Then functions* $u \in Y$ *have traces* $u|_{\Sigma'_-} \in L^2(|b|, \Sigma'_-)$, *and the restriction of the operator* $\mathscr{B}$: $X \to Y^*$ *to the set* Y *is defined by*

$$\langle \mathscr{B}u, \eta\rangle = \int_\Omega b^i u_{x_i}\eta\, dx + \int_{(\Sigma_{2,3})_-\cup\Sigma'_-} |b|u\eta\, ds, \qquad u \in Y,\ \eta \in \tilde{C}^1_{0,\Sigma_1}(\overline{\Omega}). \tag{4.2}$$

PROOF. It follows immediately from (4.1) that if $\{u_n\}$, $u_n \in \tilde{C}^1_{0,\Sigma_1}(\overline{\Omega})$, $n = 1, 2, \ldots$, converges to $u \in Y$ in Y, then the sequence $\{u_n|_{\Sigma'_-}\}$ converges in $L^2(|b|, \Sigma'_-)$ to some function $(u)_- \in L^2(|b|, \Sigma'_-)$ which is obviously the trace of the function $u \in Y$ on Σ'_-. Using (2.22) and integrating by parts, we find that

$$\langle \mathscr{B}u_n, \eta\rangle = \int_\Omega b^i (u_n)_{x_i}\eta\, dx + \int_{(\Sigma_{2,3})_-\cup\Sigma'_-} |b|u_n\eta\, ds, \qquad n = 1, 2, \ldots, \quad \forall \eta \in \tilde{C}^1_{0,\Sigma_1}(\overline{\Omega}). \tag{4.3}$$

Passing to the limit as $n \to \infty$ in (4.3), we obtain (4.2). The lemma is proved.

LEMMA 4.2. *Suppose condition* (4.1) *is satisfied, and suppose for each of the sets* Σ' *and* Σ'_+ *condition* (4.3.2) *holds, while for the set* $(\Sigma_{2,3})_-$ *condition* (4.2.25) *is satisfied, where* $m_* = \min(m, m_1, \ldots, m_2) \geqslant 2$. *Then in order that a function* $u \in Y$ *belong to the subspace* V *it is necessary and sufficient that* $u = 0$ *on* Σ'_-.

PROOF. Because of the validity of condition (4.3.2) for Σ'_+ and Lemma 2.8 a function $u \in Y$ belongs to V if and only if condition 2) in (2.35) holds for it, i.e., $\lim_{k\to\infty}\hat{\mathscr{B}}(u, \eta_k - \eta) = 0$ for all $\eta \in \tilde{C}^1_{0,\Sigma_1\cup\Sigma'_+}(\overline{\Omega})$ and any sequence $\{\eta_k\}$, $\eta_k \in \tilde{C}^1_{0,\Sigma_1\cup\Sigma'_+}(\overline{\Omega})$, $k = 1, 2, \ldots$, converging to η in H_λ. Applying the formula for integration by parts, we rewrite $\hat{\mathscr{B}}(u, \eta_k - \eta)$ in the form

$$\hat{\mathscr{B}}(u, \eta_k - \eta) = \int_\Omega b^i u_{x_i}(\eta_k - \eta)\, dx + \int_{(\Sigma_{2,3})_-\cup\Sigma'_-} |b|u(\eta_k - \eta)\, ds, \tag{4.4}$$

where η and $\{\eta_k\}$ are as described above. In view of condition (4.2.25) for $(\Sigma_{2,3})_-$ the equality

$$\lim_{\eta_k\to\eta \text{ in } H_\lambda} \hat{\mathscr{B}}(u, \eta_k - \eta) = 0$$

for the indicated η and $\{\eta_k\}$ is equivalent to

$$\lim_{\eta_k \to \eta \text{ in } H_\lambda} \int_{\Sigma'_-} bu(\eta_k - \eta)\, ds = 0. \tag{4.5}$$

Thus, a function $u \in Y$ belongs to V if and only if (4.5) holds for any η and $\{\eta_k\}$ possessing the properties indicated above. If $u = 0$ on Σ'_-, then the validity of (4.5) is obvious, so that in this case $u \in V$. We shall now prove that (4.5) implies that $u = 0$ on Σ'_-. Let η be any fixed function in $\tilde{C}^1_{0,\Sigma_1 \cup \Sigma'_+}(\overline{\Omega})$. In view of the fact that the set Σ' satisfies condition (4.3.2) (see (2.1)) there exists a sequence $\{\eta_k\}$, $\eta_k \in \tilde{C}^1_{0,\Sigma_1 \cup \Sigma'_+}(\overline{\Omega})$, $k = 1, 2, \ldots$, converging to the selected function η in H_λ (see Lemma 2.3). Then $\int_{\Sigma'_-} bu\eta\, ds = 0$ for all $\eta \in \tilde{C}^1_{0,\Sigma_1 \cup \Sigma'_+}(\overline{\Omega})$. In view of the arbitrariness of η and the condition $b \neq 0$ on Σ'_-, this implies that $u = 0$ on Σ'_-. Lemma 4.2 is proved.

PROPOSITION 4.1. *Suppose that condition* (4.1) *is satisfied, and that condition* (4.3.2) *holds for each of the sets* Σ' *and* Σ'_+ *while for the set* $(\Sigma_{2,3})_-$ *condition* (4.2.25) *is satisfied with* $m^* = \min(m, m_1, \ldots, m_n) \geqslant 2$. *Then condition* 2) *of Theorem* 4.1 *is satisfied, i.e., condition* (4.6.9) *holds for the operator* $\mathscr{B}\colon X \to Y^*$.

PROOF. In view of Lemma 4.2 and quality (4.2) the restriction of $\mathscr{B}\colon X \to Y^*$ to the set $V \cap Y$ is defined by

$$\langle \mathscr{B}u, \eta \rangle = \int_\Omega b^i u_{x_i} \eta\, dx + \int_{(\Sigma_{2,3})_-} |b| u\eta\, ds, \qquad u \in V \cap Y, \eta \in \tilde{C}^1_{0,\Sigma_1}(\overline{\Omega}),$$

from which we easily obtain

$$|\langle \mathscr{B}u, \eta \rangle| \leqslant \|u\|_Y \|\eta\|_X, \qquad u \in V \cap Y, \eta \in \tilde{C}^1_{0,\Sigma_1}(\overline{\Omega}).$$

In view of the density of $\tilde{C}^1_{0,\Sigma_1}(\overline{\Omega})$ in X this implies that the restriction of $\mathscr{B}\colon X \to Y^*$ to $V \cap Y$ is a bounded linear operator from $(V \cap Y) \subset Y$ into X^*. Proposition 4.1 is proved.

LEMMA 4.3. *Suppose that a condition of the form* (4.3.2) *is satisfied for the set* Σ'_-. *Then the set* $\tilde{C}^1_{0,\Sigma_1 \cup \Sigma'_-}(\overline{\Omega})$ *is dense in* X.

PROOF. It suffices to show that for any $u \in \tilde{C}^1_{0,\Sigma_1}(\overline{\Omega})$ there exists a sequence $\{u_n\}$, $u_n \in \tilde{C}^1_{0,\Sigma_1 \cup \Sigma'_-}(\overline{\Omega})$, $n = 1, 2, \ldots$, converging to u in X. We consider the sequence $\{u_{\delta_n}\}$, where u_δ is defined for a fixed function $u \in \tilde{C}^1_{0,\Sigma_1}(\overline{\Omega})$ by (4.3.10) with $\mathscr{P} \equiv \Sigma'_-$, and $\delta_n \to 0$ as $n \to \infty$. Exactly as in the proof of Lemma 4.3.1, we establish that $u_{\delta_n} \to u$ in H_λ. Taking into account that $u_\delta = u$ outside an n-dimensional neighborhood of Σ'_- which contracts to Σ'_- as $\delta \to 0$ (see condition 2) in (4.3.2)), we see that $u_{\delta_n} \to u$ in $L^2(\beta, \Omega_\beta)$. Taking, finally, into account that $(\Sigma_{2,3})_\pm \cup \Sigma'_+$ does not intersect Σ'_- and that $\operatorname{meas}_{n-1} \partial\Sigma'_- = 0$ (see (2.8)), we establish that $u_{\delta_n} \to u$ in $L^2(|b|, (\Sigma_{2,3})_\pm \cup \Sigma'_+)$. From what has been proved it follows that $u_{\delta_n} \to u$ in X. Lemma 4.3 is proved.

PROPOSITION 4.2. *Suppose for each of the sets* Σ', Σ'_+, *and* Σ'_- *condition* (4.3.2) *is satisfied, while condition* (4.2.25) *holds for* $(\Sigma_{2,3})_-$. *Let* $m_* = \min(m, m_1, \ldots, m_n) \geqslant 2$. *Then condition* 3) *of Theorem* 4.1 *is satisfied, i.e.,* $V \cap Y$ *is dense in* X.

PROOF. It is obvious that $\tilde{C}^1_{0,\Sigma_1 \cup \Sigma'_-}(\overline{\Omega}) \subset Y$. In view of Lemma 4.2 $\tilde{C}^1_{0,\Sigma_1 \cup \Sigma'_-}(\overline{\Omega}) \subset V$, so that $\tilde{C}^1_{0,\Sigma_1 \cup \Sigma'_-}(\overline{\Omega}) \subset V \cap Y$. By Lemma 4.3, $\tilde{C}^1_{0,\Sigma_1 \cup \Sigma'_-}(\overline{\Omega})$ is dense in X. Hence, $V \cap Y$ is dense in X. Proposition 4.2 is proved.

PROPOSITION 4.3. *Suppose that for almost all $x \in \Omega$ and any $u \in \mathbb{R}$, $q = Ap$ and $p \in \mathbb{R}^n$*

$$l'^i(x, u, q)q_i + l'_0(x, u, q)u - (1/2)(\partial b^i(x)/\partial x_i)u^2 \geqslant 0. \tag{4.6}$$

Then the operator $\mathcal{L}: X \to Y^*$ is locally coercive.

PROOF. Suppose first that $u \in \tilde{C}^1_{0,\Sigma_1}(\overline{\Omega})$. Integrating by parts, we obtain

$$\begin{aligned}\langle \mathcal{L}u, u\rangle &= \int_\Omega \left[\mathbf{l}'(x, u, A\nabla u)\cdot A\nabla u + l'_0(x, u, A\nabla u)u - \frac{\partial b^i}{\partial x_i}u^2 - b^i u u_{x_i}\right] dx \\ &\quad + \int_{\Sigma_3} \lambda u^2\, ds + \int_{(\Sigma_{2,3})_+ \cup \Sigma'_+} bu^2\, ds \\ &= \int_\Omega \left[\mathbf{l}'\cdot A\nabla u + l'_0 u - \frac{1}{2}\frac{\partial b^i}{\partial x_i}u^2\right] dx + \int_{\Sigma_3}\lambda u^2\, ds + \frac{1}{2}\int_{(\Sigma_{2,3})_\pm \cup \Sigma'_\pm} |b|u^2\, ds.\end{aligned} \tag{4.7}$$

From (4.6) and (4.7) we obtain

$$\langle \mathcal{L}u, u\rangle \geqslant 0 \quad \forall u \in \tilde{C}^1_{0,\Sigma_1}(\overline{\Omega}). \tag{4.8}$$

Taking into account that $\tilde{C}^1_{0,\Sigma_1}(\overline{\Omega})$ is dense in Y, $Y \to X$, and also that $\mathcal{L}: X \to Y^*$ is continuous, we conclude that (4.8) is valid for all $u \in Y$. Proposition 4.3 is proved.

PROPOSITION 4.4. *Suppose that for almost all $x \in \Omega$ and any $u \in \mathbb{R}$, $q = Ap$ and $p \in \mathbb{R}^n$*

$$\begin{aligned}&l'^i(x, u, q)q_i + l'_0(x, u, q)u - \frac{1}{2}\frac{\partial b^i}{\partial x_i}u^2 \\ &\qquad \geqslant \nu_1 \sum_{i=1}^n |q_i|^{m_i} + \nu_2|u|^m + \nu_3\beta(x)u^2 - \varphi(x),\end{aligned} \tag{4.9}$$

where ν_1, ν_2, ν_3 are positive constants, and $\varphi \in L^1(\Omega)$. Then the operator $\mathcal{L}: X \to Y^*$ is coercive.

PROOF. Suppose first that $u \in \tilde{C}^1_{0,\Sigma_1}(\overline{\Omega})$. For such a function (4.7) holds. Using (4.9), we compute

$$\begin{aligned}\langle \mathcal{L}u, u\rangle &\geqslant \nu_1 \sum_{i=1}^n \|A_i\nabla u\|^{m_i}_{m_i,\Omega} + \nu_2\|u\|^m_{m,\Omega} + \nu_3\|u\|^2_{L^2(\beta,\Omega_\beta)} + \|u\|^2_{L^2(\lambda,\Sigma_3)} \\ &\quad + \frac{1}{2}\int_{(\Sigma_{2,3})_\pm \cup \Sigma'_\pm} |b|u^2\, ds - \int_\Omega \varphi(x)\, dx.\end{aligned} \tag{4.10}$$

It follows from (4.10) that for any function $u \in \tilde{C}^1_{0,\Sigma_1}(\overline{\Omega})$ with $\|u\|_X \geqslant n+3$

$$\langle \mathcal{L}u, u\rangle \geqslant c\|u\|_X^{m_*} - \int_\Omega \varphi(x)\, dx, \tag{4.11}$$

where $c = \min(\nu_1, \nu_2, \nu_3, 1/2)(n+3)^{m_*}$, $m_* = \min(m_1, \dots, m_n, m, 2) > 1$. Taking account of the continuity of the operator $\mathcal{L}: X \to Y^*$, the density of $\tilde{C}^1_{0,\Sigma_1}(\overline{\Omega})$ in Y, and the imbedding $Y \to X$, we conclude that (4.11) holds also for any function $u \in Y$ with $\|u\|_X \geqslant n+3$. Since $m_* > 1$, from this the validity of a condition of the form (4.5.1) follows easily. Proposition 4.4 is proved.

PROPOSITION 4.5. *Suppose that for almost all* $x \in \Omega$ *and any* $u \in \mathbb{R}$, $q = Ap$, $p \in \mathbb{R}^n$, $\xi_0 \in \mathbb{R}$, $\eta = A\xi$ *and* $\xi \in \mathbb{R}^n$

$$\frac{\partial l''}{\partial q_j}\eta_i\eta_j + \frac{\partial l''}{\partial u}\xi_0\eta_i + \frac{\partial l'_0}{\partial q_j}\eta\xi_0 + \frac{\partial l'_0}{\partial u}\xi_0^2 - \frac{1}{2}\frac{\partial b^i}{\partial x_i}\xi_0^2 \geqslant 0. \tag{4.12}$$

Then the operator $\mathscr{L}\colon X \to Y^*$ *is monotone (and hence also has semibounded variation).*

PROOF. Suppose first that $u, v \in \tilde{C}^1_{0,\Sigma_1}(\overline{\Omega})$. We transform $\langle \mathscr{L}u - \mathscr{L}v, u - v\rangle$ as follows:

$$\begin{aligned}
\langle \mathscr{L}u &- \mathscr{L}v, u - v\rangle \\
&= \int_\Omega \Big\{ [\mathbf{l}'(x, u, A\nabla u) - \mathbf{l}'(x, v, A\nabla u)] \cdot A\nabla(u - v) \\
&\qquad + [l'_0(x, u, A\nabla u) - l'_0(x, v, A\nabla v)](u - v) - (u - v)\frac{\partial}{\partial x_i}[b^i(u - v)] \Big\}\, dx \\
&\quad + \int_{\Sigma_3} \lambda(u - v)^2\, ds + \int_{(\Sigma_{2,3})_\pm \cup \Sigma'_+} b(u - v)^2\, ds \\
&= \int_\Omega \int_0^1 \Big\{ \frac{d}{d\tau}[l''(x, v + \tau(u - v), A\nabla v + \tau A\nabla(u - v))]\, A_i\nabla(u - v) \\
&\qquad + \frac{d}{d\tau}[l'_0(x, v + \tau(u - v), A\nabla v + \tau A\nabla(u - v))](u - v) \Big\}\, d\tau\, dx \\
&\quad + \int_\Omega b^i \frac{\partial}{\partial x_i}\Big[\frac{1}{2}(u - v)^2\Big]\, dx - \int_\Omega \frac{\partial b^i}{\partial x_i}(u - v)^2\, dx \\
&\quad + \int_{(\Sigma_{2,3})_+ \cup \Sigma'_+} b(u - v)^2\, ds + \int_{\Sigma_3} \lambda(u - v)^2\, ds \\
&= \int_\Omega \int_0^1 \Big[\frac{\partial l''(x, v + \tau(u - v), A\nabla v + \tau A\nabla(u - v))}{\partial q_j} A_j\nabla(u - v) A_i\nabla(u - v) \\
&\quad + \frac{\partial l''(x, v + \tau(u - v), A\nabla v + \tau A\nabla(u - v))}{\partial u}(u - v) A_i\nabla(u - v) \\
&\quad + \frac{\partial l'_0(x, v + \tau(u - v), A\nabla u + \tau A\nabla(u - v))}{\partial q_j} A_j\nabla(u - v)(u - v) \\
&\qquad\qquad + \frac{\partial l'_0(x, v + \tau(u - v), A\nabla v + \tau A\nabla(u - v))}{\partial u}(u - v)^2 \Big]\, d\tau\, dx \\
&\quad - \frac{1}{2}\int_\Omega \frac{\partial b^i(x)}{\partial x_i}(u - v)^2\, dx + \int_{\Sigma_3} \lambda(u - v)^2\, ds \\
&\quad + \frac{1}{2}\int_{(\Sigma_{2,3})_\pm \cup \Sigma'_\pm} |b|(u - v)^2\, ds.
\end{aligned} \tag{4.13}$$

In view of condition (4.12), from (4.13) we obtain

$$\langle \mathscr{L}u - \mathscr{L}v, u - v\rangle \geqslant 0, \qquad u, v \in \tilde{C}^1_{0,\Sigma_1}(\overline{\Omega}). \tag{4.14}$$

Taking into account that $\mathscr{L}\colon X \to Y^*$ is continuous and also the density of $\tilde{C}^1_{0,\Sigma_1}(\overline{\Omega})$ in Y and the imbedding $Y \to X$, we conclude that the first condition in (4.6.4) holds for any $u, v \in Y$. Proposition 4.5 is proved.

PROPOSITION 4.6. *Suppose that for almost all* $x \in \Omega$ *and any* $u \in \mathbb{R}$, $\xi_0 \in \mathbb{R}$, $q = Ap, p \in \mathbb{R}^n$, $\eta = A\xi$ *and* $\xi \in \mathbb{R}^n$

$$\frac{\partial l''^i(x,u,q)}{\partial q_j}\eta_i\eta_j + \frac{\partial l''^i(x,u,q)}{\partial u}\xi_0\eta_j + \frac{\partial l'_0(x,u,q)}{\partial q_j}\eta_j\xi_0 + \frac{\partial l'_0(x,u,q)}{\partial u}\xi_0^2 - \frac{1}{2}\frac{\partial b^i(x)}{\partial x_i}\xi_0^2$$
$$\geqslant \alpha_0\left[\sum_{i=1}^n |q_i|^{m_i-2}\xi_i^2 + \left(|u|^{m-2} + \beta(x)\right)\xi_0^2\right], \qquad \alpha_0 = \text{const} > 0, \tag{4.15}$$

with

$$m_i \geqslant 2, \quad i = 1,\dots,n; \qquad m \geqslant 2. \tag{4.16}$$

Then the operator $\mathscr{L}: X \to Y^*$ *is strongly monotone.*

PROOF. Suppose first that $u, v \in \tilde{C}^1_{0,\Sigma_1}(\overline{\Omega})$. In proving Proposition 4.5 it was established that (4.13) holds for such functions. In view of (4.15), from (4.13) we obtain

$$\langle \mathscr{L}u - \mathscr{L}v, u-v\rangle \geqslant \alpha_0\int_\Omega\int_0^1\left[\sum_{i=1}^n |A_i\nabla v + \tau A_i\nabla(u-v)|^{m_i-2}|A_i\nabla(u-v)|^2 + |v+\tau(u-v)|^{m-2}(u-v)^2 + \beta(x)(u-v)^2\right]d\tau\,dx$$
$$+ \int_{\Sigma_3}\lambda(u-v)^2\,ds + \frac{1}{2}\int_{(\Sigma_{2,3})_\pm\cup\Sigma'_\pm}|b|(u-v)^2\,ds, \tag{4.17}$$

which is valid for any $u, v \in \tilde{C}^1_{0,\Sigma_1}(\overline{\Omega})$. Taking into account the elementary equality

$$\int_0^1 |\mathbf{a} + \tau(\mathbf{b}-\mathbf{a})|^{q-2}\,d\tau \geqslant c_1|\mathbf{b}-\mathbf{a}|^{q-2}, \tag{4.18}$$

where $\mathbf{a}$ and $\mathbf{b}$ are fixed vectors (of any finite dimension), $q \geqslant 2$, and $c_1 > 0$ is a constant depending only on q, we deduce from (4.17) that

$$\langle \mathscr{L}u - \mathscr{L}v, u-v\rangle \geqslant \alpha_0 c_1\left[\sum_{i=1}^n \|A_i\nabla(u-v)\|^{m_i}_{m_i,\Omega} + \|u-v\|^m_{m,\Omega} + \|u-v\|^2_{L^2(\beta,\Omega_\beta)}\right]$$
$$+ \int_{\Sigma_3}\lambda(u-v)^2\,ds$$
$$+ \frac{1}{2}\int_{(\Sigma_{2,3})_\pm\cup\Sigma'_\pm}|b|(u-v)^2\,ds, \qquad u, v \in \tilde{C}^1_{0,\Sigma_1}(\overline{\Omega}). \tag{4.19}$$

It follows from (4.19) that for any $u, v \in \tilde{C}^1_{0,\Sigma_1}(\overline{\Omega})$ with $\|u-v\|_X \leqslant 1$

$$\langle \mathscr{L}u - \mathscr{L}v, u-v\rangle \geqslant c_2\|u-v\|_X^{m_*}, \tag{4.20}$$

where $c_2 = \min(\alpha_0 c_1, \frac{1}{2})$ and $m^* = \max(m_1,\dots,m_n,2) \geqslant 2$, and for any $u, v \in \tilde{C}^1_{0,\Sigma_1}(\overline{\Omega})$ with $\|u-v\|_X \geqslant n+4$

$$\langle \mathscr{L}u - \mathscr{L}v, u-v\rangle \geqslant c_3\|u-v\|_X^{m_*}, \tag{4.21}$$

where $c_3 = c_2(n+4)^{-m_*}$ and $m_* = \min(m_1,\ldots,m_n, m, 2) = 2$. Finally, for any $u, v \in \tilde{C}^1_{0,\Sigma_1}(\overline{\Omega})$ with $1 \leqslant \|u - v\|_X \leqslant n+4$ both (4.20) and (4.21) obviously follow from (4.19) with constants $\tilde{c}_2$ and $\tilde{c}_3$ respectively which depend only on $n, m, m_1,\ldots,m_n$. Thus, for any $u, v \in \tilde{C}^1_{0,\Sigma_1}(\overline{\Omega})$,

$$\langle \mathscr{L}u - \mathscr{L}v, u - v\rangle \geqslant \delta(\|u - v\|_X)\|u - v\|_X, \tag{4.22}$$

where $\delta(\rho) = c\min(\rho^{m^*}, \rho^{m_*})\rho^{-1}$ and $c = \min(c_2, c_3, \tilde{c}_2, \tilde{c}_3)$. It is obvious that $\delta(\rho)$ is continuous, increasing, equal to 0 only for $\rho = 0$, and $\lim_{\rho\to+\infty}\delta(\rho) = +\infty$. In view of the density of $\tilde{C}^1_{0,\Sigma_1}(\overline{\Omega})$ in Y and the continuity of $\mathscr{L}\colon X \to Y^*$ and of $\delta(\rho)$ the estimate (4.22) remains valid also for any $u, v \in Y$. But this means precisely that the condition of strong monotonicity of the operator $\mathscr{L}\colon X \to Y^*$ is satisfied (see the second inequality in (4.6.4)). Proposition 4.6 is proved.

Other criteria for coercivity and also criteria for the variation of the operator $\mathscr{L}$: $X \to Y^*$ to be semibounded will be given below for so-called weakly degenerate $(A, \mathbf{0})$-elliptic equations.

Thus, concrete sufficient conditions have been obtained for all the assumptions of Theorems 4.1–4.4 with the exception of condition 4) of Theorem 4.1. As concerns condition 4), its verification requires distinguishing more special classes of $(A, \mathbf{b}, m, \mathbf{m})$-elliptic equations. In any case in Chapters 7 and 8 this verification will be carried out for the classes of $(A, \mathbf{0}, m, \mathbf{m})$-elliptic and $(A, \mathbf{0}, m, \mathbf{m})$-parabolic equations which are of greatest interest from the applied point of view.

§5. Linear $(A, \mathbf{b})$-elliptic equations

In this section we assume that equation (1.35) is linear, i.e., that it has the form

$$-(d/dx_i)\left(\alpha^{ij}u_{x_j} + \alpha^i u + g^i\right) + \beta^i u_{x_i} + \beta_0 u + g_0 = f, \tag{5.1}$$

where α^{ij}, α^i, g^i, β^i, β_0 and g_0 are measurable functions in Ω, $\alpha^{ij}\xi_i\xi_j \geqslant 0$ in Ω for all $\xi \in \mathbb{R}^n$, and $\alpha^{ij} = \alpha^{ji}$, $i, j = 1,\ldots,n$. We suppose that for almost all $x \in \Omega$ and any $u \in \mathbb{R}$ and $p \in \mathbb{R}^n$

$$\begin{gathered}\mathfrak{A}p + \boldsymbol{\alpha}u + \mathbf{g} = A^*(QAp + \mathbf{a}u + \mathbf{f}),\\ \boldsymbol{\beta}\cdot p + \beta_0 u + g_0 = \boldsymbol{\gamma}\cdot Ap + a_0 u + f_0 + b^i p_i,\end{gathered} \tag{5.2}$$

where

$$\mathfrak{A} \equiv \|\alpha^{ij}\|,\quad \boldsymbol{\alpha} = (\alpha^1,\ldots,\alpha^n),\quad \mathbf{g} = (g^1,\ldots,g^n),\quad \boldsymbol{\beta} = (\beta^1,\ldots,\beta^n),$$
$$A \equiv \|a^{ij}\|,\quad Q = \|q^{ij}\|,\quad \mathbf{a} = (a^1,\ldots,a^n),\quad \mathbf{f} = (f^1,\ldots,f^n),\quad \boldsymbol{\gamma} = (\gamma^1,\ldots,\gamma^n),$$

and a^{ij}, a^i, f^i, γ^i, a_0, f_0 and b^i are measurable functions in Ω. Suppose that conditions (2.1) are satisfied (wtih $m = 2$ and $\mathbf{m} = \mathbf{2}$) as well as (2.8), and

$$a^{ij} \in L^2(\Omega),\quad \frac{\partial a^{ij}}{\partial x_j} \in L^2_{\mathrm{loc}}(\Omega),\quad b^i \in C(\overline{\Omega}),\quad \frac{\partial b^i}{\partial x_i} \in C(\overline{\Omega}),\qquad i, j = 1,\ldots,n;$$

$$\begin{gathered}g^{ij} \in L^\infty(\Omega),\quad i, j = 1,\ldots,n;\qquad \mathbf{a}, \boldsymbol{\gamma} \in \mathbf{L}^\infty(\Omega),\\ f \in L^2(\Omega),\quad a_0 \in L^\infty(\Omega),\quad f_0 \in L^2(\Omega).\end{gathered} \tag{5.3}$$

The conditions (1.2) and (1.3) with $m = 2$ and $\mathbf{m} = \mathbf{2}$ are obviously satisfied for the functions $\mathbf{l}'(x, u, q) \equiv Qq + \mathbf{a}u + \mathbf{f}$ and $l_0'(x, u, q) \equiv \boldsymbol{\gamma}\cdot q + a_0 u + f_0$ for almost all $x \in \Omega$ and any $u \in \mathbb{R}$ and $q \in \mathbb{R}^n$.

In the present case the general boundary value problem takes the form

$$-(d/dx_i)\left(\alpha^{ij}u_{x_j} + \alpha^i u + g^i\right) + \beta^i u_{x_i} + \beta_0 u + g_0 = f \quad \text{in } \Omega,$$
$$u = 0 \quad \text{on } \Sigma_1 \cup \Sigma'_-, \qquad \frac{\partial u}{\partial N} + \boldsymbol{\alpha}\cdot \nu u + \mathbf{g}\cdot \nu + cu = 0 \quad \text{on } \Sigma_2, \tag{5.4}$$
$$\frac{\partial u}{\partial N} + \boldsymbol{\alpha}\cdot \nu u + \mathbf{g}\cdot \nu + (c-\lambda)u = 0 \quad \text{on } \Sigma_3,$$

where $c = 0$ on $(\Sigma_{2,3})_{0,+}$, $c = b(x)$ on $(\Sigma_{2,3})_-$, $\partial u/\partial N \equiv A\nabla u \cdot A\nu$ is the derivative of u with respect to the conormal, and the integral identity (3.3) has the form

$$\int_\Omega \left[(QA\nabla u + \mathbf{a}u + \mathbf{f})\cdot A\nabla\eta + (\boldsymbol{\gamma}\cdot A\nabla u + a_0 u + f_0)\eta - u(b^i\eta)_{x_i}\right] dx$$
$$+ \int_{\Sigma_3} \lambda u|_\Sigma \eta\, ds + \int_{(\Sigma_{2,3})_+} bu|_\Sigma \eta\, ds = \langle F, \eta\rangle \quad \forall \eta \in C^1_{0,\Sigma_1\cup\Sigma'_+}(\overline{\Omega}), \tag{5.5}$$

where $u \in H_\lambda \equiv H^{0,\Sigma_1}_{2,2}(A;\Omega;\Sigma_3,\lambda)$ and $\langle F,\eta\rangle = \int_\Omega f\eta\, dx$, $f \in L^2(\Omega)$.

We henceforth consider the identity (5.5) for all $F \in H^*_\lambda$ and call a function $u \in H_\lambda$ satisfying this identity a solution of problem (5.4) with $f \equiv F \in H^*_\lambda$. It is easy to see that in the present case the operator $\mathscr{A}\colon X \to H^*_\lambda$ (see (2.39)) is weakly compact.

THEOREM 5.1. *Suppose that for almost all $x \in \Omega$ and any $\xi \in \mathbb{R}^n$*

$$Q(x)\xi\cdot\xi \geqslant c_1|\xi|^2, \qquad c_1 = \text{const} \geqslant 0; \tag{5.6}$$
$$a_0(x) - \frac{1}{2}\frac{\partial b^i}{\partial x_i} - \frac{1}{2\varepsilon_1}|\mathbf{a}(x)|^2 - \frac{1}{2\varepsilon_2}|\boldsymbol{\gamma}(x)|^2 \geqslant c_2(1+\beta(x)), \qquad c_2 = \text{const} > 0,$$

where $\varepsilon_1 > 0$, $\varepsilon_2 > 0$ and $1/2(\varepsilon_1+\varepsilon_2) < c_1$. Then for any

$$F \in H^*_\lambda \equiv \left(H^{0,\Sigma_1}_{2,2}(A;\Omega;\Sigma_3,\lambda)\right)^*$$

the general boundary value problem (5.4) *for equation* (5.1) *has at least one generalized solution.*

PROOF. In view of the weak compactness of the operator $\mathscr{A}\colon X \to H^*_\lambda$ and Theorem 4.5, to prove Theorem 5.1 it suffices to prove coerciveness of the operator $\mathscr{L}\colon X \to Y^*$. Applying the Cauchy inequality and taking (5.3) into account, we write

$$l'^i q_i + l'_0 u - \frac{1}{2}\frac{\partial b^i}{\partial x_i}u^2 \equiv Qq\cdot q + \mathbf{a}\cdot qu + \mathbf{f}\cdot q + (\boldsymbol{\gamma}\cdot q + a_0 u + f_0)u - \frac{1}{2}\frac{\partial b^i}{\partial x_i}u^2$$
$$\geqslant \left(c_1 - \frac{\varepsilon_1}{2} - \frac{\varepsilon_2}{2} - \frac{\varepsilon_4}{2}\right)|q|^2$$
$$+ \left(a_0 - \frac{1}{2}\frac{\partial b^i}{\partial x_i} - \frac{1}{2\varepsilon_1}|\mathbf{a}|^2 - \frac{1}{2\varepsilon_2}|\boldsymbol{\gamma}|^2 - \frac{\varepsilon_2}{2}\right)u^2 - \frac{f_0^2}{2\varepsilon_3} - \frac{|\mathbf{f}|^2}{2\varepsilon_4}, \tag{5.7}$$

where $\varepsilon_3, \varepsilon_4 > 0$. An inequality of the form (4.9) with $m = 2$ and $\mathbf{m} = \mathbf{2}$ follows easily from (5.6) and (5.7). Indeed, suppose that in (5.7) values $\varepsilon_1, \varepsilon_2 > 0$ have been chosen for which $\varepsilon_1/2 + \varepsilon_2/2 < c_1$, while ε_3 and ε_4 have been chosen so that

$$\varepsilon_3/2 < c_2/2, \qquad \varepsilon_4/2 < c_1 - \varepsilon_1/2 - \varepsilon_2/2.$$

Then (5.7) gives an inequality of the form (4.9) with

$$m = 2, \quad \mathbf{m} = \mathbf{2}, \quad \nu_1 = c_1 - \varepsilon_1/2 - \varepsilon_2/2 - \varepsilon_4/2, \quad \nu_2 = \nu_3 = c_2/2,$$
$$\varphi = f_0^2/2\varepsilon_2 + |\mathbf{f}|^2/2\varepsilon_4 \in L^1(\Omega).$$

Theorem 5.1 is proved.

We note that in view of the boundedness of $\beta(x)$ in Ω the second condition in (5.6) will certainly be satisfied if for almost all $x \in \Omega$

$$a_0(x) - \frac{1}{2}\frac{\partial b^i(x)}{\partial x_i} - \frac{1}{2\varepsilon_1}|\mathbf{a}(x)|^2 - \frac{1}{2\varepsilon_2}|\gamma(x)|^2 \geqslant \tilde{c}_2 = \text{const} > 0 \tag{5.8}$$

for the same ε_1 and ε_2 as in (5.6).

We consider a special case of an equation of the form (5.1) defined by the following conditions: the matrix $\mathfrak{A} \equiv \|\alpha^{ij}(x)\|$ is symmetric and nonnegative in Ω, $\boldsymbol{\alpha}(x) \equiv \mathbf{0}$, and $\mathbf{g}(x) \equiv \mathbf{0}$ in Ω. In this case equalities (5.2) are obviously satisfied with $A = \mathfrak{A}^{1/2}$ and $Q = I$, where I is the identity matrix, $a_0 = \beta_0$, $f_0 = g_0$ and $b^i = \beta^i$, $i = 1,\dots,n$. We assume that

$$a^{ij} \in L^2(\Omega), \quad \frac{\partial a^{ij}}{\partial x_j} \in L^2_{\text{loc}}(\Omega), \quad \beta^i \in C(\overline{\Omega}), \quad \frac{\partial \beta^i}{\partial x_i} \in C(\overline{\Omega}), \qquad i, j = 1,\dots,n,$$
$$\beta_0 \in L^2(\Omega), \quad g_0 \in L^2(\Omega), \quad f \in L^2(\Omega), \tag{5.9}$$

where the a^{ij} are the elements of $A = \mathfrak{A}^{1/2}$. Theorem 5.1 then implies the following assertion.

COROLLARY 5.1. *Under conditions* (5.9) *and the condition*

$$\beta_0(x) - (1/2)(\partial\beta^i(x)/\partial x_i) \geqslant c_0 = \text{const} > 0 \quad a.e. \text{ in } \Omega \tag{5.10}$$

the general boundary value problem (5.4) *has at least one generalized solution.*

We now consider the nondivergence linear equation

$$\alpha^{ij}(x)u_{x_ix_j} + \beta^i(x)u_{x_i} + \beta(x)u = f(x) \tag{5.11}$$

with a matrix $\mathfrak{A} \equiv \|\alpha^{ij}(x)\|$ which is nonnegative and symmetric in Ω. Let

$$\alpha^{ij}, \frac{\partial \alpha^{ij}}{\partial x_j} \in C(\overline{\Omega}), \qquad \frac{\partial^2 \alpha^{ij}}{\partial x_i \partial x_j} \in C(\overline{\Omega}), \qquad i, j = 1,\dots,n;$$
$$\beta^i \in C(\overline{\Omega}), \quad \frac{\partial \beta^i}{\partial x_i} \in C(\overline{\Omega}), \quad i = 1,\dots,n; \qquad \beta \in L^\infty(\Omega), \quad f \in L^2(\Omega). \tag{5.12}$$

Conditions (5.12) make it possible to rewrite (5.11) in the divergence form

$$-(d/dx_i)(\alpha^{ij}u_{x_j}) + \hat{\beta}^i u_{x_i} + \hat{\beta}u = \hat{f}, \tag{5.13}$$

where $\hat{\beta}^i = \partial\alpha^{ij}/\partial x_j - \beta^i$, $i = 1,\dots,n$, $\hat{\beta} = -\beta$, $\hat{f} = -f$, and for (5.13) conditions of the form (5.9) are satisfied. For (5.11) we consider a general boundary value problem of the form

$$\begin{gathered} \alpha^{ij}u_{x_ix_j} + \beta^i u_{x_i} + \beta u = f \ \text{ in } \Omega, \qquad u = 0 \ \text{ on } \Sigma_1 \cup \Sigma'_-, \\ \partial u/\partial N + cu = 0 \ \text{ on } \Sigma_2, \qquad \partial u/\partial N + (c - \lambda)u = 0 \ \text{ on } \Sigma_3, \end{gathered} \tag{5.14}$$

with $c = 0$ on $(\Sigma_{2,3})_{0,+}$ and $c = b(x) \equiv -b^i\nu_i$ on $(\Sigma_{2,3})_-$, where $\partial\Omega$ is decomposed into the parts $(\Sigma_1)_{0,+,-}$, $(\Sigma_2)_{0,+,-}$, $(\Sigma_3)_{0,+,-}$ and $\Sigma'_{0,+,-}$ on the basis of the matrix $A = \mathfrak{A}^{1/2}$ and the vector $\mathbf{b} = (b^1,\dots,b^n)$, where $b^i = \beta^i - \partial\alpha^{ij}/\partial x_i$, $i = 1,\dots,n$, $\partial u/\partial N \equiv A\nabla u A\boldsymbol{\nu}$. In other words, we consider a general boundary value problem of the form (5.4) for equation (5.13).

A *generalized solution* of (5.14) is defined to be any function $u \in H_\lambda \equiv H_{2,2}^{0,\Sigma_1}(A;\Omega;\Sigma_3,\lambda)$ having a generalized limit value

$$u|_{\tilde\Sigma} \in L^1_{\mathrm{loc}}(\tilde\Sigma) \cap L^2(|b|,(\Sigma_{2,3})_\pm) \cap L^2(\lambda,\Sigma_3),$$

where $\tilde\Sigma \equiv \Sigma_1 \cup (\Sigma_{2,3})_\pm \cup \Sigma_3$, and satisfying the integral identity

$$\int_\Omega \left\{ A\nabla u \cdot A\nabla\eta + u\left[\left(\frac{\partial\alpha^{ij}}{\partial x_j} - \beta^i\right)\eta\right]_{x_i} - \beta u\eta \right\} dx + \int_{\Sigma_3} \lambda\, u|_{\tilde\Sigma}\eta\, ds$$
$$+ \int_{(\Sigma_{2,3})_+} bu|_{\tilde\Sigma}\eta\, ds = \int_\Omega f\eta\, dx, \qquad \eta \in \tilde C^1_{0,\Sigma_1\cup\Sigma'_+}(\overline\Omega), \tag{5.15}$$

where $A \equiv \mathfrak{A}^{1/2}$ and $A\nabla u$ is the generalized A-gradient of u.

Since (5.15) is the integral identity corresponding to a general boundary value problem for the divergence equation (5.13), Corollary 5.1 implies the following result.

THEOREM 5.2. *Suppose that for almost all $x \in \Omega$*

$$\frac{1}{2}\left(\frac{\partial\beta^i}{\partial x_i} - \frac{\partial^2\alpha^{ij}}{\partial x_i\partial x_j}\right) - \beta(x) \geqslant c_0 = \mathrm{const} > 0. \tag{5.16}$$

Then the general boundary value problem (5.14) *for equation* (5.11), *considered under conditions* (5.12), *has at least one generalized solution.*

We observe that the result of Theorem 5.2 is obviously preserved if the function $f \in L^2(\Omega)$ in (4.10) is replaced by an arbitrary element $F \in H^*_\lambda$, where $H_\lambda \equiv H_{2,2}^{0,\Sigma_1}(A;\Omega;\Sigma_3,\lambda)$.

CHAPTER 6
EXISTENCE THEOREMS FOR REGULAR GENERALIZED SOLUTIONS OF THE FIRST BOUNDARY VALUE PROBLEM FOR $(A,\mathbf{b})$-ELLIPTIC EQUATIONS

§1. Nondivergence $(A,\mathbf{b})$-elliptic equations

In this chapter $(A,\mathbf{b})$-elliptic equations of the form (5.1.35) are considered under conditions on the structure of the equation and the domain Ω such that the first boundary value problem is well posed:

$$\mathscr{L}u \equiv -\frac{dl^i(x,u,\nabla u)}{dx_i} + l_0(x,u,\nabla u) = f(x) \quad \text{in } \Omega, \qquad u = 0 \quad \text{on } \partial\Omega. \tag{1.1}$$

The conditions enable us to establish the existence of a generalized solution of problem (1.1) in the class of Lipschitz functions in $\overline\Omega$ having a trace on the entire boundary $\partial\Omega$. In establishing existence theorems for solutions of problem (1.1) in

Chapter 6 we go beyond the framework of divergence equations and consider so-called $(A, \mathbf{b})$-elliptic equations of nondivergence form.

Let the function u be defined in a neighborhood of a point $x \in \mathbb{R}^n$, and let $A \equiv \|a^{ij}(x)\|$ be a symmetric matrix of order n with elements a^{ij}, $i, j = 1,\dots,n$, continuous in this neighborhood; the matrix A may admit degeneracy of any rank. By definition, the *ith A-derivative* $u_{\hat{i}} \equiv \hat{\partial} u/\partial x_i \equiv A_i \nabla u$ of the function u at the point x is the derivative of u in the direction of the vector $\mathbf{a} = (a^{i1},\dots,a^{in})$ defined by the ith row of the matrix A. In particular, if $\mathbf{a}^i = \mathbf{0}$ at x, then $u_{\hat{i}} = 0$ at x. Thus, if u is differentiable (in the usual sense) at x, then $u_{\hat{i}} = \sum_{j=1}^n a^{ij}\partial u/\partial x_j$, $i = 1,\dots,n$. The A-derivatives of second and higher orders are defined similarly. For example, $u_{\hat{i}\hat{j}} = (u_{\hat{i}})_{\hat{j}}$, etc. Let $f(x, u(x), \mathbf{v}(x))$ be a composite function of x. We denote by $\hat{d}f(x, u, \mathbf{v})/dx_i$ the (total) ith A-derivative of this composite function. On the other hand, $\hat{\partial} f(x, u, \mathbf{v})/\partial x_i$ denotes the partial A-derivative of f with respect to the argument $x = (x_1,\dots,x_n)$ (with frozen values of $u = u(x)$ and $\mathbf{v} = \mathbf{v}(x)$). If $f(x, u, \mathbf{v})$ is continuous, has continuous partial derivatives

$$\frac{\partial f}{\partial u}(x, u, \mathbf{v}), \quad \frac{\partial f}{\partial v_j}(x, u, \mathbf{v}), \qquad j = 1,\dots,n,$$

and partial A-derivatives

$$\frac{\hat{\partial}}{\partial x_i} f(x, u, \mathbf{v}), \qquad i = 1,\dots,n,$$

then it also has total A-derivatives

$$\frac{\hat{d}}{dx_i} f(x, u, \mathbf{v}), \qquad i = 1,\dots,n,$$

and

$$\frac{\hat{d}}{dx_i} f(x, u, \mathbf{v}) = \frac{\hat{\partial}}{\partial x_i} f(x, u, \mathbf{v}) + \frac{\partial f(x, u, \mathbf{v})}{\partial u} u_{\hat{i}} + \frac{\partial f(x, u, \mathbf{v})}{\partial v_j} v_{j\hat{i}},$$

where $v_{ji} \equiv (v_j)_{\hat{i}} \equiv \hat{\partial} v_j/\partial x_i$.

We consider the quasilinear differential equation

$$\hat{\alpha}^{ij}(x, u, \hat{\nabla} u) u_{\hat{j}\hat{i}} - \hat{\alpha}(x, u, \hat{\nabla} u) - b^i(x) u_{x_i} = 0, \tag{1.2}$$

containing A-derivatives of first and second orders and ordinary derivatives of first order (the latter occur in the equation in linear fashion) of the unknown function $u = u(x)$, where $\hat{\nabla} u \equiv (u_{\hat{1}},\dots,u_{\hat{n}})$.

We call an equation of the form (1.2) a *nondivergence* $(A, \mathbf{b})$*-elliptic* (*strictly* $(A, \mathbf{b})$*-elliptic*) *equation in* Ω if for any $x \in \Omega$, $u \in \mathbb{R}$, $q = Ap$, and all $p \in \mathbb{R}^n$

$$\begin{gathered} \hat{\alpha}^{ij}(x, u, q)\eta_i\eta_j \geqslant 0, \qquad \eta = A\xi \quad \forall \xi \in \mathbb{R}^n \\ \left[\hat{\alpha}^{ij}(x, u, q)\eta_i\eta_j > 0, \qquad \eta \neq 0, \eta = A\xi \quad \forall \xi \in \mathbb{R}^n\right]. \end{gathered} \tag{1.3}$$

The divergence $(A, \mathbf{b})$-elliptic equations of the form (5.1.35) considered above are a special case of nondivergence $(A, \mathbf{b})$-elliptic equations, since by (2.1.2) differentiation of the first term of (5.1.35) leads to an equation of the form (1.2) with

$$\hat{\alpha}^{ij}(x,u,q)=\frac{\partial l'^i(x,u,q)}{\partial q_j},$$

$$\hat{\alpha}(x,u,q)=-\frac{\partial l'^k(x,u,q)}{\partial u}q_k-a^{ki}(x)\frac{\partial l'^k(x,u,q)}{\partial x_i}$$
$$-\frac{\partial a^{ki}(x)}{\partial x_i}l'^k(x,u,q)-f(x)+l'_0(x,u,q), \tag{1.4}$$

where $l'^i(x,u,q)$, $i=1,\dots,n$, and $l'_0(x,u,q)$ are the reduced coefficients of (5.1.35).

An equation of the form (1.2) can also be written in terms of ordinary derivatives alone, i.e., in the form

$$\alpha^{ij}(x,u,\nabla u)u_{ij}-\alpha(x,u,\nabla u)=0, \tag{1.5}$$

where $\nabla u=(u_1,\dots,u_n)$, $u_i=\partial u/\partial x_i$, $u_{ij}=\partial^2 u/\partial x_i\partial x_j$, $i,j=1,\dots,n$, and

$$\begin{aligned}\alpha^{ij}(x,u,p)&=a^{ki}(x)\hat{\alpha}^{ks}(x,u,A(x)p)a^{sj}(x),\\ \alpha(x,u,p)&=-a^{ki}(x)\hat{\alpha}^{ks}(x,u,A(x)p)\big(\partial a^{sj}(x)p_j/\partial x_i\big)\\ &\quad+\hat{\alpha}(x,u,A(x)p)+b^i(x)p_i.\end{aligned} \tag{1.6}$$

If (1.2) is generated by the divergence $(A,\mathbf{b})$-elliptic equation (1.1), then the coefficients α^{ij} and α of the corresponding equation (1.5) can also be expressed by the formulas

$$\begin{aligned}\alpha^{ij}(x,u,p)&=\frac{\partial l^i(x,u,p)}{\partial p_j},\\ \alpha(x,u,p)&=-\frac{\partial l^k(x,u,p)}{\partial u}p_k-\frac{\partial l^i(x,u,p)}{\partial x_i}-f(x)+l_0(x,u,p).\end{aligned} \tag{1.7}$$

From the invariance of the reduced coefficients l'^i and l'_0 (see Proposition 5.1.2) established in §5.1 it follows that for an equation (1.2) generated by (1.1) the coefficients $\hat{\alpha}^{ij}(x,u,q)$, $i,j=1,\dots,n$, and $\hat{\alpha}(x,u,q)$ are invariant, i.e., as a result of an arbitrary nondegenerate smooth transformation (5.1.30) an equation of the form (5.1.31) is obtained to which there corresponds a nondivergence $(\tilde{A},\tilde{\mathbf{b}})$-elliptic equation of the form

$$\hat{\tilde{\alpha}}^{ij}(\tilde{x},u,\hat{\tilde{\nabla}}u)u_{\tilde{j}\tilde{i}}-\hat{\tilde{\alpha}}(\tilde{x},u,\hat{\tilde{\nabla}}u)-\tilde{b}^i(\tilde{x})u_{\tilde{x}_j}=0, \tag{1.8}$$

where the differentiations in the first two terms correspond to the matrix $\tilde{A}=AP^*$, and where $\tilde{\mathbf{b}}=P\mathbf{b}$, P is the Jacobi matrix of the transformation (5.1.30), and for all $x\in\overline{\Omega}$, $u\in\mathbb{R}$ and $q\in\mathbb{R}^n$

$$\begin{aligned}\hat{\tilde{\alpha}}^{ij}(\tilde{x},u,q)&=\hat{\alpha}^{ij}(x(\tilde{x}),u,q),\quad i,j=1,\dots,n;\\ \hat{\tilde{\alpha}}(\tilde{x},u,q)&=\hat{\alpha}(x(\tilde{x}),u,q).\end{aligned} \tag{1.9}$$

This follows in an obvious way from (5.1.34) and (1.4). The following assertion has thus been established.

LEMMA 1.1. *The functions* $\hat{\alpha}^{ij}(x, u, q)$, $i, j = 1,\dots,n$, *and* $\hat{\alpha}(x, u, q)$ *defined by* (1.4) *on the basis of the reduced coefficients of a divergence* $(A, \mathbf{b})$*-elliptic equation of the form* (1.1) *are invariant (in the sense indicated above) under an arbitrary smooth, nondegenerate coordinate transformation.*

§2. Existence and uniqueness of regular generalized solutions of the first boundary value problem

In this section in a bounded domain $\Omega \subset \mathbb{R}^n$, $n \geq 2$, we first consider $(A, \mathbf{b})$-elliptic equations of nondivergence form (1.2) (which can also be rewritten in the form (1.5), (1.6)) under the assumption that the functions $\hat{\alpha}^{ij}(x, u, q)$, $i, j = 1,\dots,n$, and $\hat{\alpha}(x, u, q)$ belong to the class $\tilde{C}^1(\bar{\Omega} \times \mathbb{R} \times \mathbb{R}^n)$, i.e., they are continuous in $\bar{\Omega} \times \mathbb{R} \times \mathbb{R}^n$ and all their partial derivatives of first order are bounded on any compact set in $\bar{\Omega} \times \mathbb{R} \times \mathbb{R}^n$.(1) We also assume that the elements $a^{ij}(x)$, $i, j = 1,\dots,n$, of the matrix $A(x)$ and the components $b^i(x)$, $i = 1,\dots,n$, of the vector $\mathbf{b}(x)$ belong to the class $\tilde{C}^1(\bar{\Omega})$ (see the basic notation). It is obvious that the functions $\alpha^{ij}(x, u, p)$, $i, j = 1,\dots,n$, and $\alpha(x, u, p)$ defined by (1.6) then belong to $\tilde{C}^1(\bar{\Omega} \times \mathbb{R} \times \mathbb{R}^n)$.

LEMMA 2.1. *Suppose that*

$$\begin{aligned} \hat{\alpha}^{ij}(x, u, q)\eta_i\eta_j > 0 \quad &\text{for all } (x, u, q) \in \bar{\Omega} \times \mathbb{R} \times \mathbb{R}^n, \eta \in \mathbb{R}^n, \eta \neq 0, \\ a^{ij}(x)\xi_i\xi_j > 0 \quad &\text{for all } x \in \bar{\Omega}, \xi \in \mathbb{R}^n, \xi \neq 0, \end{aligned} \tag{2.1}$$

and that there exists a constant $m_0 \geq 0$ *such that for all* $x \in \bar{\Omega}$ *and any* u *with* $|u| \geq m_0$

$$\hat{\alpha}(x, u, 0)u \geq 0. \tag{2.2}$$

Let u *be a classical solution of* (1.2) (*i.e.*, $u \in C^2(\Omega) \cap C(\bar{\Omega})$) *with* $u = \varphi$ *on* $\partial\Omega$, *where* $\varphi \in C(\bar{\Omega})$. *Then*

$$\max_{\bar{\Omega}} |u| \leq \max\Big(m_0, \max_{\partial\Omega} |\varphi|\Big).$$

PROOF. Lemma 2.1 follows from a well-known classical estimate for the maximum modulus of solutions of quasilinear elliptic equations (see, for example, [83]).

REMARK 2.1. We have chosen the simplest sufficient condition under which an a priori estimate of the maximum modulus of the solution itself is valid. In place of condition (2.2) it is possible to impose any other condition of this sort. This must always be borne in mind below.

We further assume that Ω belongs at least to the class C^2. We set

$$D_\delta \equiv \{x \in \Omega: \operatorname{dist}(x, \partial\Omega) < \delta\}, \qquad \delta \in (0, K^{-1}), \tag{2.3}$$

where K is the supremum of the absolute values of the normal curvatures on $\partial\Omega$, and the number δ is so small that for each point $x \in D_\delta$ there exists a unique point $y \in \partial\Omega$ such that $\operatorname{dist}(x, \partial\Omega) = \operatorname{dist}(x, y)$.

LEMMA 2.2. *Suppose that the functions* $\alpha^{ij}(x, u, p)$, $i, j = 1,\dots,n$, *and* $\alpha(x, u, p)$ *defined by* (1.6) *are bounded together with their partial derivatives* $\partial\alpha^{ij}/\partial p_k$ *and*

(1) This definition of the class $\tilde{C}^1(\bar{\Omega} \times \mathbb{R} \times \mathbb{R}^n)$ will henceforth be used in this chapter without recalling its meaning.

$\partial\alpha/\partial p_k$, $i, j, k = 1,\ldots,n$, *on any compact set in* $\overline{D}_\delta \times [-m, m] \times \mathbf{R}^n$, *where* $m =$ const > 0, *and* $\Omega \in C^3$. *Suppose that condition* (2.1) *holds. Assume that, for all* $x \in D_\delta$, *and* $u \in [-m, m]$ *and any* $\rho \geqslant l =$ const > 0,

$$|\alpha(x, u, \rho\boldsymbol{\nu})| \leqslant \tilde{\psi}(\rho)\hat{\mathscr{E}}_1(x, u, \rho A\boldsymbol{\nu}), \tag{2.4}$$

where $\hat{\mathscr{E}}_1(x, u, q) \equiv \hat{\alpha}^{ks}(x, u, q)q_kq_s$, $\boldsymbol{\nu}$ *is the unit vector of the inner normal to* $\partial\Omega$ *at the point* $y = y(x)$, *where* $y(x)$ *is the point on* $\partial\Omega$ *nearest* $x \in D_\delta$, $A\boldsymbol{\nu} \equiv A(x)\,\boldsymbol{\nu}(y(x))$, *and* $\tilde{\psi}(\rho)$ *is a positive increasing function of* $\rho \geqslant 0$ *such that*

$$\int^{+\infty} \frac{d\rho}{\rho\tilde{\psi}(\rho)} = +\infty.$$

Suppose that in this notation the condition

$$\tilde{\psi}(\rho)\hat{\mathscr{E}}_1(x, u, \rho A\boldsymbol{\nu}) \geqslant \operatorname{Sp}\mathfrak{A}(x, u, \rho\boldsymbol{\nu})\rho \tag{2.5}$$

is also satisfied, where $\mathfrak{A}(x, u, p) = A^*(x)\hat{\mathfrak{A}}(x, u, A(x)p)A(x)$, $\mathfrak{A} \equiv \|\alpha^{ij}\|$ *and* $\hat{\mathfrak{A}} \equiv \|\hat{\alpha}^{ij}\|$. *Then for any solution* $u \in C^2(D_\delta) \cap C(\overline{D}_\delta)$ *of* (1.5) *satisfying the conditions* $u = 0$ *on* $\partial\Omega$ *and* $|u| \leqslant m$ *on* D_δ *the estimate*

$$|\partial u(y)/\partial\nu| \leqslant M_1, \qquad y \in \partial\Omega, \tag{2.6}$$

holds, with a constant M_1 *depending only on* m, l, $\tilde{\psi}(\rho)$, δ, *and* K.

PROOF. The hypotheses of Lemma 2.2 imply all the conditions of the second part of Theorem 1.4.1$'$. Indeed, because of (1.6)

$$\begin{aligned}\hat{\mathscr{E}}_1(x, u, \rho A\boldsymbol{\nu}) &= \hat{\alpha}^{ks}(x, u, \rho A\boldsymbol{\nu})\rho A_k\boldsymbol{\nu}\rho A_s\boldsymbol{\nu} \\ &= \alpha^{ij}(x, u, \rho\boldsymbol{\nu})\rho\nu_i\rho\nu_j = \mathscr{E}_1(x, u, \rho\boldsymbol{\nu}),\end{aligned} \tag{2.7}$$

where $\mathscr{E}_1(x, u, p) \equiv \alpha^{ij}(x, u, p)p_ip_j$. Therefore, condition (1.4.19) with $\delta(\rho) \equiv 0$ follows from (2.4), while (1.4.20) follows from (2.5). Hence, Lemma 2.2 follows from Theorem 1.4.1$'$.

THEOREM 2.1. *Suppose that the domain* Ω *belongs to the class* C^3 *and that condition* (2.1) *is satisfied. Assume that on the set* $\overline{D}_\delta \times [-m, m] \times (|q| > \hat{l})$, *where* $m =$ const $\geqslant 0$ *and* $\hat{l} =$ const $\geqslant 0$,

$$|q| \max_{i,j=1,\ldots,n} |\hat{\alpha}^{ij}(x, u, q)| + |\hat{\alpha}(x, u, q)| + |q| \leqslant \psi(|q|)\hat{\mathscr{E}}_1(x, u, q), \tag{2.8}$$

where $\hat{\mathscr{E}}_1(x, u, q) \equiv \hat{\alpha}^{ij}(x, u, q)q_iq_j$ *and* $\psi(\rho)$ *is a positive, nondecreasing function satisfying the condition* $\int^{+\infty}(d\rho/\rho\psi(\rho)) = +\infty$. *Suppose that for all* $y \in \partial\Omega$

$$2\gamma^{-1} \leqslant |A(y)\boldsymbol{\nu}(y)| \leqslant \tfrac{1}{2}\gamma, \tag{2.9}$$

where $\boldsymbol{\nu}(y)$ *is the unit vector of the inner normal to* $\partial\Omega$ *at the point* y, *and* $\gamma =$ const > 0. *Assume, finally, that the function* $u \in C^2(D_\delta) \cap C^1(\overline{D}_\delta)$ *satisfies* (1.2) *in* D_δ, $u = 0$ *on* $\partial\Omega$, *and* $\max_{D_\delta}|u| \leqslant m$. *Then the estimate* (2.6) *holds for the function* u, *where* M_1 *depends only on* m, $\hat{l}$, γ, $\psi(\rho)$, *and*

$$K = \sup_{\substack{i=1,\ldots,n-1\\ y\in\partial\Omega}} |k_i(y)|,$$

$k_1(y),\ldots,k_{n-1}(y)$ *being the normal curvatures of the surface* $\partial\Omega$ *at the point* y.

PROOF. Setting $q = \rho A(x)\nu(y(x))$ in (2.8), with $\rho \geqslant l \equiv \hat{l}\gamma > 0$, and assuming the boundary strip D_δ to be so narrow that $\gamma^{-1} \leqslant |A(x)\nu(y(x))| \leqslant \gamma$ in it, we deduce from (2.8) (with (1.6) taken into account) the validity for $\rho \geqslant l$ of inequalities (2.4) and (2.5) with $\tilde{\psi}(\rho) = c_1\psi(c_2\rho)$, where c_2 depends on γ, and c_1 depends on n, γ, and the bounds of $|a^{ij}|$, $|\partial a^{ij}/\partial x_k|$ and $|b^i|$ in Ω, $i, j, k = 1,\dots,n$. Theorem 2.1 then follows from Lemma 2.2.

REMARK 2.2. Condition (2.8) is satisfied, in particular, if for all $(x, u, q) \in \overline{\Omega} \times [-m, m] \times (|q| > \hat{l})$

$$\begin{gathered}\hat{a}^{ij}\xi_i\xi_j \geqslant \nu|q|^{\overline{m}-2}\xi^2, \qquad \xi \in \mathbb{R}^n, \\ |q|\,|\hat{a}^{ij}| + |\hat{a}| \leqslant \mu\psi(|q|)|q|^{\overline{m}}, \qquad i, j = 1,\dots,n,\end{gathered} \tag{2.10}$$

where $\nu = \text{const} > 0$, $\overline{m} > 1$, $\mu = \text{const} \geqslant 0$, and $\psi(\rho)$ is the same function as in (2.8). We note further that although nondegeneracy of (1.5)–(1.6) (i.e., of (1.2) rewritten in terms of ordinary derivatives) is assumed in Theorem 2.1, the constant in the estimate (2.6) does not depend on the ellipticity constant of this equation.

We shall now show that condition (2.9) of Theorem 2.1 is essential. Suppose that $\Omega \subset \mathbb{R}^2$ is contained in the strip $0 < x_1 < 3$ and has an infinitely smooth boundary consisting of a segment joining the points $(0, -2)$ and $(0, 2)$, a segment joining the points $(3, -2)$ and $(3,2)$, and arcs γ_1 and γ_2, where γ_1 joins the points $(0, -2)$ and $(3, -2)$ and is situated in the rectangle $[0, 3] \times [-3, -2]$, while γ_2 joins $(0, 2)$ and $(3,2)$ and is situated in the rectangle $[0, 3] \times [2, 3]$. Let $\xi(t)$ be a function of class $C^\infty([-3, 3])$ which is equal to 1 for $|t| \leqslant 1$ and to 0 for $|t| \geqslant 2$. We consider the function

$$u_\varepsilon(x) = \left[(x_1 + \varepsilon)^\lambda - \varepsilon^\lambda\right]\xi(x_1)\xi(x_2),$$

where $\varepsilon \in (0, 1]$ and $\lambda \in (0, 1)$. It is easy to see that u_ε satisfies the equation

$$-\frac{\partial}{\partial x_1}\left[(x_1 + \varepsilon)^2\frac{\partial u}{\partial x_1}\right] - \frac{\partial}{\partial x_2}\left[(x_1 + \varepsilon)^2\frac{\partial u}{\partial x_2}\right] + \lambda(\lambda + 1)u = f(x), \tag{2.11}$$

where

$$\begin{aligned}f(x) = {} & 2(\lambda + 1)(x_1 + \varepsilon)^{\lambda+1}\xi'(x_1)\xi(x_2) + (x_1 + \varepsilon)^{\lambda+2}\xi''(x_1)\xi(x_2) \\ & - \varepsilon^\lambda(x_1 + \varepsilon)^2\xi''(x_1)\xi(x_2) - 2\varepsilon^\lambda(x_1 + \varepsilon)\xi'(x_1)\xi(x_2) \\ & + \left[(x_1 + \varepsilon)^{\lambda+2} - \varepsilon^\lambda(x_1 + \varepsilon)^2\right]\xi(x_1)\xi''(x_2) + \lambda(\lambda + 1)\varepsilon^\lambda\xi(x_1)\xi(x_2).\end{aligned}$$

(2.11) has the structure of a divergence $(A, \mathbf{0})$-elliptic equation relative to the matrix $A = (x_1 + \varepsilon)I$, where I is the identity matrix, while the correspondingn nondivergence equation, written in terms of A-derivatives (see (1.2) and (1.4)), has the form

$$u_{\hat{1}\hat{1}} + u_{\hat{2}\hat{2}} - u_{\hat{1}} + \lambda(\lambda + 1)u = f(x). \tag{2.12}$$

In terms of ordinary derivatives the corresponding nondivergence equation has the form (see (1.5) and (1.6))

$$(x_1 + \varepsilon)^2u_{x_1x_1} + (x_1 + \varepsilon)^2u_{x_2x_2} - 2(x_1 + \varepsilon)u_{x_1} + \lambda(\lambda + 1)u = f(x). \tag{2.13}$$

Thus, for this equation we have $\hat{a}^{ij} = \delta_i^j$ and $a^{ij} = (x_1 + \varepsilon)\delta_i^j$, so that condition (2.1) is satisfied because $\varepsilon > 0$. For the solutions u_ε we have $|u_\varepsilon| \leqslant m$ in $\overline{\Omega}$ for some absolute constant $m > 0$, uniformly with respect to $\varepsilon \in (0,1]$. On the set $\overline{\Omega} \times \{|u| \leqslant m\} \times \{|q| > 1\}$ condition (2.8) is satisfied uniformly with respect to ε with $\psi(|q|) \equiv c_0$, where c_0 is an absolute constant (so that $\int^{+\infty} d\rho/\rho\psi(\rho) = +\infty$), because in the present case $\hat{\mathscr{E}}_1 = |q|^2 \equiv q_1^2 + q_2^2$, while $\hat{\alpha} \equiv -\lambda(\lambda+1)u + f(x)$ is bounded in modulus on this set by a constant not depending on ε. Thus, all the conditions of Theorem 2.1 except for (2.9) are satisfied uniformly with respect to $\varepsilon \in (0,1]$. However, condition (2.9) is not satisfied uniformly with respect to $\varepsilon \in (0,1]$, since $|A\mathbf{v}| = \varepsilon$ at the point $(0,0) \in \partial\Omega$. The derivatives $(\partial u_\varepsilon/\partial x_1)(0,0) \equiv \lambda\varepsilon^{\lambda-1}$ are also not bounded uniformly with respect to $\varphi \in (0,1]$. Thus, condition (2.9) of Theorem 2.1 is essential (of course, only the left side of (2.9) is of basic significance).

We shall now establish an estimate of $\max_{\overline{\Omega}}|\nabla u|$ in terms of $\max_{\partial\Omega}|\nabla u|$ for any equation of the form (1.5) with nonnegative characteristic form $\alpha^{ij}\xi_i\xi_j$ without requiring, generally speaking, that this equation have the structure of an $(A,\mathbf{b})$-elliptic equation (i.e., that it be determined by an equation of the form (1.2) via formula (1.6)).[2]

THEOREM 2.2. *Suppose that the functions $\alpha^{ij}(x,u,p)$, $i,j = 1,\dots,n$, and $\alpha(x,u,p)$ are differentiable on $\overline{\Omega} \times \mathbb{R} \times \mathbb{R}^n$ and that on the set $\overline{\Omega} \times [-m,m] \times \{|p| > L\}$, where $L = \text{const} \geqslant 0$ and $m = \text{const} \geqslant 0$, they satisfy the conditions*

$$\alpha^{ij}(x,u,p)\xi_i\xi_j \geqslant 0, \qquad \xi \in \mathbb{R}^n, \tag{2.14}$$

and

$$-\frac{n}{4}\sup_{|\tau|=1}\frac{|\delta\mathfrak{A}^\tau|^2}{\mathfrak{A}^\tau} + \frac{\delta\alpha}{|p|} > 0, \tag{2.15}$$

where $\mathfrak{A}^\tau \equiv \alpha^{ij}(x,u,p)\tau_i\tau_j$, $\tau \in \mathbb{R}^n$, $|\tau| = 1$, and

$$\delta \equiv \frac{p_k}{|p|}\frac{\partial}{\partial x_k} + |p|\frac{\partial}{\partial u}.$$

Then, for any function $u \in C^3(\Omega) \cap C^1(\overline{\Omega})$ satisfying (1.5) and with $|u| \leqslant m$ in Ω,

$$\max_{\overline{\Omega}}|\nabla u| \leqslant \max(L, M_1), \tag{2.16}$$

where $M_1 \equiv \max_{\partial\Omega}|\nabla u|$.

PROOF. Applying the operator $u_k(\partial/\partial x_k)$ to (1.5) and setting $v = \sum_1^n u_k^2$, we obtain

$$\tfrac{1}{2}\alpha^{ij}v_{ij} = \alpha^{ij}u_{ki}u_{kj} + \tfrac{1}{2}\left[\alpha_{p_l} - \alpha^{ij}_{p_l}u_{ij}\right]v_l + \sqrt{v}\left(\delta\alpha - \delta\alpha^{ij}u_{ij}\right),$$

where $v_l \equiv v_{x_l}$, $v_{ij} \equiv v_{x_ix_j}$, etc. Exactly as in the derivation of (1.6.13), we obtain

$$\sqrt{v}\,\delta\alpha^{ij}u_{ij} \leqslant \alpha^{ij}u_{ki}u_{kj} + \tfrac{1}{4}\left(|\delta\tilde{\alpha}^{ii}|^2/\tilde{\alpha}^{ii}\right)v,$$

where $\tilde{\mathfrak{A}} = T^*\mathfrak{A}T$, and T is the orthogonal matrix reducing the Hessian of u at the point $x \in \Omega$ to diagonal form so that $\tilde{\alpha}^{ii} = \mathfrak{A}^{\tau_i} = \mathfrak{A}\tau_j \cdot \tau_i$, $|\tau_i| = 1$, $i = 1,\dots,n$ (see §1.6). Then, by condition (2.15), at all points $x \in \Omega$, where $|\nabla u| > l$, we have

$$\frac{1}{2}\alpha^{ij}v_{ij} \geqslant b^kv_k + \left(\frac{\delta\alpha}{|p|} - \frac{n}{4}\sup_{|\tau|=1}\frac{|\delta\mathfrak{A}^\tau|^2}{\mathfrak{A}^\tau}\right)v > b^kv_k.$$

---

[2] In this case an equation of the form (1.5) may formally be considered to have the structure of an $(I,\mathbf{0})$-elliptic (but not strictly $(I,\mathbf{0})$-elliptic) equation, where $I \equiv \|\delta_i^j\|$.

From the last inequality it obviously follows that

$$\max_{\bar{\Omega}} v \leqslant \max\Big(\max_{\partial\Omega} v,\ L^2\Big),$$

whence we obtain (2.16). Theorem 2.2 is proved.

Condition (2.15) is due to the essence of the matter. For example, we consider the equation

$$-\frac{\partial}{\partial x_i}\left[r^2\frac{\partial u}{\partial x_i}\big(r|\nabla u|\big)^{m-2}\right]+\lambda^{m-2}(m-2)\big(r|\nabla u|\big)^2|u|^{m-3}$$
$$+(n\lambda+\lambda^2)u\big(r|\nabla u|\big)^{m-2}=0, \qquad (2.17)$$

where $r=\sqrt{x_1^2+\cdots+x_n^2}$, $\lambda\in(0,1)$, and m is a sufficiently large positive number (the latter assumption is to ensure sufficient smoothness of the coefficients of the equation). It is easy to verify that in the ball $\{|x|\leqslant 1\}$ such an equation has the solution $u=r^\lambda$ with unbounded first derivatives $\partial u/\partial x_i$ at $x=0$, although on the boundary $\{|x|=1\}$ these derivatives are bounded. Equation (2.17) has the form (1.5) with

$$\alpha^{ij}=r^m|p|^{m-2}\left[\delta_i^j+(m-2)\frac{p_i}{|p|}\frac{p_j}{|p|}\right],$$

$$\alpha=-2x_ip_i\big(r|p|\big)^{m-2}-mx_i\big(r|p|\big)^{m-2}p_i$$
$$+(m-2)\lambda^{m-2}\big(r|p|\big)^2|u|^{m-3}+(n\lambda+\lambda^2)u\big(r|p|\big)^{m-2},$$

so that for this equation

$$\mathfrak{A}^\tau=r^m|p|^{m-2}\left[1+\frac{(p,\tau)^2}{|p|^2}\right],\quad \delta\mathfrak{A}^\tau=m\frac{p_kx_k}{|p|}\frac{}{r}r^{m-1}|p|^{m-2}\left[1+(m-2)\frac{(p,\tau)^2}{|p|^2}\right],$$

$$\delta\alpha|_{u=0}=[-2+n\lambda+\lambda^2]\big(r|p|\big)^{m-2}.$$

Thus, for certain values of x, p, τ and for $u=0$ the left side of (2.15) does not exceed the quantity $(r|p|)^{m-2}[n\lambda+\lambda^2-2-(1/4)nm^2]$ which is negative for $\lambda\in(0,1)$ and $m\geqslant 2$. Thus, condition (2.15) is not satisfied for (2.17). Below we shall present an example showing that a condition of the form (2.15) is essential when ∇u occurs in the equation linearly.

REMARK 2.3. In the case of a linear equation

$$\alpha^{ij}(x)u_{x_ix_j}=\beta^i(x)u_{x_i}+c(x)u+f(x), \qquad (2.18)$$

where $\alpha^{ij}(x)\xi_i\xi_j\geqslant 0$ for any $x\in\Omega$ and $\xi\in\mathbb{R}^n$, condition (2.15) becomes

$$-\frac{n}{4}\sup_{|\tau|=1}\frac{|\nabla\mathfrak{A}^\tau|^2}{\mathfrak{A}^\tau}+\min_{i=1,\dots,n}\frac{\partial\beta^i}{\partial x_i}+c(x)-\sum_{i\neq k}\left|\frac{\partial\beta^i}{\partial x_k}\right|>0, \qquad (2.19)$$

where $\mathfrak{A}^\tau\equiv\alpha^{ij}(x)\tau_i\tau_j$. Indeed, it follows from (2.19) that for all $x\in\bar{\Omega}$ and any $p\neq 0$

$$-\frac{n}{4}\sup_{|\tau|=1}\frac{|\delta\mathfrak{A}^\tau|^2}{\mathfrak{A}^\tau}+\frac{p_k\big(\partial\beta^i/\partial x_k\big)p_i}{|p|^2}>0.$$

Then for all $(x, p) \in \Omega \times \{|p| > L\}$ with sufficiently large $L > 0$

$$-\frac{n}{4}\sup_{|\tau|=1}\frac{|\delta\mathfrak{A}^\tau|^2}{\mathfrak{A}^\tau} + \frac{p_k}{|p|}\frac{\partial\beta^i}{\partial x_k}\frac{p_i}{|p|} + c(x) + \frac{p_k}{|p|}\frac{\partial c(x)}{\partial x_k} - \frac{p_k}{|p|}\frac{\partial f(x)}{\partial x_k} > 0,$$

which is equivalent to (2.15) in the case where (1.5) has the form (2.18). Condition (2.19) is very close to one of the conditions of [99] which were imposed to obtain an a priori estimate of $|\nabla u|$ for solutions of linear equations of the form (2.18).

As the next example shows, a condition of the form (2.19) is due to the essence of the matter (see also [99]). In the ball $\{|x| \leqslant 1\}$ we consider an equation of the form

$$-(\partial/\partial x_i)\left(\rho^2(\partial u/\partial x_i)\right) + \left(n\lambda + \lambda^2(r^2/\rho^2)\right)u = 0, \tag{2.20}$$

where $\rho = \sqrt{x_1^2 + \cdots x_n^2 + \varepsilon^2}$, $\varepsilon, \lambda \in (0, 1)$. The nondivergence form of this equation is

$$\rho^2\Delta u + 2x_i(\partial u/\partial x_i) - \left(n\lambda + \lambda^2(r^2/\rho^2)\right)u = 0.$$

For this equation it is easy to see that a condition of the form (2.19) is not satisfied in the entire ball $\{|x| \leqslant 1\}$ uniformly with respect to $\varepsilon \in (0, 1)$, since here the left side of (2.19) is equal to

$$-n\left(|x|^2/\left(|x|^2 + \varepsilon^2\right)\right) - 2 + n\lambda + \lambda^2(r^2/\rho^2),$$

so that for all $x \neq 0$ there is an $\varepsilon \in (0, 1)$ such that this expression is strictly less than 0. At the same time the function $u = \rho^\lambda$ satisfies (2.20) at all points of the ball $\{|x| \leqslant 1\}$ and is equal to $(1 + \varepsilon)^\lambda$ on its boundary $\{|x| = 1\}$, while $|\nabla u| = \lambda\rho^{\lambda-1}[|x|/(|x|^2 + \varepsilon^2)^{1/2}]$ is not bounded in this ball uniformly with respect to $\varepsilon \in (0, 1)$. Thus, for (2.20) it is not possible to obtain an estimate of $\max_{|x|\leqslant 1}|\nabla u|$ in terms of $\max_{|x|=1}|\nabla u|$ which does not depend on ε. We observe, however, that the solution $u = \rho^\lambda$ of (2.20) in the ball $\{|x| \leqslant 1\}$ has bounded A-derivatives (of all orders) relative to the matrix $A = |X|I$, where $I \equiv \|\delta_i^j\|$, which determines the structure of an $(A, \mathbf{0})$-elliptic equations of the form (1.2) for equation (2.20). In Chapter 7 we shall distinguish a class of $(A, \mathbf{0})$-elliptic equations of the form (1.2) (which, in particular, contains (2.20)) for whose solutions estimates of the A-derivatives will be established.

We shall now use the a priori estimates (2.6) and (2.16) established above for $(A, \mathbf{b})$-elliptic equations of nondivergence form to prove the existence of a generalized solution of problem (1.1). To this end we regularize the equation (1.2) generated by an $(A, \mathbf{b})$-elliptic equation of divergence form (5.1.35); namely, as a regularized equation we consider the $(B, \mathbf{b})$-elliptic equation

$$-(d/dx_i)\hat{l}^i(x, u, \nabla u) + \hat{l}_0(x, u, \nabla u) = f(x),$$

where $B \equiv \|b^{ij}(x)\|$, $b^{ij}(x) = a^{ij}(x) + \varepsilon\delta_i^j$ (i.e., $B = A + \varepsilon I$, where I is the identity matrix), $\varepsilon > 0$, and $\mathbf{b} = (b^1(x), \ldots, b^n(x))$ is the same vector that determines the original $(A, \mathbf{b})$-elliptic equation of the form (5.1.35); we assume that the reduced coefficients of the new equation have the form

$$\hat{l}'^i(x, u, q) = \varepsilon q_i + l'^i(x, u, q), \quad i = 1, \ldots, n, \quad \hat{l}'_0(x, u, q) = l'_0(x, u, q),$$

where $l''^i(x, u, q)$, $i = 1,\dots,n$, and $l'_0(x, u, q)$ are the reduced coefficients of the original $(A, \mathbf{b})$-elliptic equation. The nondivergence form of such an equation has the form

$$\hat{\beta}^{ij}(x, u, \hat{\nabla} u) u_{\bar{j}\bar{i}} - \hat{\beta}(x, u, \hat{\nabla} u) - b^i u_{x_i} = 0, \tag{2.21}$$

where $\hat{\beta}^{ij}(x, u, q) = \varepsilon\delta_i^j + \hat{\alpha}^{ij}(x, u, q)$, $\hat{\beta}(x, u, q) = \hat{\alpha}(x, u, q) - \varepsilon(\partial a^{ki}/\partial x_i) q_k$, and the functions $\hat{\alpha}^{ij}(x, u, q)$, $i, j = 1,\dots,n$, and $\hat{\alpha}(x, u, q)$ are expressed in terms of the reduced coefficients of the original equation (5.1.35) by formula (1.4) but with the replacement of the elements of A in (1.4) by elements of the matrix $B = A + \varepsilon I$. In (2.21) the derivatives $u_{\bar{i}}$ and $u_{\bar{j}\bar{i}}$, and the gradient $\hat{\nabla} u$ correspond to the regularized matrix $B = A + \varepsilon I$. (2.21) can also be rewritten as an equation expressed only in terms of ordinary derivatives

$$\beta^{ij}(x, u, \nabla u) u_{ij} - \beta(x, u, \nabla u) = 0, \tag{2.22}$$

where

$$\beta^{ij}(x, u, p) = \frac{\partial \tilde{l}^i(x, u, p)}{\partial p_j}, \qquad i, j = 1,\dots,n,$$
$$\beta(x, u, p) = -\frac{\partial \tilde{l}^i}{\partial u} p_i - \frac{\partial \tilde{l}^i}{\partial x_i} - f(x) + \tilde{l}_0(x, u, p) \tag{2.22'}$$

(cf. (1.7)). We note that the coefficients of (2.22) can also be written in the form

$$\beta^{ij} = b^{ki}\hat{\beta}^{ks}b^{sj}, \quad \beta = -b^{ki}\hat{\beta}^{ks}(\partial b^{sj}/\partial x_i) p_j + \hat{\beta} + b^i p_i,$$

where $\hat{\beta}^{ks} = \hat{\beta}^{ks}(x, u, q)$, $\hat{\beta} = \hat{\beta}(x, u, q)$, and $q = B(x)p$ (cf. (1.6)).

For the regularized equations (2.22) we consider the auxiliary Dirichlet problems

$$\beta^{ij} u_{ij} - \beta = 0 \quad \text{in } \Omega, \qquad u = 0 \quad \text{on } \partial\Omega, \tag{2.23}$$

that correspond to values $\varepsilon \in (0, 1)$.

LEMMA 2.3. *Let $\Omega \subset \mathbb{R}^n$, $n \geqslant 2$, be a bounded domain of class C^3, and suppose that for the $(A, \mathbf{b})$-elliptic equation of the form* (1.2), *where $a^{ij} \in \tilde{C}^1(\overline{\Omega})$, $i, j = 1,\dots,n$, $b^i \in \tilde{C}^1(\overline{\Omega})$, $i = 1,\dots,n$, $\hat{\alpha}^{ij}(x, u, q) \in \tilde{C}^1(\Omega \times \mathbb{R} \times \mathbb{R}^n)$ and $\hat{\alpha}(x, u, q) \in \tilde{C}^1$, conditions* (2.2), (2.8), (2.9), *and* (2.14) *are satisfied, while for* (2.22) *condition* (2.15) *holds. Then for any solution $u \in C^3(\Omega) \cap C^1(\overline{\Omega})$ of problem* (2.23) *of the regularized $(B, \mathbf{b})$-elliptic equation* (2.21), *where $B = A + \varepsilon I$, $\hat{\beta}^{ij} = \varepsilon\delta_i^j + \hat{\alpha}^{ij}$, $i, j = 1,\dots,n$ and $\hat{\beta} = \hat{\alpha}$, the estimates*

$$\max_{\overline{\Omega}} |u| \leqslant m, \qquad \max_{\overline{\Omega}} |\nabla u| \leqslant \overline{M}_1, \tag{2.24}$$

hold, where the constants m and $\overline{M}_1$ depend only on the structural conditions (2.2), (2.8), (2.9), (2.14), *and* (2.15) *for the original equation* (1.2), *and on the domain Ω.*

PROOF. It is obvious that a condition of the form (2.1) (with $\hat{\alpha}^{ij}$ replaced by $\hat{\beta}^{ij}$ and a^{ij} by b^{ij}, $i, j = 1,\dots,n$) is satisfied for the regularized equation (2.21). The validity of the structural conditions for the original equation (1.2) enumerated in the lemma leads to the validity of the corresponding conditions for the regularized equation (2.21), as is easily seen, and the latter are satisfied with constants not depending on $\varepsilon \in (0, 1)$. Applying successively Lemma 2.1 and Theorems 2.1 and 2.2, we then obtain (2.24), where the constants m and $\overline{M}_1$ do not depend on ε. Lemma 2.3 is proved.

LEMMA 2.4. *Let $\Omega \in C^2$, and suppose that $\alpha^{ij} \in \tilde{C}^1(\overline{\Omega})$, $b^i \in \tilde{C}^1(\Omega)$, $i, j = 1,\dots,n$, $\hat{\alpha}^{ij}(x, u, q) \in \tilde{C}^1(\overline{\Omega} \times \mathbb{R} \times \mathbb{R}^n)$, $i, j = 1,\dots,n$, and $\hat{\alpha}(x, u, q) \in \tilde{C}^1(\overline{\Omega} \times \mathbb{R} \times \mathbb{R}^n)$, where the $\hat{\alpha}^{ij}$, $i, j = 1,\dots,n$, and $\hat{\alpha}$ are the coefficients of* (1.2) *corresponding to the original $(A, \mathbf{b})$-elliptic equation of the form* (5.1.35) *with $A \equiv \|a^{ij}(x)\|$ and $\mathbf{b} = (b^1(x),\dots,b^n(x))$ (see* (1.4), *in which $l'^i(x, u, q)$, $i = 1,\dots,n$, and $l'_0(x, u, q)$ are the reduced coefficients of* (5.1.35)). *Suppose that the conditions enumerated in Lemma* 2.3 *are satisfied for* (1.2). *Then for any $\varepsilon \in (0, 1)$ problem* (2.23) *has a classical solution $u_\varepsilon \in C^2(\overline{\Omega})$, and for this solution inequalities* (2.24) *hold with constants m and $\overline{M}_1$ not depending on $\varepsilon \in (0, 1)$.*

PROOF. It follows from the conditions of Lemma 2.4 that if the above equation (1.2) is rewritten in the form (1.5), (1.6), then the functions $\alpha^{ij}(x, u, p)$, $i, j = 1,\dots,n$, and $\alpha(x, u, p)$ belong to the class $\tilde{C}^1(\overline{\Omega} \times \mathbb{R} \times \mathbb{R}^n)$. Suppose first that $\Omega \in C^3$ and $\alpha^{ij}(x, u, p)$, $i, j = 1,\dots,n$ and $\alpha(x, u, p)$ belong to the class $C^2(\overline{\Omega} \times \mathbb{R} \times \mathbb{R}^n)$. By means of Schauder's theorem it is then easy to prove that any solution of (2.23) belonging to $C^2(\overline{\Omega})$ actually belongs to $C^3(\Omega)$ also (see the proof of Theorem 1.9.1). Hence, by Lemma 2.3 estimates of the form (2.24) hold for this solution, so that equation (2.23) may be considered boundedly nonlinear and uniformly elliptic. Applying known results of Ladyzhenskaya and Ural'tseva on solvability of boundary value problems for quasilinear uniformly elliptic equations (see the theorem of Ladyzhenskaya and Ural'tseva in §1.2), we establish the existence of a solution u_ε of (2.23) belonging to $C^2(\overline{\Omega})$. It is not hard to eliminate the superfluous assumptions of smoothness of Ω and of the functions a^{ij} and a by using the standard method of approximating Ω, a^{ij}, and a by the corresponding objects having the degree of smoothness indicated at the beginning of the proof (see the proof of Theorem 1.9.1). This proves Lemma 2.4.

THEOREM 2.3. *Suppose that an equation of the form* (5.1.35) *has the structure of an $(A, \mathbf{b})$-elliptic equation in a bounded domain $\Omega \subset \mathbb{R}^n$, $n \geqslant 2$, of class C^2, where $A \equiv \|a^{ij}(x)\|$ is a symmetric nonnegative-definite matrix in $\overline{\Omega}$, $a^{ij} \in \tilde{C}^1(\overline{\Omega})$, $i, j = 1,\dots,n$, and $\mathbf{b} \equiv (b^1(x),\dots,b^n(x))$, $b^i \in \tilde{C}^1(\overline{\Omega})$, $i = 1,\dots,n$. Assume that the functions $\hat{\alpha}^{ij}(x, u, q)$, $i, j = 1,\dots,n$, and $\hat{\alpha}(x, u, q)$ defined by* (1.4), *in which $l'^i(x, u, q)$, $i = 1,\dots,n$, and $l'_0(x, u, q)$ are the reduced coefficients of the above* (5.1.35), *belong to the class $\tilde{C}^1(\overline{\Omega} \times \mathbb{R} \times \mathbb{R}^n)$. Suppose that the structural conditions* (2.2), (2.8), (2.9), *and* (2.15) *are satisfied for the equations* (1.2) *and* (2.22) *generated by* (5.1.35) *by means of equalities of the form* (1.4) *and* (2.22′). *Assume, finally, that for any $x \in \overline{\Omega}$, $|u| \leqslant 2m$, $q = Bp$, $p \in \mathbb{R}^n$ and $|q| \leqslant 2\overline{M}_1$, where m and $\overline{M}_1$ are the constants determined by conditions* (2.2), (2.8), (2.9), *and* (2.15) (*see Lemma* 2.3), *for the reduced coefficients of* (5.1.35) *the inequality*

$$\frac{\partial l'^i}{\partial q_j}\xi_i\xi_j + \frac{\partial l'^i}{\partial u}\xi_0\xi_i + \frac{\partial l'_0}{\partial q_j}\xi_j\xi_0 + \frac{\partial l'_0}{\partial u}\xi_0^2 - \frac{1}{2}\frac{\partial b^i}{\partial x_i}\xi_0^2 \geqslant 0, \qquad (2.25)$$
$$\forall \xi_0 \in \mathbb{R}, \quad \xi = B\eta, \quad \forall \eta \in \mathbb{R}^n$$

holds. Then the Dirichlet problem (1.1) *for equation* (5.1.35) *has at least one regular generalized solution u, i.e., there exists a function $u \in L^\infty(\Omega) \cap \mathring{H}_m(\Omega)$ for all $m > 1$ such that $\nabla u \in L^\infty(\Omega)$ and*

$$\int_\Omega [\mathbf{l}(x, u, \nabla u) \cdot \nabla \eta + l_0(x, u, \nabla u)\eta]\, dx = \int_\Omega f\eta\, dx \quad \forall \eta \in C_0^1(\Omega). \quad (2.26)$$

If in place of (2.25) *there is the condition: for all* $x \in \Omega$ *and any* $u \in \mathbb{R}$ *and* $q \in \mathbb{R}^n$

$$\frac{\partial l'^i}{\partial q_j}\xi_i\xi_j + \frac{\partial l'^i}{\partial u}\xi_0\xi_i + \frac{\partial l'_0}{\partial q_j}\xi_j\xi_0 + \frac{\partial l'_0}{\partial u}\xi_0^2 - \frac{1}{2}\frac{\partial b^i}{\partial x_i}\xi_0^2 > 0 \tag{2.25'}$$
$$\textit{for all } (\xi, \xi_0) \neq (0,0), \quad \xi_0 \in \mathbf{R}, \quad \xi = \mathbf{R}^n,$$

then problem (1.1) *has precisely one regular generalized solution.*

PROOF. Let u_ε be the solution of (2.23) for the regularized equation (2.21) (see also (2.22)) corresponding to (1.2), with (1.4). For such solutions the estimate (2.24) holds by Lemma 2.3. We rewrite the regularized equation (2.22) in divergence form

$$-(d/dx_i)\tilde{l}^i(x, u_\varepsilon, \nabla u_\varepsilon) + \hat{l}_0(x, u_\varepsilon, \nabla u_\varepsilon) = f(x), \tag{2.27}$$

where

$$\tilde{\mathbf{l}}(x, u, p) = \varepsilon B^*Bp + B^*\mathbf{l}'(x, u, Bp), \quad \hat{l}_0(x, u, p) = l'_0(x, u, Bp) + b^i(x)p_i,$$
$$B = A + \varepsilon I, \qquad \varepsilon > 0.$$

The corresponding integral identity for u_ε has the form

$$\varepsilon\int_\Omega B\nabla u_\varepsilon \cdot B\nabla\eta\, dx + \int_\Omega\big[\mathbf{l}'(x, u_\varepsilon, B\nabla u_\varepsilon)\cdot B\nabla\eta + l'_0(x, u_\varepsilon, B\nabla u_\varepsilon)\eta - b^i u_{\varepsilon x_i}\eta\big]\, dx = \int_\Omega f\eta\, dx, \qquad \eta \in C_0^1(\Omega). \tag{2.28}$$

In view of (2.24) it may be assumed that inequalities of the form (5.1.3) for some indices m and $\mathbf{m}$, $m > 1$, $\mathbf{m} = (m_1,\dots,m_n)$, $m_i > 1$, $i = 1,\dots,n$, hold for the functions $\mathbf{l}'(x, u, q)$ and $l'_0(x, u, q)$. It may further be assumed that the sequence $\{u_\varepsilon\}$ converges weakly in $L^m(\Omega)$ to a function $u \in L^2(\Omega)$, while $\{\nabla u_\varepsilon\}$ converges weakly in $\mathbf{L}^{\mathbf{m}}(\Omega)$ to $\nabla u \in \mathbf{L}^{\mathbf{m}}(\Omega)$. In view of (2.25),

$$\varepsilon\int_\Omega B\nabla(u_\varepsilon - \xi)\cdot B\nabla(u_\varepsilon - \xi)\, dx + \int_\Omega\{[\mathbf{l}'(x, u_\varepsilon, B\nabla u_\varepsilon) - \mathbf{l}'(x, \xi, B\nabla\xi)] \times B\nabla(u_\varepsilon - \xi) + [l'_0(x, u_\varepsilon, B\nabla u_\varepsilon) - l'_0(x, \xi, B\nabla\xi)](u_\varepsilon - \xi) + b^i(u_\varepsilon - \xi)_{x_i}(u_\varepsilon - \xi)\}\, dx \geqslant 0, \quad \forall\xi \in C_0^1(\Omega). \tag{2.29}$$

Subtracting (2.28) with $\eta = u_\varepsilon - \xi$ from (2.29), we find that

$$-\varepsilon\int_\Omega B\nabla\xi\cdot B\nabla(u_\varepsilon - \xi)\, dx - \int_\Omega\big[\mathbf{l}'(x, \xi, B\nabla\xi)\cdot B\nabla(u_\varepsilon - \xi) + l'_0(x, \xi, B\nabla\xi)(u_\varepsilon - \xi) + b^i\xi_{x_i}(u_\varepsilon - \xi)\big]\, dx \geqslant -\int_\Omega f(u_\varepsilon - \xi)\, dx, \tag{2.30}$$

where $\xi \in C_0^1(\Omega)$. Letting ε tend to 0 in (2.30) and taking account of the fact that $B\nabla u_\varepsilon \to A\nabla u$ weakly in $\mathbf{L}^{\mathbf{m}}(\Omega)$ and $B\nabla\xi \to A\nabla\xi$ uniformly in $\overline{\Omega}$, we obtain

$$-\int_\Omega\left[\mathbf{l}'(x,\xi,A\nabla\xi)\cdot A\nabla(u-\xi)+l_0'(x,\xi,A\nabla\xi)(u-\xi)+b^i\xi_{x_i}(u-\xi)\right]dx$$
$$\geqslant -\int_\Omega f(u-\xi)\,dx,\qquad \xi\in C_0^1(\Omega). \tag{2.31}$$

It is obvious that (2.31) also holds for all $\xi \in \mathring{H}_{m,\mathbf{m}}(A,\Omega)\cap \mathring{H}_2^1(\Omega)$. Setting $\xi = u + \varepsilon\eta$ in (2.31), $\eta \in \mathring{C}_1(\Omega)$, $\varepsilon > 0$, dividing the inequality so obtained by ε, and passing to the limit as $\varepsilon \to +0$, we find that

$$\int_\Omega\left[\mathbf{l}'(x,u,A\nabla u)\cdot A\nabla\eta+l_0'(x,u,A\nabla u)\eta+b^iu_{x_i}\eta\right]dx$$
$$\leqslant \int_\Omega f\eta\,dx,\qquad \eta\in C_0^1(\Omega). \tag{2.32}$$

Since η is arbitrary in (2.32), we immediately obtain

$$\int_\Omega\left[\mathbf{l}'(x,u,A\nabla u)\cdot A\nabla\eta+l_0'(x,u,A\nabla u)\eta+b^iu_{x_i}\eta\right]dx$$
$$=\int_\Omega f\eta\,dx,\qquad \eta\in C_0^1(\Omega),$$

which can also be rewritten in the form (2.26). Thus, u is a generalized solution of (1.1). It is obvious that $|u| \leqslant m$ and $|\nabla u| \leqslant \overline{M}_1$ in Ω. The existence of a regular generalized solution of (1.1) has thus been demonstrated. Suppose now that condition (2.25) is replaced by (2.25′). To prove uniqueness of a regular generalized solution of problem (1.1) we use an equality of the form (5.4.13). Let u and v be two regular generalized solutions of (1.1). It is easy to see that (cf. (5.4.13))

$$0=\langle \mathscr{L}u-\mathscr{L}v,u-v\rangle$$
$$=\int_\Omega\int_0^1\left[\frac{\partial l'^i(x,v+\tau(u-v),A\nabla v+\tau A\nabla(u-v))}{\partial p_j}\right.$$
$$\times A_j\nabla(u-v)A_i\nabla(u-v)+\frac{\partial l'^i}{\partial u}(u-v)A_i\nabla(u-v)$$
$$\left.+\frac{\partial l_0'}{\partial p_j}A_j\nabla(u-v)(u-v)+\frac{\partial l_0'}{\partial u}(u-v)^2-\frac{1}{2}\frac{\partial b^i}{\partial x_i}(u-v)^2\right]d\tau\,dx. \tag{2.33}$$

We suppose that $u \not\equiv v$ in Ω. Then there exists a subset $\Omega' \subset \Omega$, meas $\Omega' > 0$, on which $u \neq v$. In view of (2.25′) the right side of (2.33) is then strictly positive, which contradicts (2.33). Theorem 2.3 is proved.

REMARK 2.4. Theorem 2.3 remains valid if the conditions that $\hat{a}^{ij}(x,u,q)$, $i,j=1,\dots,n$, and $\hat{a}(x,u,q)$ belong to the class $\tilde{C}^1(\overline{\Omega}\times\mathbb{R}\times\mathbb{R}^n)$ are replaced by the condition that these functions possess only the differential properties necessary in order that the structural conditions (2.2), (2.8), (2.9), (2.15), and (2.25) be meaningful. Indeed, averaging the coefficients $l^i(x,u,p)$, $i=1,\dots,n$, $l_0(x,u,p)$, and $f(x)$ with respect to the variables x, u and p, we consider a problem of the form (1.1) for equations

$$-(d/dx_i)l_\varepsilon^i(x, u, \nabla u) + l_{0\varepsilon}(x, u, \nabla u) = f_\varepsilon(x),$$

where l_ε^i, $l_{0\varepsilon}$, and f_ε denote the averages of the functions l^i, l_0, and f. Suppose the structural conditions (2.2), (2.8), (2.9), (2.15), and (2.25) are satisfied uniformly with respect to ε for these approximating equations. Then the following estimates hold uniformly with respect to ε for solutions of these approximating problems (which exist by Theorem 2.3):

$$\max_{\overline{\Omega}} |u_\varepsilon| \leqslant c_0, \qquad \max_{\overline{\Omega}} |\nabla u_\varepsilon| \leqslant c_1. \tag{2.34}$$

In integral identities of the form

$$\int_\Omega \left[l'_\varepsilon(x, u_\varepsilon, A\nabla u_\varepsilon) \cdot A\nabla\eta + l'_{0\varepsilon}(x, u_\varepsilon, A\nabla u_\varepsilon)\eta + b_\varepsilon^i u_{\varepsilon x_i}\eta \right] dx$$

$$= \int_\Omega f_\varepsilon \eta \, dx, \qquad \eta \in \overset{\circ}{C}{}_0^1(\Omega),$$

we can then pass to the limit as $\varepsilon \to 0$, arguing just as in the proof of Theorem 2.3. An added feature in these arguments is the consideration that the functions $l_\varepsilon'^i(x, \xi, \nabla\xi)$, $i = 1, \dots, n$, and $l'_{0\varepsilon}(x, \xi, \nabla\xi)$, where $\xi \in C_0^1(\Omega)$, as $\varepsilon \to 0$ tend uniformly on each compact portion of Ω to $l'^i(x, \xi, \nabla\xi)$, $i = 1, \dots, n$, and $l'_0(x, \xi, \nabla\xi)$ respectively, where we assume that the reduced coefficients are continuous in $\overline{\Omega}$. The limit function u satisfies an identity of the form (2.26) and possesses the property $u \in L^\infty(\Omega) \cap \overset{\circ}{H}_m(\Omega)$ for all $m > 1$ (because of the uniformity of the estimates (2.34)).

We now consider a nonregular variational problem concerning a minimum for an integral of the form

$$\int_\Omega [\mathscr{F}(x, u, A\nabla u) - f(x)u] \, dx, \qquad u = 0 \quad \text{on } \partial\Omega, \tag{2.35}$$

where Ω is a bounded domain in $\mathbb{R}^n$, $n \geqslant 2$, of class C^2, and $\mathscr{F}(x, u, q)$ satisfies the condition

$$\frac{\partial^2 \mathscr{F}(x, u, q)}{\partial q_i \partial q_j} \eta_i \eta_j \geqslant 0, \qquad x \in \overline{\Omega}, u \in \mathbb{R}, q = Ap, p \in \mathbb{R}, \eta = A\xi, \xi \in \mathbb{R}^n. \tag{2.36}$$

The Euler equation for problem (2.35) has the form

$$-\frac{d}{dx_i}\left[a^{ki}(x) \frac{\partial \mathscr{F}(x, u, A\nabla u)}{\partial q_k} \right] + \frac{\partial \mathscr{F}(x, u, A\nabla u)}{\partial u} = f(x), \tag{2.37}$$

where a^{ij}, $i, j = 1, \dots, n$, are the elements of the matrix A. It is obvious that (2.37) has the structure of an $(A, \mathbf{0})$-elliptic equation in Ω relative to the matrix A involved in (2.35), and the reduced coefficients of this equation have the form

$$l'^i = \frac{\partial \mathscr{F}(x, u, q)}{\partial q_i}, \qquad i = 1, \dots, n, \qquad l'_0 = \frac{\partial \mathscr{F}(x, u, q)}{\partial u}. \tag{2.38}$$

Taking Remark 2.4 into account, from Theorem 2.3 we thus obtain the following result.

THEOREM 2.4. *Suppose that the integral* (2.35) *is considered under the assumptions that the domain* Ω *is bounded in* $\mathbb{R}^n$, $n \geqslant 2$, *and belongs to the class* C^2, *the matrix* $A \equiv \|a^{ij}(x)\|$ *is symmetric and nonnegative-definite in* Ω *with* $a^{ij} \in \tilde{C}^1(\overline{\Omega})$, $i, j = 1,\dots,n$, *and the function* $\mathscr{F}(x, u, q)$ *satisfies condition* (2.36). *Assume that the functions* l'^i *and* l'_0 *defined by* (2.38), *the function* $f(x)$ *in* (2.35), *and the functions* $\hat{\alpha}^{ij}(x, u, q)$, $i, j = 1,\dots,n$, *and* $\hat{\alpha}(x, u, q)$ *defined by* (1.4) *and* (2.38) *ensure the validity of conditions* (2.2), (2.8), (2.9), (2.15), *and* (2.25) ((2.25′)). *Then there exists at least one* (*precisely one*) *extremal of problem* (2.35), *i.e., there exists a* (*unique*) *function* $u \in L^\infty(\Omega) \cap \overset{\circ}{H}_m(\Omega)$ *for every* $m > 1$ *such that* $\nabla u \in \mathbf{L}^\infty(\Omega)$ *for which the identity* (2.26) *holds* (*with* l'^i, $i = 1,\dots,n$, *and* l'_0 *defined by* (2.38)).

It is easy to see that under the assumptions made in Theorem 2.4 regarding the domain Ω and the matrix A (in particular, under condition (2.9)) all the other conditions of this theorem are satisfied when $\mathscr{F}(x, u, q)$ has the form

$$\mathscr{F}(x, u, q) = |q|^4 + \kappa u^2 |q|^2 + \kappa_1 u^2, \qquad \kappa, \kappa_1 = \text{const} > 0, \tag{2.39}$$

provided that $f(x) \in \tilde{C}^1(\overline{\Omega})$ and the constants κ and κ_1 are sufficiently large. In this case (2.37) has the form

$$-\operatorname{div}\left\{A\left(4|A\nabla u|^2 + 2\kappa u^2\right)A\nabla u\right\} + 2\kappa|A\nabla u|^2 u + 2\kappa_1 u = f(x). \tag{2.40}$$

A simpler example of an admissible functional (2.35) is the case of a quadratic functional corresponding to the function $\mathscr{F}(x, u, q) = |q|^2 + \kappa u^2$ with a sufficiently large constant $\kappa > 0$ and an arbitrary function $f(x) \in \tilde{C}_1(\overline{\Omega})$. In this case (2.27) is linear and has the form

$$-\operatorname{div}\{A^2 \nabla u\} + 2\kappa u = f(x). \tag{2.41}$$

§3. The existence of regular generalized solutions of the first boundary value problem which are bounded in Ω together with their partial derivatives of first order

In this section we establish the existence of generalized solutions of problem (1.1) possessing the regularity indicated in the title. It will be shown that such solutions satisfy (5.1.35) a.e. in Ω. Smoothness of order one greater than that assumed in §2 is required of the functions determining the structure of the $(A, \mathbf{b})$-elliptic equation (1.1) (in particular, of the elements of the matrix A and the components of the vector $\mathbf{b}$). Moreover, to obtain the indicated result we needed to impose rather stringent conditions on the structure of (5.1.35) which lead, in particular, to the condition of linearity of this equation in the first derivatives. It appears to us that the results obtained in this section have not previously been established even under such stringent conditions on the nonlinearity of equation (5.1.35). Similar results were obtained in [99] for the case of linear equations with a nonnegative characteristic form.

THEOREM 3.1. *Suppose that the functions* $\alpha^{ij}(x, u, p)$, $i, j = 1,\dots,n$, *and* $\alpha(x, u, p)$ *defined by* (1.6) *are twice differentiable with respect to* x, u *and* p *in* $\overline{\Omega} \times \mathbb{R} \times \mathbb{R}^n$ *and on the set* $\overline{\Omega} \times [-m, m] \times \{|p| \leqslant \overline{M}_1\} \times \{|D^2 u| > L\}$ (m, $\overline{M}_1$ *and* L *are positive constants) satisfy* (2.14) *and the condition*

$$-n \sup_{\substack{l=1,\dots,n\\|\tau|=1}} \frac{|\delta_l \mathfrak{A}^\tau|^2}{\mathfrak{A}^\tau} + \frac{\partial \alpha}{\partial u} - \frac{\partial \alpha^{ij}}{\partial u} u_{ij} + \frac{u_{kl}}{|D^2 u|^2}\left[\delta_{kl}\alpha - (\delta_{kl}\alpha^{ij})u_{ij}\right] > 0, \quad (3.1)$$

where

$$\mathfrak{A}^\tau \equiv \alpha^{ij}(x, u, p)\tau_i\tau_j, \quad \tau \in \mathbb{R}^n, \quad |\tau| = 1,$$

$$\delta_k \equiv \frac{\partial}{\partial p_s} u_{sk} + \frac{\partial}{\partial u} u_k + \frac{\partial}{\partial x_k}, \quad k = 1,\dots,n,$$

$$\delta_{kl} \equiv \frac{\partial^2}{\partial p_s \partial p_t} u_{sk} u_{tl} + \frac{\partial^2}{\partial p_s \partial u} u_{sk} u_l + \frac{\partial^2}{\partial p_s \partial x_l} u_{sk} + \frac{\partial^2}{\partial u \partial p_t} u_k u_{tl} + \frac{\partial^2}{\partial u^2} u_k u_l$$
$$+ \frac{\partial^2}{\partial u \partial x_l} u_k + \frac{\partial^2}{\partial x_k \partial p_t} u_{tl} + \frac{\partial^2}{\partial x_k \partial u} u_l + \frac{\partial^2}{\partial x_k \partial x_l},$$

$$|D^2 u| = \sum_{i,j=1}^n u_{ij}^2, \quad u_k = \frac{\partial u}{\partial x_k}, \quad u_{ij} = \frac{\partial^2 u}{\partial x_i \partial x_j},$$

where in condition (3.1) x, u, p_k, *and* u_{ij} ($u_{ij} = u_{ji}$, $i, j = 1,\dots,n$) *are considered independent variables. Then, for any function* $u \in C^4(\Omega) \cap C^2(\overline{\Omega})$ *satisfying* (1.5) *and the inequalities* $|u| \leqslant m$ *and* $|\nabla u| \leqslant \overline{M}^1$ *in* Ω,

$$\max_{\overline{\Omega}} |D^2 u| \leqslant \max\left\{\max_{\partial\Omega} |D^2 u|, L\right\}. \quad (3.2)$$

PROOF. Applying the operator $u_{kl}\partial^2/\partial x_k \partial x_l$ to (1.5) and setting $v = \sum_{k,l=1}^n u_{kl}^2$, we obtain

$$\frac{1}{2}\alpha^{ij} v_{ij} = \alpha^{ij} u_{kli} u_{klj} + \frac{1}{2}\left(\frac{\partial \alpha}{\partial p_s} - \frac{\partial \alpha^{ij}}{\partial p_s} u_{ij}\right) v_s + \left(\frac{\partial \alpha}{\partial u} - \frac{\partial \alpha^{ij}}{\partial u} u_{ij}\right) v$$
$$- (\delta_l \alpha^{ij}) u_{ijk} u_{kl} - (\delta_k \alpha^{ij}) u_{ijl} u_{kl} + \left[\delta_{kl}\alpha - (\delta_{kl}\alpha^{ij})u_{ij}\right] u_{kl} = 0, \quad (3.3)$$

which is valid at any point $x \in \Omega$. Taking into account that condition (3.1), in particular, implies that $|\delta_l \mathfrak{A}^\tau|^2 (\mathfrak{A}^\tau)^{-1}$ is finite (for any $\tau \in \mathbb{R}^n$, $|\tau| = 1$), we find the estimates (see the derivation of (1.6.13))

$$\left|(\delta_l \alpha^{ij}) u_{ijk} u_{kl}\right| \leqslant \frac{1}{2}\sum_{k=1}^n \tilde{\alpha}_k^{ii} \tilde{u}_{kii}^2 + \frac{n}{2} \sup_{i,k,l=1,\dots,n} \frac{|\delta_l \tilde{\alpha}_k^{ii}|^2}{\tilde{\alpha}_k^{ii}} v,$$
$$\left|(\delta_k \alpha^{ij}) u_{ijl} u_{kl}\right| \leqslant \frac{1}{2}\sum_{k=1}^n \tilde{\alpha}_l^{ii} \tilde{u}_{lii}^2 + \frac{n}{2} \sup_{i,k,l=1,\dots,n} \frac{|\delta_k \tilde{\alpha}_l^{ii}|^2}{\tilde{\alpha}_l^{ii}} v, \quad (3.4)$$

where $\tilde{\alpha}_k^{ii} = \mathfrak{A}^{\tau_i^k} \equiv \mathfrak{A}\tau_i^k \cdot \tau_i^k$ and τ_i^k is the ith column of the orthogonal matrix T_k transforming the Hessian of the function $u_k \equiv \partial u/\partial x_k$ to diagonal form at the point $x \in \Omega$. Since $\tilde{\alpha}_k^{ii} \tilde{u}_{kii}^2 = \alpha^{ij} u_{kli} \cdot u_{klj}$, the sum of the left sides of (3.4) can be bounded above on the set $\Omega_L \equiv \{x \in \Omega\colon |D^2 u| > L\}$ in terms of

$$\alpha^{ij} u_{kli} u_{klj} + n \sup_{l=1,\dots,n,|\tau|=1} \frac{|\delta \mathfrak{A}^\tau|^2}{\mathfrak{A}^\tau} v.$$

Taking (3.3), (3.4), and (3.1) into account, we conclude that at points $x \in \Omega_L$

$$\tfrac{1}{2}\alpha^{ij} v_{ij} > 0,$$

which shows that v cannot achieve a maximum at points $x \in \Omega_L$. Hence,

$$\max_{\bar{\Omega}} v \leqslant \max\left\{ \max_{\partial\Omega} |D^2 u|^2, L^2 \right\},$$

whence we obtain (3.2). Theorem 3.1 is proved.

We note that in the case of a linear equation of the form (2.18) it is easy to see that (3.1) goes over into the condition

$$c + \min_{s=1,\dots,n} \frac{\partial b^s}{\partial x_s} - \sum_{t \neq s} \left| \frac{\partial b^s}{\partial x_t} \right| + \max_{i,j=1,\dots,n} \frac{\partial^2 \alpha^{ij}}{\partial x_i \partial x_j}$$

$$- \sum_{(k,l) \neq (i,j)} \left| \frac{\partial^2 \alpha^{ij}}{\partial x_k \partial x_l} \right| - n \sup_{l=1,\dots,n,\, |\tau|=1} \frac{|\partial \mathfrak{A}^\tau / \partial x_l|^2}{\mathfrak{A}^\tau} > 0, \quad (3.5)$$

where $\mathfrak{A}^\tau \equiv \alpha^{ij}(x)\tau_i\tau_j$, $|\tau| = 1$. Condition (3.5) is very close to one of the conditions under which an a priori estimate of $|D^2 u|$ was established in [99] in the case of a linear equation of the form (2.17).

It will be shown below that condition (3.1) of Theorem 3.1 is essential. It is easy to see that for the validity of (3.1) it is first of all necessary that on the set $\bar{\Omega} \times [-m, m] \times \{|p| \leqslant \overline{M}_1\} \times \{|D^2 u| > L\}$ the following condition be satisfied:

$$-\frac{\partial^2 \alpha^{ij}}{\partial p_s \partial p_t} \frac{u_{lk} u_{sk} u_{tl} u_{ij}}{|D^2 u|^2} \geqslant 0. \quad (3.6)$$

Condition (3.6) introduces very stringent conditions on the occurrence of the argument p in α^{ij}. Indeed, in the disk $\Omega \equiv \{x_1^2 + x_2^2 \leqslant 1\} \subset \mathbb{R}^2$ we consider the equation

$$-\frac{d}{dx_1}\left(x_1^2 \frac{\partial u}{\partial x_1} \left(\sqrt{1 + (x_1 u_{x_1})^2} \right)^{m-2} \right)$$

$$- \frac{d}{dx_2}\left(x_1^2 \frac{\partial u}{\partial x_2} \left(\sqrt{1 + (x_1 u_{x_2})^2} \right)^{m-2} \right) + l_0(x, u, \nabla u) = 0. \quad (3.7)$$

Carrying out the differentiation in the first two terms, we obtain an equation of the form (1.6) with $n = 2$ and coefficients $\alpha^{11}(x, u, p)$ and $\alpha^{22}(x, u, p)$ of the form

$$\alpha^{ii}(x, u, p) = x_1^2 \left(\sqrt{1 + (x_1 p_i)^2} \right)^{m-2} \left[1 + (m-2) \frac{x_1 p_i^2 + x_1^2 p_i}{1 + (x_1 p_i)^2} \right], \qquad i = 1, 2. \quad (3.8)$$

From condition (3.6) applied to (3.7) we obtain, in particular, the inequalities $\partial^2 \alpha^{11}/\partial p_1^2 \leqslant 0$ and $\partial^2 \alpha^{22}/\partial p_2^2 \leqslant 0$, which must be satisfied for all $x \in \Omega$ and $|p| \leqslant \overline{M}_1$. Twofold differentiation of (3.8) with respect to p_i ($i = 1, 2$) at $p = 0$ then leads to the inequality

$$2x_1^2(m-2)(x_1^2 + x_1) \leqslant 0.$$

Since x_1 may assume arbitrarily small values of any sign in the disk $\Omega \equiv \{x_1^2 + x_2^2 \leqslant 1\}$, from this it follows that for (3.7) condition (3.6) can hold only in the case $m = 2$.

We now consider the equation

$$-\frac{d}{dx_1}\left(x_1^2\frac{\partial u}{\partial x_1}\left(\sqrt{1+(x_1u_{x_1})^2}\right)^{m-2}\right)-\frac{d}{dx_2}\left(x_1^2\frac{\partial u}{\partial x_2}\left(\sqrt{1+(x_1u_{x_2})^2}\right)^{m-2}\right)$$
$$+\frac{d}{dx_1}\left(x_1\lambda(u-\xi(x_2))\left(\sqrt{1+[\lambda(u-\xi(x_2))]^2}\right)^{m-2}\right)=f(x_1,x_2), \quad (3.9)$$

where

$$f(x_1,x_2)=\frac{d}{dx_2}\left(x_1^2(x_1^\lambda+1)\xi'(x_2)\left(\sqrt{1+x_1^2(x_1^\lambda+1)^2[\xi'(x_2)]^2}\right)^{m-2}\right),$$

and $\xi(x_2)$ is an arbitrarily smooth function such that $\xi(x_2)=1$ for $|x_2|\leqslant 1/2$, $\xi(x_2)=0$ for $|x_2|\geqslant 3/4$, and $\lambda=[2(k+1)+l]/(2k+1)$, $l=1,\dots,2k-1$, $k=1,2,\dots$. It is easy to see that in the disk $\{x_1^2+x_2^2\leqslant 1\}$ the function $u=(x_1^\lambda+1)\xi(x_2)$ satisfies (3.9). On the boundary $\{x_1^2+x_2^2=1\}$ this function assumes an infinitely differentiable boundary value. The functions forming the nondivergence equation (1.6) corresponding to (3.9) satisfy all the conditions of Theorem 3.1 in regard to smoothness. However, for $\lambda\in(1,2)$ this solution does not have a bounded second derivative $\partial^2u/\partial x_1^2$ as $x_1\to 0$. This is related to the fact that a condition of the form (3.1) is violated for this equation. Indeed, for $m\neq 2$ this was shown above, while for $m=2$ the fact that (3.1) does not hold can easily be verified directly. This example shows that condition (3.1) and, in particular, the condition that α^{ij} be independent of p, $i,j=1,\dots,n$, are essential. To a certain extent the fact that the latter condition is essential is corroborated by the well-known example of Yu. G. Kolesov (see [21], Chapter 1, §3.2). In a bounded domain $\Omega\subset\mathbf{R}^n$, $n\geqslant 2$, we consider an equation of the form

$$-(\partial/\partial x_i)\left(|\nabla u|^{m-2}(\partial u/\partial x_i)\right)+\lambda(\nabla u)=f(x),$$

where $m\geqslant 2$, $\lambda=\lambda(p_1,\dots,p_n)\in C^\infty(\mathbf{R}^n)$, $\lambda(0,\dots,0)=0$, $f\in C^\infty(\overline{\Omega})$ and $f(x)>0$ in Ω with the boundary condition $u=0$ on $\partial\Omega$. This problem cannot have a solution which is twice differentiable in Ω and continuous in $\overline{\Omega}$, since there must exist at least one extremal point in the interior of Ω at which the left side of the equation vanishes, while the right side $f(x)$ is strictly greater than zero at this point.

Returning to condition (3.1), we note that because the term

$$\frac{\partial^2\alpha}{\partial p_s\partial p_t}\frac{u_{kl}u_{sk}u_{tl}}{|D^2u|}$$

is odd in the variables u_{ij}, (3.1) implies that p cannot occur nonlinearly in the lower order term $\alpha(x,u,p)$ or, more precisely, it is necessary that $\alpha(x,u,p)$ have the form

$$\alpha(x,u,p)\equiv\sum_{k=1}^n\alpha_k(x)p_k+\alpha_0(x,u).$$

Suppose this condition is also satisfied. Then the term $(\partial\alpha^{ij}/\partial u)u_{ij}$ cannot be suppressed by any other terms of the left side of (3.1), which leads to the condition that α^{ij} be independent of u for all $i,j=1,\dots,n$.

Thus, if we intend to apply Theorem 3.1 to the $(A, \mathbf{b})$-elliptic equation (5.1.35), then we must assume that the reduced coefficients of this equation have the form

$$\begin{aligned} l'^i(x,u,q) &= \sum_{j=1}^n l'^{ij}(x)q_j, \qquad i=1,\dots,n, \\ l'_0(x,u,q) &= \sum_{k=1}^n l'^k_0(x)q_k + \hat{l}'_0(x,u). \end{aligned} \tag{3.10}$$

When conditions (3.10) hold, an equation of the form (1.5) generated by the corresponding equation (5.1.35) has the form (see (1.6), (1.4) and (3.10))

$$\alpha^{ij}(x)u_{x_ix_j} - \beta^s(x)u_{x_s} - c(x,u) - g(x) = 0, \tag{3.11}$$

where

$$\alpha^{ij} = a^{ki}l'^{ks}a^{sj}, \quad i,j=1,\dots,n, \qquad b^s = \frac{\partial}{\partial x_i}(a^{ki}l'^{kj}a^{js}) + l'_{0k}a^{ks},$$

$$c = \hat{l}'_0(x,u), \quad g = f(x).$$

For an equation of the form (3.11) condition (3.1) takes the following form: for all $x \in \overline{\Omega}$ and $u \in [-m, m]$

$$\begin{aligned} &\frac{\partial c(x,u)}{\partial u} + \min_{s=1,\dots,n} \frac{\partial \beta^s}{\partial x_s} - \sum_{t\neq s}\left|\frac{\partial \beta^s}{\partial x_t}\right| - \max_{i,j=1,\dots,n} \frac{\partial^2 \alpha^{ij}}{\partial x_i \partial x_j} \\ &\quad - \sum_{\substack{(k,l)\neq(i,j)\\ i,j=1,\dots,n}} \left|\frac{\partial^2 \alpha^{ij}}{\partial x_k \partial x_l}\right| - n \sup_{l=1,\dots,n,\,|\tau|=1} \frac{|\partial \mathfrak{A}^\tau/\partial x_l|^2}{\mathfrak{A}^\tau} > 0, \end{aligned} \tag{3.5'}$$

where $\mathfrak{A}^\tau \equiv \alpha^{ij}(x)\tau_i\tau_j$, $|\tau| = 1$.

An a priori estimate of $\max_{\partial\Omega}|D^2u|$ is established below for solutions of an equation of the form (5.1.35) under the assumption that an estimate of the form

$$\max_{\Omega}|D^2u| \leqslant \max\left\{\max_{\partial\Omega}|D^2u|, L\right\},$$

where $L = \text{const} > 0$, is already known. At this step larger classes of quasilinear equations of the form (5.1.35) than in Theorem 3.1 are admissible.

THEOREM 3.2. *Let $\Omega \in C^3$, and suppose that the reduced coefficients of the divergence $(A, \mathbf{b})$-elliptic equation* (5.1.35) *in Ω have the form*

$$l'^i(x,u,q) = \sum_{j=1}^n l'^j(x,u)q_j, \quad i=1,\dots,n; \qquad l'_0 = l'_0(x,u,q); \tag{3.12}$$

the matrix $A = \|a^{ij}(x)\|$ is symmetric and nonnegative-definite in Ω, $a^{ij} \in \tilde{C}^2(\overline{\Omega})$, $i,j = 1,\dots,n$, and $\mathbf{b} \equiv (b^1(x),\dots,b^n(x))$, $b^i(x) \in \tilde{C}^2(\overline{\Omega})$, $i = 1,\dots,n$. Assume that the coefficients of (1.2) *generated by* (5.1.35), *i.e., the functions $\hat{\alpha}^{ij}(x,u)$, $i,j = 1,\dots,n$, and $\hat{\alpha}^{ij}(x,u,q)$ defined by* (1.4) *on the basis of the functions* (3.12), *belong to the class*

$$C^{(2)}\left(\overline{D}_\delta \times [-m,m] \times \{|p| \leqslant \overline{M}_1\}\right),$$

where the domain D_δ is defined by (2.3), $m = \text{const} \geqslant 0$ *and* $\overline{M}_1 = \text{const} \geqslant 0$, *i.e., they have bounded partial derivatives of second order with respect to all their arguments in $\overline{D}_\delta \times [-m,m] \times \{|p| \leqslant \overline{M}_1\}$. Suppose that for the functions $\hat{\alpha}^{ij}$ and a^{ij}, $i,j = 1,\dots,n$,* *condition* (2.1) *is satisfied* (*with* Ω *replaced by* $\overline{D}_8$), *and, moreover, for all* $x \in \overline{D}_8$, $u \in [-m, m]$, *and* $\xi = A\eta$, *for all* $\eta \in \mathbb{R}^n$,

$$\hat{\alpha}^{ij}(x, u)\xi_i\xi_j \geqslant \nu_0|\xi|^2, \qquad \nu_0 = \text{const} > 0. \tag{3.13}$$

Suppose, further, that for all $y \in \partial\Omega$ *inequality* (2.9) *holds. Finally, suppose that the function* $u \in C^3(D_8) \cap C^2(\overline{D}_8)$ *satisfies* (5.1.35) *in* D_8 *and that* $u = 0$ *on* $\partial\Omega$, *and*

$$\max_{\overline{\Omega}}|u| \leqslant m, \qquad \max_{\overline{\Omega}}|\nabla u| \leqslant \overline{M}_1, \quad \max_{\overline{\Omega}}|D^2u| \leqslant \max\left\{\max_{\partial\Omega}|D^2u|, L\right\},$$

where $L = \text{const} \geqslant 0$. *Then*

$$\max_{\partial\Omega}|D^2u| \leqslant M_2, \tag{3.14}$$

where $|D^2u|^2 \equiv \sum_{i,j=1}^n u_{x_ix_j}^2$, *and* M_2 *depends only on* n, ν_0, *the constant* γ *in* (2.9), m, $\overline{M}_1$, L, *the bounds of the moduli of the functions* α^{ij}, $i, j = 1,\ldots,n$, α, *and their partial derivatives of first order on* $\overline{D}_8 \times \{|u| \leqslant m\} \times \{|p| \leqslant \overline{M}_1\}$, *and the* C^2*-norms of the functions describing the boundary* $\partial\Omega$.

PROOF. Let u be the solution of (5.1.35) in D_0 indicated in the formulation of the theorem. We fix a point $x_0 \in \partial\Omega$. In view of Lemma 1.1 it may be assumed that a part $\Gamma \subset \partial\Omega$ containing x_0 in its interior is flat. Indeed, condition (3.13) is again satisfied for the new $(\tilde{A}, \tilde{\mathbf{b}})$-elliptic equation of the form (1.1) obtained by rectifying $\partial\Omega$ in a neighborhood of $x_0 \in \partial\Omega$, since the functions $\hat{\alpha}^{ij}$ are invariant under nondegenerate smooth transformations of the coordinates (see Lemma 1.1), while a condition of the form (2.9) is also satisfied, since $|A\nu| = |\tilde{A}\nu|$. Moreover, it may be assume that the axes $Ox_1,\ldots,Ox_{n-1}$ are situated in the plane containing Γ, while the axis Ox_n is directed along the inner normal to Γ at x_0. It is obvious that there exists a number $r > 0$ not depending on $x_0 \in \partial\Omega$ such that the intersection $\Omega_r \equiv K_r(x_0) \cap \Omega$ is contained in D_0, while $S_r \equiv K_r(x_0) \cap \partial\Omega$ is contained in Γ. We differentiate the equation written in the form (1.5) (which is defined by (1.6), (1.2), (1.4) and (3.12) or by (1.7), (5.1.2) and (3.12)) with respect to any variable x_τ, $\tau = 1,\ldots,n-1$. In Ω_r the function $u^\tau \equiv \partial u/\partial x_\tau$ then satisfies an equaiton of the form

$$\alpha^{ij}u^\tau_{ij} - \frac{\partial\alpha^{ij}}{\partial u}u^\tau u_{ij} - \frac{\partial\alpha^{ij}}{\partial x_\tau}u_{ij} - \frac{\partial\alpha}{\partial p_m}u^\tau_m - \frac{\partial\alpha}{\partial u}u^\tau - \frac{\partial\alpha}{\partial x_\tau} = 0. \tag{3.15}$$

For the function $u^\tau \in C^2(\Omega_r) \cap C^1(\overline{\Omega}_r)$ satisfying (3.15) in Ω_r and the condition $u^\tau|_{S_r} = 0$ the derivatives $u^\tau_{x_s}$ in the tangential directions x_s, $s = 1,\ldots,n-1$, are known, since $u^\tau_{x_s} = 0$, $\tau, s = 1,\ldots,n-1$. To estimate $u^\tau_{x_n}$ we use Theorem 1.5.1′. It is obvious that condition (1.5.7) of this theorem is satisfied at x_0. We verify (1.5.23) and (1.5.24). In view of the fact that S_r is a part of a plane we have $\nabla\rho(x) = \nu(y(x))$, where $y(x)$ is the projection of the point $x \in \Omega_r$ onto S_r. Therefore, $\mu\nabla\rho + t\nu = (\mu + t)\nu$. We recall that μ is a positive constant which in the present case is defined by $\mu = \overline{M}_1 + 2/r$, and t is a positive parameter with $t \geqslant l = \text{const} > 0$. We henceforth assume with no loss of generality that $l \geqslant \mu$. We consider (3.15) as an equation of the form $\tilde{a}^{ij}(x)u^\tau_{ij} = \tilde{a}(x, \nabla u^\tau)$, where

$$\tilde{a}^{ij} = \alpha^{ij}, \quad \tilde{a}(x, \tilde{p}) = \frac{\partial\alpha^{ij}}{\partial u}u_{x_\tau} + \frac{\partial\alpha^{ij}}{\partial x_\tau}u_{ij} + \frac{\partial\alpha}{\partial p_m}\tilde{p}_m + \frac{\partial\alpha}{\partial u}u_{x_\tau} + \frac{\partial\alpha}{\partial x_\tau},$$

and

$$\alpha^{ij},\ \frac{\partial\alpha^{ij}}{\partial p_m},\ \frac{\partial\alpha^{ij}}{\partial u}u_{x_\tau},\ \frac{\partial\alpha^{ij}}{\partial x_\tau}u_{ij},\ \frac{\partial\alpha}{\partial p_m},\ \frac{\partial\alpha}{\partial u},\ \frac{\partial\alpha}{\partial x_\tau}$$

are considered as functions of x. Therefore, in the present case $\mathscr{E}_1(x, u, \mu\nabla\rho + t\nu)$ does not depend on u and has the form $\mathscr{E}_1 = (\mu + t)^2\mathfrak{A}\nu\cdot\nu$, where $\mathfrak{A} \equiv \|\alpha^{ij}\|$ and ν is the unit vector of the inner normal to Γ. Since $\mathfrak{A}\nu\cdot\nu = \hat{\mathfrak{A}}A\nu\cdot A\nu$, where $A = \|a^{ij}(x)\|$, by (3.13) and the inequality

$$\gamma^{-1} \leqslant |A(x)\nu(y(x))| \leqslant \gamma, \qquad x \in \Omega_r, \tag{3.16}$$

which follows from (2.1) because r is sufficiently small (where the smallness depends only on the modulus of continuity of the elements of A near $\partial\Omega$), for all $x \in \Omega_r$ and $t \geqslant l \geqslant \mu$, where μ depends only on r and $\overline{M}_1$, we have

$$\mathscr{E}_1 \geqslant \nu_0\gamma^2 t^2. \tag{3.17}$$

Recalling that in Ω_r we have $|u| \leqslant m$, $|\nabla u| \leqslant \overline{M}_1$ and $|D^2u| \leqslant \max(M_2, L)$, $M_2 \equiv \sup_{\partial\Omega}|D^2u|$, and taking account of the form of the right side of (3.15), we conclude that (1.5.23) is satisfied with a function $\psi(t)$ having the form $\psi(t) = c_1t^{-1} + c_2M_2t^{-2}$, where c_1 and c_2 depend only on known quantities, and we assume that $M_2 \geqslant 1$. It is obvious that (1.5.24) is also satisfied with such a function $\psi(t)$. Applying Theorem 1.5.1′, we obtain

$$|\partial u^\tau(x_0)/\partial\nu| \leqslant \beta + c_3, \tag{3.18}$$

where c_3 is a known constant, β is determined from (see (1.3.7))

$$\int_{\bar\alpha}^{\beta}\frac{\rho\,d\rho}{(K + c_1)\rho + c_2M_2} = M_1, \tag{3.19}$$

and $\bar\alpha$ is a known quantity (see the proof of Theorem 1.5.1). It follows from (3.19) that $\beta \leqslant c_4 + c_5\sqrt{M_2}$, whence from (3.18) we obtain

$$\left|\frac{\partial^2u(x_0)}{\partial x_\tau\partial x_n}\right| \leqslant c_6\left(1 + \sqrt{M_2}\right), \qquad \tau = 1,\dots,n-1, \tag{3.20}$$

where c_6 is determined only by known quantities. To complete the proof of the theorem we note that $|\partial^2u(x_0)/\partial x_n^2|$ can be estimated in terms of the remaining derivatives of second order by (1.6) itself. Indeed, in view of (3.13)

$$\alpha^{nn} \equiv \hat\alpha^{ks}a^{kn}a^{sn} \geqslant \nu_0|\mathbf{a}^n|^2 \equiv \nu_0\sum_{i=1}^{n}(a^{in})^2,$$

while in our coordinate system $\Sigma_{i=1}^n(a^{in})^2 \equiv |A\nu|^2 \geqslant \gamma^{-2} > 0$. Therefore, the coefficient of u_{nn} is bounded away from 0 by a known quantity; this gives the assertion made above regarding the estimate of $|\partial^2u/\partial x_n^2|$ in terms of the remaining second derivatives (which have already been estimated). Because the point x_0 is arbitrary, from (3.20) it then follows that

$$\max_{\partial\Omega}|D^2u| \leqslant c_7\left(1 + \sqrt{\max_{\partial\Omega}|D^2u|}\right),$$

from which we obviously obtain (3.14). Theorem 3.2 is proved.

THEOREM 3.3. *Suppose that an equation of the form* (5.1.35) *has the structure of an* $(A, \mathbf{b})$-*elliptic equation in a bounded domain* $\Omega \subset \mathbf{R}^n$, $n \geqslant 2$, *of class* C^2, *and* $A \equiv \|a^{ij}(x)\|$ *is a symmetric nonnegative-definite matrix in* $\overline{\Omega}$, $a^{ij} \in \tilde{C}^2(\overline{\Omega})$, $i, j = 1, \dots, n$, *and* $\mathbf{b} \equiv (b^1(x), \dots, b^n(x))$, $b^i \in \tilde{C}^2(\overline{\Omega})$. *Assume that the functions* $\hat{\alpha}^{ij}(x)$, $i, j = 1, \dots, n$, *and* $\hat{\alpha}(x, u, q)$ *defined by* (1.4) *and* (3.10) *belong to* $\tilde{C}^2(\Omega)$ *and* $\tilde{C}^2(\overline{\Omega} \times \mathbf{R} \times \mathbf{R}^n)$ *respectively. Suppose that conditions* (2.2), (2.8), (2.9), (2.15), (3.5), *and* (3.13) *are satisfied uniformly in* ε *for equations* (2.21) *and* (2.22) *generated by* (5.1.35) *by means of the respective equalities* (1.4) *and* (1.7), *and suppose that condition* (2.25) (*with condition* (3.10) *taken into account*) *holds for the reduced coefficients of* (5.1.35). *Then problem* (1.1) *has at least one generalized solution u which is bounded in* $\overline{\Omega}$ *together with all its derivatives of first and second orders, i.e., there exists a function* $u \in L^\infty(\Omega) \cap \mathring{H}_m(\Omega)$ *for all* $m > 1$ *with* $\nabla u \in L^\infty(\Omega)$ *and* $D^2 u \in L^\infty(\Omega)$ *for which an identity of the form* (2.26) *holds, and u satisfies* (5.1.35) *a.e. in* Ω.

PROOF. Since the conditions of Theorem 3.3 contain as a special case the conditions of Theorem 2.3, by Theorem 2.3 there exists a function $u \in L^\infty(\Omega) \cap \mathring{H}_m(\Omega)$ such that $\nabla u \in L^\infty(\Omega)$ and (2.26) holds. Since the conditions of Theorem 3.3 imply the validity for regularized problems of the form (2.23) of all the conditions of Theorems 3.1 and 3.2, it may be assumed that for solutions u_ε of (2.23) the estimate

$$\max_{\Omega} |D^2 u_\varepsilon| \leqslant \overline{M}_2 \tag{3.21}$$

holds with a constant $\overline{M}_2$ not depending on ε. From this it obviously follows that for the limit function u we have $|D^2 u| \leqslant M_2$ a.e. in Ω. Thus, the generalized solution of (1.1) is a function having Lipschitz first derivatives in $\overline{\Omega}$. It can now be proved in standard fashion that u satisfies (5.1.35) a.e. in Ω. Theorem 3.3 is proved.

As an example related to Theorem 3.3, in a bounded domain $\Omega \subset \mathbf{R}^n$, $n \geqslant 2$, we consider a nondivergence equation of the form

$$\alpha^{ij}(x) u_{x_i x_j} - \beta^i(x) u_{x_i} - c(x, u) = g(x), \tag{3.22}$$

where $\mathfrak{A} \equiv \|\alpha^{ij}(x)\|$ is a symmetric nonnegative-definite matrix in $\overline{\Omega}$, and α^{ij}, β^i, c, and g are sufficiently smooth functions of their arguments. Equation (3.22) can be rewritten in the form

$$-\operatorname{div}_A(A \nabla u) + l_0(x, u, \nabla u) = f(x), \tag{3.23}$$

where $\operatorname{div}_A(A\nabla u) \equiv \operatorname{div}(A^* A \nabla u) \equiv \partial(\alpha^{ij} u_{x_j})/\partial x_i$, $A = \sqrt{\mathfrak{A}}$, $l_0 = c + b^i p_i$ and $b^i = \partial \alpha^{ki}/\partial x_k + \beta_i$, i.e., in the form of a divergence $(A, \mathbf{b})$-elliptic equation relative to the matrix $A = \sqrt{\mathfrak{A}}$, $\mathbf{b} \equiv (b^1(x), \dots, b^n(x))$, with reduced coefficients

$$l''(x, u, q) = q_i, \qquad l_0'(x, u, q) = c(x, u), \tag{3.24}$$

so that conditions (3.10) are satisfied for them. Therefore, under the conditions stipulated by Theorem 3.3, the problem

$$\alpha^{ij} u_{ij} - \beta^i u_i - c(x, u) = g \quad \text{in } \Omega, \qquad u = 0 \quad \text{on } \partial\Omega \tag{3.25}$$

has at least one generalized solution u which is bounded in Ω together with all its derivatives through second order, and this solution satisfies the equation a.e. in Ω and vanishes on $\partial\Omega$ as an element of the space $\mathring{H}_p^2(\Omega)$ for all $p > 1$.

We present conditions for the validity of this result in terms of the original equation (3.22). It is easy to see that all the conditions of Theorem 3.3 for equation

(3.23) generated by (3.22) are satisfied if 1) there exists $m_0 = \text{const} > 0$ such that $c(x, u)u - g(x)u \geqslant 0$ for all $x \in \overline{\Omega}$ and $|u| \leqslant m_0$; 2) $\alpha^{ij}\nu_i\nu_j > 0$ on $\partial\Omega$, where $\nu = (\nu_1, \dots, \nu_n)$ is the unit vector of the inner normal to $\partial\Omega$; 3) for all $x \in \overline{\Omega}$ and $|u| \leqslant m_0$

$$\frac{\partial c(x, u)}{\partial u} + \min_{i=1,\dots,n} \frac{\partial \beta^i}{\partial x_i} - \sum_{\substack{i \neq k \\ k=1,\dots,n}} \left| \frac{\partial \beta^i}{\partial x_k} \right| - \frac{n}{4} \sup_{|\tau|=1} \frac{|\nabla \mathfrak{A}^\tau|^2}{\mathfrak{A}^\tau} > 0, \quad (3.26)$$

where $\mathfrak{A}^\tau \equiv \alpha^{ij}(x)\tau_i\tau_j$; 4) inequality (3.5′) holds for all $x \in \overline{\Omega}$ and $|u| \leqslant m_0$; 5) $\partial c(x, u)/\partial u \geqslant 0$ for all $x \in \overline{\Omega}$ and $|u| \leqslant m_0$. The result presented above contains as a special case the corresponding result of O. A. Oleĭnik for linear equations of the form (3.22) (i.e., for the case $c(x, u) = c(x)u$) (see [99]). M. I. Freĭdlin [122] proved the existence of an analogous generalized solution (i.e., a solution having Lipschitz second derivatives in $\overline{\Omega}$) of the problem

$$\begin{gathered} \alpha^{ij}(x, u)u_{x_ix_j} + b^i(x, u)u_{x_i} - cu = 0 \quad \text{in } \Omega, \qquad u = \varphi(x) \quad \text{on } \partial\Omega, \\ \alpha^{ij}(x, u)\xi_i\xi_j \geqslant 0 \quad \alpha^{ij}\nu_i\nu_j > 0 \quad \text{on } \partial\Omega, \quad c = \text{const} > 0, \end{gathered} \quad (3.27)$$

for a sufficiently large constant $c > 0$. However, in [122] it was assumed that $\varphi(x)$ is sufficiently small together with its derivatives of first and second orders.

In conclusion we note that for equations having structure analogous to that assumed in Theorem 3.3 the existence of a solution of problem (1.1) having Lipschitz derivatives through order k for any $k = 3, 4, \dots$ can be established in a similar way.

PART III
$(A, \mathbf{0})$-ELLIPTIC AND $(A, \mathbf{0})$-PARABOLIC EQUATIONS

$(A, \mathbf{0})$-elliptic and $(A, \mathbf{0})$-parabolic equations of the form

$$\mathscr{L}u \equiv -(d/dx_i)l^i(x, u, \nabla u) + l_0(x, u, \nabla u) = f(x) \tag{1}$$

are more direct generalizations of classical elliptic and parabolic equations than general $(A, \mathbf{b})$-elliptic and $(A, \mathbf{b})$-parabolic equations of the form (1) (see the introduction to Part II). $(A, \mathbf{0})$-elliptic equations are studied in Chapter 7. At the beginning of this chapter we present results on the existence and uniqueness of a generalized solution of energy class to the general boundary value problem for $(A, \mathbf{0}, m, \mathbf{m})$-elliptic equations under the assumption of sufficient smoothness of the matrix A. In particular, the first boundary value problem for such equations has the form

$$\mathscr{L}u = f(x) \quad \text{in } \Omega, \qquad u = 0 \quad \text{on } \Sigma, \tag{2}$$

where Σ is the regular part of $\partial\Omega$ (see §§4.3 and 7.1). In addition to the requirement of a particular smoothness of the domain Ω and the matrix A, the conditions for the existence and uniqueness of a generalized solution of problem (2) contain some easily verifiable conditions on the reduced coefficients of equation (1) which have the form of algebraic inequalities (see Theorems 7.1.1 and 7.1.2) that guarantee coerciveness and monotonicity of the operator $\mathscr{L}$ corresponding to problem (2).

A corollary of these results applicable to the case of linear $(A, \mathbf{0})$-elliptic equations is presented (see Theorem 7.1.3). In Chapter 7 we also consider the case of $(A, \mathbf{0})$-elliptic equations with weak degeneracy, by which we mean $(A, \mathbf{0})$-elliptic equations corresponding to matrices A which are weakly degenerate in Ω. A matrix A is called *weakly degenerate* in Ω if for this matrix there is a continuous imbedding of the energy space $H_{m,\mathbf{m}}(A, \Omega)$ into the Sobolev space $H_{m,\mathbf{q}}(\Omega)$ (for all $m > 1$) with some $\mathbf{q} = (q_1, \dots, q_n)$, $q_i \geqslant 1$, $i = 1, \dots, n$, and hence also into some space $L^l(\Omega)$. For $(A, \mathbf{0}, m, \mathbf{m})$-elliptic, weakly degenerate equations the regular part Σ coincides with the entire boundary $\partial\Omega$, so that in this case the boundary condition is imposed on the entire boundary. In the case of weakly degenerate $(A, \mathbf{0}, m, \mathbf{m})$-elliptic equations we establish an existence theorem for a generalized solution of energy class to the general boundary value problem under weaker conditions on the structure of the equations than in the previous case (see Theorems 3.2 and 3.3).

The so-called $(A, \mathbf{0}, \overline{m}, \overline{m})$-elliptic equations, which are a special subclass of $(A, \mathbf{0}, m, \mathbf{m})$-elliptic equations with a weak degeneracy, are studied in detail. This class contains, in particular (for $\overline{m} = 2$), linear elliptic equations of the form

$$-(d/dx_i)\left(\alpha^{ij}u_{x_j} + \alpha^i u\right) + \beta^i u_{x_i} + cu = f(x), \tag{3}$$

where $\mathfrak{A} \equiv \|\alpha^{ij}\|$ is a symmetric, nonnegative-definite matrix in Ω such that the matrix $A = \mathfrak{A}^{1/2}$ is weakly degenerate in Ω, and the coefficients of (3) satisfy certain summability conditions (see (7.2.32)). Fredholm solvability of the general boundary value problem is proved for linear equations of the form (3). As an example we consider the Euler equation for the nonregular variational problem regarding a minimum of the integral

$$\int_\Omega \left[\frac{1}{\overline{m}} |A\nabla u|^{\overline{m}} + \frac{1}{m}|u|^m - f(x)u \right] dx, \qquad u|_\Sigma = 0, \tag{4}$$

which has the form

$$-\operatorname{div}\{ A^*|A\nabla u|^{\overline{m}-2} A\nabla u \} + |u|^{m-2}u = f(x), \tag{5}$$

where $\overline{m} > 1$, $m > 1$, $f \in L^{m'}(\Omega)$ and $1/m + 1/m' = 1$. Equation (5) has the structure of an $(A, \mathbf{0}, m, \mathbf{m})$-elliptic equation in the domain Ω with $\mathbf{m} = (\overline{m}, \ldots, \overline{m})$ and reduced coefficients $l'^i(x, u, q) = |q|^{\overline{m}-2} q_i$, $i = 1, \ldots, n$, and $l'_0(x, u, q) = |u|^{m-2}u$. From the results obtained in Chapter 7 it follows that under conditions of sufficient smoothness of the matrix A or its weak degeneracy in Ω the general boundary value problem for equation (5) has precisely one generalized solution of energy type which depends continuously on the right side of the equation.

We note that in the special case $A = aI$, where I is the identity matrix, the question of the existence of a generalized solution of the first boundary value problem for sufficiently smooth $a(x)$ was considered in [62]. The first boundary value problem for weakly degenerate $(A, \mathbf{0}, m, \mathbf{m})$-elliptic equations was studied in [47]. Theorems on the existence of generalized solutions of the first boundary value problem for weakly degenerate elliptic equations of higher order were established in [92]. Fredholm solvability of the first, second, and third boundary value problems for linear, weakly degenerate elliptic equations is proved in [69], [154] and [178]. The exposition in Chapter 7 reflects the author's results in [44]–[47], [49], [51] and [52].

In Chapter 7 results are also established on the solvability of problem (2) for $(A, \mathbf{0})$-elliptic equations in the class of A-regular generalized solutions (see the introduction to Part II). Analysis of the conditions of Theorems 7.3.2, 7.3.2′, 7.3.3, and 7.3.4 in which such results are formulated shows that the conditions imposed on the behavior of the reduced coefficients are of natural character, but essential restrictions arise on the character of degeneration of the matrix A (see conditions (7.3.3) and (7.3.4)). So-called uniformly degenerate matrices A characterized by the condition

$$\Lambda_A \leqslant \text{const}\, \lambda_A \quad \text{in } \Omega \tag{6}$$

are certainly admissible; here λ_A and Λ_A are the greatest and least eigenvalues of A. The basis for obtaining these results is the possibility of establishing for solutions of nondegenerate, nondivergence $(A, \mathbf{0})$-elliptic equations of the form

$$\hat{a}^{ij}(x, u, \hat{\nabla} u) u_{\hat{i}\hat{j}} - \hat{a}(x, u, \hat{\nabla} u) = 0 \tag{7}$$

(see the introduction to Part II) the a priori estimate

$$\max_\Omega (|u| + |\hat{\nabla} u|) \leqslant \hat{M}, \tag{8}$$

where $\hat{\nabla} u \equiv A\nabla u$, with a constant $\hat{M}$ not depending on the ellipticity constant of (7). To establish the estimate we develop those methods by means of which estimates of the gradients of nonuniformly elliptic equations of the form (1.1.1) were established in Chapter 1. The validity of an estimate of the form (8) makes it possible to prove the existence of solutions of the Dirichlet problems for the regularized equations (see (6.2.21)) and to pass to the limit in the integral identities corresponding to such problems. The results of Theorems 7.3.2 and 7.3.2′ are connected with the use of the condition $A\mathbf{v} \neq \mathbf{0}$ on $\partial\Omega$. Elimination of this condition leads to the more stringent condition (7.3.4) in place of (7.3.3). In the case the results obtained are given in Theorems 7.3.3 and 7.3.4. It should be noted that at one stage of the proof of the last theorems a useful role is played by a special cut-off function analogous to the one used in the paper [117], which is devoted to the proof of solvability of the first boundary value problem for a linear, degenerate, elliptic equation in the weight space $W_p^2(a, \Omega)$, where $a(x) \geqslant 0$ in Ω and $1 < p < +\infty$.

In particular, it follows from Theorem 7.3.4 that with sufficients smoothness of the domain Ω and the function $a(x) \geqslant 0$ equation (5) with $A = a(x)I$, considered under the condition $u = 0$ on $\{x \in \partial\Omega: a(x) > 0\}$, has exactly one A-regular generalized solution. By a similar method it is possible to establish the existence of A-regular generalized solutions of problem (2) possessing bounded A-derivatives of second and higher orders. In connection with these results we consider as an example the simplest equation

$$-(\partial/\partial x_i)\left(|x|^2(\partial u/\partial x_i)\right) + (n\lambda + \lambda^2)u = 0, \tag{9}$$

where $\lambda = \text{const} \in (0, 1)$ and $x \in \Omega \equiv \{|x| \leqslant 1\}$. Equation (9), which has the structure of an $(A, \mathbf{0})$-elliptic equation in Ω relative to the matrix $A = |x|I$, possesses a solution $u = |x|^\lambda$ which is bounded in Ω and equal to 1 on $\partial\Omega$ but does not have a bounded gradient ∇u in Ω (here the condition $A\mathbf{v} \neq \mathbf{0}$ on $\partial\Omega$ is satisfied for equation (9)). At the same time this solution has bounded A-derivatives of all orders in Ω. This example shows that for equations of this structure it is natural to pass from investigation of smoothness to the investigation of A-smoothness of their solutions.

In Chapter 8, which completes Part III, we study $(A, \mathbf{0})$-parabolic equations in a cylinder $Q = \Omega \times (T_1, T_2)$. At the beginning of this chapter we study parabolic analogues of the basic function spaces connected with the study of the general boundary value problem for $(A, \mathbf{b}, m, \mathbf{m})$-elliptic equations and the parabolic analogue of the operator $\mathscr{L}$ corresponding to this problem. Here also, easily verifiable conditions are established which ensure the compatibility of the operators $\mathscr{A}$ and $\mathscr{B}$ which are the components of $\mathscr{L}$ (see the introduction to Part II). One such condition is the condition that the matrix A be independent of t (although the reduced coefficients $l'^i(t, x, u, q)$, $i = 1, \ldots, n$, and $l'_0(t, x, u, q)$ may depend on all their arguments). Assuming that such conditions are satisfied, we establish existence and uniqueness theorems for generalized solutions of energy type of the general boundary value problem for $(A, \mathbf{0}, m, \mathbf{m})$-parabolic equations. As in the case of $(A, \mathbf{0}, m, \mathbf{m})$-elliptic equations, the conditions of these theorems are expressed in the requirement of sufficient smoothness of the domain Ω and the matrix $A(x)$ and certain algebraic inequalities for the reduced coefficients of these equations (see Theorems 8.2.1–8.2.5). We note that in the case of $(A, \mathbf{0})$-parabolic equations the integral identity defining a generalized solution of energy type acquires a special feature related to the fact that such a solution has a derivative with respect to t (understood in the sense of generalized functions) which belongs to the space $\mathscr{H}^*$ (see (8.2.8)). We note also that the conditions for uniquenss of a generalized solution are here relaxed somewhat as compared with the general case (see Lemma 8.2.2). In the case $m = 2$ the algebraic conditions on the reduced coefficients indicated above acquire a freer character (see Theorems 8.2.3–8.2.5).

The major part of Chapter 8 is devoted to the study of $(A, \mathbf{0}, m, \mathbf{m})$-parabolic equations with a weak degeneracy, i.e., to $(A, \mathbf{0}, m, \mathbf{m})$-parabolic equations corresponding to matrices A which are weakly degenerate in the cylinder Q. A matrix A is called weakly degenerate in a cylinder Q if for it there is a continuous imbedding of the energy space $\mathscr{A}_{m,\mathbf{m}}(A, Q)$ in the space $\mathfrak{A}_{m,\mathbf{q}}(Q)$ for some $\mathbf{q} = (q_1, \dots, q_n)$, $q_i \geqslant 1$, $i = 1, \dots, n$, for all $m > 1$ (see the beginning of §8.3, where a definition of weak degeneracy of a matrix A is given in a somewhat more general situation connected with the use of the "double" norms which are traditional for the parabolic case). It is clear that such an imbedding implies also a continuous imbedding of the space $\mathfrak{A}_{m,\mathbf{m}}(A, Q)$ into a space $L^{\tilde{l}, \tilde{l}_0}(Q)$. We note that in the case of a matrix A weakly denegerate in a cylinder Q the carrier part of the boundary of this cylinder (i.e., the set where it is necessary to prescribe the value of the desired solution) is the traditional parabolic boundary Γ of the cylinder Q, i.e.,

$$\Gamma = (\partial\Omega \times (T_1, T_2)) \cup (\Omega \times \{t = T_1\}).$$

For weakly degenerate $(A, \mathbf{0}, m, \mathbf{m})$-parabolic equations in a cylinder Q, results are established on the existence and uniqueness of generalized solutions of energy type analogous to those established for $(A, \mathbf{0}, m, \mathbf{m})$-parabolic equations under conditions of sufficient smoothness of the matrix A (see Theorem 8.3.1).

We further consider weakly degenerate $(A, \mathbf{0})$-parabolic equations of special type. The class of such equations is called the class of $(A, \mathbf{0}, 2, \mathbf{m})^{\sim}$-parabolic equations (see conditions (8.3.20) and (8.3.21)). This class is not formally a subclass of the $(A, \mathbf{0}, 2, \mathbf{m})$-parabolic equations, in connection with the fact that it involves considering "double" norms. However, it retains the basic features of $(A, \mathbf{0}, m, \mathbf{m})$-parabolic equations with weak degeneracy and is a nonlinear analogue of the class of linear parabolic equations with summable coefficients. For this class we establish existence and uniqueness theorems for generalized solutions of energy type to the general boundary value problem under less stringent conditions on the structure of the equations than in the previous cases (see Theorems 8.3.3 and 8.3.4). To prove these results we apply the method of elliptic regularization. In the case of nondegenerate parabolic equations a similar method is applied, in particular, in [20]. The results established in Theorems 8.3.3 and 8.3.4 make it possible to obtain an existence and uniqueness theorem for a generalized solution of the general boundary value problem for linear degenerate parabolic equations with summable coefficients. The case of the first boundary value problem for the latter equations was studied previously by another method in the author's papers [28] and [43]. The results of Chapter 8 enumerated above reflect the author's investigations [44]–[46], [50] and [51]. We illustrate some of the results obtained in Chapter 8 with the example of the simple equation

$$u_t - \operatorname{div}\{A^*|A\nabla u|^{\overline{m}-2} A\nabla u\} + |u|^{m-2}u = f(x, t), \tag{10}$$

where $\overline{m} = \text{const} > 1$, $m = \text{const} > 1$ and $A = A(x)$. If the matrix $A(x)$ is sufficiently smooth in $\overline{\Omega}$ or weakly degenerate in Ω and $f \in L^{m'}(Q)$, then the general boundary value problem for (10) has precisely one generalized solution (of energy type) which depends continuously on the right side.

Finally, we note that, in the case where $A = A(x)$, results on the existence and uniqueness of A-regular generalized solutions of the first boundary value problem can be established for $(A, \mathbf{0})$-parabolic equations in the cylinder Q exactly as in the case of $(A, \mathbf{0})$-elliptic equations. However, these results are not presented in this monograph.

CHAPTER 7

$(A, \mathbf{0})$-ELLIPTIC EQUATIONS

§1. The general boundary value problem for $(A, \mathbf{0}, m, \mathbf{m})$-elliptic equations

In a bounded domain $\Omega \subset \mathbb{R}^n$, $n \geqslant 2$, we consider an equation of the form (5.1.35) assuming that it has $(A, \mathbf{0}, m, \mathbf{m})$-structure in Ω, i.e., the conditions enumerated in Definition 5.1.2 are satisfied in the case $\mathbf{b} \equiv \mathbf{0}$ in Ω. We suppose also that the validity of condition (4.1.3) is ensured by conditions (4.1.1) and (4.1.4) for $m_i > 1$, $i = 1, \ldots, n$, and $m > 1$. Let $\partial\Omega = \Sigma \cup \Sigma'$ and suppose that condition (4.2.5) holds for Σ while (4.3.2) holds for Σ'. Then Σ is the regular and Σ' the singular part of $\partial\Omega$. As usual, we partition Σ into parts Σ_1, Σ_2 and Σ_3, and let λ be a function on Σ_3 which possesses the same properties as in the general case. Taking into account that $\mathbf{b} \equiv \mathbf{0}$ in Ω, we have here

$$(\Sigma_i)_+ = (\Sigma_i)_- = \varnothing, \quad i = 1, 2, 3; \qquad \Sigma'_+ = \Sigma'_- = \varnothing. \tag{1.1}$$

In view of (1.1) the remaining conditions we used in the general case in formulating the general boundary value problem (see §5.3) are automatically satisfied. The general boundary value problem for equation (5.1.35) now takes the following form

$$\begin{gathered} -dl^i/dx_i + l_0 = F \quad \text{in } \Omega; \qquad u = 0 \quad \text{on } \Sigma_1; \\ \mathbf{l}' \cdot A\boldsymbol{\nu} = 0 \quad \text{on } \Sigma_2; \qquad \mathbf{l}' \cdot A\boldsymbol{\nu} - \lambda u = 0 \quad \text{on } \Sigma_3; \end{gathered} \tag{1.2}$$

while the integral identity (5.3.3) takes the form

$$\int_\Omega [\mathbf{l}'(x, u, A\nabla u) \cdot A\nabla\eta + l'_0(x, u, A\nabla u)\eta]\, dx + \int_{\Sigma_3} \lambda u|_{\tilde{\Sigma}}\eta\, ds = \langle F, \eta\rangle, \qquad \eta \in H_\lambda, \tag{1.3}$$

where $\tilde{\Sigma} = \Sigma_1 \cup \Sigma_3$ and $H_\lambda \equiv H_{m,\mathbf{m}}(A;\ \Omega;\ \Sigma_3, \lambda)$. In the present case we see that $X \equiv Y \equiv H_\lambda$, $\mathscr{L} \equiv \mathscr{A}$ and $\mathscr{B} \equiv 0$, where 0 is the zero operator. A generalized solution of problem (1.2) can be defined here as any function $u \in H_\lambda$ satisfying (1.3). The results in the general case (see Theorems 5.4.1–5.4.4 and Propositions 5.4.3–5.4.6) imply, in particular, the following results.

THEOREM 1.1. *Suppose that for almost all $x \in \Omega$ and any $u \in \mathbb{R}$, $q = Ap$, $p \in \mathbb{R}^n$, $\xi_0 \in \mathbb{R}$, $\eta = A\xi$ and $\xi \in \mathbb{R}^n$*

$$l'^i(x, u, q)q_i + l'_0(x, u, q)u \geqslant \nu_1 \sum_{i=1}^n |q_i|^{m_i} + \nu_2|u|^m + \varphi(x), \tag{1.4}$$

where $\nu_1, \nu_2 = \text{const} > 0$, $\varphi \in L^1(\Omega)$, *and*

$$\frac{\partial l'^i(x,u,q)}{\partial q_j}\eta_i\eta_j + \frac{\partial l'^i(x,u,q)}{\partial u}\xi_0\eta_i + \frac{\partial l'_0(x,u,q)}{\partial q_j}\eta_j\xi_0 + \frac{\partial l'_0(x,u,q)}{\partial u}\xi_0^2 \geqslant 0. \tag{1.5}$$

Then for any $F \in H_\lambda^*$ *problem* (1.2) *has at least one generalized solution. If the operator* $\mathscr{L}: H_\lambda \to H_\lambda^*$ *corresponding to* (1.2) (*defined by the left side of* (1.3)) *is strictly monotone, then for all* $F \in H_\lambda^*$ *problem* (1.2) *has at most one generalized solution.*

Theorem 1.1 follows directly from Theorems 5.4.1, 5.4.2 and Propositions 5.4.4 and 5.4.5.

THEOREM 1.2. *Suppose that for almost all* $x \in \Omega$ *and any* $u \in \mathbb{R}$, $q = Ap$, $p \in \mathbb{R}^n$, $\xi_0 \in \mathbb{R}$, $\eta = A\xi$ *and* $\xi \in \mathbb{R}^n$

$$\frac{\partial l'^i(x,u,q)}{\partial q_j}\eta_i\eta_j + \frac{\partial l'^i(x,u,q)}{\partial u}\xi_0\eta_i + \frac{\partial l'_0(x,u,q)}{\partial q_j}\eta_j\xi_0 + \frac{\partial l'_0(x,u,q)}{\partial u}\xi_0^2$$
$$\geqslant \alpha_0\left(\sum_{i=1}^n |q_i|^{m_i-2}\eta_i^2 + |u|^{m-2}\xi_0^2\right), \qquad \alpha_0 = \text{const} > 0, \tag{1.6}$$

and let $m_i \geqslant 2$, $i = 1, \ldots, n$, *and* $m \geqslant 2$. *Then for any* $F \in H_\lambda^*$ *problem* (1.2) *has exactly one generalized solution. Moreover, the operator* $\mathscr{L}: H_\lambda \to H_\lambda^*$ *corresponding to problem* (1.2) *is a homeomorphism.*

Theorem 1.2 follows directly from Theorem 5.4.4. We note that all the conditions in Theorems 1.1 and 1.2 have easily verifiable character. We omit formulations of other theorems following from the results of the general case.

We consider, in particular, a linear equation of the form (5.5.1) under conditions (5.5.2) (in the case $\mathbf{b} \equiv \mathbf{0}$ in Ω) and (5.5.3) and under the assumption that conditions (4.1.4) with $m = 2$ and $\mathbf{m} = \mathbf{2}$ are valid for the matrix A in (5.5.2). As shown in §5 of Chapter 5, this equation has $(A, \mathbf{0}, 2, \mathbf{2})$-structure in Ω.

THEOREM 1.3. *Suppose that for almost all* $x \in \Omega$ *and any* $\xi \in \mathbb{R}^n$

$$Q(x)\xi\cdot\xi \geqslant c_1|\xi|^2, \qquad c_1 = \text{const} > 0,$$
$$a_0(x) - |\mathbf{a}(x)|^2/2\varepsilon_1 - |\gamma(x)|^2/2\varepsilon_2 \geqslant c_2, \qquad c_2 = \text{const} > 0, \tag{1.7}$$

where $\varepsilon_1, \varepsilon_2 > 0$ *and* $(\varepsilon_1 + \varepsilon_2)/2 < c_1$. *Then for any* $F \in H_\lambda^*$ *the general boundary value problem of the form* (5.5.4) (*with* $\mathbf{b} \equiv 0$) *has precisely one generalized solution* $u \in H_\lambda$, *and this solutions depends continuously in* H_λ *on* F *in* H_λ^*.

PROOF. Since in this case the left side of (1.6) (with $m = 2$ and $\mathbf{m} = \mathbf{2}$) has the form

$$q^{ij}\eta_i\eta_j + a^i\xi_0\eta_i + \gamma^j\eta_j\xi_0 + a_0\xi_0^2,$$

by (1.7) we obtain an inequality of the form (1.6) with a constant

$$\alpha_0 = \min(c_1 - \varepsilon_1/2 - \varepsilon_2/2, c_2).$$

Theorem 1.3 then follows from Theorem 1.2.

As the simplest example connected with the results established above, in a bounded domain $\Omega \subset \mathbb{R}^n$, $n \geqslant 2$, of class $\tilde{C}^{(1)}$ we consider the equation

$$-\operatorname{div}\{A^*|A\nabla u|^{\overline{m}-2}A\nabla u\} + |u|^{m-2}u = f(x),$$

which is the Euler equation for the variational problem of a minimum for the integral

$$\int_\Omega \left(\overline{m}^{-1}|A\nabla u|^{\overline{m}} + m^{-1}|u|^m - f(x)u\right) dx,$$

assuming that for some $m > 1$ and $\mathbf{m} = (\overline{m},\dots,\overline{m})$, $\overline{m} > 1$, the matrix $A \equiv \|a^{ij}(x)\|$ satisfies conditions (4.1.1) and (4.1.4) and that the boundary $\partial\Omega$ is partitioned into parts Σ and Σ' with condition (4.2.5) holding for Σ and (4.3.2) for Σ' relative to $\mathbf{m} = (\overline{m},\dots,\overline{m})$. This equation has the structure of an $(A, \mathbf{0}, m, \mathbf{m})$-elliptic equation in Ω with reduced coefficients of the form $l'^i = |q|^{\overline{m}-2}q_i$, $i = 1,\dots,n$, $l'_0 = |u|^{m-2}u$, and they satisfy condition (1.6) with $\alpha_0 = 1$ and $m_i = \overline{m}$, $i = 1,\dots,n$. It then follows from Theorem 1.2 that for all $f \in L^{m'}(\Omega)$ the general boundary value problem of the form (1.2) for this equation has precisely one generalized solution, and this solution depends continuously on the right side $f(x)$.

§2. $(A,0)$-elliptic equations with weak degeneracy

1. $(A, \mathbf{0}, m, \mathbf{m})$-elliptic equations with weak degeneracy. Let $\Omega \subset \mathbb{R}^n$, $n \geqslant 2$, be a bounded strongly Lipschitz domain, and let $A \equiv \|a^{ij}(x)\|$ be a square matrix of order n with

$$a^{ij} \in L^{m_i}(\Omega), \qquad m_i > 1, \qquad i, j = 1,\dots,n. \tag{2.1}$$

DEFINITION 2.1. We call a matrix A *weakly degenerate in* Ω if $\det A \neq 0$ for almost all $x \in \Omega$ and there exists an index $\mathbf{q} = (q_1,\dots,q_n)$, $q_i \geqslant 1$, $i = 1,\dots,n$, such that

$$\|\nabla u\|_{\mathbf{q},\hat{\Omega}} \leqslant c\|A\nabla u\|_{\mathbf{m},\hat{\Omega}}, \quad \forall u \in \tilde{C}^1(\overline{\Omega}), \tag{2.2}$$

where $\hat{\Omega}$ is an arbitrary measurable subset of Ω, $\mathbf{m} = (m_1,\dots,m_n)$, and the constant c depends only on n, $\mathbf{m}$, $\mathbf{q}$, and Ω.

LEMMA 2.1. *Suppose that the matrix A (satisfying condition* (2.1)) *for almost all $x \in \Omega$ has a bounded inverse matrix $A^{-1} \equiv B \equiv \|b^{ij}(x)\|$ with*

$$b^{ij} \in L^{r_{ij}}(\Omega), \qquad r_{ij} \geqslant 1, \qquad i, j = 1,\dots,n;$$
$$\max_{k=1,\dots,n} (1/m_k + 1/r_{ik}) \leqslant 1, \qquad i = 1,\dots,n. \tag{2.3}$$

Then A is weakly degenerate in Ω, and inequality (2.2) *holds with an index $\mathbf{q} = (q_1,\dots,q_n)$ such that*

$$1/q_i = \max_{k=1,\dots,n} (1/m_k + 1/r_{ik}), \qquad i = 1,\dots,n, \tag{2.4}$$

and the constant c in (2.2) *depends only on n, m_k, r_{ik}, ω, and $\|b^{ik}\|_{r_{ik}}$, Ω, $i, k = 1,\dots,n$.*

PROOF. We estimate the norm $\|u_{x_i}\|_{q_i,\Omega}$ for some $i \in \{1,\dots,n\}$. Taking into account that $u_{x_i} = (BA)_i\nabla u = B_iA\nabla u = B(A\nabla u)\cdot \mathbf{e}_i = A\nabla u \cdot B^*\mathbf{e}_i$, where $\mathbf{e}_i = (0,\dots,0,1,0,\dots,0)$, and applying the Hölder inequality, for any subdomain $\hat{\Omega} \subset \Omega$

we estimate

$$\|u_{x_i}\|_{q_i,\hat{\Omega}} = \left(\int_{\hat{\Omega}} |A\nabla u \cdot B^* \mathbf{e}_i|^{q_i}\, dx\right)^{1/q_i} \leqslant c \sum_{k=1}^{u} \|A_k \nabla u\|_{m_k,\hat{\Omega}} \|B_k^* \mathbf{e}_i\|_{r_{ik},\hat{\Omega}}, \quad (2.5)$$

where we have used the fact that $q_i \leqslant m_k$ and $q_i m_k/(m_k - q_i) \leqslant r_{ik}$, $i, k = 1, \ldots, n$ (see (2.4)). It is obvious that the constant c in (2.5) depends only on q_i, m_k, and meas Ω. Taking account of the fact that $B_k^* \mathbf{e}_i = b^{ik}$, we conclude that Lemma 2.1 follows from (2.5).

Lemma 2.1 thus gives sufficient conditions for weak degeneracy of A.

LEMMA 2.2. *If the matrix A is weakly degenerate in Ω, then condition* (4.1.3) *holds for it.*

PROOF. If $u_n \to 0$ weakly in $L^m(\Omega')$ and $A\nabla u_n \to \mathbf{v}$ weakly in $\mathbf{L}^{\mathbf{m}}(\Omega')$, where $u_n \in \tilde{C}^1_{\text{loc}}(\Omega)$, $n = 1, 2, \ldots$, and Ω' is a fixed, strictly interior subdomain of Ω, then, applying the Banach-Saks theorem, it is easy to prove the existence of a sequence $\{\hat{u}_n\}$ whose elements are the arithmetic means of elements of a subsequence extracted from the original sequence $\{u_n\}$ such that $\hat{u}_n \to 0$ in $L^m(\Omega')$ and $A\nabla \hat{u}_n \to \mathbf{v}$ in $\mathbf{L}^{\mathbf{m}}(\Omega')$ (strongly). But then $A\nabla \hat{u}_n \to \mathbf{v}$ a.e. in Ω', and hence $\nabla \hat{u}_n \to A^{-1}\mathbf{v}$ a.e. in Ω'. On the other hand, using (2.2) we conclude that $\nabla(\hat{u}_n - \hat{u}_m) \to 0$ in $\mathbf{L}^{\mathbf{q}}(\Omega')$ as n, $m \to \infty$. Then $\nabla \hat{u}_n \to \mathbf{w}$ in $\mathbf{L}^{\mathbf{q}}(\Omega')$, where $\mathbf{w} \in \mathbf{L}^{\mathbf{q}}(\Omega')$, and it is obvious that $\mathbf{w} = A^{-1}\mathbf{v}$ a.e. in Ω'. Recalling that $\hat{u}_n \to 0$ in $L^m(\Omega')$, we conclude that $\mathbf{w} = \mathbf{0}$ a.e. in Ω, and hence $\mathbf{v} = \mathbf{0}$ a.e. in Ω. Lemma 2.2 is proved.

The next assertion is proved similarly.

LEMMA 2.3. *Suppose that the matrix A is weakly degenerate in Ω. If a function $u \in L^m_{\text{loc}}(\Omega)$ ($u \in L^m(\Omega)$) has a generalized A-gradient $A\nabla u \in \mathbf{L}^{\mathbf{m}}_{\text{loc}}(\Omega)$ ($A\nabla u \in \mathbf{L}^{\mathbf{m}}(\Omega)$), then it also has an (ordinary) generalized gradient $\nabla u \in \mathbf{L}^{\mathbf{q}}_{\text{loc}}(\Omega)$ ($\nabla u \in \mathbf{L}^{\mathbf{q}}(\Omega)$), and the generalized A-gradient $A\nabla u$ is equal to the vector obtained by the action of the matrix on the vector ∇u.*

Taking account of S. L. Sobolev's familiar theorem on the continuous imbedding of $H_{\mathbf{q}}(\Omega)$ in $L^1(\partial\Omega)$, we establish the following result.

LEMMA 2.4. *Suppose that the matrix A is weakly degenerate in Ω. Then the entire boundary $\partial\Omega$ is regular relative to the matrix A and the indices m and $\mathbf{m}$ (see Definition 4.3.1). Any function $u \in H_{m,\mathbf{m}}(A, \Omega)$ has a generalized limit value on $\partial\Omega$ which coincides with the trace on $\partial\Omega$ of this function considered as an element of $H_{m,\mathbf{q}}(\Omega)$ with $\mathbf{q} = (q_1, \ldots, q_n)$ given by* (2.4).

We consider equation (5.1.35) in Ω assuming that it has $(A, \mathbf{0}, m, \mathbf{m})$-structure in Ω relative to a weakly degenerate matrix A. In view of Lemma 2.4 the general boundary value problem for this equation has the form

$$\mathscr{L}u = f \quad \text{in } \Omega; \qquad u = 0 \quad \text{on } \Sigma_1;$$

$$\mathbf{l}' \cdot A\mathbf{v} = 0 \quad \text{on } \Sigma_2; \qquad \mathbf{l}' \cdot A\mathbf{v} - \lambda u = 0 \quad \text{on } \Sigma_3, \quad (2.6)$$

where $\Sigma_1 \cup \Sigma_2 \cup \Sigma_3 = \partial\Omega$ and $\Sigma_i \cap \Sigma_j = \varnothing$ for $i \neq j$, $i, j = 1, 2, 3$.

A generalized solution of problem (2.6) for all $f \equiv F \in H^*$ can be defined here either by means of an identity of the form (1.3) or (taking Lemma 2.3 into account) as any function $u \in H^{0,\Sigma_1}_{m,\mathbf{m}}(A,\Omega) \cap H^{0,\Sigma_1}_{m,\mathbf{q}}(\Omega)$ satisfying the identity

$$\int_\Omega [\mathbf{l}(x,u,\nabla u)\cdot\nabla\eta + l_0(x,u,\nabla u)\eta]\,dx = \langle F,\eta\rangle, \qquad \eta \in \tilde{C}^1_{0,\Sigma_1}(\overline{\Omega}), \tag{2.7}$$

since in the present case (2.7) is also meaningful. From the results of the general case (see Theorems 5.4.1–5.4.4 and Propositions 5.4.3–5.4.6), with (1.1) and the fact that $\mathscr{B} \equiv 0$ taken into account, we obtain, in particular, the following theorem.

THEOREM 2.1. *Suppose that the reduced coefficients of an equation of the form* (5.1.35) *having* $(A,\mathbf{0},m,\mathbf{m})$*-structure in a bounded strongly Lipschitz domain* $\Omega \subset \mathbb{R}^n$, $n \geqslant 2$, *relative to a weakly degenerate matrix* A *and some indices* $m > 1$ *and* $\mathbf{m} = (m_1,\dots,m_n)$, $m_i > 1$, $i = 1,\dots,n$, *satisfy conditions* (1.4) *and* (1.5). *Then for every* $F \in H^*$ *problem* (2.6) *has at least one generalized solution. If in place of* (1.4) *and* (1.5) *condition* (1.6) *holds for the reduced coefficients, then for every* $F \in H^*$ *problem* (2.6) *has precisely one generalized solution, and the operator* $\mathscr{L}\colon H \to H^*$ *corresponding to this problem is a homeomorphism.*

In the case of weakly degenerate $(A,\mathbf{0})$-elliptic equations it is possible to obtain a number of supplementary results, to which we now proceed. We introduce some new function spaces. We denote by $\tilde{H} \equiv \widetilde{H^{0,\Sigma_1}_m}(A,\Omega)$ the completion of the set $\tilde{C}^1_{0,\Sigma_1}(\overline{\Omega})$ in the norm $\|u\|_{\tilde{H}} \equiv \|A\nabla u\|_{\mathbf{m},\Omega} + \delta(\Sigma_1)\|u\|_{1,\Omega}$, where

$$\delta(\Sigma_i) = \begin{cases} 0, & \text{if } \Sigma_i \neq \varnothing,\ i = 1,2,3,\\ 1, & \text{if } \Sigma_i = \varnothing, \end{cases}$$

where in considering a partition of $\partial\Omega$ into parts Σ_1, Σ_2, Σ_3, we always assume that $\operatorname{meas}_{n-1}\Sigma_i > 0$ if $\Sigma_i \neq \varnothing$. We denote by $\tilde{H}_\lambda \equiv \widetilde{H^{0,\delta_1}_{\mathbf{m}}}(A;\Omega;\Sigma_3,\lambda)$ the completion of the set $\tilde{C}^1_{0,\Sigma_1}(\overline{\Omega})$ in the norm

$$\|u\|_{\tilde{H}_\lambda} = \|A\nabla u\|_{\mathbf{m},\Omega} + \delta(\Sigma_1)\delta(\Sigma_3)\|u\|_{1,\Omega} + [1-\delta(\Sigma_3)]\,\|u\|_{L^2(\lambda,\Sigma_3)}, \tag{2.8}$$

where λ is a given positive function on Σ_3 and $\lambda \in L^1(\Sigma_3)$. We note that by (2.2) and Lemma 4.4.1 the expressions $\|\cdot\|_{\tilde{H}}$ and $\|\cdot\|_{\tilde{H}_\lambda}$ are actually norms, so that $\tilde{H}$ and $\tilde{H}_\lambda$ are Banach spaces. If $\lambda \in L^\kappa(\Sigma_3)$, where

$$\kappa > r/(r-2) \geqslant 1, \quad (n-1)/r = n/q^* - 1 \quad \text{for } q_* = \min(q_1,\dots,q_n) < r,$$
$$r \in [2,+\infty) \quad \text{for } q_* \geqslant n, \tag{2.9}$$

and the indices $q_1,\dots,q_n$ are related to $\mathbf{m}$ by condition (2.2), then we need not require positivity of the function λ, since by means of Sobolev's familiar imbedding theorem and condition (2.2) for all $u \in \tilde{C}^1_{0,\Sigma_1}(\overline{\Omega})$ it is easy to establish the estimate

$$\left(\int_{\Sigma_3} \lambda u^2\,ds\right)^{1/2} \leqslant \varepsilon\|u\|_{\tilde{H}} + c(\varepsilon)\|u\|_{1,\Omega}, \quad \forall\varepsilon > 0,$$

in view of which the sign of λ plays no role. In this case the last term in the definition of the norm (2.8) is to be omitted.

LEMMA 2.5. *Suppose that the matrix A is weakly degenerate in Ω, and that the index $\hat{l} > 1$ satisfies the conditions*

$$\frac{1}{\hat{l}} = \frac{1}{n}\left(\sum_{i=1}^{n}\frac{1}{q_i} - 1\right) \text{ for } \sum_{i=1}^{n}\frac{1}{q_i} > 1, \quad \hat{l} \in (2, +\infty) \text{ for } \sum_{i=1}^{n}\frac{1}{q_i} \leqslant 1, \tag{2.10}$$

where $\mathbf{q} = (q_1, \ldots, q_n)$ is the index in the condition of weak degeneracy of the matrix A (see (2.2)). Then (see Remark 4.4.3) there are the imbeddings $\tilde{H}_\lambda \to \tilde{H} \to \widetilde{H_{\mathbf{q}}^{0,\Sigma_1}}(\Omega) \to L^{\hat{l}}(\Omega)$. In particular, for any function $u \in \tilde{H}$

$$\|u\|_{\hat{l},\Omega} \leqslant c_0\|u\|_{\tilde{H}}, \tag{2.11}$$

where the constant c_0 depends only on n, q, Ω and the constant c in (2.2). *If*

$$\frac{1}{l} > \frac{1}{n}\left(\sum_{i=1}^{n}\frac{1}{q_i} - 1\right) \text{ for } \sum_{i=1}^{n}\frac{1}{q_i} > 1, \quad l \in (1 + \infty) \text{ for } \sum_{i=1}^{n}\frac{1}{q_i} \leqslant 1, \tag{2.12}$$

then the imbedding $\tilde{H} \to L^l(\Omega)$ is compact. Here for any function $u \in \tilde{H}$ and any $\varepsilon > 0$

$$\|u\|_{l,\Omega} \leqslant \varepsilon\|A\nabla u\|_{\mathbf{m},\Omega} + c_1\varepsilon^{-\theta}\|u\|_{1,\Omega}, \tag{2.13}$$

where c_1 depends only on n, l, $\mathbf{q}$, Ω, and the constant c in (2.2), *while $\theta > 0$ depends only on n, l, and $\mathbf{q}$. In the case $\Sigma_1 \equiv \partial\Omega$ the constants c_0 and c_1 in* (2.12) *and* (2.13) *do not depend on Ω.*

PROOF. From (2.2), (2.10), and Lemma 4.4.1 we obtain

$$\|u\|_{\hat{l},\Omega} \leqslant c_1\|u\|_{\tilde{H}_{\mathbf{q}}(\Omega)} \leqslant c_2\|u\|_{\tilde{H}_{\mathbf{m}}(A,\Omega)}, \quad \forall u \in \tilde{C}^1_{0,\Sigma_1}(\overline{\Omega}), \tag{2.14}$$

where $c_1 = c_1(n, m, \mathbf{q}, \Omega)$, $c_2 = c_2(n, m, \mathbf{q}, \Omega, c)$, and c is the constant in (2.2). From this it follows easily that $\tilde{H} \equiv \widetilde{H_{\mathbf{m}}^{0,\Sigma_1}}(A, \Omega)$ can be identified with a subspace of $\widetilde{H_{\mathbf{q}}^{0,\Sigma_1}}(\Omega)$, and $\widetilde{H_{\mathbf{q}}^{0,\Sigma_1}}(\Omega)$ can be identified with a subspace of $L^{\hat{l}}(\Omega)$, where the inequalities (2.14) are preserved for all $u \in \tilde{H}$. In the case of conditions (2.12) the compactness of the imbedding $\tilde{H} \to L^l(\Omega)$ follows from the compactness of the imbedding $\widetilde{H_{\mathbf{q}}^{0,\Sigma_1}}(\Omega) \to L^l(\Omega)$ (see Lemma 4.4.1), while (2.12) follows from (4.4.6) with $s = 1$ and (2.2). Lemma 2.5 is proved.

Thus, under condition (2.10) the spaces

$$\overset{0,\Sigma_1}{H}_{\hat{l},\mathbf{m}}(A, \Omega) \quad \text{and} \quad \widetilde{\overset{0,\Sigma_1}{H}}_{\mathbf{m}}(A, \Omega)$$

$$\left(\overset{0,\Sigma_1}{H}_{\hat{l},\mathbf{m}}(A; \Omega; \Sigma_3, \lambda) \quad \text{and} \quad \widetilde{\overset{0,\Sigma_1}{H}}_{\mathbf{m}}(A; \Omega; \Sigma_3, \lambda)\right)$$

are isomorphic.

2. $(A, 0, \hat{l}, \mathbf{m})$-elliptic equations with weak degeneracy. Below we consider a problem of the form (2.6) for equations having $(A, \mathbf{0}, \hat{l}, \mathbf{m})$-structure in Ω, where the index $\hat{l}$ satisfies (2.10); we assume that the function λ is defined on Σ_3, is positive, and belongs to $L^1(\Omega)$. If $\lambda \in L^\kappa(\Sigma_3)$ with an index κ satisfying (2.9), then the condition of positivity of λ is removed. Progress in investigating the solvability of problem (2.6) for such equations is connected with the possibility of giving for these equations new algebraic criteria for coercivity and strong monotonicity and also semibounded variation of the operator $\mathscr{L}$ corresponding to the problem. We note that the operator can therefore be written in the form $\mathscr{L}$: $\tilde{H}_\lambda \to \tilde{H}_\lambda^*$ where $\tilde{H}_\lambda \equiv \widetilde{H_{\mathbf{m}}^{0,\Sigma_3}}(A; \Omega; \Sigma_3, \lambda)$ is isomorphic to the space $H_{\hat{l},m}^{0,\Sigma_1}(A; \Omega; \Sigma_3, \lambda)$.

PROPOSITION 2.1. *Suppose that for almost all* $x \in \Omega$ *and any* $u \in \mathbb{R}$, $q = Ap$ *and* $p \in \mathbb{R}^n$

$$l'^i(x,u,q)q_i + l'_0(x,u,q)u \geqslant \nu_1\left[\sum_{i=1}^n |q_i|^{m_i} + \delta(\Sigma_1)\delta(\Sigma_3)|u|^\kappa\right] - \nu_2|u|^m - \varphi(x), \tag{2.15}$$

where $\nu_1, \nu_2 = \text{const} > 0$, $\kappa > 1$, $\varphi \in L^1(\Omega)$, $m > 1$, *and*

$$m < m_* = \min(m_1,\dots,m_n), \qquad \delta(\Sigma_1)\delta(\Sigma_3)m < \kappa, \qquad [1-\delta(\Sigma_3)]m < 2, \qquad m \leqslant \hat{l}, \tag{2.16}$$

or

$$\begin{gathered} m \leqslant m_*, \quad \delta(\Sigma_1)\delta(\Sigma_3)m < \kappa, \qquad [1-\delta(\Sigma_3)]m < 2, \quad m \leqslant \hat{l}, \\ \hat{\nu}_1\Delta^{-m_*} > c_0|\nu_2|(\operatorname{meas}\Omega)^{(\hat{l}-m)/\hat{l}}, \qquad \hat{\nu}_1\Delta^{-\kappa} > \delta(\Sigma_1)c_0^m(\operatorname{meas}\Omega)^{(\hat{l}-m)/\hat{l}}, \\ \hat{\nu}_1\Delta^{-2} > [1-\delta(\Sigma_3)]c_0^m|\nu_2|(\operatorname{meas}\Omega)^{(\hat{l}-m)/\hat{l}}, \\ \Delta \equiv n + \delta(\Sigma_1)\delta(\Sigma_3) + 1 - \delta(\Sigma_3), \\ \hat{\nu}_1 \equiv \min(1,\nu)\cdot\min\left(1,(\operatorname{meas}\Omega)^{-\kappa/\kappa'}\right), \qquad 1/\kappa + 1/\kappa' = 1. \end{gathered} \tag{2.17}$$

Then the operator $\mathscr{L}\colon \tilde{H}_\lambda \to \tilde{H}_\lambda^*$ *is coercive.*

PROOF. Suppose first that conditions (2.15) and (2.16) hold. In view of (2.15), for all $u \in \tilde{H}_\lambda$ we have

$$\langle \mathscr{L}u, u\rangle \geqslant \hat{\nu}_1\left[\sum_{i=1}^n \|A_i\nabla u\|_{m_i,\Omega}^{m_i} + \delta(\Sigma_1)\delta(\Sigma_3)\|u\|_{\kappa,\Omega}^\kappa + \|u\|_{L^2(\lambda,\Sigma_3)}^2\right] - |\nu_2|\,\|u\|_{m,\Omega}^m - \int_\Omega \varphi\,dx.$$

Taking into account that $\|u\|_{1,\Omega}^\kappa \leqslant \|u\|_{\kappa,\Omega}^\kappa(\operatorname{meas}\Omega)^{\kappa/\kappa'}$ and using (2.11), from this we obtain

$$\langle \mathscr{L}u, u\rangle \geqslant \hat{\nu}_1\left[\sum_{i=1}^n \|A_i\nabla u\|_{m_i,\Omega}^{m_i} + \delta(\Sigma_1)\delta(\Sigma_3)\|u\|_{1,\Omega}^\kappa + \|u\|_{L^2(\lambda,\Sigma_3)}^2\right] - |\nu_2|c_0^m(\operatorname{meas}\Omega)^{(\hat{l}-m)/\hat{l}}\|u\|_{\tilde{H}_\lambda}^m - \int_\Omega \varphi(x)\,dx.$$

Suppose that $\|u\|_{\tilde{H}_\lambda} = \rho \geqslant \Delta \equiv n + \delta(\Sigma_1)\delta(\Sigma_3) + 1 - \delta(\Sigma_3)$. From the last inequality we then obtain

$$\langle \mathscr{L}u, u\rangle \geqslant \hat{\nu}_1\rho^q\Delta^{-q} - |\nu_2|c_0^m(\operatorname{meas}\Omega)^{(\hat{l}-m)/\hat{l}}\rho^m - \int_\Omega \varphi(x)\,dx, \tag{2.18}$$

where q is one of the numbers of the collection $\{m_1,\dots,m_n, \delta(\Sigma_1)\delta(\Sigma_3)\kappa, 2(1-\delta(\Sigma_3))\}$ and $q > m > 1$ by (2.16). Now a condition of the form (4.5.1) (in the case $X \equiv \tilde{H}_\lambda$) follows easily from (2.18), i.e., the operator $\mathscr{L}\colon \tilde{H}_\lambda \to \tilde{H}_\lambda^*$ is coercive. In the case of conditions (2.15) and (2.17) we again obtain (2.18) but now with $q \geqslant m$. Taking (2.17) into account, we deduce from (2.18) that

$$\langle \mathscr{L}u, u\rangle \geqslant c\rho^m - \int_\Omega \varphi(x)\,dx, \quad \forall u \in \tilde{H}_\lambda, \qquad \|u\|_{\tilde{H}_\lambda} = \rho > \Delta, \tag{2.19}$$

where $c = c(c_0, \nu_2, \operatorname{meas}\Omega, m, \hat{l})$. Since $m > 1$, coerciveness of the operator $\mathscr{L}$: $\tilde{H}_\lambda \to \tilde{H}_\lambda^*$ follows immediately from (2.19). Proposition 2.1 is proved.

PROPOSITION 2.2. *Suppose that the following conditions are satisfied:*
1) *The reduced coefficients of equation* (5.1.35) *have the form*

$$l'^i = \bar{l}'^i(x, u, q) + \bar{\bar{l}}'^i(x, u), \qquad i = 1, \dots, n,$$

$$l'_0 = \bar{l}'_0(x, u, q) + \bar{\bar{l}}'_0(x, u, q). \tag{2.20}$$

2) *The operator* $\bar{\mathscr{L}}$: $\tilde{H}_\lambda \to \tilde{H}_\lambda^*$ *defined by*

$$\langle \bar{\mathscr{L}} u, \eta \rangle = \int_\Omega (\bar{\mathbf{l}}' \cdot A\nabla\eta + \bar{l}'_0\eta)\, dx + \int_{\Sigma_2} \lambda u|_\Sigma \, ds, \qquad u, \eta \in \tilde{H}_\lambda, \tag{2.21}$$

satisfies a condition of strong monotonicity of the form

$$\langle \bar{\mathscr{L}} u - \bar{\mathscr{L}} v, u - v \rangle$$

$$\geqslant \nu_0 \left[\sum_{i=1}^n \|A_i \nabla(u - v)\|_{m_i, \Omega}^{m_i} + \delta(\Sigma_1)\delta(\Sigma_3)\|u - v\|_{\kappa, \Omega}^\kappa + \|u - v\|_{L^2(\lambda, \Sigma_3)}^2 \right]; \tag{2.22}$$

3) *For almost all* $x \in \Omega$ *and any* $u \in \mathbf{R}$ *and* $q \in \mathbf{R}^n$

$$\left| \frac{\partial \bar{\bar{l}}'^i}{\partial u} \right| \leqslant \mu_1 |u|^{l/m'_i - 1} + \varphi_i;$$

$$\left| \frac{\partial \bar{\bar{l}}'_0}{\partial q_j} \right| \leqslant \mu_2 \left(\sum_{k=1}^n |q_k|^{m_k(1/m'_i - 1/l)} + |u|^{l/m'_i - 1} + \tilde{\varphi}_i \right), \quad i = 1, \dots, n;$$

$$\left| \frac{\partial \bar{\bar{l}}'_0}{\partial u} \right| \leqslant \mu_3 \left(\sum_{k=1}^n |q_k|^{m_k(1 - 2/l)} + |u|^{l-2} + \varphi_0 \right), \tag{2.23}$$

where $\mu_i = \mathrm{const} \geqslant 0$, $i = 1, 2, 3$, $\varphi_i, \tilde{\varphi}_i \in L^{(1/m'_i - 1/l)^{-1}}(\Omega)$, $1/m_i + 1/m'_i = 1$, $i = 1, \dots, n$, $\varphi_0 \in L^{l/(l-2)}(\Omega)$, $l \geqslant 2$, *and* $\max(m'_1, \dots, m'_n) \leqslant l < \hat{l}$.
Then the operator $\mathscr{L}$: $\tilde{H}_\lambda \to \tilde{H}_\lambda^*$ *has semibounded variation.*

PROOF. We denote by $\bar{\bar{\mathscr{L}}}$: $\tilde{H}_\lambda \to \tilde{H}_\lambda^*$ the operator defined by

$$\langle \bar{\bar{\mathscr{L}}} u, \eta \rangle = \int_\Omega (\bar{\bar{l}}' \cdot A\nabla\eta + \bar{\bar{l}}'_0\eta)\, dx, \qquad u, \eta \in \tilde{H}_\lambda.$$

Using conditions 1) and 2) of the proposition and arguing as in the proof of Proposition 5.4.5, we obtain

$$\langle \mathscr{L} u - \mathscr{L} v, u - v \rangle = \langle \bar{\mathscr{L}} u - \bar{\mathscr{L}} v, u - v \rangle + \langle \bar{\bar{\mathscr{L}}} u - \bar{\bar{\mathscr{L}}} v, u - v \rangle$$

$$\geqslant \tilde{\nu}_0 \left[\sum_{i=1}^n \|A_i \nabla(u - v)\|_{m_i, \Omega}^{m_i} + \delta(\Sigma_1)\delta(\Sigma_3)\|u - v\|_{\kappa, \Omega}^\kappa \right] + I_1 + I_2 + I_3, \tag{2.24}$$

where $\tilde{\nu}_0 = \min(\nu, 1)$ and

$$I_1 = \int_\Omega \int_0^1 \frac{\partial l\|^{\prime i}(x, v + \tau(u - v))}{\partial u}(u - v) A_i \nabla(u - v)\, d\tau\, dx,$$

$$I_2 = \int_\Omega \int_0^1 \frac{\partial l\|'_0(x, v + \tau(u - v), A\nabla v + \tau A\nabla(u - v))}{\partial u}(u - v)^2\, d\tau\, dx,$$

$$I_3 = \int_\Omega \int_0^1 \frac{\partial l\|'_0(\dots)}{\partial q_j} A_j \nabla(u - v)(u - v)\, d\tau\, dx.$$

Let $\|u\|_{\tilde{H}_\lambda} \leqslant \rho$ and $\|v\|_{\tilde{H}_\lambda} \leqslant \rho$. Taking account of condition 3) of the proposition and applying the Hölder inequality and Lemma 2.5, we obtain

$$\begin{gathered}
|I_1| \leqslant \varepsilon \sum_{i=1}^n \|A_i \nabla(u - v)\|_{m_i,\Omega}^{m_i} - \sum_{i=1}^n \beta_i \|u - v\|_{l,\Omega}^{m'_i} + \beta_0 \|u - v\|_{l,\Omega}^l, \\
|I_2| \leqslant \beta\left(\|u - v\|_{l,\Omega}^2 + \|u - v\|_{l,\Omega}^l\right), \\
|I_3| \leqslant \varepsilon \sum_{i=1}^n \|A_i \nabla(u - v)\|_{m_i,\Omega}^{m_i} + \sum_{i=1}^n \gamma_i \|u - v\|_{l,\Omega}^{m'_i} + \gamma \|u - v\|_{l,\Omega}^l,
\end{gathered} \tag{2.25}$$

where $\varepsilon > 0$, $\beta_i = \beta_i(\varepsilon, \rho) > 0$, $\beta_0 = \beta_0(\varepsilon, \rho) > 0$, $\beta = \beta(\varepsilon, \rho) > 0$, $\gamma_i = \gamma_i(\varepsilon, \rho) > 0$, $\gamma = \gamma(\varepsilon, \rho) > 0$, and these functions depend continuously on $\varepsilon > 0$ and $\rho > 0$. From (2.24) and (2.25) we obtain, in an obvious manner,

$$\begin{aligned}
&\langle \mathscr{L}u - \mathscr{L}v, u - v\rangle \\
&\quad \geqslant \frac{\tilde{\nu}_0}{2}\left[\sum_{i=1}^n \|A_i \nabla(u - v)\|_{m_i,\Omega}^{m_i} + \delta(\Sigma_1)\delta(\Sigma_3)\|u - v\|_{\kappa,\Omega}^{\kappa} + \|u - v\|_{L^2(\lambda,\Sigma_3)}^2\right] \\
&\quad\quad - \gamma(\rho, \|u - v\|_{l,\Omega}),
\end{aligned} \tag{2.26}$$

where $\gamma(\rho, \tau)$ has the form $\gamma(\rho, \tau) = \sum_1^n d_i(\rho)\tau^{m'_i} + d_0(\rho)\tau^l$, and $d_i(\rho)$, $i = 1, \dots, n$, and $d_0(\rho)$ are continuous nonnegative functions of $\rho > 0$. Since $m'_i > 1$, $i = 1, \dots, n$ and $l > 1$, from this it follows that $\gamma(\rho, \tau)\tau^{-1} \to 0$ as $\tau \to +0$. In view of Lemma 2.5 the norm $\|\cdot\|_{l,\Omega}$ is compact relative to the norm $\|\cdot\|_{\tilde{H}_\lambda}$. From (2.26) it then follows that the operator $\mathscr{L}\colon \tilde{H}_\lambda \to \tilde{H}_\lambda^*$ has semibounded variation. Proposition 2.2 is proved.

PROPOSITION 2.3. *Suppose that for almost all $x \in \Omega$ and any $u \in \mathbb{R}$, $q \in \mathbb{R}^n$, $\xi_0 \in \mathbb{R}$ and $\eta \in \mathbb{R}^n$*

$$\begin{aligned}
&\frac{\partial l^{\prime i}(x, u, q)}{\partial q_j}\eta_i\eta_j + \frac{\partial l^{\prime i}(x, u, q)}{\partial u}\eta_i\xi_0 + \frac{\partial l'_0(x, u, q)}{\partial q_j}\eta_j\xi_0 + \frac{\partial l'_0(x, u, q)}{\partial u}\xi_0^2 \\
&\qquad \geqslant c\left[\sum_{i=1}^n |q_i|^{m_i - 2}\eta_i^2 + \delta(\Sigma_1)\delta(\Sigma_3)|u|^{\kappa - 2}\xi_0^2\right],
\end{aligned} \tag{2.27}$$

where $c = \text{const} > 0$, and the indices κ and $m_1, \dots, m_n$ satisfy the condition $\kappa \geqslant 2$, $m_i \geqslant 2$, $i = 1, \dots, n$. Then the operator $\mathscr{L}\colon \tilde{H}_\lambda \to \tilde{H}_\lambda^*$ is strongly monotone.

PROOF. Proposition 2.3 is established in exactly the same way as Proposition 5.4.6 with the additional fact that the norms of the spaces

$$\tilde{H}_\lambda \equiv \overset{\widetilde{0,\Sigma_1}}{H_{\mathbf{m}}}(A;\Omega;\Sigma_3,\lambda) \quad \text{and} \quad \overset{0,\Sigma_1}{H_{l,\mathbf{m}}}(A;\Omega;\Sigma_3,\lambda)$$

are equivalent taken into account.

The next results on the solvability of the general boundary value problem of the form (2.6) for weakly degenerate $(A, \mathbf{0}, \hat{l}, \mathbf{m})$-elliptic equations follow directly from Theorems 5.4.1–5.4.4 and Propositions 5.4.4–5.4.6.

THEOREM 2.2. *Suppose that conditions* (2.15), (2.16) *or* (2.15), (2.17) *and conditions* 1)– 3) *of Proposition* 2.2 *are satisfied. Then for all* $F \in \tilde{H}^*_\lambda$ *problem* (2.6) *has at least one generalized solution.*

THEOREM 2.3. *Suppose condition* (2.27) *is satisfied. Then for any* $F \in \tilde{H}^*_\lambda$ *problem* (2.6) *has precisely one generalized solution. Here the operator* $\mathscr{L}: \tilde{H}_\lambda \to \tilde{H}^*_\lambda$ *corresponding to problem* (2.6) *is a homeomorphism.*

3. $(A, \mathbf{0}, \overline{m}, \overline{m})^{\sim}$**-elliptic equations with weak degeneracy.** We suppose that conditions (5.1.2) relative to a weakly degenerate matrix A and the vector $\mathbf{b} \equiv \mathbf{0}$ in Ω are satisfied for an equation of the form (5.1.35). Let $\mathbf{m} = (\overline{m}, \dots, \overline{m})$, where $\overline{m} > 1$, and let $\mathbf{q} = (q_1, \dots, q_n)$ be the index connected with this $\mathbf{m}$ by condition (2.2) (the existence of such $\mathbf{q}$ follows from Lemma 2.1); suppose that the index $\hat{l}$ is determined on the basis of $\mathbf{q}$ by condition (2.10). We assume that $\overline{m} < \hat{l}$. Suppose that for the functions l' and l'_0 in (5.1.2) we have

$$\begin{aligned} |l'^i(x,u,q)| &\leqslant \tilde{\mu}_1 |q|^{\overline{m}-1} + \tilde{a}_1(x)|u|^{\overline{m}-1} + \psi_i(x), \qquad i = 1, \dots, n, \\ |l'_0(x,u,q)| &\leqslant \tilde{a}_2(x)|q|^{\overline{m}-1} + \tilde{a}_3(x)|u|^{\overline{m}-1} + \psi_0(x), \end{aligned} \tag{2.28}$$

where $\tilde{\mu}_1 = \text{const} \geqslant 0$, $\tilde{a}_i(x) \geqslant 0$, $a_1^{\overline{m}'}, a_2^{\overline{m}} \in L^{\hat{s}}(\Omega)$, $a_3 \in L^{\hat{s}}(\Omega)$, $1/\hat{s} + \overline{m}/\hat{l} = 1$, $\psi_i \in L^{\overline{m}'}(\Omega)$, $1/\overline{m} + 1/\overline{m}' = 1$, and $\psi_0 \in L^{\hat{l}'}(\Omega)$, $1/\hat{l} + 1/\hat{l}' = 1$. With the help of Young's inequality it is easy to see that conditions (2.28) imply (5.1.3) with $m_1 = \cdots = m_n = \overline{m}$ and $m = \hat{l}$, which expresses the fact that the equation in question has isotropic $(A, \mathbf{0}, \hat{l}, \mathbf{m})$-structure with $\mathbf{m} = (\overline{m}, \dots, \overline{m})$. Equations of the form (5.1.35) possessing the properties just enumerated will be called *equations having* $(A, \mathbf{0}, \overline{m}, \overline{m})^{\sim}$*-structure* in the domain Ω. The following assertions hold for equations having this structure.

PROPOSITION 2.4. *Suppose that the conditions indicated above are satisfied, and that for almost all* $x \in \Omega$ *and any* $u \in \mathbb{R}$ *and* $q \in \mathbb{R}^n$

$$l'^i q_i + l'_0 u \geqslant \nu\left[|q|^m + \delta(\Sigma_1)\delta(\Sigma_3)|u|^\kappa\right] - a_4(x)|u|^{\overline{m}} - \psi(x), \tag{2.29}$$

where $\nu = \text{const} > 0$, $\kappa > 1$, $a_4 \in L^{\hat{s}}(\Omega)$, $\overline{m}/\hat{l} + 1/\hat{s} = 1$, $\psi \in L^1(\Omega)$, and

$$\begin{gathered} \delta(\Sigma_1)\delta(\Sigma_3)\overline{m} \leqslant \kappa, \quad [1 - \delta(\Sigma_3)]\,\overline{m} \leqslant 2, \quad \hat{\nu}\Delta^{-\overline{m}} - c_0^{\overline{m}}\|(a_4)_+\|_{\hat{s},\Omega} > 0, \\ \Delta \equiv n + \delta(\Sigma_1)\delta(\Sigma_3) + 1 - \delta(\Sigma_3), \quad \hat{\nu} \equiv \min(1,\nu)\min\left(1, (\text{meas}\,\Omega)^{-\kappa/\kappa'}\right), \end{gathered} \tag{2.30}$$

where c_0 *is the constant in* (2.11), *and* f_+ *denotes the positive part of the function* f, *i.e.*, $f_+(x) = \max(0, f(x))$. *Then the operator* $\mathscr{L}: \tilde{H}_\lambda \to \tilde{H}^*_\lambda$ *corresponding to problem* (2.6) *is coercive.*

PROOF. Obviously, for all $u \in \tilde{H}_\lambda$ we have (see the proof of Proposition 2.1)

$$\langle \mathscr{L}u, u\rangle \geqslant \hat{\nu}\Big[\|A\nabla u\|^{\overline{m}}_{\overline{m},\Omega} + \delta(\Sigma_1)\delta(\Sigma_3)\|u\|^{\kappa}_{\kappa,\Omega} + \|u\|^2_{L^2(\lambda,\Sigma_3)}\Big]$$
$$= \int_\Omega (a_4)_+|u|^{\overline{m}}\,dx - \int_\Omega \psi(x)\,dx.$$

Applying Hölder's inequality and using an estimate of the form (2.11) and condition (2.29), we obtain for all $u \in \tilde{H}_\lambda$ such that $\|u\|_{\tilde{H}_\lambda} \equiv \rho > n + \delta(\Sigma_1)\delta(\Sigma_3) + 1 - \delta(\Sigma_3) \equiv \Delta$ the inequality

$$\langle \mathscr{L}u, u\rangle \geqslant \left[\nu\Delta^{-\overline{m}} - c_0^{\overline{m}}\|(a_4)_+\|_{\hat{s},\Omega}\right]\rho^{\overline{m}} - \left|\int_\Omega \psi(x)\,dx\right|$$

from which coerciveness of the operator $\mathscr{L}$: $\tilde{H}_\lambda \to \tilde{H}^*_\lambda$ obviously follows by (2.30). Proposition 2.4 is proved.

PROPOSITION 2.5. *Suppose that for an equation* (5.1.35) *having* $(A, \mathbf{0}, \overline{m}, \overline{m})^\sim$-*structure in* Ω *the following conditions are satisfied*:

1) *The reduced coefficients* l'^i, $i = 1,\dots,n$, *and* l'_0 *have the form* (2.20).

2) *The operator* $\overline{\mathscr{L}}$: $\tilde{H}_\lambda \to \tilde{H}^*_\lambda$ *defined by* (2.21) *satisfies a condition of the form* (2.22) *with* $m_1 = \cdots = m_n = \overline{m}$.

3) *For almost all* $x \in \Omega$ *and any* $u \in \mathbb{R}$ *and* $q \in \mathbb{R}^n$

$$\left|\partial\bar{\bar{l}}'^i/\partial u\right| \leqslant a_5|u|^{\overline{m}-2} + \psi_i;$$
$$\left|\partial\bar{\bar{l}}'_0/\partial q_i\right| \leqslant a_6|q|^{\overline{m}-2} + a_7|u|^{\overline{m}-2} + \tilde{\psi}_i, \qquad i = 1,\dots,n;$$
$$\left|\partial\bar{\bar{l}}'_0/\partial u\right| \leqslant a_8|q|^{\overline{m}-2} + a_9|u|^{\overline{m}-2} + \psi_0, \tag{2.31}$$

where $a_i = a_i(x) \geqslant 0$, $i = 5,\dots,9$, $\psi_i = \psi_i(x) \geqslant 0$, $\tilde{\psi}_i = \tilde{\psi}_i(x) \geqslant 0$, $i = 1,\dots,n$, $\psi_0 = \psi_0(x) \geqslant 0$, $a_5^{\overline{m}'}$, $a_6^{\overline{m}}$, $a_7^{\overline{m}'}$, $a_8^{\overline{m}/2}$, $a_9 \in L^s(\Omega)$, $1/s + \overline{m}/l = 1$, $2 \leqslant \overline{m} \leqslant l < \hat{l}$, ψ_i, $\tilde{\psi}_i \in L^{(1/\overline{m}'-1/l)^{-1}}(\Omega)$ *and* $\psi_0 \in L^{l/(l-2)}(\Omega)$.

Then the operator $\mathscr{L}$: $\tilde{H}_\lambda \to \tilde{H}^*_\lambda$ *corresponding to problem* (2.6) *has semibounded variation.*

PROOF. Proposition 2.5 is a special case of Proposition 2.2, since from (2.31) by means of Young's inequality it is easy to derive conditions of the form (2.23) with

$$m_1 = \cdots = m_n = \overline{m}, \qquad \varphi_i = \psi_i + a_5^{(l-\overline{m}')/(l-\overline{m})},$$
$$\tilde{\varphi}_i = \psi_i + a_6^{(l\overline{m}-l-\overline{m})/(l-\overline{m})} + a_7^{(l-\overline{m}')/(l-\overline{m})}, \ \varphi_0 = \psi_0 + a_8^{\overline{m}(l-2)/2(l-\overline{m})} + a_9^{(l-2)/(l-\overline{m})},$$

where the functions φ_i, $\tilde{\varphi}_i$ and φ_0 satisfy the conditions required in Proposition 2.2. This proves Proposition 2.5.

The next assertion obviously follows from Theorem 5.4.1 and Propositions 2.4 and 2.5.

THEOREM 2.4. *Suppose that conditions* (2.28)–(2.30) *and conditions* 1)–3) *of Proposition* 2.5 *are satisfied. Then for any* $F \in \tilde{H}^*_\lambda$ *problem* (2.6) *has at least one generalized solution.*

4. Linear elliptic equations with weak degeneracy. In a bounded, strongly Lipschitz domain $\Omega \subset \mathbb{R}^n$, $n \geqslant 2$, we consider a linear equation of the form (5.5.1) where the (in general, nonsymmetric) matrix $\mathfrak{A} \equiv \|\alpha^{ij}(x)\|$ is positive definite for almost all $x \in \Omega$. We henceforth always assume that there exists a constant $k_0 > 0$ such that for almost all $x \in \Omega$ and any $\xi, \eta \in \mathbb{R}^n$

$$|\alpha^{ij}(x)\xi_i\eta_j| \leqslant k_0\big(\alpha^{ij}(x)\xi_i\xi_j\big)^{1/2}\big(\alpha^{ij}(x)\eta_i\eta_j\big)^{1/2}.$$

(It is obvious that this condition is satisfied for $k_0 = 1$ in the case of a symmetric matrix $\mathfrak{A}$.) We set $A \equiv \|a^{ij}(x)\| \equiv [\frac{1}{2}(\mathfrak{A} + \mathfrak{A}^*)]^{1/2}$. It is obvious that $\alpha^{ij}\xi_i\xi_j = |A\xi|^2$. We suppose that the matrix A is weakly degenerate in Ω (see Definition 2.1). Suppose that the index $\hat{l}$ satisfies condition (2.10), where $\mathbf{q} = (q_1,\ldots,q_n)$ is the index connected with $\mathbf{m} = \mathbf{2} = (2,\ldots,2)$ by condition (2.2); we assume here that $\hat{l} > 2$. It is easy to see that under these conditions and the conditions

$$a^{ij} \in L^2(\Omega), \quad i,j = 1,\ldots,n; \qquad A^{-1}\alpha,\ A^{-1}\beta \in \mathbf{L}^{2\hat{l}/(\hat{l}-2)}(\Omega);$$
$$A^{-1}g \in \mathbf{L}^2(\Omega); \qquad \beta_0 \in L^{\hat{l}/(\hat{l}-2)}(\Omega), \quad g_0 \in L^{\hat{l}'}(\Omega), \quad 1/\hat{l} + 1/\hat{l}' = 1, \tag{2.32}$$

equation (5.5.1) has $(A,\mathbf{0},2,2)^{\sim}$-structure in Ω. Indeed, equalities of the form (5.1.2) hold for the functions $\mathbf{l} \equiv \mathfrak{A}p + \alpha\mathbf{u} + \mathbf{g}$ and $l_0 \equiv \beta\cdot p + \beta u + g_0$ with $\mathbf{l}' \equiv Qq + \mathbf{a}u + \mathbf{f}$ and $l_0' \equiv \boldsymbol{\gamma}\cdot q + a_0 u + f_0$ for $A = ((\mathfrak{A} + \mathfrak{A}^*)/2)^{1/2}$, $Q = A^{-1}\mathfrak{A}A^{-1}$, $\mathbf{a} = A^{-1}\boldsymbol{\alpha}$, $\mathbf{f} = A^{-1}\mathbf{g}$, $\boldsymbol{\gamma} = A^{-1}\boldsymbol{\beta}$, $\mathbf{b} = 0$, $a_0 = \beta_0$ and $f_0 = g_0$ (where we have taken into account that the matrix A is symmetric). For such functions l'^i, $i = 1,\ldots,n$, and l_0' inequalities of the form (2.28) hold with $\overline{m} = 2$, $\tilde{a}_1 = |A^{-1}\boldsymbol{\alpha}|$, $\psi_i = |A^{-1}\mathbf{g}|$, $i = 1,\ldots,n$, $\tilde{a}_2 = |A^{-1}\boldsymbol{\beta}|$, $\tilde{a}_3 = |\beta_0|$ and $\psi_0 = |g_0|$ with a constant $\tilde{\mu}_1$ depending on k_0, since it is easy to see that $\|A^{-1}\mathfrak{A}A^{-1}\| \leqslant$ const in Ω. The validity of the last inequality follows from the estimate

$$|A^{-1}\mathfrak{A}A^{-1}p\xi| = |\mathfrak{A}q\cdot\eta| \leqslant k_0|Aq|\,|A\eta| = k_0|p|\,|\xi|,$$

where $p = Aq$, $\xi = A\eta$ and $q, \eta \leftarrow \mathbb{R}^n$, and from the fact that $\mathbb{R}^n$ coincides with the set $\{p = A(x)q,\ q \in \mathbb{R}^n\}$ for almost all $x \in \Omega$. Thus, equation (5.5.1) indeed has $(A,0,2,2)^{\sim}$-structure in Ω.

THEOREM 2.5. *Suppose conditions* (2.32) *are satisfied, and suppose there exist positive numbers* $\varepsilon_1, \varepsilon_2, \varepsilon_3, \varepsilon_4$, $(\varepsilon_1 + \varepsilon_2 + \varepsilon_3)/2 < 1$, *and* θ *such that*

$$\delta(\Sigma_1)\delta(\Sigma_3)\theta\beta_0(x) \geqslant c_1 = \text{const} > 0$$

for almost all $x \in \Omega$, *and*

$$\nu\Delta^{-2} - c_0^2\left\|\left[(\theta - 1)\beta_0 + \frac{|A^{-1}\alpha|^2}{2\varepsilon_1} + \frac{|A^{-1}\beta|^2}{2\varepsilon_3} + \frac{\varepsilon_4}{2}|f_0|^{(l-2)/(l-1)}\right]_+\right\|_{\hat{s},\Omega} \geqslant c_2$$
$$= \text{const} > 0, \qquad \frac{1}{\hat{s}} + \frac{2}{\hat{l}} = 1, \tag{2.33}$$

where $\hat{\nu} = \min(1,\nu)$ $(\min(1,\text{meas}^{-1}(\Omega)))$, $\nu = \min(1 - (\varepsilon_1 + \varepsilon_2 + \varepsilon_3)/2, c_1)$, $\Delta = n + \delta(\Sigma_1)\delta(\Sigma_3) + 1 - \delta(\Sigma_3)$, *and* c_0 *is the constant in* (2.11). *Then for all* $f \equiv F \in \tilde{H}_\lambda \equiv H_2^{0,\Sigma_1}(A;\ \Omega;\ \Sigma_3,\ \lambda)$ *the general boundary value problem of the form* (5.5.4) (*in the case* $\Sigma = \partial\Omega$, $\Sigma' = \varnothing$) *has precisely one generalized solution, and the operator* $\mathscr{L}$: $\tilde{H}_\lambda \to \tilde{H}_\lambda^*$ *corresponding to this problem is a homeomorphism.*

PROOF. Applying the Cauchy and Young inequalities, we find that for almost all $x \in \Omega$ and any $u \in \mathbb{R}$ and $q \in \mathbb{R}^n$

$$\begin{aligned} l'^i q_i + l'_0 u &= Qq \cdot q + \mathbf{a}u \cdot q + \mathbf{f} \cdot q + \gamma \cdot qu + a_0 u^2 + f_0 u \\ &\geqslant \left(1 - \frac{\varepsilon_1 + \varepsilon_2 + \varepsilon_3}{2}\right)|q|^2 + \delta(\Sigma_1)\delta(\Sigma_3)\theta a_0 u^2 \\ &\quad + c_0^2\left[(1-\theta)a_0 - \frac{1}{2\varepsilon_1}|\mathbf{a}|^2 - \frac{1}{2\varepsilon_3}|\gamma|^2 - \frac{\varepsilon_4}{2}|f_0|^{(l-2)/(l-1)}\right]u^2 \\ &\quad - \frac{1}{2\varepsilon_2}|\mathbf{f}|^2 - \frac{1}{2\varepsilon_4}|f_0|^{l/(l-1)}. \end{aligned}$$

In view of (2.33) this implies that the conditions of Proposition 2.4 are satisfied with

$$\bar{m} = 2, \qquad \kappa = 2, \qquad \psi = -(1/2\varepsilon_2)|\mathbf{f}|^2 - (1/2\varepsilon_4)|f_0|^{l/(l-1)}.$$

It follows from the proof of Proposition 2.4 that in the case where $\psi \equiv 0$ in Ω in (2.29) we have $\langle \mathscr{L}u, u\rangle \geqslant c_2\|u\|^2_{\tilde{H}_\lambda}$ for all $u \in \tilde{H}_\lambda$. In view of the linearity of the operator $\mathscr{L}\colon \tilde{H}_\lambda \to \tilde{H}^*_\lambda$ this immediately implies its strong monotonicity, since in bounding the expression $\langle \mathscr{L}(u-v), u-v\rangle$ from below we may assume that $|\mathbf{f}| \equiv |f_0| \equiv 0$ in Ω; therefore, condition (2.29) is satisfied with $\psi \equiv 0$ in Ω. Theorem 2.5 then follows from Theorem 5.4.4.

REMARK 2.1. Condition (2.33) is certainly satisfied if for almost all $x \in \Omega$

$$\delta(\Sigma_1)\delta(\Sigma_3)\beta_0(x) \geqslant \hat{c}_1 = \text{const} > 0; \qquad \beta_0(x) \geqslant |A^{-1}\alpha|^2 + |A^{-1}\beta|^2. \tag{2.34}$$

Indeed, setting $\varepsilon_1 = \varepsilon_2 = 1/2(1-\theta)$, $\theta = 1/4$ and $\varepsilon_3 = 1/3$ and taking into account that $(a+b)_+ \leqslant a_+ + b_+$, we see that condition (2.33) is satisfied if we choose $\varepsilon_+ > 0$ so small that

$$\tilde{\nu}/2 \geqslant c_0(\varepsilon_4/2)\|\,|g_0|^{(\hat{l}-2)/(\hat{l}-1)}\|_{\hat{s},\Omega},$$

where c_0 is the constant in (2.11).

We now suppose that in (2.32) the limit index $\hat{l}$ is replaced by any index $l \in [\,2, \hat{l}\,)$, and by $\mathscr{L}\colon \tilde{H}_\lambda \to \tilde{H}^*_\lambda$ we henceforth mean the operator defined by

$$\langle \mathscr{L}u, \eta\rangle = \int_\Omega [(\mathfrak{A}\nabla u + u\alpha)\cdot\nabla\eta + (\beta\cdot\nabla u + \beta_0 u)\eta]\,dx + \int_{\Sigma_3}\lambda u\eta\,ds, \qquad u, \eta \in \tilde{H}_\lambda. \tag{2.35}$$

Let $\mathscr{L}_{\tau,h}u \equiv \mathscr{L}u + \tau R_h u$, $\tau \in \mathbb{R}$, where $h = |A^{-1}\alpha|^2 + |A^{-1}\beta|^2 + |\beta_0| + \delta(\Sigma_1)\delta(\Sigma_3)$, and $R_h\colon \tilde{H}_\lambda \to \tilde{H}^*_\lambda$ is defined by

$$R_h u = \int_\Omega hu\eta\,dx, \qquad u, \eta \in \tilde{H}_\lambda.$$

We observe that R_h actually does act from $\tilde{H}_\lambda$ into $\tilde{H}^*_\lambda$; this follows from the inequality

$$\left|\int_\Omega hu\eta\,dx\right| \leqslant \|h\|_{s,\Omega}\|u\|_{l,\Omega}\|\eta\|_{l,\Omega},$$

where $u, \eta \in \tilde{H}_\lambda$ and $1/s + 2/l = 1$, and the imbeddings $\tilde{H}_\lambda \to L^l(\Omega) \to L^2(\Omega) \to \tilde{H}_\lambda^*$ following from Lemma 2.5. Moreover, from what has been said it follows that the operator R_h is compact, since in the case of a nonlimit index l the imbedding $\tilde{H}_\lambda \to L^l(\Omega)$ is compact. By Theorem 2.5 and Remark 2.1 the operator $\mathscr{L}_{1,h}$: $\tilde{H}_\lambda \to \tilde{H}_\lambda^*$ is a homeomorphism. We consider the one-parameter family of equations

$$\mathscr{L}_{\tau,h} u \equiv \mathscr{L} u + \tau R_h u = F, \qquad F \in \tilde{H}_\lambda^*. \tag{2.36}$$

Equation (2.36) is equivalent to

$$u + (\tau - 1)\mathscr{L}_{1,h}^{-1} R_h u = \mathscr{L}_{1,h}^{-1} F, \tag{2.37}$$

which has the form of an equation $u + (\tau - 1)Tu = \Phi$ with a compact operator $T \equiv \mathscr{L}_{1,h}^{-1} \circ R_h$ acting in the Hilbert space $\tilde{H}_\lambda \equiv \widetilde{H_2^{0,\Sigma_1}}(A; \Omega; \Sigma_3, \lambda)$. We denote by $\mathscr{L}_{\tau,h}^*$: $\tilde{H}_\lambda \to \tilde{H}_\lambda^*$ the operator formally adjoint to $\mathscr{L}_{\tau,h}$. The operators $\mathscr{L}_{\tau,h}$ and $\mathscr{L}_{\tau,h}^*$ are thus defined by

$$\langle \mathscr{L}_{\tau,h} u, \eta \rangle = \int_\Omega [\mathfrak{A} \nabla u \cdot \nabla \eta + \alpha \cdot \nabla \eta u + \beta \cdot \nabla u \eta + (\beta_0 + \tau h) u \eta]\, dx + \int_{\Sigma_3} \lambda u \eta \, ds \tag{2.38}$$

and

$$\langle \mathscr{L}_{\tau,h}^* u, \eta \rangle = \int_\Omega [\mathfrak{A}^* \nabla u \cdot \nabla \eta + \beta \cdot \nabla \eta u + \alpha \cdot \nabla u \eta + (\beta_0 + \tau h) u \eta]\, dx + \int_{\Sigma_3} \lambda u \eta \, ds, \tag{2.39}$$

where $u, \eta \in \tilde{H}_\lambda$. Applying well-known results of the Riesz-Schauder theory, we obtain the following assertion.

THEOREM 2.6. *Suppose that condition* (2.32) *is satisfied with $\hat{l}$ replaced by $l \in [2, \hat{l})$. Then there exists a countable isolated set $\mathfrak{M} \subset \mathbf{R}$ such that for all $\tau \notin \mathfrak{M}$ the operator $\mathscr{L}_{\tau,h}$: $\tilde{H}_\lambda \to \tilde{H}_\lambda^*$ is a bijection. For all $\tau \in \mathfrak{M}$ the kernels of $\mathscr{L}_{\tau,h}$ and $\mathscr{L}_{\tau,h}^*$ have positive, finite dimension. The range of $\mathscr{L}_{\tau,h}$ in $\tilde{H}_\lambda^*$ is the orthogonal complement of the kernel of $\mathscr{L}_{\tau,h}^*$.*

In the case $\Sigma_2 = \varnothing$ or in the case of any set Σ_2 under the additional condition that the function β_0 is bounded above in Ω we also consider the traditional one-parameter family $\mathscr{L}_\tau u \equiv \mathscr{L} u + \tau R u$, where $\mathscr{L}_\tau$: $\tilde{H}_\lambda \to \tilde{H}_\lambda^*$ is defined by

$$\langle \mathscr{L}_\tau u, \eta \rangle = \langle \mathscr{L} u, \eta \rangle + \tau \langle R u, \eta \rangle;$$

here $\langle \mathscr{L} u, \eta \rangle$ has the form (2.35), and $\langle Ru, \eta \rangle = \int_\Omega u \eta \, dx$. It is easy to see that $\mathscr{L}_\tau$ acts from $\tilde{H}_\lambda$ into $\tilde{H}_\lambda^*$ and is compact. It is also obvious that there exists a number $\tau_0 \in \mathbf{R}_+$ for which the quantity

$$\| [-\tau_0 + \beta_0 + |A^{-1}\alpha|^2 + |A^{-1}\beta|^2]_+ \|_{s,\Omega}$$

is sufficiently small; and, furthermore, in the case where the function β_0 is bounded above in Ω we have the inequality

$$\delta(\Sigma_1)\delta(\Sigma_3)(\tau_0 - \beta_0(x)) \geqslant c_1 = \text{const} > 0$$

a.e. in Ω. We further assume that either $\delta(\Sigma_1)\delta(\Sigma_3) = 0$ or β_0 is bounded above in Ω. It then follows from Theorem 2.5 that the operator $\mathscr{L}_{\tau_0}: \tilde{H}_\lambda \to \tilde{H}_\lambda^*$ is a homeomorphism, and the operator $\mathscr{L}_{\tau_0}^{-1} \circ R: \tilde{H}_\lambda \to \tilde{H}_\lambda^*$ is compact. We consider the family of equations $\mathscr{L}u + \tau Ru = F$, $F \in \tilde{H}_\lambda^*$, which can also be rewritten in the form $u + (\tau - \tau_0)\mathscr{L}_{\tau_0}^{-1}Ru = \mathscr{L}_{\tau_0}^{-1}F$. The following analogue of Theorem 2.6 then holds.

THEOREM 2.7. *Under the conditions indicated above there exists a countable isolated set* $\mathfrak{M} \subset \mathbb{R}$ *such that for each* $\tau \notin \mathfrak{M}$ *the operator* $\mathscr{L}_\tau: \tilde{H}_\lambda \to \tilde{H}_\lambda^*$ *is a bijection. For all* $\tau \in \mathfrak{M}$ *the kernels of* $\mathscr{L}_\tau: \tilde{H}_\lambda \to \tilde{H}_\lambda^*$ *and the operator* $\mathscr{L}_\tau^*: \tilde{H}_\lambda \to \tilde{H}_\lambda^*$ *formally adjoint to it, defined by* (2.38) *and* (2.39) *in the case* $h \equiv 1$ *in* Ω, *have positive finite dimension. The range of* $\mathscr{L}_\tau$ *in* $\tilde{H}_\lambda^*$ *is the orthogonal complement of the kernel of* $\mathscr{L}_\tau^*$.

§3. Existence and uniqueness of A-regular generalized solutions of the first boundary value problem for $(A,0)$-elliptic equations

In a bounded domain $\Omega \subset \mathbb{R}^n$, $n \geqslant 2$, we first consider an arbitrary $(A,0)$-elliptic equation of nondivergence form

$$\hat{a}^{ij}(x,u,\hat{\nabla}u)u_{\hat{j}\hat{i}} - \hat{a}(x,u,\hat{\nabla}u) = 0, \tag{3.1}$$

where $u_{\hat{i}}$ and $u_{\hat{i}\hat{j}}$ are respectively the derivatives of first and second orders (see §6.1) relative to the matrix $A \equiv \|a^{ij}(x)\|$, which is symmetric and nonnegative-definite in Ω, and $\hat{\nabla}u = (u_{\hat{1}},\dots,u_{\hat{n}})$.

THEOREM 3.1. *Let* $a^{ij} \in \tilde{C}^1(\overline{\Omega})$, $i,j = 1,\dots,n$, *and suppose that the functions* $\hat{a}^{ij}(x,u,q)$, $i,j = 1,\dots,n$, *and* $\hat{a}(x,u,q)$ *are continuous, differentiable with respect to the variables* u *and* q, *and have partial* A*-derivatives* $\hat{\partial}\hat{a}^{ij}/\partial x_k$ *and* $\hat{\partial}\hat{a}/\partial x_k$, $i,j,k = 1,\dots,n$, *in* $\overline{\Omega} \times \mathbb{R} \times \mathbb{R}^n$ (*see* § 6.1). *Suppose that on the set* $\overline{\Omega} \times [-m,m] \times \{|q| > \hat{L}\}$, *where* m, $\hat{L} = \text{const} \geqslant 0$, *the inequalities*

$$\hat{\mathscr{E}}_1 > 0, \quad \Lambda \leqslant \sqrt{\sigma_0\lambda\hat{\mathscr{E}}_1}, \qquad \max_{i,j=1,\dots,n} |\hat{a}^{ij}| \leqslant \mu_0\Lambda,$$

$$\left(\sum_{i,j=1}^n \left(q\hat{a}^{ij}_q\right)^2\right)^{1/2} \leqslant \sqrt{\mu_1\lambda\hat{\mathscr{E}}_1}\,|q|^{-1}, \qquad \left(\sum_{i,j=1}^n (\hat{\delta}\hat{a}^{ij})^2\right)^{1/2} \leqslant \sqrt{\sigma_1\lambda\hat{\mathscr{E}}_1},$$

$$|\hat{a} - q\hat{a}_q| \leqslant \mu_2\hat{\mathscr{E}}_1, \qquad \hat{\delta}\hat{a} \geqslant -\sigma_2\hat{\mathscr{E}}_1|q| \tag{3.2}$$

hold, where

$$\hat{\mathscr{E}}_1 \equiv \hat{a}^{ij}(x,u,q)q_iq_j, \qquad \hat{\delta} \equiv |q|(\partial/\partial u) + (q_n/|q|)(\hat{\partial}/\partial x_k),$$

and $\lambda = \lambda(x,u,q)$ *and* $\Lambda = \Lambda(x,u,q)$ *are respectively the least and greatest eigenvalues of the matrix* $\frac{1}{2}(\hat{\mathfrak{A}} + \hat{\mathfrak{A}}^*)$, $\hat{\mathfrak{A}} \equiv \|\hat{a}^{ij}(x,u,q)\|$, $\mu_0, \mu_1, \mu_2 = \text{const} \geqslant 0$, *and* σ_0, σ_1 *and* σ_2 *are nonnegative constants which are sufficiently small, depending on* n, μ_1, μ_2, *and* m. *Suppose that for the matrix* A *the following condition is satisfied: for all* $x \in \Omega$, $\xi \in \mathbb{R}^n$ *and* $\eta \in \mathbb{R}^n$

$$|\Gamma^r_{ik}\xi_r\eta_k| + |a^{jt}(\Gamma^r_{ik})_{,t}\xi_r\eta_k| + |\Gamma^r_{ik}(\partial a^{jt}/\partial x_r)\xi_t\eta_k| \leqslant \mu_3|A\xi|\,|\eta|, \quad i,j = 1,\dots,n, \tag{3.3}$$

where

$$\Gamma^r_{ij} = (\partial a^{ir}/\partial x_s)a^{js} - a^{is}(\partial a^{ir}/\partial x_s), \qquad \mu_3 = \text{const} \geqslant 0.$$

Then, for any solution $u \in C^3(\Omega) \cap C^1(\overline{\Omega})$ *of* (3.1) *such that* $|u| \leqslant m$ *in* Ω,

$$\max_{\overline{\Omega}} |\hat{\nabla} u| \leqslant \overline{\hat{M}_1},$$

where $\overline{\hat{M}_1}$ *depends only on* m, $\hat{M}_1 \equiv \max_{\partial\Omega} |\hat{\nabla} u|$, L, n, μ_0, μ_1, μ_2 *and* μ_3.

REMARK 3.1. Condition (3.3) is satisfied, in particular, if the matrix A satisfies the condition $a^{ij} \in \tilde{C}^2(\overline{\Omega})$, $i, j = 1, \ldots, n$, and

$$\Lambda_A \leqslant \mu_4 \lambda_A, \tag{3.4}$$

where $\lambda_A \equiv \lambda_A(x)$ and $\Lambda_A \equiv \Lambda_A(x)$ are respectively the least and greatest eigenvalues of A, and $\mu_4 = \text{const} \geqslant 1$.

Indeed, taking into account that $|a^{ij}| \leqslant c\Lambda_A \leqslant c\mu_4\lambda_A$, $c = \text{const} \geqslant 1$, we obtain the inequalities $|a^{ij}\xi_k| \leqslant c\mu\lambda_A|\xi| \leqslant c\mu|A\xi|$, $i, j, k = 1, \ldots, n$. Taking into account the boundedness in Ω of the functions a^{ij}, $\partial a^{ij}/\partial x_k$ and $\partial^2 a^{ij}/\partial x_k \partial x_l$, $i, j, k, l = 1, \ldots, n$, and the form of the left sides of (3.3), we then easily establish (3.3). If condition (3.4) is satisfied we say that A is uniformly degenerate in Ω. We further note that (3.3) contains the condition of differentiability of Γ^r_{ik}, $i, k, r = 1, \ldots, n$.

PROOF OF THEOREM 3.1. Applying the operator $u_{\hat{k}}\,\hat{\partial}/\partial x_k$ (see §6.1) to equation (3.1), we obtain

$$\hat{\alpha}^{ij} u_{\hat{j}\hat{i}\hat{k}} u_{\hat{k}} = \left((\partial\hat{\alpha}/\partial q_l) - \partial\hat{\alpha}^{ij} u_{\hat{j}\hat{i}}/\partial q_l\right) u_{\hat{l}\hat{k}} u_{\hat{k}} + \sqrt{\hat{v}}\left(\hat{\delta}\hat{\alpha} - \hat{\delta}\hat{\alpha}^{ij} u_{\hat{j}\hat{i}}\right), \tag{3.5}$$

where $\hat{v} \equiv \sum_1^n u_{\hat{k}}^2$. We first transform the left side of (3.5). Since

$$u_{\hat{i}\hat{j}} = a^{jr} a^{is} u_{rs} + a^{js}(\partial a^{ir}/\partial x_s) u_r,$$

it follows that

$$u_{\hat{i}\hat{j}} - u_{\hat{j}\hat{i}} = \Gamma^r_{ij} u_{\hat{r}}, \qquad \Gamma^r_{ij} \equiv (\partial a^{ir}/\partial x_s) a^{js} - a^{is}(\partial a^{jr}/\partial x_s). \tag{3.6}$$

Taking account of (3.6) and the equality

$$u_{\hat{i}r} = u_{r\hat{i}} + (\partial a^{it}/\partial x_r) u_t, \tag{3.7}$$

we find that

$$u_{\hat{j}\hat{i}\hat{k}} = u_{\hat{k}\hat{j}\hat{i}} + Y_{jik},$$

$$Y_{jik} = \Gamma^r_{jk} u_{\hat{i}\hat{r}} + \Gamma^r_{ik} u_{\hat{j}\hat{r}} + \left[a^{it}(\Gamma^r_{jk})_{,t} u_r - \Gamma^r_{jk}(\partial a^{it}/\partial x_r) u_t\right]. \tag{3.8}$$

Taking further into account that $(u_{\hat{k}}^2)_{\hat{j}\hat{i}} = 2u_{\hat{k}\hat{i}} u_{\hat{k}\hat{j}} + 2u_{\hat{k}} u_{\hat{k}\hat{j}\hat{i}}$, and hence

$$u_{\hat{k}\hat{j}\hat{i}} u_{\hat{k}} = \frac{1}{2}\left(\sum_{k=1}^n u_{\hat{k}}^2\right)_{\hat{j}\hat{i}} - u_{\hat{k}\hat{i}} u_{\hat{k}\hat{j}} = \frac{1}{2}\hat{v}_{\hat{j}\hat{i}} - u_{\hat{k}\hat{i}} u_{\hat{k}\hat{j}},$$

we find that

$$\hat{\alpha}^{ij} u_{\hat{j}\hat{i}\hat{k}} u_{\hat{k}} = \hat{\alpha}^{ij} \hat{v}_{\hat{j}\hat{i}}/2 + \hat{\alpha}^{ij} u_{\hat{k}\hat{j}} u_{\hat{k}\hat{i}} + \hat{\alpha}^{ij} Y_{jik} u_{\hat{k}}.$$

From (3.5) we then obtain

$$\begin{aligned} \hat{\alpha}^{ij} v_{\hat{j}\hat{i}}/2 &= \hat{\alpha}^{ij} u_{\hat{k}\hat{i}} u_{\hat{k}\hat{j}} - \hat{\alpha}^{ij} Y_{jik} u_{\hat{k}} \\ &\quad + \tfrac{1}{2}\left(\partial\hat{\alpha}/\partial q_l - (\partial\hat{\alpha}^{ij}/\partial q_l) u_{\hat{j}\hat{i}}\right)\hat{v}_{\hat{l}} + \sqrt{\hat{v}}\left(\hat{\delta}\hat{\alpha} - \hat{\delta}\hat{\alpha}^{ij} u_{\hat{j}\hat{i}}\right), \end{aligned} \tag{3.9}$$

in deriving which we have also taken into account that $u_{\hat{l}\hat{k}} u_{\hat{k}} = \frac{1}{2}\hat{v}_{\hat{l}}$. Let $z = z(u)$ be a positive function which is twice differentiable on the interval $[-m, m]$. We introduce the function w defined by $\hat{v} = z(u)w$. Taking into account that $\hat{v}_i = z'u_{\hat{i}}w + zw_{\hat{i}}$ and $\hat{v}_{\hat{j}\hat{i}} = z''u_{\hat{i}}u_{\hat{j}}w + z'u_{\hat{j}\hat{i}}w + z'u_{\hat{j}}w_{\hat{i}} + z'u_{\hat{i}}w_{\hat{j}} + zw_{\hat{j}\hat{i}}$, from (3.9) we derive

$$\begin{aligned} z\hat{\alpha}^{ij}w_{\hat{j}\hat{i}} + b^k w_{\hat{k}} &= -z''\hat{\mathscr{E}}_1 w + 2\alpha^{ij}u_{\hat{k}\hat{i}}u_{\hat{k}\hat{j}} - 2\hat{\alpha}^{ij}Y_{jik}u_{\hat{k}} \\ &+ z'(q_l\hat{\alpha}_{ql} - \hat{\alpha})w + 2(\hat{\delta}\hat{\alpha}/|q|)\hat{v} - z'\hat{\alpha}^{ij}_{ql}q_l u_{\hat{j}\hat{i}}w - 2\sqrt{\hat{v}}\,\hat{\delta}\hat{\alpha}^{ij}u_{\hat{j}\hat{i}}, \end{aligned} \tag{3.10}$$

where $|q|^2 \equiv \hat{v}$, $q_l = u_{\hat{l}}$, $l = 1,\dots,n$, and the form of the functions b^k is irrelevant for subsequent considerations. We henceforth consider and transform (3.10) only on the set $\hat{\Omega}_L \equiv \{x \in \Omega\colon |\hat{\nabla} u| > \hat{L}\}$, which enables us to use condition (3.2). Taking the conditions on $q\hat{\alpha}^{ij}_q$ and $\hat{\delta}\hat{\alpha}^{ij}$ in (3.2) into account, we obtain

$$\begin{aligned} |q\hat{\alpha}^{ij}_q z'u_{\hat{j}\hat{i}}w| &\leqslant \frac{1}{2}\lambda \sum_{i,j=1}^n u^2_{\hat{i}\hat{j}} + \frac{1}{2}\frac{z'^2}{z}\mu_1\hat{\mathscr{E}}_1 w, \\ |2\sqrt{\hat{v}}\,\hat{\delta}\hat{\alpha}^{ij}u_{\hat{j}\hat{i}}| &\leqslant \frac{1}{2}\lambda \sum_{i,j=1}^n u^2_{\hat{i}\hat{j}} + 2\sigma_1 z\hat{\mathscr{E}}_1 w. \end{aligned} \tag{3.11}$$

Taking the conditions on $\max_{i,j=1,\dots,n}\hat{\alpha}^{ij}$ and Λ in (3.2) and condition (3.3) into account, we obtain (assuming that $\hat{L} \geqslant 1$)

$$\begin{aligned} |2\alpha^{ij}Y_{jik}u_{\hat{k}}| &\leqslant 2\mu_0\Lambda\left[2n\mu_3\left(\sum_{i,j=1}^n u^2_{\hat{i}\hat{j}}\right)^{1/2}\hat{v}^{1/2} + 2n^2\mu_3\hat{v}\right] \\ &\leqslant \frac{1}{2}\lambda \sum_{i,j=1}^n u^2_{\hat{i}\hat{j}} + 8n^2\mu_0^2\mu_3^2 z\hat{\mathscr{E}}_1 w + \sqrt{\sigma_0}\,L^{-1}2n^2\mu_3 z\hat{\mathscr{E}}_1 w, \end{aligned} \tag{3.12}$$

where we have also take account of the fact that $\lambda \leqslant \hat{\mathscr{E}}_1|q|^{-2}$ on $\hat{\Omega}_{\hat{L}}$.

In view of (3.10)–(3.12) and the conditions on $\hat{\alpha} - q\hat{\alpha}_q$ and $\hat{\delta}\hat{\alpha}$ in (3.2) we have

$$z\hat{\alpha}^{ij}w_{\hat{j}\hat{i}} + b^k w_{\hat{k}} \geqslant \left[-z'' - (\mu_1/2)(z'^2/z) - \mu_2|z'| - \nu z\right]\hat{\mathscr{E}}_1 w, \qquad x \in \hat{\Omega}_{\hat{L}}, \tag{3.13}$$

where $\nu = 8n^2\mu_0^2\mu_3^2\sigma_0 + 4\mu_0 n^2\sqrt{\sigma_0}\,L^{-1}\mu_3 + 2(\sigma_1 + \sigma_2)$. Suppose that for $z(u)$ we take the function $z(u) = (\gamma + 1)e^{\kappa m} - e^{\kappa u}$, where $\gamma = \text{const}$ and $\kappa = \text{const} > 0$. Taking into account that $z' = -\kappa e^{\kappa u}$, $z'' = -\kappa^2 e^{\kappa u}$ and $z'^2/z \leqslant \kappa^2 e^{\kappa u}/\gamma$, we observe that the square bracket in (3.13) is bounded below by

$$s \equiv \left(\kappa^2 - \mu_1\kappa^2/2\gamma - \mu_2\kappa\right)e^{-\kappa m} - \nu(\gamma + 1)e^{\kappa m}.$$

Choosing first $\gamma = \mu_1$, then $\kappa = 4\mu_2$ and requiring further that the constants σ_0, σ_1 and σ_2 ensure for ν the inequality $\nu(\gamma + 1)e^{\kappa m} < 1/4\kappa^2 e^{-\kappa m}$, we observe that with this choice $s > 0$. Then in place of (3.13) we have $z\hat{\alpha}^{ij}w_{\hat{j}\hat{i}} + b^k w_{\hat{k}} > 0$ on $\hat{\Omega}_{\hat{L}}$, from which it obviously follows that w does not assume its maximum in $\overline{\Omega}$ on $\hat{\Omega}_{\hat{L}}$. Hence,

$$\max_{\overline{\Omega}} w \leqslant \max\left\{\max_{\partial\Omega}(\hat{v}/z),\, \hat{L}^2/z\right\},$$

and so, obviously,

$$\max_{\overline{\Omega}} \hat{v} \leqslant (\max z/\min z)\max\left\{\max_{\partial\Omega}\hat{v},\, \hat{L}^2\right\} \leqslant ((\gamma + 1)/\gamma)\left\{\max_{\partial\Omega}\hat{v},\, \hat{L}^2\right\}.$$

Hence

$$\max_{\overline{\Omega}}|\hat{\nabla} u| \leqslant ((\gamma + 1)/\gamma)^{1/2}\max\left(\hat{M}_1,\, \hat{L}\right); \qquad \hat{M}_1 \equiv \max_{\partial\Omega}|\hat{\nabla} u|. \tag{3.14}$$

Theorem 3.1 is proved.

It is easy to see that if, for example, on $\bar{\Omega}\times\{|u|\leqslant m\}\times\{|q|>\hat{L}\}$ the conditions

$$\hat{\alpha}^{ij}=\hat{\alpha}^{ji},\qquad i,j=1,\dots,n;$$

$$\hat{\nu}|q|^{m-2}\xi^2\leqslant\hat{\alpha}^{ij}(x,u,q)\xi_i\xi_j\leqslant\hat{\mu}|q|^{m-2}\xi^2,\quad \hat{\nu},\hat{\mu}=\text{const}>0,\quad m>1;$$

$$|q\hat{\alpha}^{ij}_q|\leqslant\tilde{\mu}_1|q|^{m-2},\quad |\hat{\delta}\hat{\alpha}^{ij}|\leqslant\tilde{\sigma}_1|q|^{m-1},\quad |\hat{\alpha}-q\hat{\alpha}_q|\leqslant\mu_2|q|^m,\quad \hat{\delta}\hat{\alpha}\geqslant-\tilde{\sigma}_2|q|^{m+1}$$

hold, where $\tilde{\mu}_1$ and $\tilde{\mu}_2$ are arbitrary while $\tilde{\sigma}_1$ and $\tilde{\sigma}_2$ are sufficiently small constants, then conditions (3.2) are satisfied.

As an example in connection with Theorem 3.1 we now consider equation (6.2.17). As was shown in §6.2, this equation has the solution $u=r^\lambda$, $\lambda\in(0,1)$, which does not have a bounded gradient in the region $\{|x|\leqslant 1\}$. It is easy to see that (6.2.17) has the structure of an $(A,\mathbf{0})$-elliptic equation relative to the matrix $A=rI$, where I is the identity matrix. In terms of A-derivatives the nondivergence form of this equation has the form

$$|\hat{\nabla}u|^{m-2}\left(\hat{\Delta}u+\frac{x_i}{r}u_{\hat{i}}\right)-(m-2)\lambda^{m-3}|u|^{m-3}|\hat{\nabla}u|^2$$
$$-(n\lambda+\lambda^2)u|\hat{\nabla}u|^{m-2}=0,\qquad \hat{\Delta}u\equiv\sum_{i=1}^n u_{\hat{i}\hat{i}}.\quad (3.15)$$

It is obvious that the solution $u=r^\lambda$, $\lambda\in(0,1)$, of this equation has a bounded A-gradient $A\nabla u$ relative to the matrix $A=rI$, which is in agreement with Theorem 3.1.

We distinguish a special case of Theorem 3.1 pertaining to a linear $(A,\mathbf{0})$-elliptic equation.

THEOREM 3.1′. *Suppose that the coefficients of the linear $(A,\mathbf{0})$-elliptic equation*

$$\hat{\alpha}^{ij}(x)u_{\hat{j}\hat{i}}-\hat{\beta}^i(x)u_{\hat{i}}-c(x)u-f(x)=0\qquad(3.16)$$

are continuous in $\bar{\Omega}$ and have bounded A-derivatives $\hat{\partial}\hat{\alpha}^{ij}/\partial x_k$, $\hat{\partial}\hat{\beta}^i/\partial x_k$, $\hat{\partial}c/\partial x_k$, and $\hat{\partial}f/\partial x_k$, $i,j,k=1,\dots,n$, in Ω. Let $\nu\leqslant\lambda(x)<\Lambda(x)\leqslant\mu$, where $\lambda(x)$ and $\Lambda(x)$ are respectively the least and greatest eigenvalues of the matrix $\frac{1}{2}(\hat{\mathfrak{A}}+\hat{\mathfrak{A}}^*)$, $\hat{\mathfrak{A}}\equiv\|\hat{\alpha}^{ij}(x)\|$, and ν, $\mu=\text{const}>0$. Suppose that for the matrix $A\equiv\|a^{ij}(x)\|$ relative to which differentiation is performed in (3.16) condition (3.3) is satisfied. Then for any solution $u\in C^3(\Omega)\cap C^1(\bar{\Omega})$ of (3.16) such that $|u|\leqslant m$ in Ω the estimate $\max_{\bar{\Omega}}|\hat{\nabla}u|\leqslant\hat{M}_1$ holds with a constant $\hat{M}_1$ depending only on n, m, $\hat{M}_1\equiv\max_{\partial\Omega}|\hat{\nabla}u|$, ν, μ, μ_3, and the upper bounds in Ω for the absolute values of the coefficients of (3.16) and their A-derivatives of first order.

PROOF. It is easy to see that under the conditions of Theorem 3.1′ the conditions (3.2) of Theorem 3.1 are satisfied on $\bar{\Omega}\times\{|u|\leqslant m\}\times\{|q|>\hat{L}\}$ with a sufficiently large $\hat{L}>0$ depending only on ν, μ, and the bounds in Ω for the absolute values of the coefficients of (3.16) and their A-derivatives of first order. Therefore, Theorem 3.1′ follows from Theorem 3.1.

We consider, for example, equation (6.2.20), which can also be rewritten in the form

$$\hat{\Delta}u+\frac{x_i}{\rho}u_{\hat{i}}-\left(n\lambda+\lambda^2\frac{|x|^2}{\rho^2}\right)u=0,\qquad \hat{\Delta}u\equiv\sum_{i=1}^n u_{\hat{i}\hat{i}},\ \lambda\in(0,1),\qquad(3.17)$$

where $\rho = \sqrt{x_1^2 + \cdots + x_n^2 + \varepsilon^2}$, $\varepsilon \in (0,1)$, and the derivatives in (3.17) correspond to the matrix $A = \rho I$. It was noted in §6.2 that the solutions $u = \rho^\lambda$ of (3.17) do not possess an estimate of $|\nabla u|$ in the region $\{|x| \leqslant 1\}$ which is uniform with respect to $\varepsilon \in (0,1)$. It is easy to see that these solutions have a bound for the A-gradient $A\nabla u$ (relative to the matrix $A = \rho I$) which is uniform with respect to $\varepsilon \in (0,1)$; this is in complete agreement with Theorem 3.1, all conditions of which are satisfied for (3.17) and its solution $u = \rho^\lambda$.

We shall now establish an existence theorem for a solution of problem (6.1.1) for $(A,\mathbf{0})$-elliptic equations of the form (5.1.35) on the basis of Theorems 6.2.1 and 3.1, which will be applied to regularized equations of the form (6.2.21). This existence theorem is proved in analogy to the proof of Theorem 6.2.3. Indeed, we consider auxiliary divergence $(B,\mathbf{0})$-elliptic equations of the form (6.2.27), where $B = A + \varepsilon I$, $\varepsilon \in (0,1)$, and I is the identity matrix, which have reduced coefficients of the form

$$\tilde{l}'^i(x,u,q) = \varepsilon q_i + l'^i(x,u,q), \quad \tilde{l}'_0(x,u,q) = l'_0(x,u,q) + \varepsilon\big(\partial l'^i/\partial x_i\big). \tag{3.18}$$

The nondivergence form of (6.2.27) has the form (6.2.21), where

$$\hat{\beta}^{ij} = \varepsilon\delta_i^j + \hat{\alpha}^{ij}, \qquad \hat{\alpha}^{ij} = \partial l'^i/\partial q_j, \qquad \hat{\beta} = \hat{\alpha} - \varepsilon\big(\partial a^{ki}/\partial x_i\big)q_k,$$
$$\hat{\alpha} = -\big(\partial l'^k/\partial u\big)q_k - a^{ki}(\partial/\partial x_i)l'^k - \big(\partial a^{ki}/\partial x_i\big)l'^k - f + l'_0. \tag{3.19}$$

Equations (6.2.27) can also be written in the form (6.2.22), where

$$\beta^{ij} = \partial\tilde{l}^i(x,u,p)/\partial p_j, \qquad \beta = -\big(\partial\tilde{l}^i/\partial u\big)p_i - \partial\tilde{l}^i/\partial x_i - f(x) + \tilde{l}_0(x,u,p).$$

For equations of the form (6.2.22) we consider the Dirichlet problems (6.2.23) corresponding to values $\varepsilon \in (0,1)$.

LEMMA 3.1. *Let $\Omega \in C^2$, and suppose that $a^{ij} \in \tilde{C}^1(\overline{\Omega})$, $i,j = 1,\ldots,n$, and $\hat{\alpha}^{ij}(x,u,q)$, $\hat{\alpha}(x,u,q) \in C^{(1)}(\overline{\Omega} \times \mathbb{R} \times \mathbb{R}^n)$, $i,j = 1,\ldots,n$, where the $\hat{\alpha}^{ij}$, $i,j = 1,\ldots,n$, and $\hat{\alpha}$ are the coefficients of an equation of the form* (3.1) *corresponding to the original $(A,\mathbf{0})$-elliptic equation of the form* (5.1.35) *relative to the matrix $A \equiv \|a^{ij}(x)\|$* (*see* (6.1.4)). *Assume that for equation* (3.1) *conditions* (6.2.2), (6.2.8), (6.2.9), (3.2), *and* (3.3) *are satisfied. Then for any $\varepsilon \in (0,1)$ problem* (6.2.23) *has a classical solution $u_\varepsilon \in C^2(\overline{\Omega})$, and for this solution the inequalities*

$$\max_{\overline{\Omega}}|u\varepsilon| \leqslant m, \qquad \max_{\overline{\Omega}}|A\nabla u_\varepsilon| \leqslant \overline{\hat{M}}_1, \tag{3.20}$$

hold, where the constants m and $\overline{\hat{M}}_1$ do not depend on $\varepsilon \in (0,1)$.

PROOF. Lemma 3.1 is proved in exactly the same way as Lemmas 6.2.3 and 6.2.4. We first suppose that $\Omega \in C^3$ and the coefficients of (3.1) are so smooth that any solution of (6.2.22) belonging to the class $C^2(\overline{\Omega})$ automatically belongs also to $C^3(\Omega)$. Since from conditions (6.2.2), (6.2.8), (6.2.9), (3.2), and (3.3) the validity of analogous conditions for the regularized equations follows easily (and these conditions are satisfied uniformly with respect to $\varepsilon \in (0,1)$), uniform a priori estimates of the form (3.20) hold for solutions $u_\varepsilon \in C^2(\overline{\Omega})$ of (6.2.23). Taking the structure of equations (6.2.22) into account, we may then assume that these equations are uniformly elliptic and boundedly nonlinear. Applying the theorem of Ladyzhenskaya and Ural'tseva of §1.2, we establish the existence of the required solutions of (6.2.22) for all $\varepsilon \in (0,1)$. Lemma 3.1 is proved.

THEOREM 3.2. *Suppose that an equation of the form* (5.1.35) *has the structure of an* $(A, \mathbf{0})$*-elliptic equation in a bounded domain* $\Omega \subset \mathbb{R}^n$, $n \geqslant 2$, *of class* C^2, *where* $A \equiv \|a^{ij}(x)\|$ *is a symmetric nonnegative-definite matrix in* Ω *with elements* $a^{ij} \in \tilde{C}^1(\bar{\Omega})$, $i, j = 1,\dots,n$. *Assume that the reduced coefficients* $l'^i(x, u, q)$, $i = 1,\dots,n$, $l'_0(x, u, q)$ *and the right side* $f(x)$ *of* (5.1.35) *are continuous in* $\bar{\Omega} \times \mathbb{R} \times \mathbb{R}^n$, *and the functions* $\hat{\alpha}^{ij}(x, u, q)$, $i, j = 1,\dots,n$, *and* $\hat{\alpha}(x, u, q)$ *defined by* (6.1.4) *ensure the validity of conditions* (6.2.2), (6.2.8), (6.2.9), (3.2), (3.3), *and* (6.2.25) ((6.2.25′)). *Then the Dirichlet problem of the form* (6.1.1) *for equation* (5.1.35) *has at least one* (*precisely one*) *A-regular generalized solution u, i.e., there exists a* (*unique*) *function* $u \in L^\infty(\Omega) \cap \mathring{H}_m(A, \Omega)$ *for every* $m > 1$ *having A-gradient* $A\nabla u \in \mathbf{L}^\infty(\Omega)$ *and satisfying the identity*

$$\int_\Omega [\mathbf{l}'(x, u, A\nabla u) \cdot A\nabla\eta + l'_0(x, u, A\nabla u)\eta]\, dx = \int_\Omega f\eta\, dx, \quad \forall \eta \in C_0^1(\Omega). \tag{3.21}$$

PROOF. Theorem 3.2 is proved in exactly the same way as Theorem 6.2.3, with Remark 6.2.2 taken into account.

We distinguish specially the case of linear nondivergence $(A, \mathbf{0})$-elliptic equations reducing to an equation of the form (5.1.35) under conditions of sufficient smoothness of the matrix of leading coefficients. We shall consider the general linear equation of the form (6.2.18) with nonnegative characteristic form $\alpha^{ij}(x)\xi_i\xi_j$ in Ω, assuming that $\alpha^{ij} = \alpha^{ji}$, $i, j = 1,\dots,n$, and $\alpha^{ij} \in \tilde{C}^1(\bar{\Omega})$. It is obvious that this equation automatically has the structure of a nondivergence $(A, \mathbf{b})$-elliptic equation relative to $\|a^{ij}\| \equiv A = \mathfrak{A}^{1/2}$ and $\mathbf{b} = (b^1,\dots,b^n)$, where $b^i = \partial\alpha^{ij}/\partial x_j + \beta^i$, $i = 1,\dots,n$, and in the present case $\hat{\alpha}^{ij} = \delta_i^j$ and $\hat{\alpha} = -(\partial a^{ik}/\partial x_i)q_k + cu + f$. In terms of A-derivatives this equation has the form

$$\hat{\Delta} u + (\partial a^{ik}/\partial x_i)u_{\hat{k}} - cu - f - b^i u_{x_i} = 0, \tag{3.22}$$

where $\hat{\Delta} u \equiv \sum_1^n u_{\hat{i}\hat{i}}$. (3.22) can be rewritten in the form of a divergence $(A, \mathbf{b})$-elliptic equation

$$-\operatorname{div}(A^* A\nabla u) + cu + f + b^i u_{x_i} = 0. \tag{3.23}$$

In order that (3.22), and hence also (6.2.18), be an $(A, \mathbf{0})$-elliptic (relative to $A = \mathfrak{A}^{1/2}$) equation of the form (3.1) it is obviously necessary and sufficient that the vector $\mathbf{b}$ defined above (i.e., $\mathbf{b} = (b^1,\dots,b^n)$, $b^i = \partial\alpha^{ij}/\partial x_i + \beta^i$, $i = 1,\dots,n$) be an A-vector, i.e., that there exist a vector $\boldsymbol{\gamma} = \boldsymbol{\gamma}(x)$ such that $\mathbf{b} = A\boldsymbol{\gamma} \equiv A^*\boldsymbol{\gamma}$ in Ω. This condition can be written in the form: there exists a vector $\boldsymbol{\gamma} = (\gamma'(x),\dots,\gamma^n(x))$ such that

$$\beta^i + \partial x^{ij}/\partial x_j = a^{ik}\gamma_k, \qquad i = 1,\dots,n, \tag{3.24}$$

where $\|\alpha^{ij}\| \equiv \mathfrak{A}$ and $\|a^{ij}\| \equiv A \equiv \mathfrak{A}^{1/2}$. We note that (3.24) is trivially satisfied in the case of a linear equation of the form

$$-(\partial/\partial x_i)(\alpha^{ij}(\partial u/\partial x_j)) + cu + f = 0; \qquad \alpha^{ij} = \alpha^{ji}, i, j = 1,\dots,n, \tag{3.25}$$

where $\alpha^{ij}\xi_i\xi_j \geqslant 0$ in Ω. If condition (3.24) is satisfied, then (3.23) assumes the form

$$-\operatorname{div}(A^* A\nabla u) + \gamma A\nabla u + cu + f = 0. \tag{3.26}$$

The nondivergence form of (3.26) is

$$\hat{\Delta}u + \hat{\beta}^i u_{\hat{i}} - cu - f = 0, \tag{3.27}$$

where $\hat{\beta}^i = -\gamma^i + \partial a^{ik}/\partial x_k$, $i = 1,\dots,n$. If condition (3.24) is satisfied we call the original linear equation of the form (6.2.18) $(A, \mathbf{0})$-*elliptic* (relative to $A = \mathfrak{A}^{1/2}$). The next result follows from Theorem 3.2 with Theorems 3.1 and 6.2.1 taken into account.

THEOREM 3.2′. *Suppose the linear equation of the form* (6.2.18) *is* $(A, \mathbf{0})$-*elliptic* (*in the sense indicated above*) *relative to the matrix* $A = \mathfrak{A}^{1/2}$, *where* $\mathfrak{A} \equiv \|\alpha^{ij}(x)\|$ *is symmetric and nonnegative-definite in* Ω *and* $\alpha^{ij} \in \tilde{C}^1(\overline{\Omega})$, $i, j = 1,\dots,n$. *Suppose the condition* $c(x) \geqslant c_0$ *in* Ω, *where* $c_0 = \text{const} > 0$, *and also condition* (6.2.9) *for the matrix* $A \equiv \mathfrak{A}^{1/2}$ *and* $\partial\Omega$, *are satisfied. Assume that the coefficients of equations* (3.26) *and* (3.27) *generated by* (6.2.18) *are continuous in* $\overline{\Omega}$ *and have bounded A-derivatives* $\hat{\partial}\beta^i/\partial x_k$, $\hat{\partial}c/\partial x_k$ *and* $\hat{\partial}f/\partial x_k$, $i, j, k = 1,\dots,n$, *in* Ω. *Suppose that condition* (3.3) *is satisfied for the matrix A. Then the Dirichlet problem of the form* (6.1.1) *for equation* (6.2.18) *has at least one A-regular generalized solution, i.e., there exists a function* $u \in L^\infty(\Omega) \cap \overset{\circ}{H}_m(A, \Omega)$ *for every* $m > 1$ *such that* $A\nabla u \in \mathbf{L}^\infty(\Omega)$ *and*

$$\int_\Omega [A\nabla u \cdot A\nabla\eta + (\boldsymbol{\gamma} \cdot A\nabla u + cu + f)\eta]\, dx = 0, \quad \forall \eta \in C_0^1(\Omega). \tag{3.28}$$

If for all $x \in \Omega$, $\eta = A\xi$, $\xi \in \mathbb{R}^n$ *and* $\xi_0 \in \mathbb{R}$, $(\xi_0, \eta) \neq (0, 0)$,

$$\sum_{i=1}^n \eta_i^2 + \sum_{i,j=1}^n \gamma_j \eta_i \xi_0 + c\xi_0^2 > 0 \tag{3.29}$$

then problem (6.1.1) *for equation* (6.2.18) *has precisely one A-regular generalized solution.*

PROOF. Theorem 3.2′ is proved in exactly the same way as Theorem 3.2 with the facts taken into account that conditions of the form (6.2.8) are certainly satisfied for equation (3.27) and that it is not necessary to assume a monotonicity condition of the form (6.2.25) to pass to the limits in the integral identities (6.2.29)–(6.2.31), in view of the linearity of the equation. Theorem 3.2′ is proved.

We consider the variational problem of a minimum for an integral of the form (6.2.35) under the condition (6.2.36). The Euler equation for this problem, which has the form (6.2.37), is an $(A, \mathbf{0})$-elliptic equation in Ω (see also (6.2.38)). The next result obviously follows from Theorem 3.2.

THEOREM 3.2″. *Suppose that the integral* (6.2.35) *is considered under the assumptions that the domain* Ω *is bounded in* $\mathbb{R}^n$, $n \geqslant 2$, *and belongs to the class* C^2, *the matrix* $A \equiv \|a^{ij}(x)\|$ *is symmetric and nonnegative-definite in* Ω *with* $a^{ij} \in \tilde{C}^1(\overline{\Omega})$, $i, j = 1,\dots,n$, *and the function* $F(x, u, q)$ *satisfies condition* (6.2.36). *Assume that the functions* l'^i, $i = 1,\dots,n$, *and* l'_0 *defined by* (6.2.38), *the function* $f(x)$ *in* (6.2.35) *and the functions* $\hat{\alpha}^{ij}$, $i, j = 1,\dots,n$, *and* $\hat{\alpha}$ *defined by* (6.1.4) *and* (6.2.38) *ensure the validity of conditions* (6.2.2), (6.2.8), (6.2.9), (3.2), (3.3), *and* (6.2.25) ((6.2.25′)). *Then there exists at least one* (*precisely one*) *extremal of problem* (6.2.35), *i.e., there exists a* (*unique*) *function* $u \in L^\infty(\Omega \cap \overset{\circ}{H}_m(A, \Omega))$ *for every* $m > 1$ *such that* $A\nabla u \in \mathbf{L}^\infty(\Omega)$ *for which the identity* (3.21) *is satisfied* (*with* l'^i, $i = 1,\dots,n$, *and* l'_0 *defined by* (6.2.38)).

Examples of more concrete functionals of the form (6.2.35) for which the result of Theorem 3.2 holds will be presented below.

In establishing the existence of an A-regular generalized solution of the first boundary value problem for $(A, \mathbf{0})$-elliptic equations it is possible to relinquish condition (6.2.9). Elimination of this condition leads to consideration of a first boundary value problem of the form

$$\mathscr{L}u = f(x) \quad \text{in } \Omega, \qquad u = 0 \quad \text{on } \Sigma \equiv \{x \in \partial\Omega : A\nu \neq \mathbf{0}\}. \tag{3.30}$$

In considering a problem of the form (3.30), we assume that (5.1.35) has the structure of an $(A, \mathbf{0})$-elliptic equation relative to a matrix A satisfying condition (3.4). Condition (3.4) enables us to reduce consideration of problem (3.30) to the case of (5.1.35), which has the structure of an $(aI, \mathbf{0})$-elliptic equation in Ω, where $a = a(x) = (1/n)\operatorname{Tr} A$. Indeed, any equation of the form (5.1.35) having the structure of an $(A, \mathbf{0})$-elliptic equation with reduced coefficients $l'^i(x, u, q)$, $i = 1, \dots, n$, and $l'_0(x, u, q)$ also has the structure of an $(aI, \mathbf{0})$-elliptic equation with reduced coefficients

$$\tilde{\mathbf{l}}'(x, u, \mathbf{q}) = (A^*/a)\mathbf{l}'(x, u, A\tilde{q}/a), \qquad \tilde{l}'_0(x, u, q) = l'_0(x, u, A\tilde{q}/a). \tag{3.31}$$

In view of (3.4) the conditions on the reduced coefficients $\tilde{\mathbf{l}}'(x, u, \tilde{q})$ and $\tilde{l}'_0(x, u, \tilde{q})$ which are imposed below to ensure the solvability of problem (3.30) in the class of (aI)-regular generalized solutions can easily be rewritten in terms of the original reduced coefficients $\mathbf{l}'(a, u, q)$ and $l'_0(x, u, q)$, and the resulting solution of problem (3.30) can be interpreted as an A-regular generalized solution relative to the original matrix A. In view of what has been said we shall henceforth simply assume that $A = a(x)I$, where $a(x) \geqslant 0$ in Ω, in order to shorten the exposition. We note that condition (3.3) is trivially satisfied for the matrix $A = a(x)I$, since in this case $\lambda_A \equiv \Lambda_A \equiv a(x)$ in Ω.

THEOREM 3.3. *Suppose* (5.1.35) *has the structure of an* $(aI, \mathbf{0})$*-elliptic equation, where* $a \in \check{C}^1(\overline{\Omega}) \cap \operatorname{Lip}(\mathbf{R}^n)$ *and* $\inf_{\mathbf{R}^n} a(x) > 0$. *Suppose that the functions* $\hat{a}^{ij}(x, u, q)$, $i, j = 1, \dots, n$, *and* $\hat{a}(x, u, q)$ *defined by* (6.1.4) *belong to the class* $C^{(1)}(\overline{\Omega} \times \mathbf{R} \times \mathbf{R}^n)$. *Suppose for equation* (3.1) *generated by the above equation* (5.1.35) *by means of equalities* (6.1.4) *conditions* (3.2) *are satisfied as well as the following condition: on the set* $D_\delta \times \{|u| \leqslant m\} \times \mathbf{R}^n$ *(see* (6.2.3)) *the inequalities*

$$|\hat{a}^{ij}| \leqslant \mu_5 \lambda(\varepsilon_0 |q| + 1), \quad i, j = 1, \dots, n; \qquad |\hat{a}| \leqslant \mu_6 \lambda(\varepsilon_0 |q|^2 + 1), \tag{3.32}$$

hold, where $\lambda = \lambda(x, u, q) > 0$ *is the least eigenvalue of the matrix* $\frac{1}{2}(\hat{\mathfrak{A}} + \hat{\mathfrak{A}}^*)$, $\hat{\mathfrak{A}} \equiv \|\hat{a}^{ij}(x, u, q)\|$, $\mu_5, \mu_6 = \text{const} \geqslant 0$, *and* ε_0 *is a sufficiently small constant.*(¹) *Then, for any function* $u \in C^2(D_\delta) \cap C^1(\overline{D_\delta})$ *such that* $u = 0$ *on* Σ *and* $|u| \leqslant m$ *in* D_δ,

$$\max_{\partial\Omega} |A \nabla u| \leqslant \hat{M}_1 \tag{3.33}$$

with a constant $\hat{M}_1$ *depending only on* n *and* $m \equiv \max_{\overline{\Omega}} |u|$, *on the known quantities determined by conditions* (3.2) *and* (3.32), *and on the* C^2*-norms of the functions describing the boundary of* Ω.

(¹) The nature of the smallness of ε_0 will be specified in the proof.

PROOF. The function u satisfies (3.1) and (6.1.5), whose coefficients in the present case are connected by the relations

$$\alpha^{ij}(x, u, p) = a^2(x)\hat{\alpha}^{ij}(x, u, q), \qquad i, j = 1, \dots, n,$$
$$\alpha(x, u, p) = \hat{\alpha}(x, u, q) - \hat{\alpha}^{ij}(x, u, q)(\partial a(x)/\partial x_i)q_j, \quad q = a(x)p. \tag{3.34}$$

We remark that it follows from (3.32), in particular, that the matrix $\hat{\mathfrak{A}}(x, u, q)$ is nondegenerate on $D_\delta \times \{|u| \leqslant m\} \times \mathbb{R}^n$. Let $x_0 \in \partial\Omega$ be a point at which the maximum of $a|\partial u/\partial \nu|$ on $\partial\Omega$ is realized, where ν is the unit vector of the inner normal to $\partial\Omega$. We set $a_0 = a(x_0)$. We introduce the following sets:

$$F_i \equiv \{x \in \mathbb{R}^n: a_0/2^{i+1} \leqslant a(x) \leqslant 2^{i+1}a_0\}, \qquad i = 0, 1, 2. \tag{3.35}$$

We denote by $\zeta = \zeta(x)$ the average with a fixed smooth kernel of the characteristic function of the set F_1, assuming that the radius h of averaging is

$$h = \min\{\operatorname{dist}(F_0, \mathbb{R}^n \setminus F_1), \operatorname{dist}(F_1, \mathbb{R}^n \setminus F_2)\}. \tag{3.36}$$

It is easy to see (see also [117]) that

$$h^{-1} \leqslant cKa_0^{-1}, \tag{3.37}$$

where K is the Lipschitz constant of the function $a(x)$ in $\mathbb{R}^n$ and $c > 0$ is an absolute constant. Indeed, let $x_1 \in F_0$ and $x_2 \in \mathbb{R}^n \setminus F_1$. Then

$$a_0/4 \leqslant |a(x_2) - a(x_1)| \leqslant K|x_2 - x_1|,$$

whence

$$\operatorname{dist}(F_0, \mathbb{R}^n \setminus F_1) = \min_{x_1 \in F_0,\, x_2 \in \mathbb{R}^n \setminus F_1} |x_2 - x_1| \geqslant \left(\frac{1}{4}K\right)a_0.$$

In a completely analogous way we find that

$$\operatorname{dist}(F_1, \mathbb{R}^n \setminus F_2) \geqslant (1/8K)a_0.$$

Inequality (3.37) with $c = 8$ follows from what has been proved. In view of (3.36) and (3.37) we have $\zeta(x) = 1$ on the set F_0, $\zeta(x) = 0$ on $\mathbb{R}^n \setminus F_2$. and there exists a constant c_0 depending only on K such that

$$|\nabla\zeta| \leqslant c_0 a_0^{-1}, \qquad |D^2\zeta| \leqslant c_0 a_0^{-2}. \tag{3.38}$$

We set $v = u\zeta$. Taking into account that u satisfies an equation of the form (6.1.5), we find for v the identity

$$\alpha^{ij}v_{ij} - 2\alpha^{ij}u_i\zeta_j - \alpha^{ij}u\zeta_{ij} - \alpha\zeta = 0, \qquad x \in \Omega. \tag{3.39}$$

Making the change of variables $\tilde{x} = a_0^{-1}x$ and taking (3.34) into account, we reduce (3.39) to

$$\hat{\alpha}^{ij}v_{\tilde{i}\tilde{j}} - 2\hat{\alpha}^{ij}(a_0/a)(au_i)(a_0\zeta_j) - \hat{\alpha}^{ij}u\left(a_0^2\zeta_{ij}\right) - (a_0/a)^2\alpha\zeta = 0, \quad \tilde{x} \in \tilde{\Omega}, \tag{3.40}$$

where $\tilde{\Omega}$ is the image of Ω under this coordinate transformation. We note that (3.40) is nontrivial only on the set $\tilde{F}_2 \equiv \{\tilde{x} \in \mathbb{R}^n: 1/8 \leqslant a(x)/a_0 \leqslant 8\} \cap \tilde{\Omega}$. Taking the conditions of Theorem 3.3, the estimates (3.38), and Theorem 3.1 into account, we conclude that for the function v and (3.40), considered as an equation of the form $A^{ij}(\tilde{x})v_{\tilde{x}_i\tilde{x}_j} - A(\tilde{x}) = 0$, where

$$\tilde{x} \in \tilde{\Omega}, \qquad A^{ij}(\tilde{x}) = \alpha^{ij}(x(\tilde{x}), u, A\nabla u), \qquad i, j = 1, \dots, n,$$
$$A(\tilde{x}) = 2\hat{\alpha}^{ij}(a_0/a)A_i\nabla u(a_0\zeta_j) + \hat{\alpha}^{ij}u\left(a_0^2\zeta_{ij}\right) + (a_0/a)^2\alpha\zeta,$$

all the conditions of Theorem 1.4.1 are satisfied; in particular, the inequality of the form (1.4.1) is satisfied with $l = 0$, $\delta(t) \equiv 0$, and $\psi(t) \equiv c_1(1 + \varepsilon_0 \hat{M}_1^2)t^{-2}$, where c_1 depends only on m and the known quantities determined by (3.1), (3.32), and (3.14) with $\hat{M}_1 \equiv \max_{\partial\Omega} a|\nabla u|$. From Theorem 1.4.1 (with Remark 1.5.1 taken into account) we then obtain the estimate $|\partial v(\tilde{x}_0)/\partial\tilde{\nu}| \leqslant \beta$, or

$$|a_0(\partial v(x_0)/\partial\nu)| \leqslant \beta, \tag{3.41}$$

where β is determined from (1.3.7) with

$$\Phi(t) = K/t^2 + 2\psi(t)/t$$

and a number $\bar{\alpha}$ depending only on known quantities. Taking into account that $\zeta(x_0) = 1$, from (3.41) and (3.32) we obtain

$$|a_0(\partial u(x_0)/\partial\nu)| \leqslant \beta + c_2 m.$$

Since the derivatives in tangential directions are equal to 0 at the point $x_0 \in \partial\Omega$, this implies that

$$\hat{M}_1 \equiv \max_{\partial\Omega} |a\nabla u| \leqslant \beta + c_2 m. \tag{3.42}$$

From (1.3.7) and the form of $\psi(t)$ we easily deduce the inequality $\beta \leqslant c_3\sqrt{\varepsilon_0}\,\hat{M}_1 + c_4$, where the constants c_3 and c_4 depend only on known quantities. Let ε_0 be chosen so that $1 - c_3\sqrt{\varepsilon_0} = 1/2$. The estimate (3.33) then follows from (3.42). Theorem 3.3 is proved.

In formulating below the theorem on the solvability of problem (3.30) for an $(A, \mathbf{0})$-elliptic equation of the form (5.1.35) under the condition (3.4) of uniform degeneracy of the matrix A, we take into account the possibility of qualifying such an equation also as an $(aI, \mathbf{0})$-elliptic equation relative to $a = (1/n)\operatorname{Tr} A$ (see (3.31)).

THEOREM 3.4. *Suppose that an equation of the form* (5.1.35) *has the structure of an* $(A, \mathbf{0})$*-elliptic equation in a bounded domain* $\Omega \subset \mathbb{R}^n$, $n \geqslant 2$, *of class* C^2 *relative to a matrix* $A = a(x)I$, $a(x) \geqslant 0$ *in* $\mathbb{R}^n$, $a \in \tilde{C}^1(\bar{\Omega}) \cap \operatorname{Lip}(\mathbb{R}^n)$. *Suppose that the reduced coefficients* $l''(x, u, q)$, $i = 1, \dots, n$, *and* $l_0'(x, u, q)$ *and the right side* $f(x)$ *of* (5.1.35) *are continuous in* $\bar{\Omega} \times \mathbb{R} \times \mathbb{R}^n$ *and the functions* $\hat{\alpha}^{ij}(x, u, q)$, $i, j = 1, \dots, n$, *and* $\hat{\alpha}(x, u, q)$ *defined by* (6.1.4) *ensure the validity of conditions* (6.2.2), (3.2), (3.32), *and* (6.2.25) ((6.2.25′)). *Then the problem of the form* (3.30) *has at least one* (*precisely one*) *A-regular generalized solution, i.e., there exists a function* $u \in L^\infty(\Omega) \cap H_m^{0,\Sigma}(A, \Omega)$, $\forall m > 1$, *such that* $A\nabla u \in L^\infty(\Omega)$ *and*

$$\int_\Omega [\mathbf{l}'(x, u, A\nabla u) \cdot A\nabla\eta + l_0'(x, u, A\nabla u)\eta]\, dx = \int_\Omega f\eta\, dx,$$

$$\forall \eta \in \tilde{C}^1_{0,\Sigma}(\bar{\Omega}). \tag{3.43}$$

PROOF. Theorem 3.4 is proved by means of Theorems 3.2 and 3.3 in exactly the same way as Theorem 6.2.3 was proved by means of Theorems 6.2.1 and 6.2.2 (see also Remark 6.2.2).

As an example related to Theorems 3.2 and 3.4, we consider an equation of the form (1.7) with $\bar{m} \geqslant 2$ and $m \geqslant 2$. The next assertion, in particular, follows from Theorems 3.2 and 3.4.

THEOREM 3.5. *Suppose that the matrix* $A \equiv \|a^{ij}(x)\|$ *is symmetric and nonnegative-definite in a bounded domain* $\Omega \subset \mathbb{R}^n$, $n \geqslant 2$, *of class* C^2, *where* $a^{ij} \in \tilde{C}^1(\overline{\Omega})$, $i, j = 1, \dots, n$, *and suppose that* $f(x)$ *is bounded and has bounded A-derivatives* $\hat{\partial} f(x)/\partial x_k$, $k = 1, \dots, n$, *in* Ω. *If conditions* (3.3) *and* (6.2.9) *are satisfied, then the problem of the form* (6.1.1) *for equation* (1.7) *has precisely one A-regular generalized solution, i.e., there exists a unique function* $u \in L^\infty(\Omega) \cap \overset{\circ}{H}_m(A, \Omega)$ *for every* $m > 1$ *such that* $A\nabla u \in \mathbf{L}^\infty(\Omega)$ *and*

$$\int_\Omega \left[|A\nabla u|^{\overline{m}-2} A\nabla u \cdot A\nabla\eta + |u|^{m-2} u\eta \right] dx = \int_\Omega f\eta\, dx, \quad \forall \eta \in \tilde{C}_0^1(\Omega). \tag{3.44}$$

If the condition $A = a(x)I$, $a(x) \in \tilde{C}^1(\overline{\Omega}) \cap \mathrm{Lip}(\mathbb{R}^n)$, $a(x) \geqslant 0$ *in* $\mathbb{R}^n$, *holds and also under the same assumptions regarding the right side* $f(x)$ *as in the first part of the theorem, the problem of the form* (3.30) *has precisely one A-regular generalized solution, i.e., there exists a unique function* $u \in L^\infty(\Omega) \cap \overset{\circ}{H}_m(A, \Omega)$ *for every* $m > 1$ *such that* $A\nabla u \in \mathbf{L}^\infty(\Omega)$ *and* (3.44) *holds for all* $\eta \in \tilde{C}^1_{0,\Sigma}(\overline{\Omega})$, *where* $\Sigma \equiv \{x \in \partial\Omega\colon A\mathbf{\nu} \neq \mathbf{0}\}$.

PROOF. Since (6.2.25′) is satisfied, Theorem 3.5 follows directly from Theorems 3.2 and 3.4 and the fact that the results of Theorems 3.2 and 3.4 are obviously preserved if the right side $f(x)$ is bounded in Ω together with its A-derivatives $\hat{\partial} f(x)/\partial x_k$, $k = 1, \dots, n$.

The existence of A-regular generalized solutions of problem (3.30) possessing bounded A-derivatives of second order in Ω can be established in an analogous way. For brevity we assume forthwith that $A = a(x)I$, where $a(x) \geqslant 0$ in Ω, since the more general case determined by condition (3.4) can easily be reduced to this. Moreover, it should be noted that at the stage of obtaining an a priori estimate of $\max_{\overline{\Omega}} |D^2 u|$ in terms of $\max_{\partial\Omega} |D^2 u|$ we admit a larger class of matrices (here there are conditions of approximately the same character as condition (3.3) on the matrix A).

THEOREM 3.6. *Suppose that* (5.1.35) *has the structure of an* $(A, \mathbf{0})$*-elliptic equation in a bounded domain* $\Omega \subset \mathbb{R}^n$, $n \geqslant 2$, *of class* C^2 *relative to the matrix* $A = a(x)I$, *where*

$$a(x) \in C^2(\overline{\Omega}) \cap \mathrm{Lip}(\mathbb{R}^n), \qquad \inf_{\mathbb{R}^n} a(x) \geqslant 0. \tag{3.45}$$

Suppose that the reduced coefficients $l'^i(x, u, q)$, $i = 1, \dots, n$, *and* $l'_0(x, u, q)$ *and the right side* $f(x)$ *of* (5.1.35) *are continuous in* $\Omega \times \mathbb{R} \times \mathbb{R}^n$ *and the functions* $\hat{\alpha}^{ij}(x, u, q)$, $i, j = 1, \dots, n$, *and* $\hat{\alpha}(x, u, q)$ *defined by* (6.1.4) *ensure that conditions* (6.2.2), (3.2), (3.32) *and the following condition are satisfied: on the set* $\overline{\Omega} \times [-m, m] \times \{|q| \leqslant \hat{M}_1\}$, *where the constants* m *and* $\hat{M}_1$ *are determined respectively by conditions* (6.2.2) *and* (3.2), (3.32), *the inequalities*

$$\hat{\alpha}^{ij}(x, u, q)\eta_i\eta_j \geqslant \nu|\eta|^2, \qquad \eta = A\xi,\ \xi \in \mathbb{R}^n,\ \nu = \text{const} > 0,$$

$$\left| \frac{\partial \hat{\alpha}^{ij}(x, u, q)}{\partial q_s} \right|, \left| \frac{\partial^2 \hat{\alpha}^{ij}(x, u, q)}{\partial q_s \partial q_t} \right| \leqslant \sigma_1, \qquad i, j, s, t = 1, \dots, n, \tag{3.46}$$

hold, where σ_1 *is a nonnegative constant which is sufficiently small, depending on* $\hat{M}_1$ *and* ν. *Assume also that for all* $x \in \overline{\Omega}$, $u \in [-2m, m]$, $q \in \mathbb{R}^n$, $|q| \leqslant 2\hat{M}_1$, $\eta \in \mathbb{R}^n$ *and* $\xi_0 \in \mathbb{R}$, *where* m *and* $\hat{M}_1$ *are the same constants as above, inequality* (6.2.25) ((6.2.25′))

holds. Then problem (3.30) *has at least one* (*precisely one*) *A-regular generalized solution possessing bounded A-derivatives of second order in* Ω. *This solution satisfies the corresponding equation* (3.1) *a.e. in* Ω.

PROOF. We first estimate $\max_{\overline{\Omega}}|D^2u|$ in terms of $\max_{\partial\Omega}|D^2u|$. Applying to equation (3.1) corresponding to the $(A, \mathbf{0})$-elliptic equation (5.1.35) under consideration the operator

$$u_{\hat{k}\hat{l}}\frac{\hat{\partial}^2}{\partial x_k \partial x_l} \equiv \sum_{k,l=1}^{n} u_{\hat{k}\hat{l}}\frac{\hat{\partial}}{\partial x_l}\left(\frac{\hat{\partial}}{\partial x_k}\right),$$

introducing the notation $v \equiv \sum_{k,l=1}^{n} u_{\hat{k}\hat{l}}^2$, and arguing as in the proof of Theorems 6.3.1 and 3.3, we establish the estimate

$$\frac{1}{2}\hat{\alpha}^{ij}v_{\hat{j}\hat{i}} \geqslant \frac{\nu}{2}|\hat{D}^3u|^2 - c_1\sigma_1\left(v^2 + \sqrt{v}\,|\hat{\nabla}v|\right) - c_2\left(v^{3/2} + |\hat{\nabla}v|\right), \qquad x \in \Omega_L, \quad (3.47)$$

where $\Omega_L \equiv \{x \in \Omega: |\hat{D}^2u| > L\}$ and c_1 and c_2 are constants depending only on known quantities. Making the substitution $v = zw \equiv (1 + \overline{\hat{M}}_1^2 - \sum_1^n u_{\hat{k}}^2)w$ in (3.47), where $\hat{M}_1 \equiv \max_{\overline{\Omega}}|\hat{\nabla}u|$, we find that w cannot assume its maximum in $\overline{\Omega}$ in Ω_L, whence we easily obtain

$$\max_{\overline{\Omega}}|\hat{D}^2u| \leqslant \left(1 + \overline{\hat{M}}_1^2\right)^{1/2}\max\left\{\max_{\partial\Omega}|\hat{D}^2u|, L\right\}, \qquad (3.48)$$

where $L = L(n, \overline{\hat{M}}_1, \nu, c_1, c_2)$. Having obtained the qualified estimate (3.48), we proceed to estimate $\max_{\partial\Omega}|\hat{D}^2u|$. This estimate is obtained by means of the same arguments as in the proof of Theorems 6.3.2 and 3.3. We thus establish the a priori estimate $\max_{\overline{\Omega}}|D^2u| \leqslant c$ with a constant c depending only on known quantities. The remainder of the proof of Theorem 3.6 is altogether analogous to the proof of Theorems 3.2 and 6.3.3.

We illustrate Theorems 3.4 and 3.6 with the example of a linear equation of the form (6.2.18) having a nonegative characteristic form $\alpha^{ij}(x)\xi_i\xi_j$ in Ω. Suppose that (6.2.18) is $(A, \mathbf{0})$-elliptic in Ω relative to the matrix $A = \mathfrak{A}^{1/2}$, i.e., condition (3.24) holds for this equation. We assume that A is uniformly degenerate in Ω, i.e., condition (3.4) holds. Equation (6.2.18) can then be qualified as an (aI)-elliptic equation in Ω relative to $a = (\operatorname{Tr} A)/n$, while its factorization can be realized in a somewhat different way than in the general situation described above (see (3.26) and (3.27)). It is obvious that under these assumptions (6.2.18) can always be written in the form

$$a^2\hat{\alpha}^{ij}(x)u_{ij} - a\hat{\gamma}^i(x)u_i - c(x)u - f(x) = 0, \qquad (3.49)$$

where

$$\hat{\alpha}^{ij} = \frac{\alpha^{ij}}{a^2}, \qquad i, j = 1, \ldots, n,$$

$$\hat{\gamma}^i = \frac{a^{ik}\gamma^k}{a} - 2\frac{\alpha^{ij}(\partial a/\partial x_j)}{a^2}, \qquad i = 1, \ldots, n,$$

$$a = \frac{\operatorname{Tr} A}{n} \geqslant 0 \quad \text{in } \Omega,$$

and the coefficients γ^k, $k = 1,\dots,n$, are determined by (3.24). It is obvious that $\hat{\alpha}^{ij}\xi_i\xi_j \geqslant c_0\xi^2$ for all $\xi \in \mathbb{R}^n$ and the constant $c_0 > 0$ depends only on n and the constant μ_4 in the condition for uniform degeneracy of A. In terms of (aI)-derivatives (6.2.18) has the form

$$\hat{\alpha}^{ij}u_{\hat{j}\hat{i}} - \hat{\beta}^i u_{\hat{i}} - cu - f = 0, \tag{3.50}$$

where $\hat{\beta}^i = -\hat{\alpha}^{ij}(\partial a/\partial x_u) + \hat{\gamma}^i$, $i = 1,\dots,n$, and its divergence form is

$$-\operatorname{div}\{(aI)\hat{\mathfrak{A}}(a\nabla u)\} + \hat{\alpha}^i u_{\hat{i}} + cu + f = 0, \tag{3.51}$$

where $\hat{\alpha}^i = \hat{\beta}^i + a(\partial\hat{\alpha}^{ij}/\partial x_j)$, $i = 1,\dots,n$. Taking the linearity of equation (3.51) into account, from Theorems 3.4 and 3.6 we obviously obtain the following result.

THEOREM 3.7. *Let Ω be a bounded domain in $\mathbb{R}^n$, $n \geqslant 2$, of class C^2, and suppose that condition* (3.45) *is satisfied. Assume that the functions $\hat{\alpha}^{ij}(x)$, $\hat{\beta}^i(x)$, $c(x)$, and $f(x)$ are continuous in $\overline{\Omega}$ and have bounded (aI)-derivatives of first (and second) order in Ω. Suppose, for all $x \in \overline{\Omega}$, $\eta = (aI)\xi$ and $\xi \in \mathbb{R}^n$, that $\nu\eta^2 \leqslant \hat{\alpha}^{ij}(x)\eta_i\eta_j \leqslant \mu\eta^2$, where ν, $\mu = \mathrm{const} > 0$, and that $c(x) \geqslant c_0 = \mathrm{const} > 0$ in Ω. Then the problem*

$$\hat{\alpha}^{ij}u_{\hat{j}\hat{i}} - \hat{\beta}^i u_{\hat{i}} - cu - f = 0 \quad \text{in } \Omega,$$
$$u = 0 \quad \text{on } \Sigma \equiv \{x \in \partial\Omega : a(x) > 0\}, \tag{3.52}$$

where $u_{\hat{i}}$ and $u_{\hat{j}\hat{i}}$ are the A-derivatives relative to the matrix $A = a(x)I$, has at least one (aI)-regular generalized solution u (possessing bounded (aI)-derivatives of second order in Ω), i.e., there exists a function $u \in L^\infty(\Omega) \cap \mathring{H}_m(aI, \Omega)$ for every $m > 1$ such that $a\nabla u \in \mathbf{L}^\infty(\Omega)$ $(a^2\mathscr{D}^2 u \in L^\infty(\Omega))$ and

$$\int_\Omega [\hat{\mathfrak{A}}\hat{\nabla} u \cdot \hat{\nabla}\eta + \boldsymbol{\alpha}\hat{\nabla} u\eta + cu\eta + f\eta]\, dx = 0, \quad \forall \eta \in \tilde{C}^1_{0,\Sigma}(\overline{\Omega}), \tag{3.53}$$

where $\hat{\mathfrak{A}} \equiv \|\hat{\alpha}^{ij}(x)\|$ and $\boldsymbol{\alpha} \equiv (\alpha^1,\dots,\alpha^n)$ (and the function u satisfies (3.50) *a.e. in Ω). Under the additional condition*

$$\hat{\alpha}^{ij}\eta_i\eta_j + \hat{\alpha}^j\eta_i\xi_0 + c\xi_0^2 > 0,$$
$$x \in \overline{\Omega}, \quad \eta = (aI)\xi, \quad \xi \in \mathbb{R}^n, \quad \xi_0 \in \mathbb{R}, \quad (\xi_0, \eta) \neq (0,0), \tag{3.54}$$

problem (3.52) *has precisely one (aI)-regular generalized solution.*

We note that the solvability of the first boundary value problem for a linear degenerate equation of the form (3.49) in the weighted space $W_p^2(a(x), \Omega)$, where $a(x) \geqslant 0$ in Ω and $1 < p < +\infty$, was established in [117]. This result does not follow from our Theorem 3.7, nor does Theorem 3.7 follow from the results of [117].

CHAPTER 8
$(A, 0)$-PARABOLIC EQUATIONS

§1. The basic function spaces connected with the general boundary value problem for $(A, 0, m, \mathbf{m})$-parabolic equations

In this chapter the domain in which the boundary value problems are considered is always the cylinder $Q \equiv \Omega \times (T_1, T_2)$, where Ω is a bounded domain in $\mathbb{R}^n$, $n \geqslant 1$, and T_1 and T_2 are fixed numbers. In the case $T_1 = 0$, $T_2 = T$ the corresponding cylinder is denoted by $Q_T \equiv \Omega \times (0, T)$. The variables $x_1, \ldots, x_n$ are called space variables, and t is called the time.

Let $A = \|a^{ij}(t, x)\|$ be a square matrix of order n satisfying the condition

$$a^{ij} \in L^{m_i, m_{0i}}(Q), \quad i, j = 1, \ldots, n, \quad m_i \geqslant 1, \quad m_{0i} \geqslant 1, \quad i = 1, \ldots, n. \tag{1.1}$$

To each function $u \in \tilde{C}^1_{\text{loc}}(Q) \subset L^{m, m_0}_{\text{loc}}(Q)$ we assign the vector $A\nabla u \equiv (A_1\nabla u, \ldots, A_n \nabla u)$, where $A_i \nabla u \equiv a^{ij}(t, x)\partial u/\partial x_j$, $i = 1, \ldots, n$, considered as an element of $\mathbf{L}^{\mathbf{m}, \mathbf{m}_0}_{\text{loc}}(Q)$. This mapping we call the operator of taking the spatial A-gradient in Q. We assume that the following condition is satisfied:

$$\text{the operator of taking the spatial } A\text{-gradient admits weak closure.} \tag{1.2}$$

If condition (1.2) is satisfied we say that the function $u \in L^{m, m_0}_{\text{loc}}(Q)$ has a *generalized spatial A-gradient* $A\nabla u \in \mathbf{L}^{\mathbf{m}, \mathbf{m}_0}_{\text{loc}}(Q)$ in Q if u belongs to the domain of the weak closure of the operator of taking the spatial A-gradient; the vector-valued function $A\nabla u$ is the value of this operator at the function u. The components $A_1\nabla u, \ldots, A_n\nabla u$ of $A\nabla u$ we call the *generalized spatial A-derivatives* of u in Q; here $A_i\nabla u \in L^{m_i, m_{0i}}(Q)$, $i = 1, \ldots, n$.

Since the generalized spatial A-derivatives of a function u are generalized $\tilde{A}$-derivatives of this function in the domain $\tilde{\Omega} \equiv Q \subset \mathbb{R}^{n+1}$ relative to the matrix $\tilde{A}$ of order $n + 1$ having the form

$$\tilde{A} = \begin{Vmatrix} 0 & \cdots & 0 \\ \vdots & A & \\ 0 & & \end{Vmatrix},$$

the facts presented below are established in exactly the same way as the corresponding propositions in §4.1.

LEMMA 1.1. *Under conditions* (1.1) *and the conditions*

$$a^{ij}, \frac{\partial a^{ij}}{\partial x_j} \in L^{m', m_0'}_{\text{loc}}(Q), \qquad \frac{1}{m} + \frac{1}{m'} = 1, \qquad \frac{1}{m_0} + \frac{1}{m_0'} = 1, \tag{1.3}$$

the operator of taking the spatial A-gradient admits weak closure (*i.e. condition* (1.2) *is satisfied*).

It will be shown below that condition (1.3) is ensured also by a condition of sufficiently weak degeneracy of the matrix A in Q.

LEMMA 1.2. *Under conditions* (1.1) *and* (1.3), *for any function* $u \in L^{m, m_0}_{\text{loc}}(Q)$ *having generalized spatial A-gradient* $A\nabla u \in \mathbf{L}^{\mathbf{m}, \mathbf{m}_0}_{\text{loc}}(Q)$ *the following identities hold*:

$$\iint_Q u\left(A_i \nabla \eta + \frac{\partial a^{ij}}{\partial x_j}\eta\right) dt\, dx = -\iint_Q A_i \nabla u \eta\, dt\, dx,$$
$$\forall \eta \in \tilde{C}^1_{0, \partial\Omega \times (T_1, T_2)}(Q), \quad i = 1, \ldots, n. \tag{1.4}$$

LEMMA 1.3. *Suppose that conditions* (1.1) *and* (1.3) *are satisfied and that for any compact subregion* Ω' *in* Ω

$$|a^{ij}(t, x) - a^{ij}(t, y)| \leqslant K|x - y|, \quad x, y \in \Omega', t \in [T_1, T_2], i, j = 1, \ldots, n, \tag{1.5}$$

where K depends only on Ω'. Assume that for functions $u \in L_{\mathrm{loc}}^{m,m_0}(Q)$ and $\mathbf{v} \in \mathbf{L}_{\mathrm{loc}}^{\mathbf{m},\mathbf{m}_0}(Q)$

$$\iint_Q u\left(A_i\nabla\eta + \frac{\partial a^{ij}}{\partial x_j}\eta\right)dt\,dx = -\iint_Q v_i\eta\,dt\,dx,$$

$$\eta \in \tilde{C}^1_{0,\partial\Omega\times(T_1,T_2)}(Q), \quad i,\,j = 1,\dots,n. \tag{1.6}$$

Then the function $u \in L_{\mathrm{loc}}^{m,m_0}(Q)$ has generalized spatial A-gradient $A\nabla u \in \mathbf{L}_{\mathrm{loc}}^{\mathbf{m},\mathbf{m}_0}(Q)$, and $A\nabla u = \mathbf{v}$.

Suppose that conditions (1.1) and (1.2) are satisfied, and let Γ be an arbitrary part of the boundary ∂Q of the cylinder Q. We denote by $\mathscr{H} \equiv \mathscr{H}^{0,\Gamma}_{m,\,m_0;\mathbf{m},\mathbf{m}_0}(A,Q)$ the completion of the set $\tilde{C}^1_{0,\Gamma}(Q)$ in the norm

$$\|u\|_{\mathscr{H}} \equiv \|u\|_{m,m_0,Q} + \|A\nabla u\|_{\mathbf{m},\mathbf{m}_0,Q}, \tag{1.7}$$

where

$$\|A\nabla u\|_{\mathbf{m},\mathbf{m}_0,Q} \equiv \sum_{i=1}^n \|A_i\nabla u\|_{m_i,m_{0i},Q}.$$

In the case $\Gamma = \varnothing$ we omit the upper indices in the notation for this space.

To each function $u \in \tilde{C}^1(Q)$ we assign its value $u|_\Pi$ on the set $\Pi \subset \partial Q$. We consider this mapping as a linear operator acting from $\tilde{C}^1(Q) \subset \mathscr{H}_{m,\,m_0;\mathbf{m},\mathbf{m}_0}(A,Q)$ into $L^1_{\mathrm{loc}}(\Pi)$, and call it the *operator of taking the limit value on* Π.

We suppose that for some set $\Pi \subset \partial Q$ the following condition is satisfied:

the operator of taking the limit value on the set Π admits closure. (1.8)

If condition (1.8) is satisfied we say that a function $u \in \mathscr{H} \equiv \mathscr{H}_{m,m_0;\mathbf{m},\mathbf{m}_0}(A,Q)$ has *generalized limit value* $u|_\Pi$ on the set Π if u belongs to the domain of the closure of the operator of taking the limit value on Π and $u|_\Pi$ is the value of this operator at the given function u.

Sufficient conditions that condition (1.8) be satisfied, which involve the assumption of sufficient regularity of the surface Π, smoothness of the elements of the matrix A in a neighborhood of Π, and nontangency of the vector $A^*A\nu$ to Π, are determined by the proofs of Lemmas 4.2.1–4.2.3. In particular, the next assertions follow from these propositions and the form of the norm (1.7) in an obvious way.

LEMMA 1.4. *Let $\Pi = \pi \times (T_1,T_2)$, where $\pi \subset \partial\Omega$, and suppose that $A = A(x)$ (i.e., the elements of the matrix A do not depend on t). Assume that conditions* (4.1.1), (4.1.3), *and* (4.2.5) *are satisfied for the set π and the matrix A. Then for any $m \geqslant 1$, $m_0 \geqslant 1$, $\mathbf{m} = (m_1,\dots,m_n)$ and $\mathbf{m}_0 = (m_{01},\dots,m_{0n})$, $m_i \geqslant 1$, $m_{0i} \geqslant 1$, $i = 1,\dots,n$, the operator of taking the limit value on the set Π admits closure. Any function $u \in \mathscr{H}$ has a generalized limit value $u|_\Pi$ on Π, and $u|_\Pi \in L^{m_*}_{\mathrm{loc}}(\Pi)$, where $m_* = \min(m, m_0, m_1,\dots,m_n, m_{01},\dots,m_{0n})$. For any $x_0 \in \operatorname{int}\pi$ there exists a neighborhood $\hat{p}_{x_0} \subset \pi$ such that*

$$\lim_{h\to 0}\int_{T_1}^{T_2}\int_{\hat{p}_{x_0}} |u((t,y) + hA^*(y)A(y)\nu(t,y)) - u|_\Pi(t,y)|^{m_*}\,dt\,ds = 0, \tag{1.9}$$

where $\nu(t, y)$ is the unit vector of the inner normal to Π at the point $(t, y) \in \hat{p}_{x_0} \times (T_1, T_2)$, and ds is the area element on $\hat{p}_{x_0}$. Moreover,

$$\int_{T_1}^{T_2} \int_{\hat{p}_{x_0}} |u|^{m_*}\, dt\, ds \leqslant c \int_{T_1}^{T_2} \int_{\hat{\omega}_{x_0}} (|u|^{m_*} + |A \nabla u|^{m_*})\, dt\, dx, \tag{1.10}$$

where $\hat{\omega}$ is a part of Ω abutting $\hat{p}_{x_0}$, and the constant c does not depend on u.

LEMMA 1.5. *Suppose that condition (4.2.25) is satisfied for some set $\Pi = \pi \times (T_1, T_2)$ and a matrix $A = A(x)$. Then the operator of taking the limit value on the set Π admits closure. Any function $u \in \mathscr{H}$ has a generalized limit value $u|_\Pi \in L^{m_*}(\Pi)$, where $m_* = \min(m, m_0, m_1, m_{01}, \dots, m_n, m_{0n})$, and*

$$\lim_{h \to 0} \int_\Pi |u((t, y) + hA^*(y)A(y)\nu(t, y)) - u|_\Pi(t, y)|^{m_*}\, dt\, ds = 0 \tag{1.11}$$

and

$$\int_\Pi |u|^{m_*}\, dt\, ds \leqslant c \int_{T_1}^{T_2} \int_\Pi (|u|^{m_*} + |A \nabla u|^{m_*})\, dt\, dx, \tag{1.12}$$

where c does not depend on u.

Sufficient conditions for the validity of (1.8) for the entire lateral surface of the cylinder Q involving assumptions of another kind (weak degeneracy of the matrix $A(t, x)$ in Q) will be given in §3. We note that condition (1.8) is certainly not satisfied for the upper and lower bases of Q (as we shall see below).

For the rest of this section we always assume that conditions (1.1) and (1.2) are satisfied. In analogy to Definition 4.3.1 we introduce the concepts of regular and singular parts of the boundary ∂Q relative to the matrix A (of order n) and the indices m, m_0, $\mathbf{m}$ and $\mathbf{m}_0$. We call a set $\Pi \subset \partial Q$ regular if the operator of taking the limit value on Π admits closure (i.e., condition (1.8) holds for Π). We call a set $\mathscr{P} \subset \partial Q$ a singular part of ∂Q if

$$\text{the set } \tilde{C}^1_{0,\mathscr{P}}(\overline{Q}) \text{ is dense in } \mathscr{H}_{m, m_0; \mathbf{m}, \mathbf{m}_0}(A, Q). \tag{1.13}$$

Sufficient conditions for regularity of sets of the form $\Pi = \pi \times (T_1, T_2)$, $\pi \subset \partial\Omega$, are given above. We shall present conditions guaranteeing singularity of sets $\mathscr{P} \subset \partial Q$. The facts stated below follow easily from the proof of the general Lemma 4.3.1.

LEMMA 1.6. *The sets $\tilde{C}^1_{0,\Omega_{T_1}}(\overline{Q})$, $\tilde{C}^1_{0,\Omega_{T_2}}(\overline{Q})$, and $\tilde{C}^1_{0,\Omega T_1 \cup \Omega T_2}(\overline{Q})$ are dense in $\mathscr{H}$, so that the lower base Ω_{T_1} and the upper base Ω_{T_2} are singular parts of the boundary ∂Q (for any indices m, m_0, $\mathbf{m}$ and $\mathbf{m}_0$).*

PROOF. Since for the set $\mathscr{P} = \Omega_{T_1}$ or $\mathscr{P} = \Omega_{T_2}$ the function $\zeta_\delta(t, x)$, constructed in analogy with (4.3.4) is independent of x, the analogue of the third term on the right side of (4.3.6) is equal to 0, whence the assertion of Lemma 1.6 follows.

LEMMA 1.7. *Let $\mathscr{P} = p \times (T_1, T_2)$, where $p \subset \partial\Omega$, and let $A = A(x)$. Assume that for the set p and the matrix A condition (4.3.2) is satisfied. Then $\mathscr{P}$ is a singular part of the boundary ∂Q.*

PROOF. Let $u_\delta = u\zeta_\delta$, where the cut-off function ζ_δ is constructed according to a formula of the form (4.3.3), so that $u_\delta \in \tilde{C}^1_{0,\mathscr{P}}(\overline{Q})$. Then (cf. (4.3.5))

$$\|u - u_\delta\|_{\mathscr{H}} \leqslant \|u(1-\zeta_\delta)\|_{m,m_0,Q} + \sum_{i=1}^n \|A_i\nabla u(1-\zeta\delta)\|_{m_i,m_{0i},Q} + \sum_{i=1}^n \|uA_i\nabla\zeta_\delta\|_{m_i,m_{0i},q_{\delta/2,\delta}}, \tag{1.14}$$

where $q_{\delta/2,\delta} = \omega_{\delta/2,\delta} \times (T_1, T_2)$, and the set $\omega_{\delta/2,\delta}$ is defined as in (4.3.3). Since

$$\|uA_i\nabla\zeta_\delta\|_{m_i,m_{0i},q_{\delta/2,\delta}} = \left(\int_{T_1}^{T_2} \|uA_i\nabla\zeta_\delta\|^{m_{0i}/m_i}_{m_i,\omega_{\delta/2,\delta}}\,dt\right)^{1/m_{0i}}, \tag{1.15}$$

from formulas of the form (4.3.6), (4.3.7) we obtain

$$\|uA_i\nabla\zeta_\delta\|_{m_i,m_{0i}q_{\delta/2,\delta}} \leqslant c\delta^{\varepsilon_i}, \qquad i = 1,\dots,n, \tag{1.16}$$

where the constant c does not depend on δ, and $\varepsilon_i = \alpha_i - 1/m_i' > 0$, $i = 1,\dots,n$. It follows easily from (1.14) and (1.16) that $\lim_{\delta\to 0}\|u - u_\delta\|_{\mathscr{H}} = 0$. Lemma 1.7 is proved.

For the rest of this chapter we always assume that the following condition is satisfied:

$$\partial Q = \Sigma \cup \Sigma', \Sigma = \sigma \times (T_1, T_2), \Sigma' = (\sigma' \times (T_1, T_2)) \cup \Omega_{T_1} \cup \Omega_{T_2},$$
where Σ is the regular and Σ' the singular part of ∂Q relative to the given matrix A and indices m, m_0, $\mathbf{m}$ and $\mathbf{m}_0$, $m, m_0, m_i, m_{0i} > 1$, $i = 1,\dots,n$.
(1.17)

We recall that conditions (1.1) and (1.2) are satisfied for the matrix A, and the sets Ω_{T_1} and Ω_{T_2} are certainly singular parts of ∂Q. Suppose that the set Σ is partitioned into parts $\Sigma_1 \equiv \sigma_1 \times (T_1, T_2)$, $\Sigma_2 \equiv \sigma_2 \times (T_1, T_2)$ and $\Sigma_3 \equiv \sigma_3 \times (T_1, T_2)$, so that $\Sigma_1 \cup \Sigma_2 \cup \Sigma_3 = \Sigma$ and $\Sigma_i \cap \Sigma_j = \varnothing$ for $i \neq j$, $i, j = 1, 2, 3$, where we assume that $\operatorname{meas}_n \partial\Sigma_i = 0$, $i = 1, 2, 3$. Suppose that on Σ_3 there is given a piecewise continuous, bounded, positive function λ.

The completion of the set $\tilde{C}^1_{0,\Sigma_1}(Q)$ in the norm

$$\|u\|_{\mathscr{H}_\lambda} \equiv \|u\|_{m,m_0,Q} + \|A\nabla u\|_{\mathbf{m},\mathbf{m}_0,Q} + \|u\|_{L^2(\lambda,\Sigma_3)}, \tag{1.18}$$

where

$$\|A\nabla u\|_{\mathbf{m},\mathbf{m}_0,Q} \equiv \sum_{i=1}^n \|A_i\nabla u\|_{m_i,m_{0i},Q}, \qquad A_i\nabla u \equiv a^{ij}u_{x_j}, \quad i = 1,\dots,n,$$

is denoted by

$$\mathscr{H}_\lambda \equiv \mathscr{H}^{0,\Sigma_1}_{m,m_0;\mathbf{m},\mathbf{m}_0}(A; Q; \Sigma_3, \lambda).$$

In the case $\Sigma_3 = \varnothing$ the space $\mathscr{H}_\lambda$ coincides with the space

$$\mathscr{H} \equiv \mathscr{H}^{0,\Sigma_1}_{m,m_0;\mathbf{m},\mathbf{m}_0}(A, Q)$$

introduced above.

The next assertions are proved in exactly the same way as Lemmas 5.2.1–5.2.3.

LEMMA 1.8. *The space $\mathscr{H}_\lambda$ is separable and reflexive. Any linear functional $\mathscr{F}$ in $\mathscr{H}_\lambda$ can be defined by*

$$\langle \mathscr{F}, \eta\rangle = \iint_Q (f_0\eta + \mathbf{f}\cdot A\nabla\eta)\,dt\,dx + \int_{\Sigma_3} \lambda\psi\eta\,ds, \qquad \eta \in \tilde{C}^1_{0,\Sigma_1}(Q), \tag{1.19}$$

where $f_0 \in L^{m',m'_0}(Q)$, $1/m + 1/m' = 1$, $1/m_0 + 1/m'_0 = 1$, $\mathbf{f} = (f^1,\dots,f^n)$, $f^i \in L^{m'_i,m'_{0i}}(Q)$, $1/m_i + 1/m'_i = 1$, $1/m_{0i} + 1/m'_{0i} = 1$, $i = 1,\dots,n$, $\psi \in L^2(\lambda, \Sigma_3)$, and f_0, f, and ψ can be chosen so that

$$\|\mathscr{F}\|_{\mathscr{H}^*_\lambda} = \sup\Big(\|f_0\|_{m',m'_0,Q}, \|f^1\|_{m'_1,m'_{01},Q},\dots,\|f^n\|_{m'_n,m'_{0n},Q}, \|\psi\|_{L^2(\lambda,\Sigma_3)}\Big). \tag{1.20}$$

Any expression of the form (1.19), *considered under the conditions on f_0, $\mathbf{f}$, and ψ indicated above, defines a linear functional $\mathscr{F}$ in $\mathscr{H}_\lambda$ with norm $\|\mathscr{F}\|_{\mathscr{H}^*_\lambda}$ not exceeding the quantity on the right side of* (1.20).

LEMMA 1.9. *There is the dense imbedding $\mathscr{H}_\lambda \to \mathscr{H}$. If $A = A(x)$, $m \geqslant 2$, $m_0 \geqslant 2$, $m_i \geqslant 2$, $m_{0i} \geqslant 2$, $i = 1,\dots,n$, and condition* (4.2.25) *is satisfied for the set σ_3 and the matrix A, then the spaces $\mathscr{H}_\lambda$ and $\mathscr{H}$ are isomorphic.*

LEMMA 1.10. *The sets $\tilde{C}^1_{0,\Sigma_1\cup\Omega_{T_1}\cup\Omega_{T_2}}(\overline{Q})$, $\tilde{C}^1_{0,\Sigma_1\cup\Omega_{T_1}}(\overline{Q})$, and $\tilde{C}^1_{0,\Sigma_1\cup\Omega_{T_2}}(\overline{Q})$ are dense in $\mathscr{H}_\lambda$.*

We now introduce analogues of the spaces X and Y (see (5.2.10) and (5.2.11)). Taking into account that the equations considered below will have $(\tilde{A}, \tilde{\mathbf{b}})$-structure in the cylinder Q relative to the matrix

$$\tilde{A} = \begin{Vmatrix} 0 & \cdots & 0 \\ \vdots & A & \\ 0 & & \end{Vmatrix}$$

and vector $\tilde{\mathbf{b}} = (1, 0, \dots, 0)$, we consider the following spaces $\mathscr{X}$ and $\mathscr{Y}$.

The completion of the set $\tilde{C}^1_{0,\Sigma_1}(Q)$ in the norm

$$\|u\|_{\mathscr{X}} = \|u\|_{\mathscr{H}_\lambda} + \|u\|_{2,Q} + \|u\|_{2,Q_{T_2}} \tag{1.21}$$

we denote by

$$\mathscr{X} \equiv \mathscr{X}^{0,\Sigma_1}_{m,m_0;\mathbf{m},\mathbf{m}_0}(A; Q; \Sigma_3, \lambda).$$

The completion of the set $\tilde{C}^1_{0,\Sigma_1}(Q)$ in the norm

$$\|u\|_{\mathscr{Y}} = \|u\|_{\mathscr{X}} + \|u_t\|_{2,Q} \tag{1.22}$$

we denote by

$$\mathscr{Y} \equiv \mathscr{Y}^{0,\Sigma_1}_{m,m_0;\mathbf{m},\mathbf{m}_0}(A; Q; \Sigma_3, \lambda).$$

The following results are proved in exactly the same way as Lemmas 5.2.4–5.2.6.

LEMMA 1.11. *The space $\mathscr{X}$ is separable and reflexive. Any linear functional $\mathscr{F}$ in $\mathscr{X}$ can be defined by*

$$\langle \mathscr{F}, \eta\rangle = \iint_Q (f_0\eta + \mathbf{f}\cdot A\nabla\eta + g_0\eta)\,dt\,dx + \int_{\Omega_{T_2}} q\eta\,dx$$
$$+ \int_{\Sigma_3} \lambda\psi\eta\,ds, \qquad \eta \in \tilde{C}^1_{0,\Sigma_1}(Q), \tag{1.23}$$

where f_0, $\mathbf{f}$, *and* ψ *are the same functions as in* (1.19), $g_0 \in L^2(Q)$, $q \in L^2(\Omega)$, *and these functions can be chosen so that*

$$\|\mathscr{F}\|_{\mathscr{X}^*} = \sup\left\{ \|f_0\|_{m',m'_0,Q}, \dots, \|f^n\|_{m'_n,m'_{0n},Q}, \|g_0\|_{2,Q}, \|q\|_{2,\Omega_{T_2}}, \|\psi\|_{L^2(\lambda,\Sigma_3)} \right\}. \tag{1.24}$$

Any expression of the form (1.24), *considered under the conditions indicated above on* f_0, $\mathbf{f}$, ψ, g_0, *and* q, *defines a linear functional* $\mathscr{F}$ *in* $\mathscr{X}$ *with norm* $\|\mathscr{F}\|_{\mathscr{X}^*}$ *not exceeding the quantity on the right side of* (1.24).

LEMMA 1.12. *The space* $\mathscr{Y}$ *is separable and reflexive. Any linear functional* $\mathscr{F}$ *in* $\mathscr{Y}$ *can be defined by*

$$\langle \mathscr{F}, \eta \rangle = \iint_Q (f_0\eta + \mathbf{f} \cdot A\nabla\eta + g_0\eta + h_0\eta_t)\, dt\, dx + \int_{\Omega_{T_2}} q\eta\, dx$$

$$+ \int_{\Sigma_3} \lambda\psi\eta\, ds, \qquad \eta \in \tilde{C}^1_{0,\Sigma_1}(Q), \tag{1.25}$$

where f_0, $\mathbf{f}$, g_0, q *and* ψ *are the same functions as in* (1.23), $h_0 \in L^2(Q)$, *and these functions can be chosen so that*

$$\|F\|_{\mathscr{Y}^*} = \sup\left\{ \|f_0\|_{m',m'_0,Q}, \dots, \|f^n\|_{m'_n,m'_{0n},Q}, \right.$$

$$\left. \|g_0\|_{2,Q}, \|h_0\|_{2,Q}, \|q\|_{2,\Omega_{T_2}}, \|\psi\|_{L^2(\lambda,\Sigma_3)} \right\}. \tag{1.26}$$

Any expression of the form (1.25), *considered under the conditions on* f_0, $\mathbf{f}$, ψ, g_0, q, *and* h_0 *indicated above, defines a linear functional* $\mathscr{F}$ *in* $\mathscr{Y}$ *with norm* $\|\mathscr{F}\|_{\mathscr{Y}^*}$ *not exceeding the quantity on the right side of* (1.26).

We denote by $\hat{\mathscr{H}}_\lambda$ the completion of $\tilde{C}^1_{0,\Sigma_1}(Q)$ in the norm $\|u\|_{\mathscr{H}_\lambda} + \|u\|_{2,Q}$. It is obvious that $\hat{\mathscr{H}}_\lambda \to \mathscr{H}_\lambda$.

LEMMA 1.13. $\mathscr{X}$ *can be identified with a subspace of* $\hat{\mathscr{H}}_\lambda \times L^2(\Omega_{T_2})$.

REMARK 1.2. $\mathscr{X}$ cannot be identified with a subspace of $\hat{\mathscr{H}}_\lambda$ (see the proof of Remark 5.2.1).

LEMMA 1.14. $\mathscr{Y}$ *can be identified both with a subspace of* $\hat{\mathscr{H}}_\lambda$ *and with a subspace of* $\mathscr{X}$.

In view of Lemma 1.13 elements $\mathbf{u} \in \mathscr{X}$ can be written as pairs $\mathbf{u} = (u, \varphi)$, where $u \in \hat{\mathscr{H}}_\lambda$ and $\varphi \in L^2(\Omega)$. In view of Lemma 1.14 elements $\mathbf{u} \in \mathscr{Y}$ can also be written as pairs $\mathbf{u} = (u, \varphi)$, $u \in \hat{\mathscr{H}}_\lambda$, $\varphi \in L^2(\Omega)$, but in this case the second component φ is uniquely determined by the first component u by the formula

$$\int_{\Omega_{T_2}} \varphi\eta\, dx = -\iint_Q (u_t\eta + u\eta_t)\, dt\, dx, \qquad \eta \in \tilde{C}^1_{0,\Omega_{T_1}}(Q), \tag{1.27}$$

which is derived in the same was as (5.2.20). Therefore, we henceforth agree to write the component φ of an element $\mathbf{u} = (u, \varphi) \in \mathscr{Y}$ as $[u]_{\Omega_{T_2}}$. Then elements $\mathbf{u} \in \mathscr{Y}$ can be written either in the form $\mathbf{u} = (u, [u]_{\Omega_{T_2}})$ or simply as a function $u \in \hat{\mathscr{H}}_\lambda$.

Lemmas 1.8, 1.11, and 1.12 obviously imply the following imbeddings:

$$\mathscr{X} \to \hat{\mathscr{H}}_\lambda \times L^2(\Omega_{T_2}), \qquad \mathscr{Y} \to \mathscr{X}, \qquad \mathscr{Y} \to \hat{\mathscr{H}}_\lambda \tag{1.28}$$

and

$$L^{m',m'_0}(Q) \to \mathscr{H}^*_\lambda \to \mathscr{X}^* \to \mathscr{Y}^*. \tag{1.29}$$

In view of (1.29) and the existence of a common dense set $\tilde{C}^1_{0,\Sigma_1}(\overline{Q})$ in $\mathscr{H}$, $\mathscr{X}$, and $\mathscr{Y}$ it is possible to use the same notation $\langle\,\cdot\,,\cdot\rangle$ for the dualities between $\mathscr{H}_\lambda$ and $\mathscr{H}^*_\lambda$, $\mathscr{X}$ and $\mathscr{X}^*$, and $\mathscr{Y}$ and $\mathscr{Y}^*$.

A remark analogous to Remark 5.2.2 is valid, but we omit the explicit formulation.

Finally, we formulate the following obvious assertion.

LEMMA 1.15. *Any function $u \in \mathscr{Y}$ belongs to the space $C([T_1, T_2]; L^2(\Omega))$, and*

$$\|u\|_{C([T_1,T_2];L^2(\Omega))} \leqslant c\big(\|u\|_{2,Q} + \|u_t\|_{2,Q}\big), \tag{1.30}$$

where the constant c does not depend on $u \in \mathscr{Y}$; for all $u \in \mathscr{Y}$ the formula for integration by parts

$$\iint_Q u_t\eta\,dt\,dx = -\iint_Q u\eta_t\,dt\,dx + \int_{\Omega_{T_2}} u\eta\,dx - \int_{\Omega_{T_1}} u\eta\,dx, \quad \forall\eta \in \tilde{C}^1(Q),$$

holds, so that for every $u \in \mathscr{Y}$ the value of $[u]_{\Omega_{T_2}}$ defined by (1.27) coincides with the value of $u(x, T_2) \in L^2(\Omega)$.

We consider the linear operator $\mathscr{B}$: $\tilde{C}^1_{0,\Sigma_1}(\overline{Q}) \subset \mathscr{X} \to \mathscr{Y}^*$ defined by

$$\langle \mathscr{B}u, \eta\rangle = -\iint_Q u\eta_t\,dt\,dx + \int_{\Omega_{T_2}} u\eta\,dx, \qquad u, \eta \in \tilde{C}^1_{0,\Sigma_1}(\overline{Q}). \tag{1.31}$$

We establish the following propositions in analogy with the results of the general case (see §2.2).

LEMMA 1.16. *For any $u, \eta \in \tilde{C}^1_{0,\Sigma_1}(\overline{Q})$*

$$|\langle \mathscr{B}u, \eta\rangle| \leqslant \|u\|_{\mathscr{X}}\|\eta\|_{\mathscr{Y}}, \tag{1.32}$$

so that the operator $\mathscr{B}$: $\tilde{C}^1_{0,\Sigma_1}(\overline{Q}) \subset \mathscr{X} \to \mathscr{Y}^*$ can be extended by continuity to the entire space $\mathscr{X}$. The restriction of the extended operator $\mathscr{B}$: $\mathscr{X} \to \mathscr{Y}^*$ to $\mathscr{Y}$ is a bounded linear operator acting from $\mathscr{Y}$ to $\mathscr{X}^*$.

In the proof of Lemma 1.16 the fact that the set $\tilde{C}^1_{0,\Sigma\cup\Omega_{T1}}(\overline{Q})$ is dense in $\mathscr{X}$ is taken into account, and this assertion is proved in exactly the same way as Lemma 1.10 (see also the proof of Lemma 5.2.3). The next assertion is proved with Lemma 1.10 taken into account.

LEMMA 1.17. *The subspace $V \equiv \{\mathbf{u} \in \mathscr{X}\colon \mathscr{B}\mathbf{u} \in \mathscr{H}^*\}$ can be identified with a subspace of $\hat{\mathscr{H}}_\lambda$.*

In the proof of Lemma 1.17 (which proceeds in complete analogy to the proof of Lemma 5.2.7) it is established that the second component $\varphi \in L^2(\Omega)$ of an element

$\mathbf{u} = (u, \varphi) \in \mathscr{V}$ is uniquely determined by its first component $u \in \hat{\mathscr{H}}_\lambda$ by the formula

$$\int_{\Omega_{T_2}} \varphi\eta\, dx = -\lim_{\eta_n \to \eta \text{ in } \mathscr{H}_\lambda} \iint_Q u(\eta_n - \eta)_t\, dt\, dx, \tag{1.33}$$

where η is an arbitrary function in $\tilde{C}^1_{0,\Sigma_1}(Q)$, and $\{\eta_n\}$ is any sequence formed from functions $\eta_n \in \tilde{C}^1_{0,\Sigma_1 \cup \Omega_{T_2}}(Q)$, $n = 1, 2, \ldots$, which converges to η in $\mathscr{H}_\lambda$. The value of φ for elements $(u, \varphi) \in \mathscr{V}$ defined by (1.33) we agree to denote by $(u)_{T_2}$. Lemma 1.17 establishes the identification of the notation $(u, (u)_{T_2})$ with u for elements of $\mathscr{V}$. In the proof of Lemma 1.17 it is also established that the restriction of the operator $\mathscr{B}\colon \mathscr{X} \to \mathscr{Y}^*$ to the set $\mathscr{V}$ is completely determined by the formula

$$\langle \mathscr{B}u, \eta\rangle = -\iint_Q u\eta_t\, dt\, dx, \qquad u \in \mathscr{V}, \eta \in \tilde{C}^1_{0,\Sigma_1 \cup \Omega_{T_2}}(Q). \tag{1.34}$$

The following lemmas are analogues of Lemmas 5.2.8 and 5.4.2.

LEMMA 1.18. *A function $u \in \hat{\mathscr{H}}_\lambda$ belongs to the subspace $\mathscr{V}$ if and only if the following conditions are satisfied*:

1) *There exists a sequence $\{u_k\}$, $u_k \in \tilde{C}^1_{0,\Sigma_1}(Q)$, $k = 1, 2, \ldots$, such that*
$$\lim_{k\to\infty} \|u_k - u\|_{\mathscr{H}_\lambda} = 0, \qquad \lim_{k,\delta\to\infty} \|u_k - u_s\|_{\mathscr{H}} = 0.$$

2) *For any $\eta \in \tilde{C}^1_{0,\Sigma_1 \cup \Omega_{T_2}}(Q)$ and any sequence $\{\eta_k\}$, $\eta_k \in \tilde{C}^1_{0,\Sigma_1 \cup \Omega_{T_2}}(Q)$, $k = 1, 2, \ldots$, converging to η in $\mathscr{H}_\lambda$ the following equality holds*:

$$\lim_{k\to\infty} \iint_Q u(\eta_k - \eta)_t\, dt\, dx = 0. \tag{1.35}$$

LEMMA 1.19. *In order that a function $u \in \tilde{C}^1_{0,\Sigma_1}(Q)$ belong to the subspace $\mathscr{V}$ it is necessary and sufficient that $u = 0$ on Ω_{T_1}.*

Taking Lemma 1.15 into account, we establish the following generalization of Lemma 1.19 in an analogous way.

LEMMA 1.20. *In order that a function $u \in \mathscr{Y}$ belong to the subspace $\mathscr{V}$ it is necessary and sufficient that $u = 0$ on Ω_{T_1}.*

For the rest of this section we assume that the following condition is satisfied:

the matrix A does not depend on t, and condition (4.1.3) holds for the matrix $A \equiv A(x)$ in the domain Ω; $m \geqslant 2$, $m_0 \geqslant 2$, $m_i \geqslant 2$, $m_{0i} \geqslant 2$, $i = 1, \ldots, n$; the spaces $\mathscr{H}_\lambda$ and $\mathscr{H}$ are isomorphic. (1.36)

We observe that the last condition in (1.36) regarding the isomorphism of $\mathscr{H}_\lambda$ and $\mathscr{H}$ is certainly satisfied if for the set $\sigma_3 \subset \partial\Omega$ (we recall that $\Sigma_3 = \sigma_3 \times (T_1, T_2)$) the matrix A, and the indices m and $\mathbf{m}$ condition (4.2.25) is satisfied. We consider the realization of the space U defined in §7 of Chapter 4 which is obtained with the following choice of the spaces $B_0, B_1, \ldots, B_n$ $(N = n)$, set G, operators $l_k\colon G \subset B_0 \to B_k$, $k = 1, \ldots, n$, and indices $p_0, p_1, \ldots, p_k$: $B_0 = L^m(\Omega)$, $B_k = L^{m_k}(\Omega)$, $k = 1, \ldots, n$, $G = \tilde{C}^1_{0,\sigma_1}(\Omega) \subset L^m(\Omega)$, the operators $l_k\colon \tilde{C}^1_{0,\sigma_1}(\Omega) \subset L^m(\Omega) \to L^{m_k}(\Omega)$, $k = 1, \ldots, n$, are defined by the formula $l_k u = A_k \nabla u$, $u \in \tilde{C}^1_{0,\sigma_1}(\Omega)$, $p_0 = m_0$, $p_1 = m_1, \ldots, p_n = m_n$; by (1.36) the operators l_k, $k = 1, \ldots, n$, admit closure. We denote by $\hat{B}$ (see Chapter 4, §7) the closure of $\tilde{C}^1_{0,\sigma_1}(\overline{\Omega})$ in the norm $\|u\|_{\hat{B}} \equiv \|u\|_H + \sum_{k=1}^n \|u\|_{m_k,\Omega}$ where $H \equiv H^{0,\sigma_1}_{m,\mathbf{m}}(A, \Omega)$. It is easy to see that all conditions needed for the construction of the space U are satisfied. Thus, in the present case U is the Banach space of functions in $([T_1, T_2] \to H)$ equipped with the norm

$$\|u\|_U = \|u\|_{m, m_0, Q} + \|A \nabla u\|_{\mathbf{m}, \mathbf{m}_0, Q}.$$

LEMMA 1.21. $U \equiv \mathscr{H}$.

PROOF. The set $\tilde{C}^1_{0,\Sigma_1}(\overline{Q})$ is dense in $\mathscr{H}$. We shall prove that $\tilde{C}^1_{0,\Sigma_1}(\overline{Q})$ is also dense in U. It follows from Corollary 4.7.1 that the set $C^\infty([T_1, T_2], H)$, where $H \equiv H^{0,\sigma_1}_{m,\mathbf{m}}(A, \Omega)$, is dense in U. Now it is easy to see that any function $u \in C^\infty([T_1, T_2], H)$ can be approximated by polynomials of the form $u_N = \sum_{k=0}^{M_n} v_{N,k} t^k$, where $v_{N,k} \in \tilde{C}^1_{0,\sigma_1}(\overline{\Omega})$, in the norm of U. Since $u_N \in \tilde{C}^1_{0,\Sigma_1}(\overline{Q})$, it follows from what has been proved that $\tilde{C}^1_{0,\Sigma_1}(\overline{Q})$ is dense in U. Thus, the spaces $\mathscr{H}$ and U coincide, since they have the same norms and a common dense set. Lemma 1.21 is proved.

Thus, $U = \mathscr{H}$ and $U^* = \mathscr{H}^*$. In view of Lemma 1.8 (in the case $\mathscr{H}_\lambda \equiv \mathscr{H}$) any element $\mathscr{F} \in \mathscr{H}^*$ is determined by a function

$$F(t) = F_0(t) + \sum_{k=1}^n D_k^* F_k(t),$$

where

$$F_0(t) \in L^{m_0'}([T_1, T_2]; L^{m'}(\Omega)) \to L^{m_0'}([T_1, T_2]; H^*),$$
$$F_k(t) \in L^{m_{0k}'}([T_1, T_2]; L^{m_k'}(\Omega)),$$
$$D_k^* F_k(t) \in L^{m_{0k}'}([T_1, T_2]; H^*), \qquad k = 1, \ldots, n,$$

and

$$\langle \mathscr{F}, \eta \rangle = \int_{T_1}^{T_2} (F(t), \eta(t))\, dt, \qquad \eta \in \mathscr{H},$$

$$\begin{gathered} (f(t), \psi) = (F_0(t), \psi) + \sum_{k=1}^n (D_k^* F_k(t), \psi), \\ (F_0(t), \psi) \equiv \int_\Omega F_0(t, x) \psi(x)\, dx, \end{gathered} \tag{1.37}$$

$$(D_k^* F_k(t), \psi) = \int_\Omega F_k(t, x) A_k \nabla \psi\, dx, \qquad k = 1, \ldots, n, \psi \in H.$$

We denote (in correspondence with (4.7.22)) by $\mathscr{W}$ the following subspace of $\mathscr{H}$:

$$\mathscr{W} = \{u \in \mathscr{H} : u' \in \mathscr{H}^*\}, \tag{1.38}$$

where u' denotes the derivative of u in the sense of distributions on $[T_1, T_2]$ with values in the Banach space H^*, where $H \equiv H^{0,\sigma_1}_{m,\mathbf{m}}(A, \Omega)$, i.e., the mapping $\varphi(t) \to -\int_{T_1}^{T_2} u(t) \varphi'(t)\, dt$, $\varphi \in \mathscr{D}([T_1, T_2])$.

LEMMA 1.22. $\mathscr{V} \subset \mathscr{W}$.

PROOF. Let $u \in \mathscr{V}$. Taking account of Lemma 1.17, we conclude that $u \in \mathscr{H}$. The function u may then be considered as an element of the space $\mathscr{D}^*([T_1, T_2]; H^*)$. Let

u' be the derivative of this element. To prove the lemma it suffices to establish the existence of a function $F(t) \in ([T_1, T_2] \to H^*)$ of the form (1.37) such that

$$-\int_{T_1}^{T_2} u(t)\varphi'(t)\,dt = \int_{T_1}^{T_2} F(t)\varphi(t)\,dt, \qquad \varphi \in \mathscr{D}([T_1, T_2]), \tag{1.39}$$

where the integrals are understood as Bochner integrals of functions on $[T_1, T_2]$ with values in H. The identity (1.39) is equivalent to

$$-\int_{T_1}^{T_2}(u(t), \varphi'(t)\psi)\,dt = \int_{T_1}^{T_2}(F(t), \varphi(t)\psi)\,dt,$$

$$\varphi \in \mathscr{D}([T_1, T_2]), \psi \in \tilde{C}^1_{0,\sigma_1}(\overline{\Omega}), \tag{1.40}$$

since $\tilde{C}^1_{0,\sigma_1}(\Omega)$ is dense in H. Now from the definition of the subspace $\mathscr{V}$ and Lemma 1.17 it follows that there exist $f_0 \in L^{m',m'_0}(Q)$ and $\mathbf{f} \in \mathbf{L}^{\mathbf{m}',\mathbf{m}'_0}(Q)$ such that for any $\eta = \varphi(t)\psi(x)$, $\psi \in \tilde{C}^1_{0,\sigma_1}(\Omega)$ and $\varphi \in \mathscr{D}([T_1, T_2])$ (see (1.37))

$$-\iint_Q u\eta_t\,dt\,dx = \iint_Q (f_0\eta + \mathbf{f}\cdot A\nabla\eta)\,dt\,dx, \tag{1.41}$$

i.e., (1.40) is satisfied with $F(t) = f_0(t) + D_k^* f_k(t)$. This proves Lemma 1.22.

The next assertion follows from the properties of elements of the space $\mathscr{Y}$ enumerated above.

LEMMA 1.23. $\mathscr{Y} \subset \mathscr{W}$.

The next assertion follows from Lemmas 1.22 and 1.23 and (4.7.29).

COROLLARY 1.1. *For any* $u, v \in \mathscr{V} \cup \mathscr{Y}$

$$(u, v)\big|_{t_1}^{t_2} = \int_{t_1}^{t_2}[(u', v) + (u, v')]\,dt, \qquad t_1, t_2 \in [T_1, T_2], \tag{1.42}$$

where $(\cdot,\cdot)$ *denotes the duality between* H *and* H^* *and the inner product in* $L^2(\Omega)$, *and* u' *and* v' *are the derivatives (in the sense of distributions on* $[T_1, T_2]$ *with values in* H^*) *of* u *and* v.

LEMMA 1.24. *For any function* $u \in \mathscr{V}$

$$u(x, T_2) = (u)_{T_2}(x), \qquad u(x, T_1) = 0 \quad a.e.\ in\ \Omega, \tag{1.43}$$

where the function $\varphi = (u)_{T_2}$ *is defined by* (1.33).

PROOF. It follows from (1.33) that

$$\int_{\Omega_{T_2}} (u)_{T_2}\eta\,dx = -\lim_{\eta_n \to \eta \text{ in } \mathscr{H}} \iint_Q u(\eta_n - \eta)_t\,dt\,dx, \qquad \eta \in \tilde{C}^1_{0,\sigma_1}(Q), \tag{1.44}$$

where $\{\eta_n\}$, $\eta_n \in \tilde{C}^1_{0,\Sigma_1\cup\Omega_{T_2}}(Q)$, $n = 1, 2, \dots$, is a sequence converging to η in $\mathscr{H}$. Using Lemma 1.22 and (1.42), we rewrite (1.44) in the form

$$((u)_{T_2}, \eta(T_2)) = \lim_{\eta_n \to \eta}\Bigg[\int_{T_1}^{T_2}(u', \eta_n - \eta)\,dt - (u(T_2), (\eta - \eta_n)(T_2))$$

$$-(u(T_1), (\eta - \eta_n)(T_1))\Bigg]. \tag{1.45}$$

Taking into account that $\int_{T_1}^{T_2}(u', \eta_n - \eta)\,dt = \langle u', \eta_n - \eta\rangle$, where $u' \in \mathscr{H}^*$ and $\eta_n \to \eta$ in $\mathscr{H}$, and also that $\eta_n(T_2) = 0$, $n = 1, 2, \dots$, we deduce from (1.45) that

$$((u)_{T_2}, \eta(T_2)) = (u(T_2), \eta(T_2)) - \lim_{\eta_n \to \eta \text{ in } \mathscr{H}}(u(T_1), (\eta - \eta_n)(T_1)). \tag{1.46}$$

Since η and $\{\eta_n\}$ are arbitrary, it follows from (1.46) that $(u)_{T_2} = u(T_2, x)$ for almost all $x \in \Omega$. From (1.46) it then follows that

$$\lim_{\eta_n \to \eta \text{ in } \mathscr{H}} (u(T_1), (\eta - \eta_n)(T_1)) = 0. \tag{1.47}$$

Using again the fact that the choice of η and $\{\eta_n\}$ is arbitrary, from (1.47) we deduce that $u(x, T_1) = 0$ a.e. in Ω. Lemma 1.24 is proved.

LEMMA 1.25. *For any functions $u \in \mathscr{V}$ and $\eta \in \mathscr{V} \cup \mathscr{Y}$*

$$\langle \mathscr{B}u, \eta \rangle = -\int_{T_1}^{T_2} (u, \eta')\,dt + (u(T_2), \eta(T_2)) = \int_{T_1}^{T_2} (u', \eta)\,dt, \tag{1.48}$$

where $u' \in \mathscr{H}^*$ is the derivative of u considered as an element of the space $\mathscr{D}^*([T_1, T_2]; H^*)$.

PROOF. Let $u \in \mathscr{V}$ and $\eta \in \tilde{C}^1_{0,\Sigma_1}(Q)$. From the definition of the operator $\mathscr{B}$: $\mathscr{X} \to \mathscr{Y}^*$ and Lemma 1.24 it follows easily that

$$\begin{aligned}\langle \mathscr{B}u, \eta \rangle &= -\iint_Q u\eta_t\,dt\,dx + \int_{\Omega_{T_2}} (u)_{T_2}\eta\,dx \\ &= -\int_{T_1}^{T_2} (u, \eta')\,dt + (u(T_2), \eta(T_2)).\end{aligned} \tag{1.49}$$

In view of Lemma 1.23 formula (1.49) also holds for all $u \in \mathscr{Y}$ with $u(T_1) = 0$. Using (1.42) with $u = v$, $t_1 = T_1$ and $t_2 = T_2$, and taking (1.43) into account, we see that the right side in (1.43) can be rewritten in the form $\int_{T_1}^{T_2} (u', \eta)\,dt$. Thus, equalities (1.48) have been established for any $u \in \mathscr{V}$ and $\eta \in \tilde{C}_{0,\Sigma_1}(Q)$. Suppose now that $u \in \mathscr{V}$ and $\eta \in \mathscr{V} \cup \mathscr{Y}$. Since $\mathscr{V} \cup \mathscr{Y} \subset \mathscr{H}$, there exists a sequence $\{\eta_n\}$, $\eta_n \in \tilde{C}^1_{0,\Sigma_1}(Q)$, $n = 1, 2, \ldots$, converging to η in $\mathscr{H}$. Passing to the limit in the equalities

$$\langle \mathscr{B}u, \eta_n \rangle = \int_{T_1}^{T_2} (u', \eta_n)\,dt, \qquad n = 1, 2, \ldots, \tag{1.50}$$

and taking into account that $\mathscr{B}u \in \mathscr{H}^*$ and $u' \in \mathscr{H}^*$, we then obtain

$$\langle \mathscr{B}u, \eta \rangle = \int_{T_1}^{T_2} (u', \eta)\,dt. \tag{1.51}$$

Applying (1.42) and taking account of (1.43), we can rewrite the right side of (1.51) also in the form

$$-\int_{T_1}^{T_2} (u, \eta')\,dt + (u(T_2), \eta(T_2)).$$

This proves Lemma 1.25.

LEMMA 1.26. *For any function $u \in \mathscr{V}$*

$$\langle \mathscr{B}u, u \rangle = \frac{1}{2}\int_\Omega u^2(x, T_2)\,dx. \tag{1.52}$$

PROOF. Let $u \in \mathscr{V}$. It follows from Lemma 1.25 that

$$\langle \mathscr{B}u, u \rangle = \int_{T_1}^{T_2} (u', u)\,dt, \tag{1.53}$$

and from (1.42) and (1.43) we obtain

$$\int_{T_1}^{T_2}(u', u)\,dt = \frac{1}{2}(u(T_2), u(T_2)). \tag{1.54}$$

(1.52) follows from (1.53) and (1.54). Lemma 1.26 is proved.

COROLLARY 1.2. *For the operator* $\mathscr{B}\colon \mathscr{X} \to \mathscr{Y}^*$ *a condition of the form* (4.6.11) *holds, i.e., the function* $v \to \langle \mathscr{B}v, v\rangle$, $v \in \mathscr{V}$, *is continuous relative to the norm of the space* $\mathscr{X}$.

PROOF. We note first of all that for a function $v \in \mathscr{V}$

$$\|v\|_{\mathscr{X}} = \|v\|_{m,m_0,Q} + \|A\nabla v\|_{\mathbf{m},\mathbf{m}_0,Q} + \|v(x, T_2)\|_{2,\Omega}. \tag{1.55}$$

Indeed, let $v \in \mathscr{V}$ and let $\{v_n\}$, $v_n \in \tilde{C}^1_{0,\Sigma_1}(Q)$, $n = 1, 2, \ldots$, be a sequence which is in $\mathscr{X}$ and converges to v in $\mathscr{H}$. Then

$$\|v\|_{\mathscr{X}} = \lim_{n\to\infty} \|v_n\|_{\mathscr{X}} = \|v\|_{m,m_0,Q} + \|A\nabla v\|_{\mathbf{m},\mathbf{m}_0,Q} + \|(v)_{T_2}\|_{2,\Omega}. \tag{1.56}$$

Now by (1.43) $(v)_{T_2} = v(x, T_2)$ a.e. in Ω, whence we obtain (1.55). The validity of a condition of the form (1.6.22) follows immediately from (1.52) and (1.55). The corollary is proved.

In the sequel we shall also use the following proposition.

LEMMA 1.27. *The average* u_h *of an arbitrary function* $u \in \mathscr{W}$ *defined by*

$$u_h(t, x) = \int_{-\infty}^{+\infty} \omega_h(t - \tau)\hat{u}(\tau, x)\,d\tau, \tag{1.57}$$

where the kernel $\omega_h(\eta)$ *is defined by* (4.7.4) *and* $\hat{u}(t, x)$ *by*

$$\hat{u}(t, x) = \begin{cases} u(T_1 + \tau, x) & \text{for } t = T_1 - \tau, 0 \leqslant \tau \leqslant T, x \in \Omega, \\ u(t, x) & \text{for } t \in [T_1, T_2], x \in \Omega, \\ u(T_2 - \tau, x) & \text{for } t = T_2 + \tau, 0 \leqslant \tau \leqslant T, x \in \Omega, \\ 0 & \text{for } t \notin [T_1 - T, T_2 + T], \end{cases} \tag{1.58}$$

with $T = T_2 - T_1$, *belongs to the space* $\mathscr{Y} \subset \mathscr{W}$ *and converges to* u *in* $\mathscr{W}$ *and in* $C([T_1, T_2]; L^2(\Omega))$, *i.e.*,

$$\lim_{h\to 0}\|u_h - u\|_{\mathscr{H}} = \lim_{h\to 0}\|u_h' - u'\|_{\mathscr{H}^*} = \lim_{h\to 0}\|u_h - u\|_{C([T_1,T_2];L^2(\Omega))} = 0. \tag{1.59}$$

PROOF. In view of Lemma 4.7.5 and Corollary 4.7.2 it suffices to verify that $u_h \in \mathscr{Y}$. To prove this we consider a sequence $\{u_n\}$, $u_n \in \tilde{C}^1_{0,\Sigma_1}(Q)$, $n = 1, 2, \ldots$, converging to u in $\mathscr{X}$, and we show that the sequence $\{u_{nh}\}$, where u_{nh} obviously belongs to $\tilde{C}^1_{0,\Sigma_1}(Q)$, converges as $n \to \infty$ to u_h in $\mathscr{Y}$ (for any fixed $h \in (0, T)$). Taking into account that the matrix A does not depend on t, we obtain

$$\|A_i\nabla(u_{nh} - u_h)\|_{m_i,Q}^{m_i} \leqslant c \sup_{\substack{\tau'\in[-h,h] \\ i=1,\ldots,n}} \iint_Q |A_i\nabla(\hat{u}_n(t - \tau', x) - \hat{u}(t - \tau', x))|\,dt\,dx, \tag{1.60}$$

where c is a constant not depending on n. Because of the convergence of u_n to u in $\mathscr{H}$, it follows from (1.60) that

$$\lim_{n\to\infty} \|A_i\nabla(u_{nh} - u_h)\|_{m_i,m_{0i},Q} = 0.$$

The equality

$$\lim_{n\to\infty} \|u_{nh} - u_h\|_{m,m_0,Q} = 0$$

is proved in a completely analogous way. We now prove that

$$\lim_{n\to\infty} \|u_{nht} - u_{ht}\|_{2,Q} = 0. \tag{1.61}$$

We have

$$\|u_{nht} - u_{ht}\|_{2,Q}^2 = \iint_Q dt\,dx \left| \int_{-h}^{h} \omega_h(\tau')[\hat{u}_h(t-\tau',x) - \hat{u}(t-\tau',x)]\,d\tau' \right|^2$$

$$\leqslant \int_{-h}^{h} d\tau' \iint_Q [\hat{u}_h(t-\tau',x) - \hat{u}(t-\tau',x)]^2\,dt\,dx$$

$$\leqslant c(h) \sup_{\tau'\in[-h,h]} \iint_Q [\hat{u}_n(t-\tau',x) - \hat{u}(t-\tau',x)]^2\,dt\,dx. \tag{1.62}$$

Equality (1.61) follows from (1.62) because of the convergence of u_n to u in $L^2(Q)$, which follows from the convergence of u_n to u in $\mathscr{X}$. Finally, we prove that

$$\lim_{n\to\infty} \|u_{nh} - u_h\|_{2,\Omega_{T_2}} = 0. \tag{1.63}$$

This follows easily from the estimate

$$\|u_{nh} - u_h\|_{C([T_1,T_2])} \leqslant c\big(\|u_{nh} - u_h\|_{2,Q} + \|u_{nht} - u_{ht}\|_{2,Q}\big), \tag{1.64}$$

where c does not depend on u_{nh} or u_h, and the facts proved above (in particular, we note that $m \geqslant 2$ and $m_0 \geqslant 2$). From this it obviously follows that $u_h \in \mathscr{Y}$ for any $h \in (0, T)$. Lemma 1.27 is proved.

COROLLARY 1.3. *The set $\tilde{C}^1_{0,\Sigma_1}(Q)$ is dense in $\mathscr{W}$.*

PROOF. It follows from Lemma 1.27 that $\mathscr{Y}$ is dense in $\mathscr{W}$. From the definition of $\mathscr{Y}$ it follows that $\tilde{C}^1_{0,\Sigma_1}(Q)$ is dense in $\mathscr{Y} \subset \mathscr{W}$. A fortiori $\tilde{C}^1_{0,\Sigma_1}(Q)$ is dense in $\mathscr{Y}$ in the norm of $\mathscr{W}$. Therefore, $\tilde{C}^1_{0,\Sigma_1}(\overline{Q})$ is also dense in $\mathscr{W}$. Corollary 1.3 is proved.

The next assertion, in particular, obviously follows from Corollary 1.3.

COROLLARY 1.4. *The set $\tilde{C}^1_{0,\Sigma_1}(\overline{Q})$ is dense in the set $\mathscr{W} \supset \mathscr{V}$ relative to the norm*

$$\|u\| \equiv \|u\|_{C([T_1,T_2];L^2(\Omega))} + \|u\|_{\mathscr{H}}.$$

§2. The general boundary value problem for $(A, \mathbf{0}, m, \mathbf{m})$-parabolic equations

In the cylinder $Q = \Omega \times (T_1, T_2)$, where Ω is a bounded domain of class $\tilde{C}^{(1)}$ in $\mathbb{R}^n$, $n \geqslant 1$, we consider the equation

$$u_t - (d/dx_i)l^i(t,x,u,\nabla u) + l_0(t,x,u,\nabla u) = f(t,x). \tag{2.1}$$

We say that an equation of the form (2.1) has spatial $(A, \mathbf{0}, m, \mathbf{m})$-structure in Q if there exist a square matrix $A \equiv \|a^{ij}(x,t)\|$ of order n satisfying conditions (1.1) and (1.2) for $m > 1$, $m_0 = m$, $\mathbf{m} = \mathbf{m}_0 = (m_1, \ldots, m_n)$, $m_i > 1$, $i = 1, \ldots, n$, and functions $l'^i(t,x,u,q)$, $i = 1, \ldots, n$, and $l'_0(t,x,u,q)$ satisfying in $Q \times \mathbb{R} \times \mathbb{R}^n$ the Carathéodory condition that for almost all $(t,x) \in Q$ and any $u \in \mathbb{R}$ and $p \in \mathbb{R}^n$

$$\mathbf{l}(t,x,u,p) = A^*\mathbf{l}'(t,x,u,Ap), \qquad l_0(t,x,u,p) = l'_0(t,x,u,Ap), \tag{2.2}$$

where A^* is the matrix adjoint to A, $\mathbf{l} \equiv (l^1,\dots,l)$ and $\mathbf{l}'(l'^1,\dots,l'^n)$, and the inequalities

$$|l'^i(t, x, u, q)| \leqslant \mu_1\left(\sum_{k=1}^{n} |q_k|^{m_k/m_i'} + |u|^{m/m_i'} + \varphi_i(t, x)\right), \quad i = 1,\dots,n,$$

$$|l_0'(t, x, u, q)| \leqslant \mu_2\left(\sum_{k=1}^{n} |q_k|^{m_k/m'} + |u|^{m/m'} + \varphi_0(t, x)\right), \tag{2.3}$$

where μ_1, $\mu_2 = \mathrm{const} \geqslant 0$, $\varphi_i \in L^{m_i'}(Q)$, $1/m_i + 1/m_i' = 1$, $i = 1,\dots,n$, and $\varphi_0 \in L^{m'}(Q)$, $1/m + 1/m' = 1$.

We call an equation of the form (2.1) having spatial $(A, \mathbf{0}, m, \mathbf{m})$-structure in Q $(A, \mathbf{0}, m, \mathbf{m})$-parabolic in Q if for almost all $(t, x) \in Q$ and any $u \in \mathbb{R}$, $q = Ap$ and $p \in \mathbb{R}^n$

$$\frac{\partial l'^i(t, x, u, q)}{\partial q_j}\xi_i\xi_j \geqslant \nu \sum_{i=1}^{n} |q_i|^{m_i-2}\xi_i^2, \qquad \nu = \mathrm{const} > 0, \tag{2.4}$$

where the indices $m_1,\dots,m_n$ in (2.3) and (2.4) coincide.

It is obvious that any equation of the form (2.1) which in the cylinder $Q \subset \mathbb{R}^{h+1}$ has spatial $(A, \mathbf{0}, m, \mathbf{m})$-structure (is $(A, \mathbf{0}, m, \mathbf{m})$-parabolic in Q) also has $(\tilde{A}, \tilde{\mathbf{b}}, \tilde{m}, \tilde{\mathbf{m}})$-structure in the domain $\tilde{\Omega} \equiv Q \subset \mathbb{R}^{n+1}$ (is also $(\tilde{A}, \tilde{\mathbf{b}}, \tilde{m}, \tilde{\mathbf{m}})$-elliptic in $\tilde{\Omega} \equiv Q \subset \mathbb{R}^{n+1}$) relative to the matrix

$$\tilde{A} = \begin{Vmatrix} 0 & \cdots & 0 \\ \vdots & A & \\ 0 & & \end{Vmatrix}$$

of order $n + 1$, the $(n + 1)$-dimensional vector $\tilde{\mathbf{b}} = (1, 0,\dots,0)$, the indices $\tilde{m} = m$, $\tilde{\mathbf{m}} = (m_0, m_1,\dots,m_n)$, $m_0 = 2$, and the reduced coefficients $\tilde{l}'^i(\tilde{x}, u, \tilde{q})$, $i = 1,\dots,n + 1$, and $\tilde{l}_0'(\tilde{x}, u, \tilde{q})$, where $\tilde{x} = (t, x)$, $\tilde{q} = (q_0, q_1,\dots,q_n)$, $\tilde{l}'^1 = q_0$, $\tilde{l}'^i = l'^{(i-1)}(t, x, u, q)$, $i = 2,\dots,n + 1$, and $\tilde{l}_0' = l_0'(t, x, u, q)$.

For an equation (2.1) having spatial $(A, \mathbf{0}, m, \mathbf{m})$-structure in Q we consider the general boundary value problem of the form (5.3.1), which here takes the following form:

$$u_t - (d/dx_i)l^i(t, x, u, \nabla u) + l_0(t, x, u, \nabla u) = \mathscr{F} \quad \text{in } Q,$$
$$u = 0 \quad \text{on } \Sigma_1 \cup \Omega_{T_1}, \qquad \mathbf{l}' \cdot A\nu = 0 \quad \text{on } \Sigma_2, \qquad \mathbf{l}' \cdot A\nu - \lambda u = 0 \quad \text{on } \Sigma_3, \tag{2.5}$$

where $\Sigma = \Sigma_1 \cup \Sigma_2 \cup \Sigma_3$, $\Sigma_i \cap \Sigma_j = \varnothing$ for $i \neq j$, is the regular part of ∂Q (see (1.17) with $m = m_0$ and $m_i = m_{0i}$, $i = 1,\dots,n$). There are no boundary conditions on the part $\Sigma_{0,-}' \equiv (\sigma' \times (T_1, T_2)) \cup \Omega_{T_2}$. We henceforth understand $\mathscr{H}_\lambda$, $\mathscr{X}$, and $\mathscr{Y}$ to be the spaces (see §1) constructed on the basis of the matrix A, the indices m, m_0, $\mathbf{m}$ and $\mathbf{m}_0$ (with $m_0 = m$ and $\mathbf{m}_0 = \mathbf{m}$) characterizing the structure of equation (2.5), the sets Σ_1 and Σ_3, and also the function λ defined on Σ_3. It is assumed that λ is piecewise continuous, bounded, and positive on Σ_3. We further recall that the sets Σ_i have the form $\Sigma_i = \sigma_i \times (T_1, T_2)$, $i = 1, 2, 3$. According to the general case (see §5.2) the operator $\mathscr{L}: \mathscr{H} \to \mathscr{Y}^*$ corresponding to problem (2.5) has the form

$$\mathscr{L}=\mathscr{A}+\mathscr{B},\qquad \mathscr{A}:\mathscr{X}\to\mathscr{H}_\lambda^*\subset\mathscr{Y}^*,\qquad \mathscr{B}:\mathscr{X}\to\mathscr{Y}^*,$$

$$\langle\mathscr{A}u,\eta\rangle=\iint_Q[\mathbf{l}'(t,x,u,A\nabla u)\cdot A\nabla\eta + l_0'(t,x,u,A\nabla u)\eta]\,dt\,dx+\int_{\Sigma_3}\lambda u\eta\,ds,$$

$$\langle\mathscr{B}u,\eta\rangle=-\iint_Q u\eta_t\,dt\,dx+\int_{\Omega_{T_2}}u\eta\,dx, \tag{2.6}$$

where $u\in\tilde{C}^1_{0,\Sigma_1}(Q)$ and $\eta\in\tilde{C}^1_{0,\Sigma_1}(Q)$ (it is obvious that the operator $\mathscr{L}:\mathscr{X}\to\mathscr{Y}^*$ is completely determined by the values of $\langle\mathscr{L}u,\eta\rangle$ for such u and η). For the rest of this section we always assume that condition (1.36) in the case $m_0=m$, $\mathbf{m}_0=\mathbf{m}$ is satisfied. In this case the operator $\mathscr{L}:\mathscr{X}\to\mathscr{Y}^*$ can be defined in the form (see Lemmas 1.17 and 1.25)

$$\langle\mathscr{L}u,\eta\rangle=\int_{T_1}^{T_2}(u',\eta)\,dt+\iint_Q(\mathbf{l}'(t,x,u,A\nabla u)\cdot A\nabla\eta + l_0'(t,x,u,A\nabla u)\eta)\,dt\,dx+\int_{\Sigma_3}\lambda u\eta\,ds,\quad u\in V,\ \eta\in\tilde{C}^1_{0,\Sigma_1}(Q), \tag{2.7}$$

where $(\cdot,\cdot)$ is the duality pairing between $H\equiv H_{m,\mathbf{m}}(A,\Omega)$ and H^*, and u' is the derivative of u as an element of $\mathscr{D}^*([T_1,T_2];H^*)$.

A generalized solution of problem (2.5) can then be defined (see §5.3 and, in particular, Proposition (5.3.1)) as any function $u\in\mathscr{W}_0$, where $\mathscr{W}_0\equiv\{u\in\mathscr{H}: u'\in\mathscr{H}^*,\ u(T_1)=0\}$, satisfying the identity

$$\int_{T_1}^{T_2}(u',\eta)\,dt+\iint_Q(\mathbf{l}'\cdot A\nabla\eta+l_0'\eta)\,dt\,dx+\int_{\Sigma_3}\lambda u|_{\tilde\Sigma}\eta\,ds=\langle\mathscr{F},\eta\rangle,\qquad \forall\eta\in\tilde{C}^1_{0,\Sigma_1}(Q), \tag{2.8}$$

where $\tilde\Sigma=\Sigma_1\cup\Sigma_3$ and $\mathscr{F}\in\mathscr{H}^*$. We note that because of the imbedding $\mathscr{W}\subset\mathscr{H}$ and the fact that $\mathscr{H}$ and $\mathscr{H}_\lambda$ are isomorphic, functions $u\in\mathscr{W}_0$ automatically have generalized limit values on the set $\tilde\Sigma\equiv\Sigma_1\cup\Sigma_3$ (see §5.3), and $u|_{\tilde\Sigma}\in L^1_{\mathrm{loc}}(\tilde\Sigma)\cap L^2(\lambda,\Sigma_3)$. We note that the definition of a generalized solution of problem (2.5) given above is equivalent to the definition which follows from the general definition given in Chapter 5, §3, since we have the following assertion.

PROPOSITION 2.1. $\mathscr{W}_0=\mathscr{V}$.

PROOF. Because of Lemmas 1.22 and 1.24, we have $\mathscr{V}\subset\mathscr{W}_0$. We prove the reverse inclusion $\mathscr{W}_0\subset\mathscr{V}$. Let $u\in\mathscr{W}_0\subset\mathscr{H}$. In view of Corollary 1.3 there exists a sequence $\{u_n\}$, $u_n\in\tilde{C}^1_{0,\Sigma_1}(Q)$, $n=1,2,\ldots$, which converges to u in $\mathscr{W}$ and hence also in $\mathscr{X}$. On the other hand, in view of an equality of the form (4.7.29) for elements of $\mathscr{W}$ for any $\eta\in\tilde{C}^1_{0,\Sigma_1\cup\Omega_{T_2}}(Q)$ and $\{\eta_k\}$, $\eta_k\in\tilde{C}^1_{0,\Sigma_1\cup\Omega_{T_2}}(Q)$, $k=1,2,\ldots$, such that $\eta_k\to\eta$ in $\mathscr{H}$, we have

$$\lim_{k\to\infty}\iint_Q u(\eta_k-\eta)_t\,dt\,dx=\lim_{k\to\infty}\int_{T_1}^{T_2}(u',\eta_k-\eta)\,dt=0.$$

It then follows from Lemma 1.18 that $u\in\mathscr{V}$. Proposition 2.1 is proved.

In place of $u|_{\tilde\Sigma}$ in the integral over Σ_3 we henceforth write simply u.

LEMMA 2.1. *Let $u \in \mathscr{W}_0$ be a generalized solution of problem* (2.5). *Then for any* $\tau_1, \tau_2 \in [T_1, T_2]$

$$\int_{\tau_1}^{\tau_2}(u', \eta)\,dt + \int_{\tau_1}^{\tau_2}\int_{\Omega}(\mathrm{l}' \cdot A\nabla\eta + l_0'\eta)\,dt\,dx + \int_{\sigma_3\times(\tau_1,\tau_2)}\lambda u\eta\,ds = \int_{\tau_1}^{\tau_2}(F(t), \eta)\,dt, \qquad \forall \eta \in \mathscr{H}_{m,\mathrm{m}}^{0,\Sigma_1}(A, Q), \tag{2.9}$$

where $F(t) \equiv \mathscr{F} \in \mathscr{H}^*$.

PROOF. Let $u \in \mathscr{V}$ be a solution of (2.5) with $\mathscr{F} \in \mathscr{H}^*$. An identity of the form (2.9) holds for this function with $\tau_1 = T_1$ and $\tau_2 = T_2$. In (2.9) we set $\eta = \Phi(t, x)\zeta_k(t)$, where $\Phi(t, x) \in \tilde{C}^1_{0,\sigma_1\times(T_1,T_2)}(\overline{Q})$, $\zeta_k \in C_0^1(T_1, T_2)$, $k = 1, 2, \dots,$ and the sequence $\{\zeta_k\}$ converges on $[T_1, T_2]$ to a function $\zeta(t) = \zeta_{\varepsilon,\tau_1,\tau_2}(t)$ continuous on $[T_1, T_2]$, equal to 1 in $(\tau_1 + \varepsilon/2, \tau_2 - \varepsilon/2)$, linear in $(\tau_1 - \varepsilon/2, \tau_1 + \varepsilon/2)$ and $(\tau_2 - \varepsilon/2, \tau_2 + \varepsilon/2)$, and equal to 0 in $(T_1, \tau_1 - \varepsilon/2)$ and $(\tau_2 + \varepsilon/2, T_2)$, where $T_1 < \tau_1 - \varepsilon/2 < \tau_1 + \varepsilon/2 < \tau_2 - \varepsilon/2 < \tau_2 < \tau_2 + \varepsilon/2 < T_2$, $\varepsilon > 0$. We further assume that $\partial\zeta_k/\partial t$ tends to $\partial\zeta/\partial t$ in $L^{m'}([T_1, T_2])$. (For the ζ_k it is possible to take, for example, averages of the function ζ with increment $h = c/k$, $c = \text{const.}$) From (2.9) we then obtain

$$\iint_Q(\mathrm{l}' \cdot A\nabla\Phi + l_0'\Phi)\zeta_k\,dt\,dx + \int_{\sigma_3\times(T_1,T_2)}\lambda u\Phi\zeta_k\,ds - \iint_Q(u\Phi_t\zeta_k + u\Phi\zeta_{kt}')\,dt\,dx = \langle\mathscr{F}, \Phi\zeta_k\rangle, \qquad k = 1, 2, \dots. \tag{2.10}$$

Letting k tend to ∞ and applying the Lebesque theorem, we obtain

$$\iint_Q(\mathrm{l}' \cdot A\nabla\Phi + l_0'\Phi)\zeta\,dt\,dx + \int_{\Sigma_3}\lambda u\Phi\zeta\,ds - \iint_Q(u\Phi_t\zeta + u\Phi\zeta')\,dt\,dx = \int_{T_1}^{T_2}(F, \Phi\zeta)\,dt, \qquad \Phi \in \tilde{C}^1_{0,\Sigma_1}(Q). \tag{2.11}$$

Letting ε tend to 0 in this identity, again using the Lebesgue theorem, and taking into account that by the continuity of u in $L^2(\Omega)$ on $[T_1, T_2]$

$$\lim_{\varepsilon\to 0} 1/\varepsilon\int_{\tau-\varepsilon/2}^{t+\varepsilon/2}u\Phi\,dt\,dx = \int_{\Omega}u\Phi\,dx\Big|^{t=\tau}, \qquad \tau \in [T_1, T_2],$$

we find that (2.9) holds for all $\eta \in \tilde{C}^1_{0,\Sigma_1}(Q)$. Taking the density of $\tilde{C}^1_{0,\Sigma_1}(Q)$ in $\mathscr{H}$ and the linearity of the expression (2.9) in η into account, from this we deduce the validity of (2.9) for all $\eta \in \mathscr{H}$ also. Lemma 2.1 is proved.

The results on solvability of the general boundary value problem of the form (2.5) are determined by Theorems 5.4.1–5.4.5 for the general case, by Lemma 1.20 and Corollary 1.2 leading to the validity of conditions of the form (4.6.10) and (4.6.11), by Lemmas 1.10 and 1.25 guaranteeing the validity of conditions (5.2.25) and (5.4.1), and also by the algebraic criteria for coercivity and monotonicity of the operator $\mathscr{L}$: $\mathscr{X} \to \mathscr{T}^*$ which follow from Propositions 5.4.3 and 5.4.6. In addition we shall use the following assertion.

LEMMA 2.2. *Suppose that for all $t \in [T_1, T_2]$ the operator A_t: $H \to H^*$, where $H = H^{0,\sigma_1}_{m,\mathbf{m}}(A; \Omega, \Sigma_3, \lambda)$, defined by*

$$\langle A_t u, \eta \rangle = \int_\Omega [\mathbf{l}'(t, x, u, A\nabla u) \cdot A\nabla\eta + l'_0(t, x, u, A\nabla u)\eta]\, dx$$

$$+ \int_{\sigma_3} \lambda u \eta\, ds, \qquad \eta \in H, \tag{2.12}$$

is monotone. Then for every $\mathscr{F} \in \mathscr{H}^*$ problem (2.5) *has at most one generalized solution.*

PROOF. Let $u_1, u_2 \in \mathscr{W}$ be any two generalized solutions of (2.5). For each of the functions u_1 and u_2 there is an identity of the form (2.9) with $\tau_1 = T_1$ and $\tau_2 = t$, $t \in (T_1, T_2\,]$. Setting $\eta = u_i$ $(i = 1, 2)$ in the identity for u_i, subtracting from the equality obtained for u_1 the analogous equality obtained for u_2, and taking into account Lemmas 2.1 and 1.25, we obtain

$$\int_0^t \langle A_\tau u_1 - A_\tau u_2, u_1 - u_2 \rangle d\tau + \|u_1 - u_2\|_{L^2(\lambda, (\Sigma_3)_t)} + \frac{1}{2}\|u_1 - u_2\|_{2,\Omega_t} = 0, \tag{2.13}$$

where $(\Sigma_3)_t = \sigma_3 \times (T_1, t)$. Because of the monotonicity of the operators A_τ, $\tau \in [T_1, T_2]$, it follows from (2.13) that $\|u_1 - u_2\|_{2,\Omega_t} = 0$, $t \in [T_1, T_2]$, i.e., $u_1 = u_2$ a.e. in Q. Lemma 2.2 is proved. In particular, this enables us to formulate the following results.

THEOREM 2.1. *Let conditions* (1.1), (1.2), (1.17), *and* (1.36) *be satisfied, and suppose that for almost all $(t, x) \in Q$ and any $u \in \mathbb{R}$, $\xi_0 \in \mathbb{R}$, $q = Ap$, $p \in \mathbb{R}^n$, $\eta = A\xi$ and $\xi \in \mathbb{R}^n$*

$$l'^i(t, x, u, q) q_i + l'_0(t, x, u, q) u \geqslant \nu_1 \sum_{i=1}^n |q_i|^{m_i} + \nu_2 |u|^m - \varphi(t, x), \tag{2.14}$$

where $\nu_1, \nu_2 = \text{const} > 0$, $\varphi \in L^1(Q)$, and

$$\frac{\partial l'^i(t, x, u, q)}{\partial q_j} \eta_i \eta_j + \frac{\partial l'^i(t, x, u, q)}{\partial u} \xi_0 \eta_i$$

$$+ \frac{\partial l'_0(t, x, u, q)}{\partial q_j} \eta_j \xi_0 + \frac{\partial l'_0(t, x, u, q)}{\partial u} \xi_0^2 \geqslant 0. \tag{2.15}$$

Then for every $\mathscr{F} \in \mathscr{H}^*$ problem (2.5) *has precisely one generalized solution.*

PROOF. Since, as already noted, in the present case conditions of the form (4.6.9), (4.6.10), and (4.6.11) are satisfied, the result of Theorem 2.1 follows from Theorem 5.4.1 and Lemma 2.2, on taking into account that (2.15) implies the monotonicity of operators of the form (2.12) (see Proposition 5.4.5), while (2.14) implies the coercivity of the operator $\mathscr{L}$: $\mathscr{X} \to \mathscr{Y}^*$ (see Proposition 5.4.4). Theorem 2.1 is proved.

THEOREM 2.2. *Suppose that conditions* (1.1), (1.2), (1.17), *and* (1.36) *are satisfied, and suppose that for almost all* $(t, x) \in Q$ *and any* $u \in \mathbb{R}$, $\eta = A\xi$, $\xi \in \mathbb{R}^n$, $q = Ap$, $p \in \mathbb{R}^n$ *and* $\xi_0 \in \mathbb{R}$

$$\frac{\partial l'^i(t,x,u,q)}{\partial q_j}\eta_i\eta_j + \frac{\partial l'^i(t,x,u,q)}{\partial u}\xi_0\eta_i + \frac{\partial l'_0(t,x,u,q)}{\partial q_j}\eta_j\xi_0 + \frac{\partial l'_0(t,x,u,q)}{\partial u}\xi_0^2$$

$$\geqslant \alpha_0\left[\sum_{i=1}^n |q_i|^{m_i-2}\eta_i^2 + \left(|u|^{m-2}+1\right)\xi_0^2\right], \qquad \alpha_0 = \text{const} > 0. \qquad (2.16)$$

Then for every $\mathscr{F} \in \mathscr{H}^*$ *problem* (2.5) *has precisely one generalized solution. Moreover, the restriction* $\hat{\mathscr{L}}$: $\mathscr{V} \subset \mathscr{X} \to \mathscr{H}^* \subset \mathscr{Y}^*$ *of the operator* $\mathscr{L}$: $\mathscr{X} \to \mathscr{Y}^*$ *to the set* $\mathscr{V}$ *is a homeomorphism.*

PROOF. This follows directly from Theorem 5.4.4 and the analogue of Proposition 5.4.6.

THEOREM 2.3. *For* $m = 2$ *the result of Theorem* 2.1 *is preserved if the positive constant* ν_2 *in* (2.14) *is replaced by any negative constant.*

THEOREM 2.4. *For* $m = 2$ *the result of Theorem* 2.2 *is preserved if* (2.16) *is replaced by the inequality*

$$\frac{\partial l'^i}{\partial q_j}\eta_i\eta_j + \frac{\partial l'^i}{\partial u}\eta_i\xi_0 + \frac{\partial l'^i}{\partial q_j}\eta_j\xi_0 + \frac{\partial l'_0}{\partial u}\xi_0^2 \geqslant \alpha_0\sum_{i=1}^n |q_i|^{m_i-2}\eta_i^2 - \beta_0\xi_0^2, \quad (2.17)$$

where α_0 *and* β_0 *are positive constants.*

Theorems 2.3 and 2.4 follow from Theorems 2.1 and 2.2 in view of the fact that the conditions of Theorems 2.3 and 2.4 reduce to those of Theorems 2.1 and 2.2 by introducing a new unknown function according to the formula $u = e^{\gamma t}\bar{u}$, $\gamma = \text{const} > 0$.

Admissibility of the matrix A, regularity of the part $\Sigma = \sigma \times (T_1, T_2)$, and singularity of the part $\Sigma' = \sigma' \times (T_1, T_2)$ were postulated in Theorems 2.1–2.4. Sufficient conditions for these assumptions to be satisfied were presented in §1. Taking these conditions into account, for the readers' convenience we here present the following version of Theorems 2.1–2.4.

THEOREM 2.5. *Suppose that equation* (2.1) *has spatial* $(A, \mathbf{0}, m, \mathbf{m})$*-structure in the cylinder* Q *and the matrix* $A = A(x)$ *satisfies the conditions*

$$a^{ij} \in L^{m_i}(\Omega), \qquad i, j = 1,\dots,n; \quad a^{ij}, \frac{\partial a^{ij}}{\partial x_j} \in L^{m'}(\Omega), \quad \frac{1}{m} + \frac{1}{m'} = 1. \quad (2.18)$$

Assume also that condition (1.36) *is satisfied. Let* $\widetilde{\partial\Omega} = \sigma \cup \sigma'$, *where condition* (4.2.5) *holds for the set* σ *and condition* (4.3.2) *holds for* σ' (*relative to the matrix* $A = A(x)$, m, *and* $\mathbf{m}$). *Suppose, finally, that conditions* (2.14) *and* (2.15) *hold. Then for every* $\mathscr{F} \in \mathscr{H}^*$ *problem* (2.5) (*with* $\Sigma = \sigma \times (T_1, T_2)$ *and* $\Sigma' = (\sigma' \times (T_1, T_2)) \cup \Omega_{T_1} \cup \Omega_{T_2}$) *has precisely one generalized solution. If inequality* (2.16) *is satisfied in place of* (2.14) *and* (2.15), *then this solution depends continuously in* $\mathscr{V} \subset \mathscr{X}$ *on* $\mathscr{F} \in \mathscr{H}^* \subset \mathscr{Y}^*$. *For* $m = 2$ *in condition* (2.14) *it may be assumed that* ν_2 *is any constant and* (2.16) *may be replaced by the less stringent condition* (2.17).

§3. $(A, \mathbf{0})$-parabolic equations with weak degeneracy

1. $(A, \mathbf{0}, m, \mathbf{m})$-parabolic equations with weak degeneracy. Suppose a square matrix $A = \|a^{ij}(t, x)\|$ of order n is defined in the cylinder $Q = \Omega \times (T_1, T_2)$, where Ω is a bounded, strongly Lipschitz domain in $\mathbb{R}^n$, $n \geqslant 1$, and

$$a^{ij} \in L^{m_i, m_{0i}}(Q), \qquad m_i > 1, \qquad m_{0i} > 1, \qquad i, j = 1, \ldots, n. \tag{3.1}$$

We call the matrix A weakly degenerate in the cylinder Q if $\det A \neq 0$ in Q and there exist indices $\mathbf{q} = (q_1, \ldots, q_n)$ and $\mathbf{q}_0 = (q_{01}, \ldots, q_{0n})$, $q_i \geqslant 1$, $q_{0i} \geqslant 1$, $i = 1, \ldots, n$, such that

$$\|\nabla u\|_{\mathbf{q}, \mathbf{q}_0, \hat{Q}} \leqslant c \|A \nabla u\|_{\mathbf{m}, \mathbf{m}_0, \hat{Q}}, \quad \forall u \in \tilde{C}^1(Q), \tag{3.2}$$

where $\hat{Q} = \hat{\Omega} \times (T_1, T_2)$, $\hat{\Omega} \subseteq \Omega$, and the constant c in (3.2) depends only on n, $\mathbf{m}$, $\mathbf{m}_0$, $\mathbf{q}$, $\mathbf{q}_0$, and Q.

LEMMA 3.1. *Suppose the matrix A (satisfying condition (3.1)) for almost all $(t, x) \in Q$ has an inverse matrix $A^{-1} \equiv B \equiv \|b^{ij}\|$, where*

$$b^{ij} \in L^{r_{ij}, r^0_{ij}}(Q), \qquad r_{ij} \geqslant 1, \qquad r^0_{ij} \geqslant 1, \qquad i, j = 1, \ldots, n;$$

$$\max_{k=1,\ldots,n} \left(\frac{1}{m_k} + \frac{1}{r_{ik}} \right) \leqslant 1, \qquad \max_{k=1,\ldots,n} \left(\frac{1}{m_{0k}} + \frac{1}{r^0_{ik}} \right) \leqslant 1, \qquad i = 1, \ldots, n. \tag{3.3}$$

Then the matrix A is weakly degenerate in the cylinder Q, and inequality (3.2) is satisfied with indices $\mathbf{q} = (q_1, \ldots, q_n)$ and $\mathbf{q}_0 = (q_{01}, \ldots, q_{0n})$, where

$$1/q_i = \max_{k=1,\ldots,n} (1/m_k + 1/r_{ik}),$$

$$1/q_{0i} = \max_{k=1,\ldots,n} (1/m_{0k} + 1/r^0_{ik}), \qquad i = 1, \ldots, n. \tag{3.4}$$

PROOF. Lemma 3.1 is proved in exactly the same way as Lemma 7.2.1.

In the case where the matrix A depends only on x it will certainly be weakly degenerate in the cylinder Q if it is weakly degenerate in Ω in the sense of Definition 7.2.1.

LEMMA 3.2. *Suppose the matrix $A \equiv \|a^{ij}(t, x)\|$ is weakly degenerate in the cylinder Q. Then condition (1.2) holds for this matrix. If a function $u \in L^{m, m_0}_{\text{loc}}(Q)$ ($u \in L^{m, m_0}(Q)$), m, $m_0 \geqslant 1$, has a generalized spatial A-gradient $A \nabla u \in \mathbf{L}^{\mathbf{m}, \mathbf{m}_0}_{\text{loc}}(Q)$ ($A \nabla u \in \mathbf{L}^{\mathbf{m}, \mathbf{m}_0}(Q)$), $\mathbf{m}$, $\mathbf{m}_0 \geqslant \mathbf{1}$, then it also has an ordinary (Sobolev) generalized spatial gradient $\nabla u \in \mathbf{L}^{\mathbf{q}, \mathbf{q}_0}_{\text{loc}}(Q)$ ($\nabla u \in \mathbf{L}^{\mathbf{q}, \mathbf{q}_0}(Q)$) (see (3.2)), and the vector $A \nabla u$ is equal to the vector obtained by the action of the matrix A on the vector ∇u.*

PROOF. Lemma 3.2 is proved in exactly the same way as Lemmas 7.2.2 and 7.2.3.

LEMMA 3.3. *Suppose the matrix $A \equiv \|a^{ij}(t, x)\|$ is weakly degenerate in the cylinder Q. Then there is a compact imbedding*

$$\mathscr{H}_{m, m_0; \mathbf{m}, \mathbf{m}_0}(A, Q) \to L^1(\partial\Omega \times (T_1, T_2))$$

such that the entire lateral surface is the regular part of ∂Q relative to the matrix A and the indices m, m_0, $\mathbf{m}$ and $\mathbf{m}_0$ (see §1). Any function $u \in \mathscr{H}_{m, m_0; \mathbf{m}, \mathbf{m}_0}(A, Q)$ has a generalized limit value $u|_{\partial\Omega \times (T_1, T_2)}$ which coincides with the trace on $\partial\Omega \times (T_1, T_2)$ of this function considered as an element of the space $\mathscr{H}_{m, m_0; \mathbf{q}, \mathbf{q}_0}(Q)$.

PROOF. Lemma 3.3 follows in an obvious way from the definition of the space $\mathscr{H}_{m,m_0;\,\mathbf{m},\mathbf{m}_0}(A,Q)$, condition (3.2), and the familiar imbedding theorem of S. L. Sobolev.(1)

In the cylinder Q we consider an equation of the form (2.1). We suppose first that this equation has spatial $(A,\mathbf{0},m,\mathbf{m})$-structure in Q relative to a matrix A and indices $m>1$ and $\mathbf{m}>\mathbf{1}$. Suppose that A is weakly degenerate in Q, and that condition (3.1) holds for it with $\mathbf{m}_0=\mathbf{m}$, while (3.2) holds with $\mathbf{m}_0=\mathbf{m}$ and some $\mathbf{q}$ and $\mathbf{q}_0$. We consider the general boundary value problem of the form (2.5). Under the assumptions of the present section $\Sigma_1\cup\Sigma_2\cup\Sigma_3=\partial\Omega\times(T_1,T_2)$, since the entire lateral surface of Q is the regular part of ∂Q. Suppose that a condition of the form (1.36) is satisfied in the case $m_0=m$, $\mathbf{m}_0=\mathbf{m}$. As in §2, a generalized solution of problem (2.5) is, by definition, any function $u\in\mathscr{W}_0$ satisfying (2.8) with $\Sigma=\partial\Omega\times(T_1,T_2)$. It is obvious that the results on solvability of problem (2.5) obtained in §2 (see Theorems 2.1–2.4) are also applicable to the case of the equations considered in this section; namely, we have the following assertion.

THEOREM 3.1. *Suppose that equation* (2.1) *has spatial* $(A,\mathbf{0},m,\mathbf{m})$*-structure in the cylinder* Q, *and conditions* (3.1), (3.26) *and* (1.36) *with* $m_0=m$ *and* $\mathbf{m}_0=\mathbf{m}$, *and also conditions* (2.14) *and* (2.15), *are satisfied. Then for every* $\mathscr{F}\in\mathscr{H}^*$ *problem* (2.5) (*where* $\Sigma=\partial\Omega\times(T_1,T_2)$ *and* $\Sigma'=\Omega_{T_1}\cup\Omega_{T_2}$) *has precisely one generalized solution. If condition* (2.16) *is satisfied in place of* (2.14) *and* (2.15), *then this solution depends continuously in* $\mathscr{V}\subset\mathscr{X}$ *on* $\mathscr{F}\in\mathscr{H}^*\subset\mathscr{Y}^*$. *For* $m=2$ *in condition* (2.14) *it may be assumed that* ν_2 *is any constant, and* (2.16) *may be replaced by the less stringent condition* (2.17).

PROOF. Theorem 3.1 follows directly from Theorems 2.1–2.4, since the weak degeneracy of A in Q implies condition (1.2) and also the regularity of $\partial\Omega\times(T_1,T_2)$.

2. The space $\tilde{\mathscr{H}}$. Below we shall establish solvability theorems for a problem of the form (2.5) for several other classes of $(A,\mathbf{0})$-parabolic equations with weak degeneracy in Q. In connection with this we first consider some questions of function theory. Let $\Gamma=\gamma\times(T_1,T_2)$, $\gamma\subset\partial\Omega$, where we assume that $\operatorname{meas}_{n-1}\gamma>0$ if $\gamma\neq\varnothing$. Suppose that A is weakly degenerate in Q in the general sense indicated at the beginning of this section. We denote by $\tilde{\mathscr{H}}\equiv\widetilde{\mathscr{H}^{0,\Gamma}_{\mathbf{m},\mathbf{m}_0}}(A,Q)$ the completion of $\tilde{C}^1_{0,\Gamma}(Q)$ in the norm

$$\|u\|_{\tilde{\mathscr{H}}}\equiv\|u\|_{2,\infty,Q}+\|A\nabla u\|_{\mathbf{m},\mathbf{m}_0,Q}. \tag{3.5}$$

In the case $\mathbf{m}=\mathbf{m}_0$ we denote this space by $\tilde{\mathscr{H}}\equiv\widetilde{\mathscr{H}^{0,\Gamma}_{\mathbf{m}}}(A,Q)$. If $\gamma=\varnothing$ we omit the upper indices in the notation for this space.

LEMMA 3.4.

$$\widetilde{\overset{0,\Gamma}{\mathscr{H}}}_{\mathbf{m},\mathbf{m}_0}(A,Q)\to\widetilde{\overset{0,\Gamma}{\mathscr{H}}}_{\mathbf{q},\mathbf{q}_0}(Q)\to L^{\bar{l},\bar{l}_0}(Q), \tag{3.6}$$

where the indices $\mathbf{q}$ *and* $\mathbf{q}_0$ *are related to the indices* $\mathbf{m}$ *and* $\mathbf{m}_0$ *by condition* (3.2), $\widetilde{\mathscr{H}^{0,\Gamma}_{\mathbf{q},\mathbf{q}_0}}(Q)$ *is the space defined in Chapter* 4, §5, *and the indices* $\bar{l}$ *and* $\bar{l}_0$ *are defined by*

(1) It is actually possible to prove an imbedding $\mathscr{H}\to L^r(\partial\Omega\times(T_1,T_2))$ for some r depending on the indices $\mathbf{q}$ and $\mathbf{q}_0$ in (3.2). However, the imbedding $\mathscr{H}\to L^1(\partial\Omega\times(T_1,T_2))$ suffices for our purposes.

(4.5.11) *and* (4.5.6) *relative to the indices* $\mathbf{q}$ *and* $\mathbf{q}_0$ *in condition* (3.2) (*see Remark* 4.5.3). *In particular, for any* $u \in \widetilde{\mathcal{H}^{0,\Gamma}_{\mathbf{m},\mathbf{m}_0}}(A, Q)$

$$\|u\|_{\bar{l},\bar{l}_0,Q} \leqslant c_0\|u\|_{\tilde{\mathcal{H}}}, \tag{3.7}$$

where the constant c_0 *depends only on* n, $\mathbf{q}$, $\mathbf{q}_0$, α, Ω, *and the constant in condition* (3.2). *If for some* $\alpha \in (0,1)$ *and* $\beta \in (0,1)$ *the indices* l *and* l_0 *satisfy conditions* (4.5.13), *then for any* $\varepsilon > 0$

$$\|u\|_{l,l_0,Q} \leqslant \varepsilon\|u\|_{\tilde{\mathcal{H}}} + c_1\varepsilon^{-\lambda}\|u\|_{2,1,Q}. \tag{3.8}$$

If, however, for some $\alpha \in (0,1)$ *and* $\beta \in (0,1)$ *the indices* l *and* l_0 *satisfy conditions* (4.5.15) *instead of* (4.5.13), *then*

$$\|u\|_{l,l_0,Q} \leqslant \varepsilon\|u\|_{\tilde{\mathcal{H}}} + c_2\varepsilon^{-\lambda}\|u\|_{2,Q}. \tag{3.9}$$

The constants c_1 *and* c_2 *in* (3.8) *and* (3.9) *depend only on* n, $\mathbf{q}$, $\mathbf{q}_0$, α, β, *and* Ω, *while* $\lambda > 0$ *depends only on* α *and* β. *In the case* $\Gamma = \partial\Omega \times (T_1, T_2)$ *and* $\Sigma_1^n 1/q_i > 1$ *the constants* c_0, c_1 *and* c_2 *in* (3.7)–(3.9) *do not depend on* Ω.

PROOF. Lemma 3.4 follows directly from Lemma 4.5.3 and condition (3.2).

REMARK 3.1. From Lemma 3.4 we obtain, in particular, the following results:

1) There is the imbedding $\widetilde{\mathcal{H}^{0,\Gamma}_{\mathbf{m},\mathbf{m}_0}}(A, Q) \to L^{\bar{l}}(Q)$, where $\bar{l} = 2 + \hat{l}_0 - 2\hat{l}_0/\hat{l}$ and $\hat{l}$ and $\hat{l}_0$ are defined in (4.5.6) on the basis of the indices $\mathbf{q}$ and $\mathbf{q}_0$ in (3.2); in particular, for all $u \in \widetilde{\mathcal{H}^{0,\Gamma}_{\mathbf{m},\mathbf{m}_0}}(A, Q)$

$$\|u\|_{\bar{l},Q} \leqslant c_0\|u\|_{\tilde{\mathcal{H}}}. \tag{3.10}$$

2) For any l satisfying (for some $\beta \in (0,1)$) condition (4.5.28), and any $\varepsilon > 0$,

$$\|u\|_{l,Q} \leqslant \varepsilon\|u\|_{\tilde{\mathcal{H}}} + c_1\varepsilon^{-\lambda}\|u\|_{1,Q}, \quad \forall u \in \tilde{\mathcal{H}}. \tag{3.11}$$

3) For any l satisfying (for some $\beta \in (0,1)$) condition (4.5.30), and any $\varepsilon > 0$,

$$\|u\|_{l,Q} \leqslant \varepsilon\|u\|_{\tilde{\mathcal{H}}} + c_2\varepsilon^{-\lambda}\|u\|_{2,Q}, \quad \forall u \in \tilde{\mathcal{H}}. \tag{3.12}$$

Henceforth in this section the space $\widetilde{\mathcal{H}^{0,\Gamma}_{\mathbf{m},\mathbf{m}_0}}(A, Q)$ is considered with $\Gamma = \Sigma_1$, where Σ_1 is the portion of $\partial\Omega \times (T_1, T_2)$ on which the first (homogeneous) boundary condition is prescribed. We further introduce the space

$$\tilde{\mathcal{H}}_\lambda \equiv \widetilde{\mathcal{H}^{0,\Sigma_1}_{\mathbf{m},\mathbf{m}_0}}(A, Q; \Sigma_3, \lambda)$$

as the completion of $\tilde{C}^1_{0,\Sigma_1}(Q)$ in the norm

$$\|u\|_{\tilde{\mathcal{H}}_\lambda} \equiv \|u\|_{\tilde{\mathcal{H}}} + \|u\|_{L^2(\lambda,\Sigma_3)}, \tag{3.13}$$

where $\lambda \in L^1(\Sigma_3)$ is a positive function on Σ_3. Since $\tilde{\mathcal{H}}_\lambda \to \tilde{\mathcal{H}}$, the results of Lemma 3.4 also pertain directly to the space $\tilde{\mathcal{H}}_\lambda$. We consider the space

$$\mathcal{W} \equiv \{u \in \mathcal{H}: u' \in \mathcal{H}^*\}, \tag{3.14}$$

where $\mathcal{H} \equiv \mathcal{H}^{0,\Sigma_1}_{\bar{l},\bar{l}_0;\mathbf{m},\mathbf{m}_0}(A, Q)$.

PROPOSITION 3.1. *If condition* (1.36) *is satisfied, then* $\mathcal{W} \subset \tilde{\mathcal{H}}$.

PROOF. By Corollary 1.4 the set $\tilde{C}^1_{0,\Sigma_1}(Q)$ is dense in $\mathcal{V}$ relative to the norm

$$\|u\| = \|u\|_{C([T_1,T_2];L^2(\Omega))} + \|u\|_{\bar{l},\bar{l}_0,Q} + \|A\nabla u\|_{\mathbf{m},\mathbf{m}_0,Q},$$

whence, with Corollary 1.3 taken into account, we obtain the imbedding $\mathscr{W} \subset \tilde{\mathscr{H}}$.

LEMMA 3.5. *Suppose that*

$$\frac{1}{l} = \frac{\alpha}{\hat{l}} + \frac{1-\alpha}{2}; \qquad \frac{1}{l_0} = \frac{\alpha}{\hat{l}_0} + \frac{(1-\alpha)\beta}{2}; \qquad \alpha, \beta \in (0,1); \hat{l} > 2, \qquad (3.15)$$

where $\hat{l}$ and $\hat{l}_0$ are defined as functions of $\mathbf{q}$ and $\mathbf{q}_0$ in (3.2) *by* (4.5.6). *Then there is the compact imbedding $\mathscr{W}_0 \to L^{l,l_0}(Q)$.*

PROOF. In view of Proposition 3.1 and Lemma 3.4 (see (3.9)) it suffices to prove compactness of the imbedding $\mathscr{W}_0 \to L^2(Q)$. We shall first prove the compactness of the imbedding $\mathscr{W}_0 \to L^{2,1}(Q)$. For this we use Lemma 4.4.6. Let $\{u_n\}$ be a sequence converging weakly to u in $\mathscr{W}$, where u_n, $u \in \mathscr{W}_0$, $n = 1, 2, \dots$. It may then be assumed that $\{u_n\}$ converges weakly to u in $\mathscr{H}$ and $L^2(Q)$, while $\{u_n'\}$ converges weakly to u' in $\mathscr{H}^*$. We shall prove that a subsequence $\{u_\nu\}$ converges strongly to u in $L^{2,1}(Q)$. Let $\{\psi_k(x)\}$ be an orthonormal basis in $L^2(\Omega)$, where $\psi_k \in \widetilde{H_{\mathbf{m}}^{0,\Sigma_1}}(A, \Omega)$, $k = 1, 2, \dots$. In view of Proposition 3.2 and condition (3.2) the functions u_n and u for almost all $t \in [T_1, T_2]$ belong to $\widetilde{H_{\mathbf{q}}^{0,\sigma_1}}(\Omega)$, where $\mathbf{q}$ and $\mathbf{q}_0$ are related to $\mathbf{m}$ and $\mathbf{m}_0 = \mathbf{m}$ by condition (3.2). Taking the condition $\hat{l} > 2$ (see (3.15)), we conclude that for the values of t indicated above inequalities of the form (4.4.22) are satisfied for the differences $u_n - u$, $n = 1, 2 \dots$. Integrating such inequalities with respect to t from T_1 to T_2, we obtain

$$\begin{aligned} \|u_n - u\|_{1,Q} &\leqslant c\|u_n - u\|_{2,1,Q} \\ &\leqslant c(T_2 - T_1) \sum_{k=1}^{N} \max_{t \in [T_1, T_2]} |(u_n - u, \psi_k)| + c\varepsilon, \end{aligned} \qquad (3.16)$$

since $\int_{T_1}^{T_2} \|u_n - u\|_{\tilde{\mathscr{H}}_{\mathbf{q}}(\Omega)}\, dt$ can obviously be estimated in terms of $\|u_n - u\|_{\tilde{\mathscr{H}}_{\mathbf{q},\mathbf{q}_0}(Q)}$, while in view of condition (3.2) $\|u_n - u\|_{\tilde{\mathscr{H}}_{\mathbf{q},\mathbf{q}_0}(Q)}$ can be estimated in terms of $\|u_n - u\|_{\tilde{\mathscr{H}}_{\mathbf{m}}(A,Q)}$ and the latter norms are uniformly bounded by virtue of the weak convergence of $\{u_n\}$ in $\mathscr{H}$.

It follows from (3.16) that to prove strong convergence of $\{u_n\}$ to u in $L^{2,1}(Q)$ it suffices to show that the sequence $\{(u_n - u, \psi_k)\}$ tends to 0 uniformly in $t \in [T_1, T_2]$ for all $k = 1, 2, \dots$. We observe that because of the imbedding $\mathscr{W} \to C([T_1, T_2]); L^2(\Omega))$ the functions $l_{n,k}(t) = (u_n - u, \psi_k)$ are continuous in t on $[T_1, T_2]$. Because of the weak convergence of $\{u_n\}$ to u in $L^2(Q)$ the functions $l_{n,k}(t)$ tend to 0 as $n \to \infty$ for almost all $t \in [T_1, T_2]$ (for each fixed $k = 1, 2, \dots$). It is thus obvious that to prove uniform convergence of a subsequence of the sequence $\{l_{n,k}(t)\}$ to 0 on $[T_1, T_2]$ it suffices to show that the functions $l_{n,k}(t)$ are equicontinuous with respect to $n = 1, 2, \dots$ on $[T_1, T_2]$ and are uniformly bounded with respect to $n = 1, 2, \dots$ on $[T_1, T_2]$ (for any $k = 1, 2, \dots$).

Applying (1.42), we find that for any t, $t + \Delta t \in [T_1, T_2]$,

$$|l_{n,k}(t + \Delta t) - l_{n,k}(t)| = \left| \int_t^{t+\Delta t} (u_n' - u', \psi_k)\, dt \right|, \qquad k = 1, 2, \dots. \qquad (3.17)$$

We take into account that the difference $u_n' - u' \in \mathscr{H}^*$ is a function in $([T_1, T_2] \to H^*)$ of the form $F(t) = F_0(t) + \sum_1^n D_k^* F_k(t)$, whose action on elements of H is realized for almost all $t \in [T_1, T_2]$ by formulas of the form (1.37) in which $F_0 \in L^{\bar{l}', \bar{l}_0'}(Q)$ and $F_i \in L^{m_i', m_{0i}'}(Q)$, $i = 1, \dots, n$, so that for almost all $t \in [T_1, T_2]$ we have $F_0(t) \in L^{\hat{l}'}(\Omega)$ and $F_i(t) \in L^{m_i'}(\Omega)$, $i = 1, \dots, n$. We recall that the functions F_0 and F_i, $i = 1, \dots, n$, can be chosen so that

$$\|u_n' - u'\|_{\mathscr{W}} = \sup\left\{ \|F_0\|_{\bar{l}', \bar{l}_0', Q}, \|F_1\|_{m_1', m_0', Q}, \dots, \|F_n\|_{m_n', m_{0n}', Q} \right\}.$$

Applying Hölder's inequality, we then have

$$\left| \int_t^{t+\Delta t} (u_n' - u', \psi_k)\, dt \right| = \left| \int_t^{t+\Delta t} \int_\Omega \left(F_0(t,x)\psi_k + \sum_{s=1}^n F_s(t,x) A_s \nabla \psi_k \right) dt\, dx \right|$$

$$\leqslant \|u_n' - u'\|_{\mathscr{W}} \left(\operatorname{meas}^{1/\bar{l}} \Omega |\Delta t|^{1/\bar{l}_0} \|\psi_k\|_{\bar{l}, \Omega} + \sum_{s=1}^n \operatorname{meas}^{1/m_s} \Omega |\Delta t|^{1/m_{0s}} \|A_s \nabla \psi_k\|_{m_s, \Omega} \right),$$

$$k = 1, 2, \dots.$$

The equicontinuity on $[T_1, T_2]$ of the functions $l_{n,k}(t)$ for all $k = 1, 2, \dots$ follows from this and from (3.17) in view of the boundedness of $\|u_n' - u'\|_{\mathscr{W}}$. Taking into account that $l_{n,k}(T_1) = 0$ (since $u_n(T_1) = u(T_1) = 0$), in a similar way we obtain the estimate

$$|l_{n,k}(t)| \leqslant \|u_n' - u'\|_{\mathscr{W}} \left(\operatorname{meas}^{1/l}(t - T_1)^{1/l_0} \|\psi_k\|_{l, \Omega} \right.$$

$$\left. + \sum_{s=1}^n \operatorname{meas}^{1/m_s} \Omega (t - T_1)^{1/m_{0s}} \|A_s \nabla \psi_k\|_{m_s, \Omega} \right),$$

from which it follows that the functions $l_{n,k}(t)$ are uniformly bounded with respect to $n = 1, 2, \dots$ on $[T_1, T_2]$.

The compactness of the imbedding $\mathscr{W}_0 \to L^{2,1}(Q)$ has thus been proved. Because of (3.8) there is the compact imbedding $\mathscr{W} \to L^{l, l_0}(Q)$ for any l and l_0 satisfying conditions (4.5.13) for some $\alpha, \beta \in (0, 1)$. We choose $\alpha, \beta > 0$ so small that the indices l and l_0 in (4.5.13) satisfy the inequalities $l > 2$ and $l_0 > 2$ (this is possible in view of the condition $\hat{l} > 2$). The imbedding $\mathscr{W} \to L^2(Q)$ is then also compact. Using (3.9) and taking account of the compactness of the imbedding $\mathscr{W} \to L^2(Q)$, we finally establish Lemma 3.5.

3. $(A, \mathbf{0}, 2, \bar{m})^{\sim}$**-parabolic equations with weak degeneracy.** In this subsection we consider equation (2.1) in a cylinder Q, assuming that identities (2.2) relative to a matrix $A \equiv A(x)$ satisfying conditions (7.2.1) and (7.2.2) with $\mathbf{m} = (\bar{m}, \dots, \bar{m})$, $\bar{m} \geqslant 2$, are valid. For the matrix A condition (3.2) then also holds with $\mathbf{q}$ and $\mathbf{m} = (\bar{m}, \dots, \bar{m})$ from (7.2.2) and $\mathbf{q}_0 = \mathbf{m}_0 = (\bar{m}, \dots, \bar{m})$. We recall that (7.2.2) is satisfied, in particular, if (see Lemma 7.2.1) for the elements b^{ij} of the matrix $B = A^{-1}$ inverse to A the following conditions hold:

$$b^{ij} \in L^{r_i}(\Omega), \quad r_i > 1, \quad i = 1, \dots, n; \qquad 1/\bar{m} + 1/r_i \leqslant 1, \quad i = 1, \dots, n.$$

In this case condition (3.2) is satisfied with indices $\mathbf{q}$, $\mathbf{q}_0$ defined by

$$1/q_i = 1/\bar{m} + 1/r_i, \qquad 1/q_{0i} = 1/\bar{m}, \qquad i = 1, \dots, n. \tag{3.18}$$

Throughout this subsection we always assume that condition (1.36) is satisfied with $m = \bar{l}$, $m_0 = \bar{l}_0$ (see (3.21)) and $\mathbf{m} = \mathbf{m}_0 = (\overline{m}, \dots, \overline{m})$. Under the assumptions made above this condition is certainly satisfied if

$$\left(\int_{T_1}^{T_2}\int_{\Sigma_3} \lambda u^2\, ds\right)^{1/2} \leqslant c\|u\|_{\tilde{\mathscr{H}}(A,Q)}, \quad \forall u \in \tilde{C}^1(Q),$$

with a constant c not depending on $u \in \tilde{C}^1(Q)$. We denote by $\tilde{\mathscr{H}} \equiv \widetilde{\mathscr{H}_{\overline{m}}^{0,\Sigma_1}}(A, Q)$ the space $\widetilde{\mathscr{H}_{\mathbf{m}}^{0,\Sigma_1}}(A, Q)$ with $\mathbf{m} = (\overline{m}, \dots, \overline{m})$. The estimate above is certainly valid if we suppose, for example, that

$$\lambda \in L^{\kappa,\kappa_0}(\Sigma_3) \quad \left(\text{i.e., } \left[\int_{T_1}^{T_2}\left(\int_{\sigma_3} |\lambda|^\kappa\, ds\right)^{\kappa_0/\kappa} dt\right]^{1/\kappa_0} < +\infty\right),$$

where the indices κ and κ_0 are defined by

$$\kappa = \frac{r}{r-2}, \quad \kappa_0 = \frac{\overline{m}}{\overline{m}-2}, \quad \frac{n-1}{r} = \frac{n}{q_*} - 1$$

$$\text{for } q_* \equiv \min(q_1, \dots, q_n) < n,$$

$$r \in [2, +\infty) \quad \text{for } q_* \geqslant n, \tag{3.19}$$

and it is assumed that $q_* \geqslant 2n/(n+1)$ (so that $r \geqslant 2$) and that the indices $q_1, \dots, q_n$ in (3.19) are defined by (3.18). Indeed, applying the Hölder inequality and the familiar Sobolev imbedding theorem, for any function $u \in \tilde{C}^1(Q)$ we obtain

$$\left(\int_{T_1}^{T_2}\int_{\sigma_3} \lambda u^2\, ds\right)^{1/2} \leqslant c_1\|\lambda\|_{\kappa,\kappa_0,\Sigma_3}^{1/2}\|u\|_{\widetilde{\mathscr{H}_{\overline{m}}^{0,\Sigma_1}}(A,Q)},$$

where c_1 does not depend on $u \in \tilde{C}^1(Q)$.

We further suppose that for the functions $l'^i(x, u, q)$, $i = 1, \dots, n$, and $l'_0(x, u, q)$ in (2.2), for almost all $(t, x) \in Q$ and any $u \in \mathbb{R}$ and $q \in \mathbb{R}^n$

$$|l'^i(t, x, u, q)| \leqslant \mu_1 |q|^{\overline{m}/m|'} + a_1(t, x)|u|^{2/\overline{m}'} + \psi(t, x), \quad i = 1, \dots, n,$$

$$|l'_0(t, x, u, q)| \leqslant a_2(t, x)|q|^{\overline{m}/2} + a_3(t, x)|u| + \psi_0(t, x), \tag{3.20}$$

where $\overline{m} \geqslant 2$, $1/\overline{m} + 1/\overline{m}' = 1$, $\mu_1, \mu_2 = \text{const} \geqslant 0$, $a_1, a_2, a_3 \geqslant 0$, $\psi, \psi_0 \geqslant 0$, $a_1^{\overline{m}'}$, a_2^2, $a^3 \in L^{\bar{s},\bar{s}_0}(Q)$, $1/\bar{s} + 2/\bar{l} = 1$, $1/\bar{s}_0 + 2/\bar{l}_0 = 1$, $\psi \in L^{\overline{m}'}(Q)$, $\psi_0 \in L^{\bar{l}',\bar{l}'_0}(Q)$, and the indices $\bar{l}$ and $\bar{l}_0$ are defined by

$$\frac{1}{\bar{l}} = \frac{\alpha}{\hat{l}} + \frac{(1-\alpha)}{2}; \ \frac{1}{\bar{l}_0} = \frac{\alpha}{\hat{l}_0} < \frac{1}{2};$$

$$\hat{l} = \frac{n}{\sum_{i=1}^n 1/q_i - 1} > 2 \quad \text{for } \sum_{i=1}^n 1/q_i > 1,\ n \geqslant 2;$$

$$\hat{l} \in (2, +\infty) \quad \text{for } \sum_{i=1}^n 1/q_i = 1,\ n \geqslant 2; \tag{3.21}$$

$$\hat{l} \in (2, +\infty)] \quad \text{for } \sum_{i=1}^n 1/q_i < 1,\ n \geqslant 2 \text{ and for } n = 1;$$

$$\hat{l}_0 = n\left(\sum_{i=1}^n 1/q_{0i}\right)^{-1},$$

where $\mathbf{q}$ and $\mathbf{q}_0$ are the indices in condition (3.2) (see Remark 4.5.3). We observe that the inequalities $\bar{l} > 2$ and $\bar{l}_0 > 2$ follow from (3.21). Equations of the form (2.1) possessing the properties enumerated above are called *equations having* $(A,\mathbf{0},2,\overline{m})^{\sim}$ *-structure* in the cylinder Q.

For an equation (2.1) having $(A,\mathbf{0},2,\overline{m})^{\sim}$-structure in Q we consider a problem of the form (2.5) in the case $\Sigma = \partial\Omega \times (T_1, T_2)$, $\Sigma' = \Omega_{T_1} \cup \Omega_{T_2}$. Suppose that on the part $\Sigma_3 \subset \Sigma$ there is given a positive function $\lambda \in L^1(\Sigma_3)$. (If $\lambda \in L^{\kappa,\kappa_0}(Q)$), where $\kappa > r/(r-2)$ and $\kappa_0 = \overline{m}/(\overline{m}-2)$ with the same r as in (3.19), then it is easy to see that the condition that λ be positive on Σ_3 can be dropped, and in this case the term $\|u\|_{L^2(\lambda,\Sigma_3)}$ is to be dropped in the definition of the norm of $\tilde{\mathscr{H}}_\lambda$ (see (3.13).) We denote by

$$\mathscr{H} \equiv \mathscr{H}^{0,\Sigma_1}_{\bar{l},\bar{l}_0,\overline{m}}(A,Q) \qquad \text{and} \qquad \mathscr{Y} \equiv \mathscr{Y}^{0,\Sigma_1}_{\bar{l},\bar{l}_0,\overline{m}}(A;Q;\Sigma_3,\lambda)$$

respectively the spaces

$$\mathscr{H}^{0,\Sigma_1}_{\bar{l},\bar{l}_0;\mathbf{m},\mathbf{m}_0}(A,Q) \qquad \text{and} \qquad \mathscr{H}^{0,\Sigma_1}_{\bar{l},\bar{l}_0;\mathbf{m},\mathbf{m}_0}(A;Q;\Sigma_3,\lambda)$$

with $\mathbf{m} = (\overline{m},\dots,\overline{m})$ and $\mathbf{m}_0 = \mathbf{m}$.

In correspondence with the definition given in the general situation, by a generalized solution of problem (2.5) we mean any function $u \in \mathscr{W}_0$, where $\mathscr{W}_0 \equiv \{u \in \mathscr{H}: u' \in \mathscr{H}^*,\ u(T_1) = 0\}$, satisfying an identity of the form (2.8). We note that by Proposition 3.1 $\mathscr{W}_0 \subset \mathscr{W} \subset \tilde{\mathscr{H}}$, so that the estimates (3.7)–(3.9) hold for functions in $\mathscr{W}_0$.

The left side of (2.8) defines an operator $\mathscr{L}\colon \mathscr{W}_0 \to \mathscr{H}^*$. Indeed, we write $\mathscr{L}$ as the sum $\mathscr{A} + \mathscr{B}$, where the operator $\mathscr{A}\colon \mathscr{H} \to \mathscr{H}^*$ (we recall that the spaces $\mathscr{H}$ and $\mathscr{H}_\lambda$ are isomorphic) is defined by

$$\langle \mathscr{A}u, \eta\rangle = \iint_Q [\mathbf{l}'(t,x,u,A\nabla u)\cdot A\nabla\eta + l_0'(t,x,u,A\nabla u)\eta]\,dt\,dx$$

$$+ \int_{\Sigma_3} \lambda u\eta\,ds, \qquad u,\eta \in \mathscr{H}, \tag{3.22}$$

while the linear operator $\mathscr{B}\colon \mathscr{W}_0 \to \mathscr{H}^*$ is defined by

$$\langle \mathscr{B}u, \eta\rangle = \int_{T_1}^{T_2} (u',\eta)\,dt, \qquad u \in \mathscr{W}_0,\ \eta \in \mathscr{H}, \tag{3.23}$$

where $(\cdot,\cdot)$ is the duality between $H \equiv H^{0,\sigma_1}_{l,\overline{m}}(A,\Omega)$ and H^*, and $\langle\cdot,\cdot\rangle$ is that between $\mathscr{H}$ and $\mathscr{H}^*$. Taking into account the inequalities

$$|\langle \mathscr{A}u,\eta\rangle| \leqslant \Big(\mu_1\|A\nabla u\|_{\overline{m},Q} + \|a_1\|_{\overline{m}'\bar{s},\overline{m}'\bar{s}_0,Q}\|u\|^{2/\overline{m}'}_{\bar{l},\bar{l}_0,Q} + \|\psi\|_{\overline{m}',Q}\Big)\|A\nabla\eta\|_{\overline{m},Q}$$

$$+\Big(\|a_2\|_{2\bar{s},2\bar{s}_0,Q}\|A\nabla u\|^{\overline{m}/2}_{\overline{m},Q} + \|a_3\|_{\bar{s},\bar{s}_0,Q}\|u\|_{\bar{l},\bar{l}_0,Q} + \|\psi_0\|_{\bar{l}',\bar{l}_0',Q}\Big)$$

$$\times\Big(\|\eta\|_{\bar{l},\bar{l}_0,Q} + \|u\|_{L^2(\lambda,\Sigma_3)}\|\eta\|_{L^2(\lambda,\Sigma_3)}\Big), \qquad u,\eta \in \mathscr{H}, \tag{3.24}$$

$$|\langle \mathscr{B}u,\eta\rangle| \leqslant \|u\|_{\mathscr{W}}\|\eta\|_{\mathscr{W}}, \qquad u \in \mathscr{W}_0,\ \eta \in \mathscr{H}, \tag{3.25}$$

we conclude that in fact $\mathscr{L}u \in \mathscr{H}^*$. It also follows from (3.24) and (3.25) that $\mathscr{L}\colon \mathscr{W}_0 \to \mathscr{H}^*$ is bounded. We shall prove that this operator is demicontinuous. Since $\mathscr{B}$ is linear and bounded, it suffices to prove that the operator $\mathscr{A}$ is demicontinuous.

Suppose that the sequence $\{u_n\}$, $u_n \in \mathscr{W}_0$, $n = 1, 2, \ldots$, converges in $\mathscr{H}$ to a function $u \in \mathscr{H}$. To prove the demicontinuity of $\mathscr{A}$ it then suffices to prove that $\{\langle \mathscr{A}u_n, \eta\rangle\}$ converges to $\langle \mathscr{A}u, \eta\rangle$ for all $\eta \in \mathscr{H}$. The convergence of the linear term $\int_{\Sigma_3} \lambda u_n \eta\, ds$ to $\int_{\Sigma_3} \lambda u \eta\, ds$ is obvious. Because of (3.24), it is also obvious that for some subsequence $\{u_\nu\}$ of $\{u_n\}$ there is also convergence $\varphi_\nu(t, x) \to \varphi(t, x)$ a.e. in Q, where

$$\varphi_\nu = \mathbf{l}'(t, x, u_\nu, A\nabla u_\nu) \cdot A\nabla\eta + l_0'(t, x, u_\nu, A\nabla u_\nu)\eta,$$
$$\varphi = \mathbf{l}'(t, x, u, A\nabla u) \cdot A\nabla\eta + l_0'(t, x, u, A\nabla u)\eta,$$

and η is a fixed function in $\mathscr{H}$. It is easy to see that the sequence $\{\varphi_\nu\}$ has absolutely continuous Lebesgue integrals in Q which are equicontinuous with respect to ν. For this we need to use the inequality obtained from (3.24) by dropping the integrals over Σ_3 and replacing Q by an arbitrary Lebesgue-measurable subset of Q, and we also need to take into account the fact that the parentheses in (3.24) are uniformly bounded with respect to ν. From the theory of the Lebesgue integral it then follows that

$$\iint_Q \varphi_\nu(t, x)\, dt\, dx \to \iint_Q \varphi(t, x)\, dt\, dx.$$

From this it follows easily that this convergence actually holds for the entire sequence $\{\varphi_n\}$. Thus, $\langle \mathscr{A}u_n, \eta\rangle \to \langle \mathscr{A}u, \eta\rangle$. We write out the property of the operator $\mathscr{A}$ just proved as the following proposition.

LEMMA 3.6. *The operator $\mathscr{A}: \mathscr{H} \to \mathscr{H}^*$, and hence also the operator $\mathscr{L}: \mathscr{W}_0 \to \mathscr{H}^*$, is demicontinuous.*

LEMMA 3.7. *Suppose conditions* (3.20) *are satisfied as well as the following condition: for almost all $(t, x) \in Q$ and any $u \in \mathbb{R}$ and $q \in \mathbb{R}^n$*

$$l'^i(t, x, u, q)q_i + l_0'(t, x, u, q)u \geqslant \nu|q|^{\overline{m}} - \hat{a}_4(t, x)u^2 - \hat{\psi}(t, x), \quad (3.26)$$

where $\nu = \text{const} > 0$, $\hat{a}_4 \in L^{s,s_0}(Q)$, $1/s + 2/l = 1$, $1/s_0 + 2/l_0 = 1$, $\hat{\psi} \in L^1(Q)$, and the indices l and l_0 are defined by

$$\frac{1}{l} = \frac{\alpha}{\hat{l}} + \frac{(1-\alpha)}{2}, \qquad \frac{1}{l_0} = \frac{\alpha}{\hat{l}_0} + \frac{(1-\alpha)\beta}{2} < \frac{1}{2},$$
$$\alpha \in (0, 1), \beta \in (0, 1), \quad (3.27)$$

in which $\hat{l}$ and $\hat{l}_0$ are determined relative to the indices $\mathbf{q}$ and $\mathbf{q}_0$ from (3.2) *as in* (3.21). *Then for all $t \in [\,T_1, T_2)$ and all $\varepsilon > 0$, for any function $u \in \mathscr{H}$,*

$$\iint_{Q_t} [\mathbf{l}'(t, x, u, A\nabla u) \cdot A\nabla u + l_0'(t, x, u, A\nabla u)u]\, d\tau\, dx$$
$$\geqslant (\nu - \varepsilon_1)\|A\nabla u\|_{\overline{m}, Q_t}^{\overline{m}} - \varepsilon_1\|u\|_{2,\infty,Q_t}^2 - c_1\varepsilon_1^{-\lambda}\|u\|_{2,Q_t}^2 - c_2, \qquad (3.28)$$

where $\varepsilon_1 > 0$, c_1 depends on n, $\mathbf{q}$, $\mathbf{q}_0$, α, β, Ω and $\|\hat{a}_4\|_{s,s_0,Q}^{\lambda}$, λ is the number in (3.9), *$c_2 = \|\hat{\psi}\|_{1,Q}$, and $Q_t = \Omega \times (T_1, t)$.*

PROOF. Lemma 3.7 follows in an obvious way from (3.26) and (3.9).

We now proceed to the direct proof of solvability of the problem of the form (2.5). We consider the regularized operator $\mathcal{L}_\varepsilon$: $\mathcal{Y}_0 \to \mathcal{Y}_0^*$, where $\mathcal{Y}_0$ is the subspace of $\mathcal{Y}$ consisting of all $u \in \mathcal{Y}$ for which $u(T_1) = 0$, defined by

$$\langle \mathcal{L}_\varepsilon u, \eta \rangle = \varepsilon \int_{T_1}^{T_2} (u', \eta')\, dt + \langle \hat{\mathcal{L}} u, \eta \rangle, \qquad u, \eta \in \mathcal{Y}_0, \tag{3.29}$$

where $\hat{\mathcal{L}}$ is the restriction of the operator $\mathcal{L}$: $\mathcal{W}_0 \to \mathcal{H}^* \subset \mathcal{Y}_0^*$ (it is obvious that $\mathcal{Y}_0$ is dense in $\mathcal{H}$) to the set $\mathcal{Y}_0 \subset \mathcal{W}_0$.

LEMMA 3.8. *Suppose conditions* (3.20) *and* (3.26) *are satisfied. Then for any* $\mathcal{F} \in \mathcal{H}^*$ *and* $\varepsilon > 0$ *for any function* $u \in \mathcal{Y}_0$ *with*

$$\langle \mathcal{L}_\varepsilon u, \eta \rangle \equiv \varepsilon \int_{T_1}^{T_2} (u', \eta')\, dt + \int_{T_1}^{T_2} (u', \eta)\, dt + \langle \mathcal{A} u, \eta \rangle = \langle \mathcal{F}, \eta \rangle, \qquad \eta \in \mathcal{Y}_0, \tag{3.30}$$

the following inequality holds:

$$\sqrt{\varepsilon}\, \|u'\|_{2,Q} + \|u\|_{2,\infty,Q} + \|A \nabla u\|_{\overline{m},Q} + \|u\|_{L^2(\lambda, \Sigma_3)} \leqslant c, \tag{3.31}$$

where the constant c does not depend either on ε *or on* $u \in \mathcal{Y}_0$. *For any* $\mathcal{F} \in \mathcal{H}^*$ *and any function* $u \in \mathcal{W}_0$ *satisfying* (3.30) *with* $\varepsilon = 0$ *an inequality of the form* (3.31) *holds with* $\varepsilon = 0$.

PROOF. We first consider the case $\varepsilon = 0$. In view of Lemma 2.1, for any $t \in (T_1, T_2]$

$$\int_{T_1}^{t} (u', \eta)\, dt + \iint_{Q_t} (\mathbf{l}' \cdot A \nabla \eta + l_0' \eta)\, dt\, dx + \int_{\sigma_3 \times (T_1, t)} \lambda u \eta\, ds = \int_{T_1}^{t} (F, \eta)\, dt, \qquad \eta \in \mathcal{W}_0. \tag{3.32}$$

In (3.32) we set $\eta = u \in \mathcal{W}_0$ and use (3.28) with $\varepsilon = \min(1/8, \nu/2)$. Then for any $t \in [T_1, \tau)$, $T_1 < \tau \leqslant t \leqslant T_2$,

$$\begin{aligned} \frac{1}{2}(u(t), u(t)) + \frac{\nu}{2} \sum_{i=1}^{n} \|A_i \nabla u\|_{\overline{m},Q_t}^{\overline{m}} + \int_{\sigma_3 \times (T_1, t)} \lambda u^2\, ds \\ \leqslant \frac{1}{8}\|u\|_{2,\infty,Q_\tau}^2 + \|\mathcal{F}\|_{\mathcal{H}^*} \|u\|_{\mathcal{H}} + c_1 4^{\nu} \|u\|_{2,Q_\tau}^2 + c_2. \end{aligned} \tag{3.33}$$

From (3.33) we easily obtain

$$\begin{aligned} \|u\|_{2,\infty,Q_\tau}^2 + \sum_{i=1}^{n} \|A_i \nabla u\|_{\overline{m},Q_\tau}^{\overline{m}} + \int_{\sigma_3 \times (T_1, \tau)} \lambda u^2\, ds \\ \leqslant c_3 \|\mathcal{F}\|_{\mathcal{H}^*} \|u\|_{\mathcal{H}} + c_4 \|u\|_{2,Q_\tau}^2 + c_5, \qquad T_1 < \tau \leqslant T_2, \end{aligned} \tag{3.34}$$

where $c_3 = 2\max(4, 2\nu^{-1})$, $c_4 = 2\max(4, 2\nu^{-1}) 4^{\lambda} c_1$ and $c_5 = 2c_2 \max(4, 2\nu^{-1})$. Applying Gronwall's lemma, we deduce from (3.34) that

$$\|u\|_{2,\infty,Q}^2 + \|A \nabla u\|_{\overline{m},Q}^{\overline{m}} + \|u\|_{L^2(\Sigma_3)}^2 \leqslant (c_6 \|\mathcal{F}\|_{\mathcal{H}^*} \|u\|_{\mathcal{H}} + c_7) e^{c_4 (T_2 - T_1)}, \tag{3.35}$$

where $c_6 = n^{\overline{m}-1}c_3$ and $c_7 = n^{\overline{m}-1}c_4$. Taking into account that

$$\|A\nabla u\|_{\overline{m},Q}^{\overline{m}} \geqslant \|A\nabla u\|_{\overline{m},Q}^{2} - c_8,$$

where $c_8 = c_8(\overline{m})$, and applying an estimate of the form (3.7), we deduce from (3.35) that

$$\|u\|_{2,\infty,Q}^2 + \|A\nabla u\|_{\overline{m},Q}^2 + \|u\|_{L^2(\lambda,\Sigma_3)}^2 \leqslant c_8\|\mathscr{F}\|_{\mathscr{H}^*}^2 + c_9, \tag{3.36}$$

where $c_8 = c_6^2c_0^2$, $c_9 = 2c_7e^{c_4(T_2-T_1)} + c_8$, and c_0 is the constant in (3.7). In the case $\varepsilon = 0$ the estimate of the form (3.31) has thus been proved. The proof of (3.31) in the case $\varepsilon > 0$ is completely analogous. Lemma 3.8 is proved.

REMARK 3.2. Suppose that the conditions of Lemma 3.8 are satisfied in the case $\overline{m} = 2$ with $\psi \equiv 0$ in (3.26). It follows easily from the proof of Lemma 3.8 that the constant c in (3.31) then has the form $c = c_1\|\mathscr{F}\|_{\mathscr{H}^*}$, where c_1 depends only on n, ν, $\mathbf{q}$, $\mathbf{q}_0$, α, β, Ω, and $\|\hat{a}_4\|_{s,s_0,Q}$. Hence, in this case there can exist only one function $u \in \mathscr{Y}_0$ ($u \in \mathscr{W}_0$) satisfying (3.29) with $\varepsilon > 0$ ((3.29) with $\varepsilon = 0$).

LEMMA 3.9. *Suppose that conditions* (3.20) *and* (3.26) *are satisfied. Assume also that the restriction of the operator* $\mathscr{A}$: $\mathscr{H} \to \mathscr{H}^*$ *(see* (3.22)) *to the set* $\mathscr{Y}_0$ *satisfies the condition of semibounded variation*

$$\langle \mathscr{A}u - \mathscr{A}v, u - v\rangle \geqslant -\gamma(\rho, \|u - v\|'_{\mathscr{Y}}), \quad u, v \in \mathscr{Y}_0, \\ \|u\|_{\mathscr{Y}} \leqslant \rho, \qquad \|v\|_{\mathscr{Y}} \leqslant \rho, \tag{3.37}$$

where the function $\gamma(\rho, \tau)$ *satisfies condition* (4.6.2), *and the norm* $\|\cdot\|'_{\mathscr{Y}}$ *is compact relative to* $\|\cdot\|_{\mathscr{Y}}$. *Then for any* $\varepsilon > 0$ *and any* $\mathscr{F} \in \mathscr{H}^*$ *there exists at least one function* $u_\varepsilon \in \mathscr{Y}_0$ *such that*

$$\langle \mathscr{L}_\varepsilon u_\varepsilon, \eta\rangle = \langle \mathscr{F}, \eta\rangle, \quad \forall \eta \in \mathscr{Y}_0. \tag{3.38}$$

Here $u'_\varepsilon \in C([T_1, T_2]; H^*)$, $u'_\varepsilon(T_2) = 0$, *and*

$$\|u'_\varepsilon\|_{\mathscr{H}^*} \leqslant K, \tag{3.39}$$

where K *does not depend on* ε.

PROOF. We fix an $\varepsilon > 0$. Because of (3.28) the operator $\mathscr{L}_\varepsilon$: $\mathscr{Y}_0 \to \mathscr{Y}_0^*$ may be assumed coercive. Indeed, setting $\eta = u \in \mathscr{Y}_0$ in (3.30), taking (1.42) and (3.28) with a suitable $\varepsilon_1 > 0$ into account, and applying Cauchy's inequality, we obtain

$$\langle \mathscr{L}_\varepsilon u, u\rangle \geqslant \frac{\varepsilon}{2}\|u'\|_{2,Q}^2 + \frac{\nu}{2}\|A\nabla u\|_{\overline{m},Q}^{\overline{m}} + \frac{1}{2}\|u\|_{2,\Omega_T}^2 - c_1\left(\frac{\nu}{2}\right)^{-\lambda}\|u\|_{2,Q}^2 - c_2. \tag{3.40}$$

Since the change of unknown function $u = e^{\gamma t}\bar{u}$, $\gamma = \text{const} > 0$, reduces (2.1) to the equation

$$\bar{u}_t - (d/dx_i)\bar{l}^i(t, x, \bar{u}, \nabla\bar{u}) + \bar{l}_0(t, x, \bar{u}, \nabla\bar{u}) = \mathscr{F}, \tag{3.41}$$

where

$$\bar{l}^i(t, x, \bar{u}, \nabla\bar{u}) \equiv l^i(t, x, e^{\gamma t}u, e^{\gamma t}\nabla u), \qquad i = 1,\dots,n,$$
$$\bar{l}_0(t, x, \bar{u}, \nabla\bar{u}) \equiv l_0(t, x, ue^{\gamma t}, \nabla ue^{\gamma t}) + \gamma\bar{u},$$

for which conditions of the form (3.20), (3.26) and (3.28) are also satisfied and for (3.41) the coefficient of $\|\bar{u}\|_{2,Q_t}^2$ in (3.28) is equal to $\gamma - c_1\varepsilon_1^{-\lambda}$, it follows that by assuming γ sufficiently large we obtain for (3.41) an inequality of the form (3.40) without the term containing $\|u\|_{2,Q}$. Since the solvability of the general boundary value problem for (3.41) is equivalent to the solvability of this problem for the original equation (2.1), it may simply be assumed that

$$\langle \mathscr{L}_\varepsilon u, u\rangle \geqslant (\varepsilon/2)\|u'\|_{2,Q}^2 + (\nu/2)\|A\nabla u\|_{\overline{m},Q}^{\overline{m}} + (1/2)\|u\|_{2,\Omega_{T_2}} - c_2, \tag{3.42}$$

from which coerciveness of the operator $\mathscr{L}_\varepsilon\colon \mathscr{Y}_0 \to \mathscr{Y}_0^*$ obviously follows. In view of (3.37) it then follows from Theorem 4.6.1 (the case $H = X = Y = \mathscr{Y}_0$, $\mathscr{L} = \mathscr{A} + \mathscr{B}$, $\mathscr{A} = \mathscr{L}_\varepsilon$, $\mathscr{B} = 0$) that there exists a function $u_\varepsilon \in \mathscr{Y}_0$ satisfying (3.38) even for all $\mathscr{F} \in \mathscr{Y}_0^*$ and not just for all $\mathscr{F} \in \mathscr{H}^* \subset \mathscr{Y}_0^*$. Moreover, from the proof of Theorem 4.6.1 we obtain

$$\|u_\varepsilon\|_{\mathscr{H}} \leqslant \|u_\varepsilon\|_{\mathscr{Y}_0} \leqslant c, \tag{3.43}$$

where c does not depend on ε; we shall prove that $u'_\varepsilon \in C([T_1, T_2]; H^*)$.

Setting $\eta = \psi\varphi$ in (3.29) with $\psi \in \tilde{C}^1_{0,\sigma_1}(\Omega)$ and $\varphi \in \mathscr{D}([T_1, T_2])$, we obtain

$$\varepsilon\int_{T_1}^{T_2} (u'_\varepsilon(t), \psi)\varphi'(t)\,dt + \int_{T_1}^{T_2} (u'_\varepsilon(t), \psi)\varphi(t)\,dt = \int_{T_1}^{T_2} (F_\varepsilon(t), \psi)\varphi(t)\,dt,$$

$$\psi \in \tilde{C}^1_{0,\sigma_1}(\Omega),\ \varphi \in \mathscr{D}([T_1, T_2]), \tag{3.44}$$

where $F_\varepsilon(t) \in ([T_1, T_2] \to H^*)$ is determined by the element $\mathscr{F}^* - \mathscr{A}u_\varepsilon \in \mathscr{H}^*$ and, according to (1.37), has the form

$$F_\varepsilon(t) = F_{\varepsilon,0}(t) + \sum_{k=1}^{n} l_k^* F_{\varepsilon,k}(t)$$

with $F_{\varepsilon,0} \in L^{\bar{l}',\bar{l}'_0}(Q)$, $F_{\varepsilon,k} \in L^{\overline{m}'}(Q)$, $k = 1, \dots, n$, and

$$\int_{T_1}^{T_2} (l_k^* F_{\varepsilon,k}, \eta)\,dt = \int_{T_1}^{T_2} (F_{\varepsilon,k}(t), A_k\nabla\eta)\,dt.$$

From (3.43) and the boundedness of $\mathscr{A}\colon \mathscr{H} \to \mathscr{H}^*$ it follows that the norms

$$\|\mathscr{F} - \mathscr{A}u_\varepsilon\|_{\mathscr{H}^*} = \sup\left(\|F_{\varepsilon,0}\|_{\bar{l}',\bar{l}'_0,Q}, \dots, \|F_{\varepsilon,n}\|_{\overline{m}',Q}\right)$$

are bounded uniformly with respect to $\varepsilon \in (0, 1]$. In view of the density of $\tilde{C}^1_{0,\sigma_1}(\Omega)$ in H and $L^2(\Omega)$, from (3.44) we obtain

$$\varepsilon\int_{T_1}^{T_2} u'_\varepsilon\varphi'\,dt + \int_{T_1}^{T_2} u'_\varepsilon\varphi\,dt = \int_{T_1}^{T_2} F_\varepsilon(t)\varphi\,dt, \quad \forall\varphi \in \mathscr{D}([T_1, T_2]), \tag{3.45}$$

where the integrals are understood as Bochner integrals of functions in $([T_1, T_2] \to H^*)$.

The identity (3.45) can be written as

$$-\varepsilon u''_\varepsilon + u'_\varepsilon = F_\varepsilon(t), \tag{3.46}$$

where u'_ε and u''_ε are the first and second derivatives of the function u_ε considered as an element of $\mathscr{D}^*([T_1, T_2]; H^*)$. From the form of $F_\varepsilon(t)$ and the condition $u'_\varepsilon \in L^2(Q) \subset L^2([T_1, T_2]; H^*)$ it follows that $u''_\varepsilon \in L^{\overline{m}'_*}([T_1, T_2]; H^*)$, $\overline{m}'_* = \min(\bar{l}'_0, \overline{m}')$, so that in any case $u''_\varepsilon \in L^1([T_1, T_2]; H^*)$. Thus, $u'_\varepsilon \in \mathscr{W}_{1,1}$, where

$$\mathscr{W}_{1,1} \equiv \{v \in L^1([T_1, T_2]; H^*)\colon v' \in L^1([T_1, T_2]; H^*)\},$$

and the norm

$$\|v\|_{\mathscr{W}_{1,1}} \equiv \|u\|_{L^1([T_1,T_2];H^*)} + \|u'\|_{L^1([T_1,T_2];H^*)}$$

is introduced in $\mathscr{W}_{1,1}$. Exactly as in the proof of the imbedding $\mathscr{W} \to C(I, B^*)$ (see the proof of Lemma 4.7.6), we establish the imbedding $\mathscr{W}_{1,1} \to C([T_1, T_2]; H^*)$, so that

$$\|u\|_{C([T_1,T_2];H^*)} \leqslant c\|u\|_{\mathscr{W}_{1,1}}. \tag{3.47}$$

In particular, it follows from (3.47) that $u'_\varepsilon \in C([T_1, T_2]; H^*)$. Using Lemma 4.7.2 and (3.47), it is easy to show that $C^\infty([T_1, T_2]; H^*)$ is dense in $\mathscr{W}_{1,1}$. Therefore, for any $u \in \mathscr{W}_{1,1}$, $u' \in \mathscr{W}_{1,1}$ there exists a sequence $\{u_n\}$, $u_n \in C^\infty([T_1, T_2]; H^*)$, $n = 1, 2, \ldots$, such that $u_n \to u$ in $L^1([T_1, T_2]; H^*)$ and $u'_n \to u'$ in $L^1([T_1, T_2]; H^*)$.

We now prove that $u'_\varepsilon(T_2) = 0$. Passing to the limit in the identity

$$\int_{T_1}^{T_2} u'_n\varphi\, dt = -\int_{T_1}^{T_2} u_n\varphi'\, dt + u_n(T_2)\varphi(T_2), \quad \forall\varphi \in C^1([T_1, T_2]), \varphi(T_1) = 0, \tag{3.48}$$

where $u_n \in C^\infty([T_1, T_2]; H^*)$, $n = 1, 2, \ldots$, $u_n \to u'_\varepsilon$ in $L^1([T_1, T_2]; H^*)$ and $u'_n \to u''_\varepsilon$ in $L^1([T_1, T_2]; H^*)$, and taking (3.47) into account, we obtain

$$\int_{T_1}^{T_2} u''_\varepsilon\varphi\, dt = -\int_{T_1}^{T_2} u'_\varepsilon\varphi'\, dt + u'_\varepsilon(T_2)\varphi(T_2), \quad \varphi \in C^1([T_1, T_2]), \varphi(T_1) = 0. \tag{3.49}$$

From (3.46) and (3.49) we then obtain

$$-\int_{T_1}^{T_2} u''_\varepsilon\varphi\, dt + \varepsilon u'_\varepsilon(T_2)\varphi(T_2) + \int_{T_1}^{T_2} u'_\varepsilon\varphi\, dt = \int_{T_1}^{T_2} F_\varepsilon(t)\, dt,$$

$$\forall\varphi \in C^1([T_1, T_2]), \qquad \varphi(T_1) = 0. \tag{3.50}$$

Multiplying (3.46) by $\varphi \in C^1([T_1, T_2])$, $\varphi(T_1) = 0$, integrating over $[T_1, T_2]$, and subtracting the equality so obtained from (3.50), since $\varphi(T_2)$ is arbitrary, we obtain $u'_\varepsilon(T_2) = 0$ as an element of H^*. Thus, the function $u'_\varepsilon(t)$ may be considered a solution of the problem

$$-\varepsilon u''_\varepsilon + u'_\varepsilon = F_\varepsilon \quad \text{in} \quad [T_1, T_2]; \qquad u'_\varepsilon(T_2) = 0, \tag{3.51}$$

where all terms are functions in $([T_1, T_2] \to H^*)$. But the unique solution of (3.51) is the function

$$u'_\varepsilon(t) = \frac{1}{\varepsilon}\int_0^{T_2-t} F_\varepsilon(T_2 - \eta)e^{-(T_2-t-\eta)/\varepsilon}\, d\eta. \tag{3.52}$$

This shows that an element $u'_\varepsilon \in \mathscr{H}^*$ acts on an element $\eta \in \mathscr{H}$ by the formula

$$\langle u'_\varepsilon, \eta\rangle = \int_{T_1}^{T_2} (\{F_\varepsilon\}(t), \eta(t))\, dt, \qquad \eta \in \mathscr{H}, \tag{3.53}$$

where, for $F \in ([T_1, T_2] \to H^*)$, $\{F\}$ denotes the function

$$\{F\} \overset{\text{def}}{=} \frac{1}{\varepsilon}\int_0^{T_2-t} F(T_2 - \eta)e^{-(T_2-t-\eta)/\varepsilon}\, d\eta. \tag{3.54}$$

Taking the form of F_ε into account, we obtain

$$u'_\varepsilon = \{F_{\varepsilon,0}\} + \sum_{k=1}^{n} l_k^*\{F_{\varepsilon,k}\}, \tag{3.55}$$

and it is easy to see that $\{F_{\varepsilon,0}\} \in L^{\bar{l}',\bar{l}'_0}(Q)$ and $\{F_{\varepsilon,k}\} \in L^{\overline{m}'}(Q)$, $k = 1,\dots,n$. Indeed, using the well-known estimate for the convolution $\|f*g\|_{p,Q} \leqslant \|f\|_{p,Q}\|g\|_{1,Q}$ and the equality $(1/\varepsilon)\int_0^{+\infty} e^{-\eta/\varepsilon}\,d\eta = 1$, we obtain

$$\|\{F_{\varepsilon,0}\}\|_{l',l'_0,Q} \leqslant \|F_{\varepsilon,0}\|_{\bar{l}',\bar{l}'_0,Q}, \qquad \|\{F_{\varepsilon,k}\}\|_{\overline{m}',Q} \leqslant \|F_{\varepsilon,k}\|_{\overline{m}',Q}, \quad k = 1,\dots,n, \tag{3.56}$$

whence the assertion follows. Since, as noted above, the norms on the right sides of (3.56) are bounded uniformly with respect to $\varepsilon \in (0,1]$, from (3.55) and (3.56) we obtain (3.39) with a constant K not depending on ε. Lemma 3.9 is proved.

THEOREM 3.2. *Suppose that conditions* (3.20) *and* (3.26) *are satisfied, and that the restriction of the operator* $\mathcal{A}: \mathcal{H} \to \mathcal{H}^*$ (*see* (3.22)) *to the set* $\mathcal{W}_0$ *satisfies the condition*

$$\langle \mathcal{A}u - \mathcal{A}v, u - v\rangle \geqslant -\gamma(\rho, \|u - v\|'_{\mathcal{W}}), \qquad u, v \in \mathcal{W}_0,$$
$$\|u\|_{\mathcal{W}} \leqslant \rho, \qquad \|v\|_{\mathcal{W}} \leqslant \rho, \tag{3.57}$$

where the norm $\|\cdot\|'_{\mathcal{W}}$ *is compact relative to* $\|\cdot\|_{\mathcal{W}}$ *and* $\gamma(\rho,\tau)$ *satisfies condition* (4.6.2). *Then for any* $\mathcal{F} \in \mathcal{X}^*$ *problem* (2.5) *has at least one generalized solution.*

PROOF. In view of Lemmas 3.8 and 3.9, for all $\varepsilon > 0$ there exists a function $u_\varepsilon \in \mathcal{Y}_0 \subset \mathcal{W}_0$ satisfying (3.38), (3.31) and (3.39). It follows easily that there exists a sequence of values ε tending to 0 for which

$$u_\varepsilon \to u \quad \text{weakly in} \quad \mathcal{H}, \qquad u'_\varepsilon \to u' \quad \text{weakly in} \quad \mathcal{H}^*, \tag{3.58}$$

where u is a function in $\mathcal{W}$. In view of the estimate

$$\|u\|_{C([T_1,T_2],L^2(\Omega))} \leqslant c\|u\|_{\mathcal{W}}, \quad \forall u \in \mathcal{W}, \tag{3.59}$$

it follows from (3.58) that $u(T_1) = 0$, so that $u \in \mathcal{W}_0$. We shall prove that u is the desired generalized solution of (2.5). Since it follows from (3.39) that

$$\varepsilon\left|\int_{T_1}^{T_2}(u'_\varepsilon, \eta')\,dt\right| \leqslant \varepsilon\|u'_\varepsilon\|_{2,Q}\|\eta'\|_{2,Q} \leqslant c\sqrt{\varepsilon}\,\|\eta'\|_{2,Q} \to 0 \quad \text{as } \varepsilon \to 0, \tag{3.60}$$

passing to the limit in (3.38), we find that for all $\eta \in \mathcal{Y}_0$

$$\int_{T_1}^{T_2}(u', \eta)\,dt + \langle f, \eta\rangle = \langle \mathcal{F}, \eta\rangle, \tag{3.61}$$

where f is the weak limit of $\mathcal{A}u_\varepsilon$ in $\mathcal{H}$ as $\varepsilon \to 0$. Since $\mathcal{Y}_0$ is dense in $\mathcal{H}$, (3.61) also holds for all $\eta \in \mathcal{H}$and, in particular for all $\eta \in \mathcal{W}_0$.

To complete the proof of the theorem it suffices to establish that $f = \mathcal{A}u$. For this we use condition (3.57). It follows from (3.57) that for all $\xi \in \mathcal{Y}_0$

$$\varepsilon\int_{T_1}^{T_2}(u'_\varepsilon - \xi', u'_\varepsilon - \xi')\,dt + \int_{T_1}^{T_2}(u'_\varepsilon - \xi', u_\varepsilon - \xi)\,dt + \langle \mathcal{A}u_\varepsilon - \mathcal{A}\xi, u_\varepsilon - \xi\rangle$$
$$\geqslant -\gamma(\rho, \|u_\varepsilon - \xi\|'_{\mathcal{W}}), \tag{3.62}$$

with $\rho = \sup_\varepsilon \|u_\varepsilon\|_{\mathscr{W}} + \|\xi\|_{\mathscr{W}}$, where in deriving (3.62) we have used Lemmas 1.25 and 1.26 and also the nonnegativity of the first term in (3.62). Subtracting (3.29) with u replaced by u_ε and η by $u_\varepsilon - \xi$ from (3.62), we obtain

$$-\varepsilon\int_{T_1}^{T_2}(\xi', u_\varepsilon' - \xi')\,dt - \int_{T_1}^{T_2}(\xi', u_\varepsilon - \xi)\,dt - \langle \mathscr{A}\xi, u_\varepsilon - \xi\rangle \geqslant -\langle \mathscr{F}, u_\varepsilon - \xi\rangle - \gamma(\rho, \|u_\varepsilon - \xi\|'_{\mathscr{W}}), \qquad \xi \in \mathscr{Y}_0. \tag{3.63}$$

Letting ε tend to 0 in (3.63) and taking into account (3.58), the properties of the function $\gamma(\cdot,\cdot)$, and the compactness of the norm $\|\cdot\|'_{\mathscr{W}}$ relative to $\|\cdot\|_{\mathscr{W}}$, we deduce from (3.63) that

$$-\int_{T_1}^{T_2}(\xi', u - \xi)\,dt - \langle \mathscr{A}\xi, u - \xi\rangle \geqslant -\langle \mathscr{F}, u - \xi\rangle - \gamma(\rho, \|u - \xi\|'_{W}), \quad \forall \xi \in \mathscr{Y}_0. \tag{3.64}$$

Adding (3.64) and (3.61) with $\eta = u - \xi \in \mathscr{W}_0$, we find that

$$\int_{T_1}^{T_2}(u' - \xi', u - \xi)\,dt + \langle f - \mathscr{A}\xi, u - \xi\rangle \geqslant -\gamma(\rho, \|u - \xi\|'_{\mathscr{W}}), \quad \forall \xi \in \mathscr{Y}_0. \tag{3.65}$$

Since $\mathscr{Y}_0$ is dense in $\mathscr{W}_0$, inequality (3.65) is also valid for all $\xi \in \mathscr{W}_0$. Setting $\xi = u - \delta\eta$, $\delta > 0$, $\eta \in \tilde{C}^1_{0,\Sigma_1 \cup \Omega_{T_1}}(Q) \subset \mathscr{W}_0$, in (3.65), we obtain

$$\delta^2\int_{T_1}^{T_2}(\eta', \eta)\,dt + \delta\langle f - \mathscr{A}(u - \delta\eta), \eta\rangle \geqslant -\gamma(\rho, \|\delta\eta\|'_{\mathscr{W}}), \quad \forall \eta \in \mathscr{W}_0. \tag{3.66}$$

Dividing both sides of (3.66) by δ, letting δ tend to 0, and taking into account the properties of γ and the continuity of the operator $\mathscr{A}\colon \mathscr{X} \to \mathscr{H}^*$, we obtain

$$\langle f - \mathscr{A}u, \eta\rangle \geqslant 0, \quad \forall \eta \in \tilde{C}^1_{0,\Sigma_1 \cup \Omega_{T_1}}(Q). \tag{3.67}$$

Since $\tilde{C}^1_{0,\Sigma_1 \cup \Omega_{T_1}}(Q)$ is dense in $\mathscr{H}$, it immediately follows from (3.67) that $\mathscr{A}u = f$. Theorem 3.2 is proved.

In view of Lemma 3.5 the next assertion follows directly from Theorem 3.2.

THEOREM 3.3. *Suppose that conditions* (3.20) *and* (3.26) *are satisfied, and that the restriction of the operator* $\mathscr{A}\colon \mathscr{H} \to \mathscr{H}^*$ *to the set* $\mathscr{W}_0$ *satisfies the condition*

$$\langle \mathscr{A}u - \mathscr{A}v, u - v\rangle \geqslant -\gamma(\rho, \|u - v\|_{l,l_0,Q}), \qquad u, v \in \mathscr{W}_0,$$
$$\|u\|_{\mathscr{W}} \leqslant \rho, \qquad \|v\|_{\mathscr{W}} \leqslant \rho, \tag{3.68}$$

where the indices l *and* l_0 *satisfy* (3.15). *Then for any* $\mathscr{F} \in \mathscr{H}^*$ *problem* (2.5) *has at least one generalized solution.*

We now consider the question of uniqueness of a generalized solution of problem (2.5). It is obvious that Lemmas 2.1 and 2.2 are also preserved for equations having $(A, \mathbf{0}, 2, \overline{m})\tilde{}$-structure in Q. Therefore, the following assertion holds.

LEMMA 3.10. *Suppose that conditions* (3.20) *are satisfied, and that the operators* A_t: $H \to H^*$, *where* $H \equiv H^{0,\sigma_1}_{l,\overline{m}}(A;\ \Omega;\ \sigma_1, \lambda)$, *defined by* (2.12) *for* $t \in [T_1, T_2]$, *are monotone. In particular, these operators are monotone if for almost all* $(t, x) \in Q$ *and any* $u, \xi_0 \in \mathbb{R}$ *and* $q, \xi \in \mathbb{R}^n$

$$\frac{\partial l'^i}{\partial q_j}\xi_i\xi_j + \frac{\partial l'^i}{\partial u}\xi_0\xi_i + \frac{\partial l'_0}{\partial q_j}\xi_j\xi_0 + \frac{\partial l'_0}{\partial u}\xi_0^2 \geqslant 0. \tag{3.69}$$

Then for every $\mathscr{F} \in \mathscr{H}^*$ *problem* (2.5) *has at most one generalized solution.*

From Remark 3.2 we obtain the following uniqueness theorem pertaining to the case $\overline{m} = 2$.

LEMMA 3.11. *Suppose that conditions* (3.26) *are satisfied with* $\overline{m} = 2$ *and* $\hat{\psi} \equiv 0$ *in* Q. *Then for any* $\mathscr{F} \in \mathscr{H}^*$ *problem* (2.5) *has at most one generalized solution.*

To conclude this section we present a simple criterion for condition (3.68) to hold.

LEMMA 3.12. *Suppose that* (3.20) *and the following conditions are satisfied*:
1) *The functions* $l'^i(t, x, u, q)$, $i = 1,\dots,n$, *and* $l'_0(t, x, u, q)$ *have the form*

$$\begin{aligned} l'^i(t, x, u, q) &= \bar{l}'^i(t, x, u, q) + \bar{\bar{l}}'^i(t, x, u), \\ l'_0(t, x, u, q) &= \bar{l}'_0(t, x, u, q) + \bar{\bar{l}}'_0(t, x, u, q). \end{aligned} \tag{3.70}$$

2) *The operator* $\bar{\mathscr{A}}: \mathscr{H} \to \mathscr{H}^*$ *defined by*

$$\langle \bar{\mathscr{A}}u, \eta\rangle = \iint_Q (\bar{\mathbf{l}}' \cdot A\nabla\eta + \bar{l}'_0)\, dt\, dx + \int_{\Sigma_3} \lambda u\eta\, ds, \qquad u, \eta \in \mathscr{H}, \tag{3.71}$$

satisfies the condition

$$\langle \bar{\mathscr{A}}u - \bar{\mathscr{A}}v, u - v\rangle \geqslant \alpha_0\|A\nabla u\|^{\overline{m}}_{\overline{m},Q}, \qquad u, v \in \mathscr{H},\ \alpha_0 = \text{const} > 0. \tag{3.72}$$

3) *For almost all* $(t, x) \in Q$ *and any* $u \in \mathbb{R}$ *and* $q \in \mathbb{R}^n$

$$\begin{gathered} \left|\partial\bar{\bar{l}}'^i/\partial u\right| \leqslant a_5|u|^{2/\overline{m}'-1} + \psi_i; \qquad \left|\partial\bar{\bar{l}}'_0/\partial q_i\right| \leqslant a_6|q|^{\overline{m}/2-1} + \tilde{\psi}_i; \\ \left|\partial\bar{\bar{l}}'_0/\partial u\right| \leqslant \psi_0, \qquad i = 1,\dots,n, \end{gathered} \tag{3.73}$$

where $a_5^{\overline{m}'}$, a_6^2, ψ_i, $\tilde{\psi}_i \in L^{q,q_0}(Q)$, $\psi_0 \in L^{s,s_0}(Q)$, $1/q = 1/\overline{m}' - 1/l$, $1/q_0 = 1/\overline{m}' - 1/l_0$, $2 \leqslant \overline{m}$, $2 \leqslant l \leqslant \bar{l}$, $2 \leqslant l_0 \leqslant \bar{l}_0$, *and* $\bar{l}$ *and* $\bar{l}_0$ *are the same indices as in* (3.21). *Then condition* (3.68) *holds.*

PROOF. We denote by $\bar{\bar{\mathscr{A}}}: \mathscr{H} \to \mathscr{H}^*$ the operator defined by

$$\langle \bar{\bar{\mathscr{A}}}u, \eta\rangle = \iint_Q \left[\bar{\bar{\mathbf{l}}}'(t, x, u) \cdot A\nabla\eta + \bar{\bar{l}}'_0(t, x, u, A\nabla u)\eta\right] dt\, dx, \qquad u, \eta \in \mathscr{H}. \tag{3.74}$$

Taking 1) and 2) into account, for any $u, v \in \mathscr{H}$ we estimate

$$\begin{aligned}
\langle \mathscr{A}u - \mathscr{A}v, u - v\rangle &= \langle \bar{\mathscr{A}}u - \bar{\mathscr{A}}v, u - v\rangle + \langle \bar{\bar{\mathscr{A}}}u - \bar{\bar{\mathscr{A}}}v, u - v\rangle \\
&\geqslant \alpha_0 \|A\nabla(u-v)\|^{\overline{m}}_{\overline{m},Q} \\
&\quad + \iint_Q \Big\{ \big[\bar{\bar{l}}'^i(t,x,u) - \bar{\bar{l}}'^i(t,x,v)\big] A_i \nabla(u-v) \\
&\qquad + \big[\bar{\bar{l}}'_0(t,x,u,A\nabla u) - \bar{\bar{l}}'_0(t,x,v,A\nabla v)\big](u-v)\Big\}\,dt\,dx \\
&\quad + \int_{\Sigma_3} \lambda (u-v)^2\,ds \\
&\geqslant \alpha_0 \|A\nabla(u-v)\|^{\overline{m}}_{\overline{m},Q} \\
&\quad + \iint_Q \int_0^1 \Big\{ \frac{\partial \bar{\bar{l}}'^i(t,x,v+\tau(u-v))}{\partial u}(u-v) A_i \nabla(u-v) \\
&\qquad + \frac{\partial \bar{\bar{l}}'_0(t,x,v+\tau(u-v), A\nabla v + \tau A\nabla(u-v))}{\partial u}(u-v)^2 \\
&\qquad + \frac{\partial \bar{\bar{l}}'_0(t,x,v+\tau(u-v), A\nabla v + \tau A\nabla(u-v))}{\partial q_j} \\
&\qquad\qquad \times A_j\nabla(u-v)(u-v)\Big\}\,d\tau\,dt\,dx \\
&\equiv \alpha_0 \|A\nabla(u-v)\|^{\overline{m}}_{\overline{m},Q} + I_1 + I_2 + I_3.
\end{aligned} \tag{3.75}$$

Taking 3) into account and applying the Hölder inequality, we obtain

$$\begin{aligned}
|I_1| &\leqslant (\alpha_0/4)\|A\nabla u\|^{\overline{m}}_{\overline{m},Q} + c_1 \|a_5\|^{\overline{m}}_{\overline{m}s,\overline{m}s_0,Q}\|u-v\|^2_{l,l_0,Q}, \\
|I_3| &\leqslant (\alpha_0/4)\|A\nabla u\|^{\overline{m}}_{\overline{m},Q} \\
&\quad + c_2\Big(\|a_6\|^2_{2s,2s_0,Q}\|u-v\|^2_{l,l_0,Q} + \|\psi\|^{\overline{m}'}_{q,q_0,Q}\|u-v\|^{\overline{m}'}_{l,l_0,Q}\Big), \\
|I_2| &\leqslant \|\psi_0\|_{s,s_0,Q}\|u-v\|^2_{l,l_0,Q},
\end{aligned} \tag{3.76}$$

where $c_1 = c_1(\alpha_0)$ and $c_2 = c_2(\alpha_0)$. It follows from (3.75) and (3.76) that for any u, $v \in \mathscr{W}_0$, $\|u\|_{\mathscr{W}} \leqslant \rho$, $\|v\|_{\mathscr{W}} \leqslant \rho$,

$$\begin{aligned}
\langle \mathscr{L}u - \mathscr{L}v, u - v\rangle &\geqslant (\alpha_0/2)\|A\nabla(u-v)\|^{\overline{m}}_{\overline{m},Q} \\
&\quad - c_3\Big(\|u-v\|^2_{l,l_0,Q} + \|u-v\|^{\overline{m}'}_{l,l_0,Q}\Big).
\end{aligned} \tag{3.77}$$

It is obvious that (3.68) follows from (3.77) and Lemma 3.5. Lemma 3.12 is proved.

From Theorem 3.3 and Lemmas 3.12 and 3.11 we obtain, in particular, the following assertion.

THEOREM 3.4. *Suppose conditions* (3.20), (3.26) *and* (3.70)–(3.73) *are satisfied. Then for every* $\mathscr{F} \in \mathscr{H}^*$ *problem* (2.5) *has at least one generalized solution. In the case* $\overline{m} = 2$ *and* $\hat{\psi} \equiv 0$ *in* (3.26) *problem* (2.5) *has precisely one generalized solution.*

§4. Linear A-parabolic equations with weak degeneracy

In the cylinder $Q = \Omega \times (T_1, T_2)$, where Ω is a bounded, strongly Lipschitz domain in $\mathbb{R}^n$, $n \geqslant 1$, we consider the linear equation

$$\partial u/\partial t - (\partial/\partial x_i)\left(\alpha^{ij}u_{x_j} + \alpha^i u + g^i\right) + \beta^i u_{x_i} + \beta_0 u + g_0 = f, \tag{4.1}$$

where the, generally speaking, nonsymmetric matrix $\mathfrak{A} \equiv \|\alpha^{ij}(t, x)\|$ is positive definite for almost all $(t, x) \in Q$, and the functions $\alpha^{ij}(t, x)$, $\alpha^i(t, x)$, $g^i(t, x)$, $\beta^i(t, x)$, $\beta_0(t, x)$ and $g_0(t, x)$, $i, j = 1, \dots, n$, are defined and measurable in Q. Suppose there exists a constant $k_0 > 0$ such that for almost all $(t, x) \in Q$ and any ξ, $\eta \in \mathbb{R}^n$

$$|\alpha^{ij}(t, x)\xi_i\eta_j| \leqslant k_0\sqrt{\alpha^{ij}(t, x)\xi_i\xi_j}\,\sqrt{\alpha^{ij}(t, x)\eta_i\eta_j}, \tag{4.2}$$

and constants $k_1, k_2 > 0$ such that for almost all $(t, x) \in Q$ and any $\xi \in \mathbb{R}^n$

$$k_1\hat{\alpha}^{ij}(x)\xi_i\xi_j \leqslant \alpha^{ij}(t, x)\xi_i\xi_j \leqslant k_2\hat{\alpha}^{ij}(x)\xi_i\xi_j, \tag{4.3}$$

where

$$\hat{\alpha}^{ij}(x) = (T_2 - T_1)^{-1}\int_{T_1}^{T_2}\alpha^{ij}(t, x)\,dt, \qquad i, j = 1, \dots, n.$$

We set

$$A(x) = \left((\hat{\mathfrak{A}} + \hat{\mathfrak{A}}^*)/2\right)^{1/2},$$

where $\hat{\mathfrak{A}} \equiv \hat{\mathfrak{A}}(x) \equiv \|\hat{\alpha}^{ij}(x)\|$. It is obvious that

$$\mathfrak{A}\xi \cdot \xi \geqslant k_1\hat{\mathfrak{A}}\xi \tag{4.4}$$

and

$$\hat{\mathfrak{A}}\xi \cdot \xi = |A\xi|^2. \tag{4.5}$$

We assume that the matrix $A = A(x)$ is weakly degenerate in Ω relative to $\mathbf{m} = \mathbf{2}$ (see Lemma 7.2.1 with $\mathbf{m} = \mathbf{2}$) and suppose that the spaces

$$\mathscr{H}_\lambda \equiv \mathscr{H}^{Q,\Sigma_1}_{l,l_0;2}(A; Q; \Sigma_3, \lambda) \qquad \text{and} \qquad \mathscr{H} \equiv \mathscr{H}^{Q,\Sigma_1}_{l,l_0;2}(A; Q)$$

(see (3.21)) are isomorphic; here $\Sigma_1 = \sigma_1 \times (T_1, T_2)$, $\Sigma_2 = \sigma_2 \times (T_1, T_2)$, $\Sigma_3 = \sigma_3 \times (T_1, T_2)$, $\sigma_1 \cup \sigma_2 \cup \sigma_2 = \partial\Omega$ and $\sigma_i \cap \sigma_j = \varnothing$ for $i \neq j$, $i, j = 1, 2, 3$. The last assumptions imply the validity of a condition of the form (1.36).

Setting

$$\mathbf{l}(t, x, u, q) \equiv \mathfrak{A}p + \boldsymbol{\alpha}u + \mathbf{g}, \qquad l_0(t, x, u, p) \equiv \boldsymbol{\beta}\cdot\mathbf{p} + \beta_0 u + g_0$$

and taking into account that

$$\mathbf{l}(t, x, u, p) = A^*(SAp + \mathbf{a}u + \mathbf{f}), \qquad l_0(t, x, u, p) = \boldsymbol{\gamma}\cdot Ap + a_0 u + f_0,$$

where $S = A^{-1}\mathfrak{A}A^{-1}$, $\mathbf{a} = A^{-1}\boldsymbol{\alpha}$, $\mathbf{f} = A^{-1}\mathbf{g}$, $\boldsymbol{\gamma} = A^{-1}\boldsymbol{\beta}$, $a_0 = \beta_0$ and $f_0 = g_0$ (here $A^* = A$), we observe that condition (2.2) is satisfied for $\mathbf{l}'(t, x, u, q) = Sq + \mathbf{a}u + \mathbf{f}$ and $l_0'(t, x, u, q) = \boldsymbol{\gamma}\cdot q + a_0 u + f_0$. Suppose that the conditions

$$\begin{gathered} A^{-1}\boldsymbol{\alpha} \in \mathbf{L}^{2s,2s_0}(Q), \qquad A^{-1}\beta \in \mathbf{L}^{2s,2s_0}(Q), \qquad \beta_0 \in L^{s,s_0}(Q), \\ A^{-1}\mathbf{g} \in \mathbf{L}^2(Q), \qquad g_0 \in L^{l',l_0'}(Q) \end{gathered} \tag{4.6}$$

are satisfied, where $1/s + 2/l = 1$, $1/s_0 + 2/l_0 = 1$, and the indices l and l_0 satisfy conditions (3.27) in which $\hat{l}$ and $\hat{l}_0$ are defined as in (3.21) relative to the indices $\mathbf{q}$ and $\mathbf{q}_0$ for which inequality (3.2) holds in the case $\mathbf{m} = \mathbf{m}_0 = 2$ (see Remark 4.5.3). Suppose, for example, that the matrix $B = \|b^{ij}(x)\|$, where $B = A^{-1}$, satisfies the conditions

$$b^{ij} \in L^{r_i}(\Omega), \quad i, j = 1,\dots,n; \qquad 1/2 + 1/r_i \leqslant 1, \quad i = 1,\dots,n, \tag{4.7}$$

and suppose that the numbers r_{0i} satisfy $1/2 + 1/r_{0i} \leqslant 1$, $i = 1,\dots,n$. Then the indices $\mathbf{q}$ and $\mathbf{q}_0$, with which (3.2) holds by Lemma 3.1, have the form

$$1/q_i = 1/2 + 1/r_i, \quad 1/q_{0_i} = 1/2 + 1/r_{0_i}, \qquad i = 1,\dots,n. \tag{4.8}$$

Finding the values of $\hat{l}$ and $\hat{l}_0$ on the basis of such $\mathbf{q} = (q_1,\dots,q_n)$ and $\mathbf{q}_0 = (q_{01},\dots,q_{0n})$ and finding the indices l and l_0 on the basis of these values of $\hat{l}$ and $\hat{l}_0$ (see (3.27)), it is possible to find the values of s and s_0 required in (4.6).

We shall show that under the above assumptions equation (4.1) has $(A, \mathbf{0}, 2, 2)\tilde{}$-structure in the cylinder Q. We note first of all that $\|S\| = \|A^{-1}\mathfrak{A}A^{-1}\| = \mu = \text{const}$ for almost all $(t, x) \in Q$. Indeed, suppose that $p = Aq$, $\xi = A\eta$, $q \in \mathbb{R}^n$ and $\xi \in \mathbb{R}^n$. Taking (4.2), (4.3), and (4.5) into account, we find that

$$\begin{aligned} |A^{-1}\mathfrak{A}A^{-1}p \cdot \xi| &= |\mathfrak{A}q \cdot \eta| \leqslant k_0(\mathfrak{A}q \cdot q)^{1/2}(\mathfrak{A}\eta \cdot \eta)^{1/2} \\ &\leqslant k_0k_2(\hat{\mathfrak{A}}q \cdot q)^{1/2}(\hat{\mathfrak{A}}\eta \cdot \eta)^{1/2} = k_0k_2|Aq|\,|A\eta| = k_0k_2|p|\,|\xi|. \end{aligned} \tag{4.9}$$

Taking into account the nondegeneracy of the matrix A for almost all $x \in \Omega$, we conclude that (4.9) implies $\|S\| \leqslant k_0k_2$ for almost all $(t, x) \in Q$. This and (4.6) obviously imply the validity of conditions (3.20) with $\overline{m} = 2$, $\mu_1 = k_0k_2$, $a_1 = |A^{-1}\boldsymbol{\alpha}|$, $\psi = |A^{-1}\mathbf{g}|$, $a_2 = |A^{-1}\boldsymbol{\beta}|$, $a_3 = |\beta_0|$ and $\psi_0 = |g_0|$, while the functions enumerated are summable over Q with indices somewhat greater than required in (3.20). Thus, (4.1) has $(A, \mathbf{0}, 2, 2)\tilde{}$-structure in Q. It is also easy to see that in the present case condition (3.26) is also satisfied (with $\overline{m} = 2$). Indeed, taking into account that here

$$l'^i q_i + l'_0 u \equiv Sq \cdot q + \mathbf{a} \cdot qu + \mathbf{f} \cdot q + \boldsymbol{\gamma} \cdot qu + a_0u^2 + f_0u, \tag{4.10}$$

where $Sq \cdot q = \mathfrak{A}(A^{-1}q) \cdot A^{-1}q \geqslant k_1|A(A^{-1}q))^2 = k_1|q|^2$ (see (4.4) and (4.5)) and estimating

$$\begin{gathered} |\mathbf{a} \cdot qu| \leqslant \frac{k_1}{6}|q|^2 + \frac{3}{2k_1}|\mathbf{a}|^2u^2, \qquad |\boldsymbol{\gamma} \cdot q|u \leqslant \frac{k_1}{6}|q|^2 + \frac{3}{2k_1}|\boldsymbol{\gamma}|^2u^2, \\ |f \cdot q| \leqslant \frac{k_1}{6}|q|^2 + \frac{3}{2k_1}|\mathbf{f}|^2, \qquad |f_0u| \leqslant \frac{1}{2}|f_0|^{2-\kappa} + \frac{1}{2}|f_0|^{\kappa}u^2, \\ |a_0u^2| \leqslant |a_0|u^2, \end{gathered} \tag{4.11}$$

with $\kappa = \min(l'/s, l'_0/s_0) \in (0, 1)$, we conclude that (3.26) is valid with $\overline{m} = 2$, $\nu = k_1/2$, $\hat{a} = 3/2k_1(|\mathbf{a}|^2 + |\boldsymbol{\gamma}|^2) + |a_0| + 1/2|f_0|^{\kappa} \in L^{s,s_0}(Q)$ and $\hat{\psi} = (3/2k_1)|\mathbf{f}|^2 + (1/2)|f_0|^{2-\kappa} \in L^1(Q)$; here we take into account that

$$\|\,|f_0|^{\kappa}\|_{s,s_0,Q} = \|f_0\|^{\kappa}_{\kappa s,\kappa s_0,Q} \leqslant c\|f_0\|^{\kappa}_{l',l'_0,Q}$$

and

$$\|\,|f_0|^{2-\kappa}\|_{1,Q} = \|f_0\|^{2-\kappa}_{2-\kappa,Q} = \|f_0\|^{2-\kappa}_{\min(l',l'_0),Q} \leqslant c\|f_0\|^{2-\kappa}_{l',l'_0,Q}.$$

Thus, a condition of the form (3.26) is satisfied.

For equation (4.1) we consider the general boundary value problem (2.5), which here assumes the form

$$\begin{gathered} u_t - (\partial/\partial x_i)\left(\alpha^{ij}u_{x_j} + \alpha^i u + g^i\right) + \beta^i u_{x_i} + \beta_0 u + g_0 = f \quad \text{in } Q, \\ u = 0 \quad \text{on } (\sigma_1 \times (T_1, T_2)) \cup \Omega_{T_1}, \\ \partial u/\partial N + \boldsymbol{\alpha} \cdot \boldsymbol{\nu} u + \mathbf{g} \cdot \boldsymbol{\nu} = 0 \quad \text{on } \sigma_2 \times (T_1, T_2), \\ \partial u/\partial N + (\boldsymbol{\alpha} \cdot \boldsymbol{\nu} - \lambda)u + \mathbf{g} \cdot \boldsymbol{\nu} = 0 \quad \text{on } \sigma_3 \times (T_1, T_2), \end{gathered} \tag{4.12}$$

where λ is a function defined on $\sigma_3 \times (T_1, T_2)$, $\partial u/\partial N \equiv \alpha^{ij}u_{x_j}\nu_i \equiv A\nabla u \cdot A\boldsymbol{\nu}$, $\sigma_1 \cup \sigma_2 \cup \sigma_3 = \partial\Omega$, and $\sigma_i \cap \sigma_j = \varnothing$ for $i \neq j$, $i, j = 1, 2, 3$. A generalized solution of problem (4.12) can be defined as any function $u \in \mathscr{W}_0$ [where

$$\mathscr{W}_0 \equiv \left\{u \in \mathscr{H} \equiv \mathscr{H}^{0,\Sigma_1}_{\bar{l},\bar{l}_0,2}(A, Q) \colon u' \in \mathscr{H}^*, u(T_1) = 0\right\};$$

here $\bar{l}$ and $\bar{l}_0$ are defined by (3.21) *for* $\mathbf{q}$ *and* $\mathbf{q}_0$ *from* (3.2) *with* $\mathbf{m} = \mathbf{m}_0 = \mathbf{2}$] satisfying the identity

$$\begin{gathered} \iint_Q [\mathfrak{A} \cdot \nabla u + \boldsymbol{\alpha} u + \mathbf{g}) \cdot \nabla\eta + (\boldsymbol{\beta} \cdot \nabla u + \beta_0 u + g_0)\eta \, dt \, dx \\ + \int_{T_1}^{T_2} (u', \eta) \, dt + \int_{\Sigma_3} \lambda u \eta \, ds = \iint_Q f\eta \, dt \, dx, \quad \forall \eta \in \tilde{C}^1_{0,\Sigma_1}(Q). \end{gathered} \tag{4.13}$$

In place of f in (4.12) it is possible to choose an arbitrary element $\mathscr{F} \in \mathscr{H}^*$. The right side of (4.13) then contains $\langle \mathscr{F}, \eta \rangle$ instead of the integral $\iint_Q f\eta \, dt \, dx$; here $\langle \cdot, \cdot \rangle$ denotes the duality between $\mathscr{H}$ and $\mathscr{H}^*$.

THEOREM 4.1. *Under the conditions enumerated in this section (in particular, under conditions* (4.2), (4.3) *and* (4.6)) *problem* (4.12) *for every* $f \equiv \mathscr{F} \in \mathscr{H}^*$, *where* $\mathscr{H} \equiv \mathscr{H}^{0,\Sigma_1}_{\bar{l},\bar{l}_0;2}(A, Q)$, *has precisely one generalized solution.*

PROOF. In view of Theorem 3.4 and the validity of conditions of the form (3.20) and (3.26) (with $\overline{m} = 2$) established above, to prove the existence of a generalized solution of problem (4.12) it suffices to verify that conditions of the form (3.70)–(3.73) (with $\overline{m} = 2$) hold. We set

$$\begin{aligned} \bar{\mathbf{l}}'(t, x, u, q) &\equiv Sq \equiv A^{-1}\mathfrak{A}A^{-1}q, \qquad \bar{l}'_0(t, x, u, q) \equiv 0, \\ \bar{\bar{\mathbf{l}}}'(t, x, u, q) &\equiv \mathbf{a}u + \mathbf{f}, \qquad \bar{\bar{l}}'_0(t, x, u, q) = \boldsymbol{\gamma} \cdot q + a_0 u + f_0. \end{aligned} \tag{4.14}$$

It is obvious that for such $\bar{\mathbf{l}}'$, $\bar{l}'_0$, $\bar{\bar{\mathbf{l}}}'$ and $\bar{\bar{l}}'_0$ condition (3.70) holds, and for the operator $\mathscr{A}\colon \mathscr{H} \to \mathscr{H}^*$ defined by (3.71) an inequality of the form (3.72) with $\overline{m} = 2$ and $\mathscr{K}_0 = k_1$ holds. Taking into account that $\partial\bar{\bar{l}}'^i/\partial u = a^i$, $\partial\bar{\bar{l}}'_0/\partial g_j = \gamma^j$ and $\partial\bar{\bar{l}}'_0/\partial u = a_0$, $i, j = 1, \ldots, n$, we conclude in view of (4.6) that conditions (3.73) are satisfied with $\overline{m} = 2$, $\psi_1 = |\mathbf{a}| = |A^{-1}\boldsymbol{\alpha}|$, $\psi_2 = |\boldsymbol{\gamma}| \equiv |A^{-1}\boldsymbol{\beta}|$ and $\psi_3 = |a_0| \equiv |\beta_0|$. Thus, the existence of a generalized solution of problem (4.12) is established.

We shall now prove the uniqueness of this solution. It is obvious that the difference $u \equiv u^1 - u^2$ of two generalized solutions of (4.12) is a generalized solution of the corresponding homogeneous problem characterized by the conditions $\mathbf{g} \equiv \mathbf{0}$, $g_0 \equiv 0$ and $f \equiv 0$ in (4.1). In this case an inequality of the form (3.36) is satisfied with $\psi \equiv 0$ (and $\overline{m} = 2$). The uniqueness of the generalized solution of (4.12) then follows from the same Theorem 3.4 (or Lemma 3.11). Theorem 4.1 is proved.

REMARK 4.1. An analogous result can also be proved if in (4.6) the indices l and l_0 are replaced by the limit indices $\bar{l}$ and $\bar{l}_0$ (the same as in (3.21) with $\mathbf{m} = \mathbf{2}$ and $\mathbf{m}_0 = \mathbf{2}$) by combining generalized solutions of problems of the form (4.12) corresponding to the partial intervals $[T_1, \tau_1]$, $[\tau_1, \tau_2], \ldots, [\tau_{n-1}, T_2]$ with lengths not exceeding a sufficiently small (fixed) number $\delta > 0$. However, this case requires alteration of the scheme of proof used above. We therefore limit ourselves here to a reference to our paper [43], in which a method is presented for proving an existence and uniqueness theorem in the case of the first boundary value problem with the indicated limit conditions on the indices of the space $\mathscr{H}$.

PART IV
ON REGULARITY OF GENERALIZED SOLUTIONS OF QUASILINEAR DEGENERATE PARABOLIC EQUATIONS

A theory of the dependence of differential properties of generalized solutions of quasilinear, nondegenerate elliptic and parabolic equations of divergence form on the properties of the functions forming these equations is expounded in the monographs of O. A. Ladyzhenskaya and N. N. Ural'tseva [80] and [83]. In this part we study questions of regularity and some qualitative questions for generalized solutions of quasilinear $(A, \mathbf{0})$-parabolic equations with a weak spatial degeneracy which contain nondegenerate parabolic equations as a special case. Here we consider questions of regularity both for generalized solutions of the general boundary value problem of the form (8.2.5) and for just generalized solutions of $(A, \mathbf{0})$-parabolic equations of the form

$$u_t - (d/dx_i)l^i(t, x, u, \nabla u) + l_0(t, x, u, \nabla u) = 0, \tag{1}$$

having $(A, \mathbf{0}, 2, 2)\tilde{}$-structure relative to a weakly degenerate matrix $A \equiv \|a^{ij}(x)\|$ in a cylinder $Q = \Omega \times (T_1, T_2)$ (see §8.3). From the results it is evident how the properties of generalized solutions improve with improvement in the regularity of the functions forming the equation. This improvement is not without bound as in the case of nondegenerate parabolic equations, however, since the presence of a weak degeneracy poses an obstacle to the improvement of the differential properties of the functions determining the structure of the equations.

Under the original assumptions regarding the structure of the equations we establish integral Hölder continuity of generalized solutions in the variable t with exponent $1/2$, and also an energy inequality for generalized solutions of the boundary value problem (8.2.5). Further improvement of the properties of generalized solutions depends on the degree of degeneracy of the matrix A. Progress in this direction is possible only after developing a special technique establishing the possibility of making substitutions of the form $\eta = f(u(t, x))\psi(t, x)$, where $f(u)$ has a uniformly Lipschitz derivative $f'(u)$ on $\mathbb{R}$ and $\psi(t, x)$ is a smooth function, in the integral identity determining the generalized solution of problem (8.2.5) or in the integral identity

$$\int_{T_1}^{T_2}(u', \eta)\, dt + \iint_Q [\mathbf{l}'(t, x, u, A\nabla u) \cdot A\nabla u + l_0'(t, x, u, A\nabla u)\eta]\, dt\, dx = 0,$$

$$\eta \in C_0^1(Q), \tag{2}$$

defining a generalized solution

$$u \in \mathscr{W} \equiv \left\{u \in \mathscr{H}_{2(\tilde{l}, \tilde{l}_0)}(A, Q) \colon u' \in \left(\mathscr{H}_{2(\tilde{l}, \tilde{l}_0)}(A, Q)\right)^*\right\}$$

(see §8.3) of equation (1) in the cylinder Q. In resolving this question we make essential use of the fact (proved in §8.1) of strong convergence of averages in the variable t of a function in $\mathscr{W}$ to this same function in the norm of $\mathscr{W}$. Substituting truncations of powers of generalized solutions as the test function η in the integral identities indicated above, using the imbedding $\mathscr{W} \to L^{l,l_0}(Q)$ with suitable l and l_0, and taking limits, we establish estimates of generalized solutions in the norms of L^{p,p_0} and L^∞ and simultaneously prove that these solutions belong to the spaces L^{p,p_0} and L^∞. Exponential summability of generalized solutions and a corresponding estimate of the integrals $\iint \exp\{c|u(t,x)|\}\,dt\,dx$ are an intermediate case of the properties indicated above. Thus, in proving the results enumerated the author proceeded along the path of developing the known iterative technique of Moser (see [149] and [150]). Within the framework of nondegenerate elliptic and parabolic equations he originally established similar results using the approach of Ladyzhenskaya and Ural'tseva to estimates of the maximum modulus of generalized solutions (see [80], [77] and [22]–[24]). However, in this monograph we shall not use this method.

In investigating the question of boundedness of generalized solutions we establish a qualified estimate of the form

$$m - c_1 k_k \leqslant u(t,x) \leqslant M + c_2 k_2, \tag{3}$$

where k_1 and k_2 depend on the structure of the equation, m and M are in a particular sense the lower and upper bounds of the solution $u(t,x)$ on the parabolic boundary of the cylinder Q, c_1 and c_2 are constants, and the inequality itself holds for almost all $(t,x) \in Q$. It follows from (3), in particular, that under more stringent conditions on the structure of the equation ($k_1 = k_2 = 0$) the generalized solution $u(t,x)$ must assume its essential infimum and supremum in $\overline{Q}$ on the parabolic boundary of the cylinder Q. Special cases of the results regarding membership of generalized solutions of equations of the form (1) in the spaces $L^{p,p_0}(Q)$ and their exponential summability have been obtained in the works [22]–[24] and [139] (linear nondegenerate equations), [25] (quasilinear nondegenerate equations), [32] and [33] (linear weakly degenerate equations), and [43] (quasilinear weakly degenerate equations).

Special cases of the results on local and global boundedness of generalized solutions of equations of the form (1) are established in [80], [98], [150], [139], [67], [68] and [22]–[24] (linear nondegenerate equations) [25], [77], [125] and [126] (quasilinear nondegenerate equations), [33]) (linear weakly degenerate equations), [70] (quasilinear $(A,\mathbf{0})$-parabolic weakly degenerate equations in the case of a diagonal matrix A under an additional assumption about the existence of a generalized derivative $u_t \in L^2(Q)$ for the generalized solution), and others. Results on the boundedness of generalized solutions of quasilinear $(A,\mathbf{0})$-parabolic weakly degenerate equations in the case of a nondiagonal matrix A are established in the author's paper [43].

We remark that the condition that the matrix A be independent of t, which is imposed throughout Chapter 9, is occasioned by the desire to work with a natural class of generalized solutions. The additional assumption of the existence of a derivative $u_t \in L^2(Q)$ for a generalized solution automatically leads to preservation of all results proved here if $A = \|a^{ij}(x)\|$ is replaced by $A = \|a^{ij}(t, x)\|$ (without any changes whatsoever in the remaining conditions guaranteeing these results). We note, however, that in the case of matrices $A = \|a^{ij}(t, x)\|$ there are no results on the existence of generalized solutions of the boundary value problems.

The restricted size of this monograph has not permitted the inclusion of the author's results connected with Harnack's inequality for generalized solutions of equations of the form (1) (see [151], [25], [125] and [171]) which is the source of many applications of qualitative and quantitative character. In particular, the estimate of the Hölder constant for generalized solutions of (1) is an important consequence of Harnacks's inequality (see [155], [80], [98], [67], [24], [78], [125], [70], and others).

CHAPTER 9
INVESTIGATION OF THE PROPERTIES OF GENERALIZED SOLUTIONS

§1. The structure of the equations and their generalized solutions

In the cylinder $Q = \Omega \times (T_1, T_2)$, where Ω is a bounded, strongly Lipschitz domain in $\mathbb{R}^n$, $n \geqslant 1$, we consider an equation of the form

$$u_t - (d/dx_i) l^i(t, x, u, \nabla u) + l_0(t, x, u, \nabla u) = 0, \tag{1.1}$$

assuming that it has $(A, \mathbf{0}, 2, 2)^{\sim}$-structure (see §8.3). Below we write out in explicit form conditions expressing this assumption regarding the structure of equation (1.1) (admitting here conditions somewhat more concrete than in §8.3). Suppose that for almost all $(t, x) \in Q$ and any $u \in \mathbb{R}$ and $p \in \mathbb{R}^n$

$$\mathbf{l}(t, x, u, p) = A^* \mathbf{l}'(t, x, u, Ap), \qquad l_0(t, x, u, p) = l_0'(t, x, u, Ap), \tag{1.2}$$

where the functions $l'^i(t, x, u, p)$, $i = 1, \dots, n$, and $l_0'(t, x, u, p)$ satisfy the Carathéodory condition in $Q \times \mathbb{R} \times \mathbb{R}^n$, and for the matrix $A = \|a^{ij}(x)\|$ the following conditions are satisfied:

1) $\det A \neq 0$ for almost all $x \in \Omega$.
2) $a^{ij} \in L^2(\Omega)$, $i, j = 1, \dots, n$.
3) For elements of the matrix $A^{-1} \equiv B \equiv \|b^{ij}(x)\|$ inverse to A we have $b^{ij} \in L^{r_i}(\Omega)$, $r_i \geqslant 2$, $i = 1, \dots, n$, and $\sum_1^n 1/r_i < 1$. (1.3)

From Lemma 8.3.4 we obtain, in particular, the following assertion.

LEMMA 1.1. *Suppose condition* (1.3) *is satisfied, and suppose the numbers* r_{0i}, $i = 1, \dots, n$, *satisfy the condition*

$$r_{0i} \geqslant 2, \qquad i = 1, \dots, n. \tag{1.4}$$

Then the imbeddings

$$\widetilde{\mathscr{H}_2^{0,\Gamma}}(A, Q) \to \widetilde{\mathscr{H}_{\mathbf{q},\mathbf{q}_0}^{0,\Gamma}}(Q) \to L^{l, l_0}(Q), \tag{1.5}$$

hold where (*cf.* *Remark* 4.5.3), *where* Γ *is any set of the form* $\Gamma = \gamma \times (T_1, T_2)$, $\gamma \subset \partial\Omega$ (*in particular*, γ *may be empty*), $\mathbf{q} = (q_1, \dots, q_n)$, $\mathbf{q}_0 = (q_{01}, \dots, q_{0n})$,

$$q/q_i = 1/2 + 1/r_i, \qquad 1/q_{0i} = 1/2 + 1/r_{0i}, \qquad i = 1, \dots, n, \tag{1.6}$$

and the indices $\bar{l}$ *and* $\bar{l}_0$ *are determined in terms of* $\mathbf{q}$ *and* $\mathbf{q}_0$ *by the conditions*

$$
\begin{gathered}
1/\bar{l} = \bar{\alpha}/\hat{l} + \frac{1-\bar{\alpha}}{2}, \qquad 1/\bar{l}_0 = \bar{\alpha}/\hat{l}_0, \qquad \bar{\alpha} \in (0, \hat{l}_0/2), \\
1/\hat{l} = \frac{\sum_{i=1}^n 1/q_i - 1}{n} \quad \textit{for} \sum_{i=1}^n 1/q_i > 1, n \geqslant 2, \\
\hat{l} \in (2, +\infty) \quad \textit{for} \sum_{i=1}^n 1/q_i = 1, n = 2, \\
\hat{l} \in (2, +\infty] \quad \textit{for } n = 1, \quad \hat{l}_0 = \frac{n}{\sum_{i=1}^n 1/q_{0i}}.
\end{gathered} \tag{1.7}
$$

In particular, for any function $u \in \tilde{\mathscr{H}} \equiv \widetilde{\mathscr{H}_2^{0,\Gamma}}(A, Q)$

$$
\|u\|_{\bar{l},\bar{l}_0,Q} \leqslant c_0 b \|u\|_{\mathscr{H}} \equiv c_0 b\big(\|u\|_{2,\infty,Q} + \|A\nabla u\|_{2,Q}\big), \tag{1.8}
$$

where c_0 *depends only on* n, r_i, r_{0i}, α, Ω, $T_2 - T_1$, *and*

$$
b \equiv b(Q) \equiv \sup_{i,j=1,\dots,n} \|b^{ij}\|_{r_i,r_{0i},Q}.
$$

If in place of $\bar{l}$ *and* $\bar{l}_0$ *indices satisfying the conditions*

$$
\begin{gathered}
1/l = \alpha/\hat{l} + \frac{1-\alpha}{2}; \; 1/l_0 = \alpha/\hat{l}_0 + \frac{(1-\alpha)\beta}{2} < 1/2, \quad \alpha \in (0,1), \beta \in (0,1); \\
1/\hat{l} = \frac{\sum_{i=1}^n 1/q_i - 1}{n} \quad \textit{for} \sum_{i=1}^n 1/q_i > 1, n \geqslant 2; \\
\hat{l} \in (2, +\infty) \quad \textit{for} \sum_{i=1}^n 1/q_i = 1, n = 2; \quad \hat{l} \in (2, +\infty] \quad \textit{for } n = 1; \\
\hat{l}_0 = \frac{n}{\sum_{i=1}^n 1/q_{0i}}; \quad \mathbf{q} \textit{ and } \mathbf{q}_0, \textit{ satisfy conditions } (1.6),
\end{gathered} \tag{1.9}
$$

are considered, then for any $\varepsilon > 0$

$$
\|u\|_{l,l_0,Q} \leqslant \varepsilon \|u\|_{\mathscr{H}} + c_1 \varepsilon^{-\lambda} \|u\|_{2,Q}, \tag{1.10}
$$

where c_1 *depends on the same quantities as* c_0 *and also on* β, *while* $\lambda > 0$ *depends only on* α *and* β.

In particular, it follows from (1.3), (1.6) and (1.7) that $\bar{l} > 2$ and $\bar{l}_0 > 2$, while from (1.3), (1.6) and (1.9) it follows that $l > 2$ and $l_0 > 2$.

Suppose that for almost all $(t, x) \in Q$ and any $u \in \mathbb{R}$ and $p \in \mathbb{R}^n$

$$
\begin{gathered}
\mathbf{l}'(t,x,u,p) \cdot p \geqslant \nu |p|^2 - a_4(t,x) u^2 - h(t,x), \\
|l_0'(t,x,u,p)| \leqslant a_2(t,x) \sum_{i=1}^n |p_i| + a_3(t,x)|u| + g(t,x),
\end{gathered} \tag{1.11}
$$

$$
|\mathbf{l}'(t,x,u,p)| \leqslant \mu \sum_{i=1}^n |p_i| + a_1(t,x)|u| + f(t,x), \tag{1.12}
$$

where $\mu = \text{const} \geqslant 0$, $\nu = \text{const} \geqslant 0$, $a_i(t, x) \geqslant 0$, $i = 1, 2, 3, 4$, $f(t, x) \geqslant 0$, $g(t, x) \geqslant 0$, $h(t, x) \geqslant 0$, and

$$a_1^2, a_2^2, a_3, a_4 \in L^{\bar{s}, \bar{s}_0}(Q), 1/\bar{s} + 2/\bar{l} = 1, \quad 1/\bar{s}_0 + 2/\bar{l}_0 = 1,$$
$$f^2, h \in L^1(Q), g \in L^{\bar{l}', \bar{l}_0'}(Q), 1/\bar{l} + 1/\bar{l}' = 1, \quad 1/\bar{l}_0 + 1/\bar{l}_0' = 1, \tag{1.13}$$
$$\bar{l}, \bar{l}_0 \text{ are the same as in (1.7)}.$$

In particular, it follows from (1.12) that $\bar{s} \in (1, +\infty)$ and $\bar{s}_0 \in (1, +\infty)$.

REMARK 1.1. It is obvious that conditions (1.11) and (1.12) are satisfied if the functions $l_i(t, x, u, p)$, $i = 1, \dots, n$, and $l_0(t, x, u, p)$ forming (1.1) satisfy the Carathéodory condition in $Q \times \mathbb{R} \times \mathbb{R}^n$ as well as the following condition: for almost all $(t, x) \in Q$ and any $u \in \mathbb{R}$ and $p \in \mathbb{R}^n$

$$\begin{gathered} \mathbf{l}(t, x, u, p) \cdot p \geqslant \nu |\tilde{A}p|^2 - a_4(t, x)u^2 - h(t, x), \\ |l_0(t, x, u, p)| \leqslant a_2(t, x) \sum_{i=1}^{h} |\tilde{A}_i p| + a_3(t, x)|u| + g(t, x), \end{gathered} \tag{1.11$'$}$$

$$\left|(\tilde{A}^*)^{-1}\mathbf{l}(t, x, u, p)\right| \leqslant \mu \sum_{i=1}^{n} |\tilde{A}_i p| + a_1(t, x)|u| + f(t, x), \tag{1.12$'$}$$

where the square matrix $\tilde{A} = \tilde{A}(t, x)$ of order n satisfies the following condition: there exist constants $k_1, k_2 > 0$ such that for almost all $x \in \Omega$, almost all $t \in [T_1, T_2]$, and all $\xi \in \mathbb{R}^n$

$$k_1|A\xi| \leqslant |\hat{A}\xi| \leqslant k_2|A\xi|, \tag{1.14}$$

where the matrix $\hat{A} \equiv \|\hat{a}^{ij}(x)\|$ is defined by

$$\hat{a}^{ij}(x) = (T_2 - T_1)^{-1} \int_{T_1}^{T_2} \tilde{a}^{ij}(t, x)\, dt, \qquad i, j = 1, \dots, n, \tag{1.15}$$

and it is assumed that condition (1.3) is satisfied for A, while conditions of the form (1.12) are satisfied for the functions a_1, a_2, a_3, a_4, f, g, and h (relative to the matrix (1.15)). In particular, if the matrix $\tilde{A} = \|a^{ij}(t, x)\|$ in (1.13) is diagonal, i.e.,

$$\tilde{A} = \begin{Vmatrix} \tilde{\lambda}_1(t, x) & & 0 \\ & \ddots & \\ 0 & & \tilde{\lambda}_n(t, x) \end{Vmatrix}, \tag{1.16}$$

then (1.11′) and (1.12′) can be replaced by the equivalent conditions

$$\begin{gathered} |l^i(t, x, u, p)| \leqslant \mu\tilde{\lambda}_i \left(\sum_{j=1}^{n} \tilde{\lambda}_j |p_j| + a_1(t, x)|u| + f(t, x) \right), \\ |l_0(t, x, u, p)| \leqslant a_2(t, x) \sum_{j=1}^{n} \tilde{\lambda}_j |p_j| + a_3(t, x)|u| + g(t, x), \\ \mathbf{l}(t, x, u, p) \cdot p \geqslant \nu \sum_{j=1}^{n} \tilde{\lambda}_j p_j^2 - a_4(t, x)u^2 - h(t, x), \end{gathered} \tag{1.17}$$

where it is assumed, of course, that conditions following from (1.14) and (1.15) for the matrix (1.16) are satisfied for the functions $\tilde{\lambda}_1, \dots, \tilde{\lambda}_n$, while the remaining functions in (1.17) satisfy conditions corresponding to (1.13).

In cases concerning properties of generalized solutions of a boundary value problem for equation (1.1) we shall always assume that the spaces

$$\mathscr{H}^{0,\Sigma_1}_{2,(\bar l_1,\bar l_0)}(A;Q;\Sigma_3,\lambda) \quad \text{and} \quad \mathscr{H}^{0,\Sigma_1}_{2,(\bar l_1,\bar l_0)}(A,Q)$$

are isomorphic, which is equivalent to the validity of the inequality

$$\|u\|_{L^2(\lambda,\Sigma_3)} \leqslant c\|u\|_{\mathscr{H}_{2(\bar l,\bar l_0)}(A,Q)}, \quad \forall u\in \tilde C^1_{0,\Sigma_1}(\overline{Q}). \tag{1.18}$$

Together with the previous assumptions condition (1.18) guarantees the validity of condition (8.1.8).

In studying local properties of generalized solutions of (1.1) (a precise definition of a generalized solution of (1.1) will be given below) we shall always assume that the following condition is satisfied:

$$a^{ij}\in L^{2\bar s,2\bar s_0}(Q),\quad i,j=1,\dots,n,\ \frac{1}{\bar s}+\frac{2}{\bar l}=1,\quad \frac{1}{\bar s_0}+\frac{2}{\bar l_0}=1,$$

$$\bar l \text{ and } \bar l_0 \text{ are as in (1.7).} \tag{1.19}$$

We recall (see §8.2) that a generalized solution of problem (8.2.5) is defined as any function u belonging to the space

$$\mathscr{V}\equiv \mathscr{W}_0^{0,\Sigma_1}\equiv \left\{u\in \mathscr{H}^{0,\Sigma_1}_{2,(\bar l_1,\bar l_0)}(A,Q) : u'\in \mathscr{H}^{0,\Sigma_1}_{2,(\bar l_1,\bar l_0)}(A,Q)^*,\ u(0)=0\right\}$$

and satisfying an identity of the form

$$\int_{T_1}^{T_2}(u',\eta)\,dt+\iint_Q(\mathbf{l}'\cdot A\nabla\eta+l_0'\eta)\,dt\,dx+\int_{\Sigma_3}\lambda u|_{\Sigma_3}\eta\,ds=\langle\mathscr{F},\eta\rangle, \tag{1.20}$$
$$\forall\eta\in\tilde C^1_{0,\Sigma_1}(\overline{Q}),$$

where u' is the derivative of u considered as an element of the space

$$\mathscr{D}^*([T_1,T_2];(H^{0,\sigma_1}_{2,\bar l}(A,\Omega))^*).$$

In view of assumption (8.3.1) a generalized solution u of problem (8.2.5) belongs to the space

$$\widetilde{\mathscr{H}^{0,\Gamma_1}_2}(A,Q)\subset C([T_1,T_2];L^2(\Omega)).$$

In what follows it is expedient to consider the genreal boundary value problem for equation (1.1) with an inhomogeneous initial condition. On the contrary, because of Lemma 8.1.8, in considering problem (8.2.5) it may always be assumed that $\mathscr{F}=0$ by suitably modifying the functions $l^i(t,x,u,p)$, $i=1,\dots,n$, and $l_0(t,x,u,p)$. Therefore, we henceforth consider a boundary value problem of the form

$$\begin{gathered}(d/dx_i)l^i(t,x,u,\nabla u)+l_0(t,x,u,\nabla u)=0 \quad \text{in } Q,\\ u=0 \quad \text{on } \Sigma_1,\quad \mathbf{l}'\cdot A\nu=0 \quad \text{on } \Sigma_2,\quad \mathbf{l}\cdot A\nu-\lambda u=0 \quad \text{on } \Sigma_3,\\ u=u_0 \quad \text{on } \Omega_{T_1},\end{gathered} \tag{1.21}$$

where $u\in L^2(\Omega)$. A generalized solution of problems (1.21) is by definition any function

$$u\in\mathscr{W}^{0,\Sigma_1}\equiv\left\{u\in\mathscr{H}^{0,\Sigma_1}_{2,(\bar l,\bar l_0)}(A,Q) : u'\in\left(\mathscr{H}^{0,\Sigma_1}_{2,(\bar l,\bar l_0)}(A,Q)\right)^*\right\}$$

equal to u_0 for $t = T_1$ and satisfying the identity

$$\int_{T_1}^{T_2}(u', \eta)\,dt + \iint_Q (\mathbf{l}' \cdot A\nabla\eta + l_0'\eta)\,dt\,dx + \int_{\Sigma_3} \lambda u|_{\Sigma_3}\eta\,ds = 0, \qquad \forall\eta \in \tilde{C}^1_{0,\Sigma_1}(Q). \tag{1.22}$$

Exactly as in the proof of Lemma 8.2.1, it can be established that (1.22) implies that for all $\tau_1, \tau_2 \in [T_1, T_2]$

$$\int_{\tau_1}^{\tau_2}(u', \eta)\,dt + \iint_{Q_{\tau_1,\tau_2}} (\mathbf{l}' \cdot A\nabla\eta + l_0'\eta)\,dt\,dx + \int_{\Sigma_3} \lambda u|_{\Sigma_3}\eta\,ds = 0, \qquad \forall\eta \in \mathscr{H}^{0,\Sigma_1}_{2,(\bar{l}',\bar{l}_0)}(A, Q). \tag{1.23}$$

We now introduce the concept of a generalized solution of (1.1) in a cylinder Q. Such a solution is by definition any function $u \in \mathscr{W}$, where

$$\mathscr{W} \equiv \left\{u \in \mathscr{H}_{2,(\bar{l},\bar{l}_0)}(A, Q) : u' \in \left(\mathscr{H}_{2,(\bar{l},\bar{l}_0)}(A, Q)\right)^*\right\},$$

satisfying the identity

$$\int_{T_1}^{T_2}(u', \eta)\,dt + \iint_Q (\mathbf{l}' \cdot A\nabla\eta + l_0'\eta)\,dt\,dx = 0, \quad \forall\eta \in \tilde{C}^1_{0,\Sigma_1}. \tag{1.24}$$

Because of Lemma 8.1.6, the identity (1.24) is also valid for any function $\eta \in \tilde{C}^1_{0,\partial\Omega\times(T_1,T_2)}(Q)$, and hence, since $\tilde{C}^1_{0,\partial\Omega\times(\tau_1,\tau_2)}(Q)$ is dense in $\mathscr{H}_{2,(\bar{l},\bar{l}_0)}(A, Q)$, also for any function

$$\eta \in \mathscr{H}^{(0)} \equiv \mathscr{H}^{0,\partial\Omega\times(T_1,T_2)}_{2,(\bar{l},\bar{l}_0)}(A, Q).$$

Further, taking Lemma 8.2.1 into account, we conclude that (1.24) is equivalent to the fact that for any $\tau_1, \tau_2 \in [T_1, T_2]$

$$\int_{\tau_1}^{\tau_2}(u', \eta)\,dt + \iint_{Q_{\tau_1,\tau_2}} (\mathbf{l}' \cdot A\nabla\eta + l_0'\eta)\,dt\,dx = 0, \qquad \forall\eta \in \mathscr{H}^{0,\partial\Omega\times(T_1,T_2)}. \tag{1.25}$$

Before introducing the concept of a local generalized solution of (1.1) in a cylinder Q, we prove the following lemma.

Lemma 1.2. *For any $u \in \mathscr{W}$ and $\xi \in \tilde{C}^1_{0,\partial\Omega\times(T_1,T_2)}(\overline{Q})$ the function $u\xi$ belongs to $\mathscr{W}^{(0)} \equiv \mathscr{W}^{0,\partial\Omega\times(T_1,T_2)}$.*

Proof. Let $u \in \mathscr{W}$. In view of Corollary 8.1.3 there exists a sequence $\{u_n\}$, $u_n \in \tilde{C}^1(Q)$, $n = 1, 2, \ldots$, converging to u in $\mathscr{W}$. We shall first prove that the sequence $\{u_n\xi\}$, $u_n\xi \in \tilde{C}^1_{0,\partial\Omega\times(T_1,T_2)}(Q)$, $n = 1, 2, \ldots$, converges to $u\xi$ in the norm of $\mathscr{H}$. Indeed, taking condition (1.19) into account, we find that

$$\begin{aligned}\|u_n\xi - u_m\xi\|_{\mathscr{H}} &\leqslant \|(u_n - u_m)\xi\|_{\bar{l},\bar{l}_0,Q} \\ &\quad + \|\xi A\nabla(u_n - u_m)\|_{2,Q} + \|A\nabla\xi(u_n - u_m)\|_{2,Q} \\ &\leqslant c\|u_n - u_m\|_{\bar{l},\bar{l}_0,Q} + c\|A\nabla(u_n - u_m)\|_{2,Q} \\ &\quad + \|A\nabla\xi\|_{2\bar{s},2\bar{s}_0,Q}\|u_n - u_m\|_{\bar{l},\bar{l}_0,Q}.\end{aligned} \tag{1.26}$$

It follows from (1.26) that the sequence $\{u_n\xi\}$ is Cauchy in $\mathscr{H}$. It is obvious that $\{u_n\xi\}$ converges to $u\xi$ in $L^{l,l_0}(Q)$. From what has been proved it now follows that $\{u_n\xi\}$ converges to $u\xi$ also in $\mathscr{H}$, whence $u\xi \in \mathscr{H}^{(0)}$. To complete the proof of Lemma 1.2 it suffices now to establish that $(u\xi)' \in (\mathscr{H}^{(0)})^*$. Taking into account that $u' \in \mathscr{H}^*$, by (8.1.19) and (1.18) we have

$$\langle u', \eta\rangle = \iint_Q \left(f_0\eta + f^k A_k \nabla\eta\right) dt\, dx, \quad \forall \eta \in \mathscr{H}^{(0)}, \tag{1.27}$$

where $f_0 \in L^{l,l_0}(Q)$ and $f^k \in L^2(Q)$, $k = 1,\dots,n$. Setting $\eta = \xi\tilde{\eta}$ in (1.27) for all $\tilde{\eta} \in \tilde{C}^1(Q)$, we obtain

$$\langle u', \xi\eta\rangle = \iint_Q \left(g_0\tilde{\eta} + g^k A_k \nabla\tilde{\eta}\right) dt\, dx, \quad \forall \tilde{\eta} \in \mathscr{H}, \tag{1.28}$$

where $g_0 = f_0\xi + g^k A_k \nabla\xi$, $g_k = f_k\xi$, $k = 1,\dots,n$, and $g_k \in L^2(Q)$, $k = 1,\dots,n$, $g_0 \in L^{\tilde{l}',\tilde{l}'_0}(Q)$, since by (1.19)

$$\|g^k A_k \nabla\xi\|_{\tilde{l}',\tilde{l}'_0,Q} \leqslant \|g^k\|_{2,Q}\|A_k\nabla\xi\|_{2\bar{s},2\bar{s}_0,Q} \leqslant \text{const}, \qquad k = 1,\dots,n.$$

From (1.28) we obtain, in particular,

$$\left.\begin{aligned} -\int_{T_1}^{T_2}(u\xi, \psi)\varphi'\, dt = \int_{T_1}^{T_2}(u\xi_t, \psi)\varphi\, dt + \int_{T_1}^{T_2}(u', \xi\psi)\varphi\, dt, \quad \varphi \in \mathscr{D}([T_1, T_2]),\\ \psi \in H^{0,\partial\Omega}, \end{aligned}\right\} \tag{1.29}$$

where $(\cdot,\cdot)$ is the duality between H and H^*. We denote by $u'\xi$ the element of $\mathscr{H}^*$ defined by (1.28), i.e., by the formula

$$\langle u'\xi, \eta\rangle \overset{\text{def}}{=} \langle u', \xi\eta\rangle, \qquad \forall \eta \in \mathscr{H}^{(0)}. \tag{1.30}$$

From (1.29) and (1.30) we then obtain

$$(u\xi)' = u\xi' + u'\xi, \tag{1.31}$$

where $u\xi' \equiv u\xi_t \in \mathscr{H}^{(0)} \subset (\mathscr{H}^{(0)})^*$; this proves that $(u\xi)' \in (\mathscr{H}^{(0)})^*$. Lemma 1.2 is proved.

We set

$$\mathscr{W}_{\text{loc}} \equiv \left\{u \in L^2_{\text{loc}}(Q)\colon \xi u \in \mathscr{W}^{(0)}, \forall \xi \in \tilde{C}^1_{0,\partial\Omega\times(T_1,T_2)}(\overline{Q})\right\}. \tag{1.32}$$

It follows from Lemma 1.2 that $\mathscr{W} \subset \mathscr{W}_{\text{loc}}$.

A local generalized solution of equation (1.1) in the cylinder Q is by definition any function $u \in \mathscr{W}_{\text{loc}}$ which is a generalized solution of (1.1) in each cylinder $\hat{Q} = \hat{\Omega} \times (\tau_1, \tau_2)$, $\hat{\Omega} \subset \Omega$, $T_1 < \tau_1 < \tau_2 < T_2$. It is obvious that (1.25) holds for a local generalized solution of (1.1) in Q for any $\tau_1, \tau_2 \in (T_1, T_2)$.

From Lemma 1.2 we obviously obtain the following assertion.

COROLLARY 1.1. *Each generalized solution of* (1.1) *in the cylinder Q is also a local generalized solution of this equation in Q.*

§2. On regularity of generalized solutions in the variable t

Here we shall prove that under the conditions indicated in the preceding section generalized solutions of (1.1) satisfy an integral Hölder condition with exponent $1/2$. An analogous result is also established for generalized solutions of problem (1.21).

THEOREM 2.1. *Let u be a generalized solution in the cylinder Q of equation* (1.1) *considered under conditions* (1.2)–(1.4), (1.11) – (1.13) *and* (1.19). *Then for any cylinder $Q' = \Omega' \times (T_1, T_2)$, $\overline{\Omega'} \subset \Omega$,*

$$\lim_{h\to 0} \frac{\|u(t,x) - u(t-h,x)\|_{2,Q'}^2}{h} = 0, \tag{2.1}$$

where the function u is assumed to be extended outside the segment $[T_1, T_2]$ according to formula (ii) *in* (4.7.2).

PROOF. In (1.25) (with $\tau_1 = T_1$ and $\tau_2 = T_2$) we set

$$\eta = \xi^2(x)\frac{1}{h}\int_{t-h}^{t} [u(x,\eta+h) - u(x,\eta)]\,d\eta, \tag{2.2}$$

where $\xi(x)$ is a smooth, nonnegative function equal to 1 in Ω' and to 0 outside a subdomain Ω'' with $\overline{\Omega'} \subset \Omega''$ and $\overline{\Omega''} \subset \Omega$. Exactly as it was proved in Lemma 8.1.27 that $u_h \in \mathscr{Y}$, we find that the function (2.2) belongs to the space

$$\mathscr{H}_{2,(l,l_0)}^{0,\partial\Omega\times(T_1,T_2)}(A, Q).$$

We set

$$\hat{u} = \xi^2 u. \tag{2.3}$$

Just as in the proof of Lemma 1.2, we establish that $\tilde{u} \in \mathscr{W}^{(0)}$. From (1.25) we then obviously obtain

$$\int_{T_1}^{T_2}\left(u', \frac{1}{h}\int_{t-h}^{t} [\tilde{u}(\eta-h,x) - \tilde{u}(\eta,x)]\,d\eta\right) dt = \left\langle Lu, \xi^2 v_h\right\rangle, \tag{2.4}$$

where

$$v_h = \frac{1}{h}\int_{t-h}^{t} [u(\eta+h,x) - u(\eta,x)]\,d\eta$$

and

$$\left\langle Lu, \xi^2 v_h\right\rangle \equiv -\iint_Q \left[\mathbf{l}' \cdot A\nabla(\xi^2 v_h) + l_0'\xi^2 v_h\right] dt\,dx. \tag{2.5}$$

It is easy to see that $\xi^2 v_h \to 0$ in $\mathscr{H}$(in particular, essential use is here made of the condition $A = A(x)$). Then

$$\lim_{h\to 0}\left\langle Lu, \xi^2 v_h\right\rangle = 0. \tag{2.6}$$

Transforming (with (8.1.42) taken into account) the left side of (2.4), we obtain

$$\int_{T_1}^{T_2}\left(u', H^{-1}\int_{t-h}^{t} [\tilde{u}(\eta+h,x) - \tilde{u}(\eta,x)]\,d\eta\right) dt = \int_\Omega \tilde{u}v_h\,dx\Big|_{T_1}^{T_2}$$
$$-\frac{1}{h}\iint_Q \{\tilde{u}(t)[\tilde{u}(t+h) - \tilde{u}(t)] - \tilde{u}(t)[\tilde{u}(t) - \tilde{u}(t-h)]\}\,dt\,dx, \tag{2.7}$$

where $v_h = h^{-1}\int_{t-h}^{t}[u(\eta + h, x) - u(\eta, x)]d\eta$. We have

$$\frac{1}{h}\iint_Q \tilde{u}(t)[\tilde{u}(t) - \tilde{u}(t-h)]\,dt\,dx = \frac{1}{h}\iint_Q \tilde{u}(t-h)[\tilde{u}(t) - \tilde{u}(t-h)]\,dt\,dx + \frac{1}{h}\iint_Q [\tilde{u}(t) - \tilde{u}(t-h)]^2\,dt\,dx. \qquad (2.8)$$

From (2.8), (2.5) and (2.4) it follows that

$$\frac{1}{h}\iint_Q [\tilde{u}(t) - \tilde{u}(t-h)]^2\,dt\,dx = \left\langle Lu, \xi^2 v_h\right\rangle - \int_\Omega \tilde{u}v_h\,dx\Big|_{T_1}^{T_2} + \frac{1}{h}\left\{\iint_Q \tilde{u}(t)[\tilde{u}(t+h) - \tilde{u}(t)]\,dt\,dx - \iint_Q \tilde{u}(t-h)[\tilde{u}(t) - \tilde{u}(t-h)]\,dt\,dx\right\}. \qquad (2.9)$$

Obviously the expression in braces in (2.9) is equal to

$$\int_{T_2-h}^{T_2}\int_\Omega \tilde{u}(t+h) - \tilde{u}(t)\,dt\,dx - \int_{-h}^{0}\int_\Omega \tilde{u}(t)[\tilde{u}(t+h) + \tilde{u}(t)]\,dt\,dx. \qquad (2.10)$$

Taking into account that $\tilde{u} \in C([T_1 - T, T_2 + T]; L^2(\Omega))$, where $T = T_2 - T_1$, it is easy to see that (2.10) tends to 0 as $h \to 0$. For this same reason

$$\lim_{h\to 0}\int_\Omega \tilde{u}v_h\,dx\Big|_{T_1}^{T_2} = 0. \qquad (2.11)$$

From (2.9), (2.6) and (2.11) we obtain

$$\lim_{h\to 0}\frac{1}{h}\iint_Q [\tilde{u}(t) - \tilde{u}(t-h)]^2\,dt\,dx = 0, \qquad (2.12)$$

from which (2.1) follows. Theorem 2.1 is proved.

THEOREM 2.2. *Suppose that conditions* (1.2)–(1.4), (1.11), (1.13) *and* (1.18) *are satisfied, and let u be a generalized solution of problem* (1.21). *Then*

$$\lim_{h\to 0}\frac{\|u(t, x) - u(t-h, x)\|_{2,Q}^2}{h} = 0, \qquad (2.13)$$

where the function u is assumed to be extended outside the segment $[T_1, T_2]$ *according to formula* (ii) *in* (4.7.2).

PROOF. In (1.23) (with $\tau_1 = T_1$, $\tau_2 = T_2$) we set

$$\eta = v_h \equiv \frac{1}{h}\int_{t-h}^{t}[u(\eta + h, x) - u(\eta, x)]\,d\eta. \qquad (2.14)$$

It is obvious that the function thus defined belongs to the space

$$\mathcal{Y} \equiv \mathcal{Y}_{2,(l,\dot{l}_0)}^{0,\Sigma_1}(A, Q).$$

It is obvious also that $v_h \to 0$ in $\mathcal{H}_{2,(l,\dot{l}_0)}^{0,\Sigma_1}(A, Q) \equiv \mathcal{H}$, and hence, since $\mathcal{H}_0$ and $\mathcal{H}_\lambda \equiv \mathcal{H}_{2,(l,\dot{l}_0)}^{0,\Sigma_1}(A; Q; \Sigma_3, \lambda)$ are isomorphic, $v_h \to 0$ in $\mathcal{H}$. Exactly as in the proof of

Theorem 2.1, we establish the equality

$$\frac{1}{h}\iint_Q [u(t)-u(t-h)]^2\,dt\,dx = \langle Lu, v_h\rangle - \int_\Omega uv_h\,dx\Big|_{T_1}^{T_2} - \int_{\Sigma_3} \lambda u|_{\Sigma_3} v_h|_\Sigma\,ds$$

$$+\frac{1}{h}\left\{\iint_Q u(t)[u(t+h)-u(t)]\,dt\,dx\right.$$

$$\left.-\iint_Q u(t-h)[u(t)-u(t-h)]\,dt\,dx\right\}, \quad (2.15)$$

where $\langle Lu, v_h\rangle$ is defined by (2.5) with $\xi \equiv 1$ in Ω. Passing to the limit as $h \to 0$ in (2.15) and arguing as in the proof of Theorem 2.1, we obtain (2.13). This proves Theorem 2.2.

§3. The energy inequality

From (8.3.33) (in the case $\overline{m} = 2$) and the proof of Lemma 8.3.8 it follows that for any generalized solution of problem (1.21) considered under conditions (8.3.26) and (8.3.27) we have the estimate

$$\|u\|_{\mathscr{H}_\lambda} \equiv \|u\|_{2,\infty,Q} + \|A\nabla u\|_{2,Q} + \|u_0\|_{L^2(\lambda,\Sigma_3)} \leqslant c\left(\|\mathscr{F}\|_{(\mathscr{H}_\lambda)^*} + \|\tilde{\psi}\|_{1,Q}\right), \quad (3.1)$$

where $c = c(\nu_0, \|\hat{a}\|_{s,s_0,Q})$, and it is not hard to see that the constant c depends on the norm $\|\hat{a}\|_{s,s_0,Q}$ in exponential fashion. We note that in conditions (8.3.27), which were used in an essential way in the derivation of (3.1), the indices s and s_0 are not limit indices.

We consider a problem of the form (1.21) under conditions (1.2)–(1.4), (1.11), (1.13) and (1.18). The following assertion holds.

PROPOSITION 3.1. *For any function* $u \in \mathscr{H}^{0,\Sigma_1}_{2,(\bar{l},\bar{l}_0)}(A, Q)$, *for any* $\varepsilon > 0$, *and for* $t_1, t_2 \in [T_1, T_2]$

$$\iint_{Q_{t_1,t_2}} [l'(t,x,u,A\nabla u)\cdot A\nabla\eta + l_0'(t,x,u,A\nabla u)\eta]\,dt\,dx$$

$$\geqslant \nu\iint_{Q_{t_1,t_2}} |A\nabla u|^2\,dt\,dx - e\left(\varepsilon, Q_{t_1,t_2}\right)$$

$$\times\left(\|u\|^2_{2,\infty,Q_{t_1,t_2}} + \|A\nabla u\|^2_{2,Q_{t_1,t_2}}\right) - E\left(\varepsilon, Q_{t_1,t_2}\right), \quad (3.2)$$

where

$$e\left(\varepsilon, Q_{t_1,t_2}\right) = c_0^2(Q)b^2(Q)$$

$$\times\left(\|a_2\|_{2\bar{s},2\bar{s}_0,Q_{t_1,t_2}} + \|a_3\|_{2\bar{s},2\bar{s}_0,Q_{t_1,t_2}} + \|a_4\|_{2\bar{s},2\bar{s}_0,Q_{t_1,t_2}} + \frac{\varepsilon}{2}\right), \quad (3.3)$$

$$E\left(\varepsilon, Q_{t_1,t_2}\right) = \frac{1}{2\varepsilon}\|g\|^2_{\bar{l}_0,\bar{l}_0',Q_{t_1,t_2}} + \|h\|_{1,Q_{t_1,t_2}},$$

$c_0(Q)$ *is the constant in inequality* (1.8) (*for the entire cylinder* Q), *and*

$$b(Q) \equiv \sup_{i,j=1,\dots,n} \|b^{ij}\|_{r_i,r_{0i},Q}.$$

PROOF. Inequality (3.2) is obtaiined by means of a Hölder inequality with conditions (1.11) and the fact that $b(Q)$ increases with increasing height of the cylinder Q taken into account.

THEOREM 3.1. *Suppose that conditions* (1.2)–(1.4), (1.11), (1.13) *and* (1.18) *are satisfied. Then, for any generalized solution of problem* (1.21),

$$\|u\|_{\mathscr{H}_\lambda} \leqslant c\left(\|u_0\|_{2,\Omega} + \|g\|_{\bar{l}',\bar{l}_0',Q} + \sqrt{\|h\|_{1,Q}}\right), \tag{3.4}$$

where c depends only on n, ν, $\|a_2\|_{2s,2s,Q}$, $\|a_3\|_{s,s_0,Q}$, $\|a_4\|_{s,s_0,Q}$, $\bar{l}_0$, *the constant* $c_0(Q)$ *from* (1.8), *and* $b(Q)$.

PROOF. Setting $\eta = u$ in (1.23) and taking (8.1.42) into account, we obtain

$$\frac{1}{2}\int_\Omega u^2\,dx\Big|_{t_1}^{t_2} + \iint\limits_{Q_{t_1,t_2}} (\mathrm{l}'\cdot A\nabla u + l_0'u)\,dt\,dx + \int_{\sigma_3\times(t_1,t_2)} \lambda\left(u|_{\Sigma_3}\right)^2 ds = 0. \tag{3.5}$$

Applying (3.2) to the middle integral in (3.5), we obtain

$$\bar{\nu}\left[\int_\Omega u^2\,dx\Big|_{t_1}^{t_2} + \iint\limits_{Q_{t_1,t_2}} |A\nabla u|^2\,dt\,dx + \int_{\sigma_3\times(t_1,t_2)} \lambda\left(u|_{\Sigma_3}\right)^2 ds\right] \leqslant e\langle\langle u\rangle\rangle^2_{Q_{t_1,t_2}} + E, \tag{3.6}$$

$$\bar{\nu} = \min(\nu, 1),$$

where $e_{t_1,t_2} = e(\varepsilon, Q_{t_1,t_2})$ and $E_{t_1,t_2} = E(\varepsilon, Q_{t_1,t_2})$ are defined by (3.3), and $\langle\langle u\rangle\rangle_{Q_{t_1,t_2}}$ denotes the expression

$$\langle\langle u\rangle\rangle^2_{Q_{t_1,t_2}} \equiv \|u\|^2_{2,\infty,Q_{t_1,t_2}} + \|A\nabla u\|^2_{2,Q_{t_1,t_2}} + \int_{\sigma_3\times(t_1,t_2)} \lambda u^2_{\Sigma_3}\,ds. \tag{3.7}$$

We decompose the interval $[T_1, T_2]$ into parts by points $T_1 = \tau_0 < \tau_1 < \cdots < \tau_{k-1} < \tau_k < \tau_{k+1} < \cdots < \tau_N = T_2$, and we set $Q_k = \Omega \times (\tau_{k-1}, \tau_k)$. In (3.6) we set $t_1 = \tau_{k-1}$ and $t_2 = \tau_{k-1} + \theta$, $k = 1,\ldots,N$, where $\theta \in [0, \tau_k - \tau_{k-1}]$. Then for any $k = 1,\ldots,N$

$$\int_\Omega u^2(x, \tau_{k-1} + \theta)\,dx + \int_{\tau_{k-1}}^{\tau_{k-1}+\theta}\int_\Omega |A\nabla u|^2\,dt\,dx + \int_{\sigma_3\times(\tau_{k-1},\tau_{k-1}+\theta)} \lambda u^2_{\Sigma_3}\,ds$$
$$\leqslant \frac{2}{\bar{\nu}}\left[e_k\langle\langle u\rangle\rangle^2_{Q_k} + E\right] + \int_\Omega u^2(x, \tau_{k-1})\,dx, \tag{3.8}$$

where $e_k = e(\varepsilon, Q_k)$ and $E = E(\varepsilon, Q)$. From (3.8) we easily obtain

$$\langle\langle u\rangle\rangle^2_{Q_k} \leqslant \frac{4}{\bar{\nu}}e_k\langle\langle u\rangle\rangle^2_{Q_k} + \frac{4}{\bar{\nu}}E + 2\int_\Omega u^2(x, \tau_{k-1})\,dx, \qquad k = 1,\ldots,N. \tag{3.9}$$

Suppose that the lengths of the intervals $[\tau_{k-1}, \tau_k]$ and the number ε are so small that

$$4e_k/\bar{\nu} \leqslant \frac{1}{2}, \qquad k = 1, 2, \ldots, N. \tag{3.10}$$

To satisfy (3.10) it suffices to set $\varepsilon = \bar{\nu}/8c_0^2b^2$, where c_0 and b are the constants in (1.8) for the entire cylinder Q, and to require that[1]

$$\|a_2\|_{2\bar{s},2\bar{s}_0,Q_k} \leqslant \frac{\nu}{48c_0^2b^2}, \quad \|a_3\|_{\bar{s},\bar{s}_0,Q_k} \leqslant \frac{\bar{\nu}}{48c_0^2b^2}, \quad \|a_4\|_{\bar{s},\bar{s}_0,Q_k} \leqslant \frac{\bar{\nu}}{48c_0^2b^2}, \qquad (3.11)$$

$$k = 1,\dots,N.$$

It is not hard to see that under condition (3.11) the number N of points of subdivision of the segment $[T_1, T_2]$ can be taken no larger than

$$\left(\frac{\bar{\nu}}{48c_0^2b^2}\right)^{-2s_0}\|a_2\|^{2\bar{s}_0}_{2\bar{s},2\bar{s}_0,Q} + \frac{\bar{\nu}}{48c_0^2b^2}\|a_3\|^{\bar{s}_0}_{\bar{s},\bar{s}_0,Q} + \frac{\bar{\nu}}{48c_0^2b^2}\|a_4\|^{\bar{s}_0}_{\bar{s},\bar{s}_0,Q} + 1$$

(see [24]). From (3.9) and (3.10) we obtain

$$z_k \leqslant 4z_{k-1} + 8E/\bar{\nu}, \qquad k = 1,\dots,N, \qquad (3.12)$$

where $z_k \equiv \langle\langle u\rangle\rangle^2_{Q_k}$, $k = 1,\dots,n$, and $z_0 = \|u(T_1, x)\|^2_{2,\Omega} \equiv \|u_0\|^2_{2,\Omega}$. From (3.10) we easily obtain

$$z_k \leqslant 4^k(z_0 + 8E/\bar{\nu}), \qquad k = 1,\dots,N. \qquad (3.13)$$

Taking into account that $\Sigma_1^N z_k = \langle\langle u\rangle\rangle^2_Q$, we obtain

$$\langle\langle u\rangle\rangle^2_Q \leqslant 4^{N+1}\|u_0\|_{2,\Omega} + \frac{32c_0^2b^2}{\bar{\nu}^2}4^{N+1}\left(\|g\|^2_{\bar{l},\bar{l}_0',Q} + \|h\|_{1,Q}\right). \qquad (3.14)$$

Replacing N by the above expression, we complete the proof of Theorem 3.1.

§4. Functions of generalized solutions

In this section we consider functions $f(u(t, x))$, where $u(t, x)$ is a generalized solution of (1.1) or of (1.21), and we establish some properties of such functions and also of functions of the form $f(u(t, x))\eta(t, x)$. It is henceforth always assumed that conditions (1.3) and (1.4) are satisfied for the matrix $A \equiv A(x)$.

LEMMA 4.1. *Suppose that a function $u(x_0, x)$, $x = (x_1,\dots,x_n)$ is defined on a set $Q \subset \mathbb{R}^{n+1}$, belongs to $L^1(Q)$, and has a generalized derivative $u_{x_i} \in L^1(Q)$, $i \in \{0,1,\dots,n\}$. Suppose the function $\omega(u)$ is uniformly Lipschitz on $\mathbb{R}$ and continuously differentiable everywhere on $\mathbb{R}$ with the exception of points $u_1,\dots,u_k$ which are corner points of ω. Then the composite function $\omega \circ u$ has a generalized derivative $\partial(\omega \circ u)/\partial x_i \in L^1(Q)$, and*

$$\frac{\partial}{\partial x_i}(\omega \circ u) = \begin{cases} \omega'(u)u_{x_i}, & (x_0, x) \in Q, u(x_0, x) \notin \{u_1,\dots,u_k\}, \\ 0, & (x_0, x) \in Q, u(x_0, x) \in \{u_1,\dots,u_k\}. \end{cases} \qquad (4.1)$$

For a proof, see, for example, [125] or [148].

COROLLARY 4.1. *Suppose a function $u \in L^1(Q)$ has a generalized derivative u_{x_i}, $i \in \{0,1,\dots,n\}$, and $u_{x_i} \in L^1(Q)$. Then for any $c \in \mathbb{R}$ such that*

$$\operatorname*{meas}_{n+1}\{u(x_0, x) = c\} > 0$$

the function u_{x_i} is equivalent to 0 on $\{u(x_0, x) = c\}$.

---

(1) We have used, in particular, the fact that the constants c_0 and b in (1.8) increase as the time interval increases.

PROOF. Suppose $c = 0$ and $\text{meas}_{n+1}\{(x_0, x)\colon u(x_0, x) = 0\} > 0$. Let $\omega_1(u) = \max(0, u)$ and $\omega_2(u) = \min(0, u)$. It is obvious that $\omega_1(u)$ and $\omega_2(u)$ satisfy the conditions of Lemma 4.1 on the function ω, and the only corner point of ω_1 and ω_2 is the point $u = 0$. Corollary 4.1 then follows from (4.1) and the equality $u = \omega_1(u) + \omega_2(u)$.

LEMMA 4.2. *Suppose that the function $\omega(u)$ is uniformly Lipschitz on $\mathbb{R}$ and has at most finitely many corner points on $\mathbb{R}$. Then for any function $u \in \mathscr{H}_{2,(\bar{l},\bar{l}_0)}(A, Q)$ the composite function $\omega \circ u$ also belongs to $\mathscr{H}_{2,(\bar{l},\bar{l}_0)}(A, Q)$. If $\omega(0) = 0$, then*

$$\omega(u) \in \mathscr{H}^{0,\Gamma}_{2,(\bar{l},\bar{l}_0)}(A, Q) \qquad \textit{for any } u \in \mathscr{H}^{0,\Gamma}_{2,(\bar{l},\bar{l}_0)}(A, Q),$$

where $\Gamma = \gamma \times [0, T]$, $\gamma \subset \partial\Omega$.

PROOF. Suppose first that $\omega \in C^1(\mathbb{R}) \cap \text{Lip}(\mathbb{R})$. We consider a sequence $\{u_n\}$, $u_n \in \tilde{C}^1(Q)$, $n = 1, 2, \ldots$, such that $u_n \to u$ in $\mathscr{H}$. It is obvious that

$$\|\omega(u_n)\|_{\mathscr{H}} = \|\omega(u_n)\|_{\bar{l},\bar{l}_0,Q} + \left(\iint_Q |\omega'(u_n)|^2 |A\nabla u_n|^2 \,dt\,dx\right)^{1/2}. \tag{4.2}$$

Taking into account that $|\omega(u_n) - \omega(u)| \leqslant K|u_n - u|$, where $K = \sup_{u\in\mathbb{R}}|\omega'(u)| < +\infty$, we easily establish that

$$\lim_{n\to\infty} \|\omega(u_n) - \omega(u)\|_{\bar{l},\bar{l}_0,Q} = 0. \tag{4.3}$$

Moreover,

$$\|\omega(u_n)\|_{\mathscr{H}} \leqslant K_1, \tag{4.4}$$

where K_1 depends on K and on $\sup_{n=1,2,\ldots}\|u_n\|_{\mathscr{H}}$. In view of the weak compactness of $\mathscr{H}$ we extract from the sequence $\{\omega(u_n)\}$ a subsequence $\{\omega(u_\nu)\}$ which converges to some function v weakly in $\mathscr{H}$ and in $L^{\bar{l},\bar{l}_0}(Q)$. It is obvious that this same subsequence converges to $\omega(u)$ strongly in $L^{\bar{l},\bar{l}_0}(Q)$. Therefore, $\omega(u) = v$, so that $\omega(u) \in \mathscr{H}_{2,(\bar{l},\bar{l}_0)}(A, Q)$. If $\omega(0) = 0$, $u \in \mathscr{H}^{0,\Gamma}_{2,(\bar{l},\bar{l}_0)}(A, Q)$, we find similarly that $\omega(u) \in \mathscr{H}^{0,\Gamma}_{2,(\bar{l},\bar{l}_0)}(A, Q)$. We shall now eliminate the assumption $\omega \in C^1(\mathbb{R})$. Let $\{u_1, \ldots, u_k\}$ be the set of all corner points of the function ω. We approximate ω by functions ω_δ, where $\delta \in (0, \bar{\delta})$, $\bar{\delta} = \frac{1}{2}\min_{i,j=1,\ldots,k}\text{dist}(u_i, u_j)$, such that

1) $\omega_\delta \in C^1(\mathbb{R}) \cap \text{Lip}(\mathbb{R})$;

2) $\omega_\delta(u) = \omega(u)$ in $\mathbb{R} \setminus \bigcup_{s=1}^{k} \theta_\delta(u_s)$, where $\theta_\delta(u_s) \equiv \{u \in \mathbb{R}\colon |u - u_s| < \delta\}$;

3) $\max\limits_{\substack{u\in\mathbb{R}\\ 0<\delta<\bar{\delta}}} |\omega_\delta'(u)| \leqslant c = \text{const} > 0$;

4) $\lim\limits_{\delta\to 0} \omega_\delta(u_s) = \omega(u_s)$, $s = 1, \ldots, k$.

By what has been proved above, $\omega_\delta(u) \in \mathscr{H}_{2,(\bar{l},\bar{l}_0)}(A, Q)$ ($\omega_\delta(u) \in \mathscr{H}^{0,\Gamma}_{2,(\bar{l},\bar{l}_0)}(A, Q)$, if $\omega(0) = 0$), and it follows from Lemma 4.1 that $\partial\omega_\delta/\partial x_i = \omega_\delta'(u)u_{x_i}$ at those points of Q where $u \notin \{u_1, \ldots, u_k\}$, and $\partial\omega_\delta/\partial x_i = 0$ on the set $\bigcup_{s=1}^k\{(t, x) \in Q\colon u(t, x) = u_s\}$. Then for an arbitrary sequence $\{\delta_m\}$, $\delta_m \to 0$ as $m \to \infty$, $\delta_m > 0$, $m = 1, 2, \ldots$, we have

$$\begin{aligned}\|\omega_{\delta_m} - \omega_{\delta_n}\|_{\mathscr{H}} &= \|\omega_{\delta_m} - \omega_{\delta_n}\|_{\bar{l},\bar{l}_0,Q} + \|A\nabla(\omega_{\delta_n} - \omega_{\delta_m})\|_{2,Q} \\ &\leqslant 2c\|u\|_{\bar{l},\bar{l}_0,q_{n,m}} + \|A\nabla u\|_{2,q_{n,m}},\end{aligned} \tag{4.5}$$

where $q_{n,m} \equiv \{(t, x) \in Q\colon\ 0 < |u - u_s| < \max(\delta_n, \delta_m),\ s \in \{1, \dots, k\}\}$. Since the right side of (4.5) tends to 0 as $n, m \to \infty$, the sequence $\{\omega_{\delta_m}\}$ tends to a function w in $\mathscr{H}$. But then some subsequence of this sequence converges to w a.e. in Q. Since it follows from the construction of ω_δ that

$$\lim_{\delta\to 0} \omega_\delta(u(t, x)) = \omega(u(t, x))$$

at any point $(t, x) \in Q$ where u is defined (i.e., for almost all $(t, x) \in Q$), we obtain the equality $w = \omega(u)$ a.e. in Q. This implies that

$$\omega(u) \in \mathscr{H}_{2,(l, l_0)}(A, Q) \qquad \left(\omega(u) \in \mathscr{H}^{0,\Gamma}_{2,(l, l_0)}(A, Q)\right).$$

Lemma 4.2 is proved.

LEMMA 4.3. *Suppose that the function $\omega(u)$ has a derivative $\omega'(u)$ satisfying a Lipschitz condition uniformly on $\mathbb{R}$, while its second derivative $\omega''(u)$ is continuous everywhere on $\mathbb{R}$ with the possible exception of finitely many points at which $\omega''(u)$ has a discontinuity of first kind. Assume that condition* (1.19) *is satisfied. Then for any $u \in \mathscr{W} \equiv \{u \in \mathscr{H}_{2,(l, l_0)}(A, Q)\colon u' \in (\mathscr{H}_{2,(l, l_0)}(A, Q))^*\}$ and any $\eta \in \tilde{C}^1_{0,\partial\Omega\times(T_1, T_2)}(\overline{Q})$ the function $\omega'(u)\eta$ belongs to the space $\mathscr{H}^{(0)}$, the function $\omega(u)$ belongs to $C([T_1, T_2]; L^2(\Omega))$, and for any $\tau_1, \tau_2 \in [T_1, T_2]$*

$$\int_{\tau_1}^{\tau_2} (u', \omega'(u)\eta)\, dt = \int_\Omega \omega(u)\eta\, dx - \int_{\tau_1}^{\tau_2} - \int_{\tau_1}^{\tau_2} (\omega(u), \eta')\, dt. \tag{4.6}$$

PROOF. We note first of all that, by Lemma 4.2, $\omega'(u) \in \mathscr{H}_{2,(l, l_0)}(A, Q)$. By Lemma 1.2 it then follows that $\omega'(u)\eta \in \mathscr{H}^{(0)}$. Further, taking into account that $u \in C([T_1, T_2]; L^2(\Omega))$ and using the inequality

$$|\omega(u(t)) - \omega(u(t'))| \leqslant K_1|u(t) - u(t')|\big(|u(t)| + |u(t')| + K_2\big), \tag{4.7}$$

following from the conditions imposed on ω, we conclude that $\omega(u(t, x)) \in c([T_1, T_2]; L^2(\Omega))$. We now prove (4.6). Since $u \in \mathscr{W}$, by Corollary 4.7.2 there exists a sequence $\{u_n\}$, $u_n \in \tilde{C}^1(Q)$, $n = 1, 2, \dots$, such that

$$u_n \to u \quad \text{in}\ \mathscr{H}, \qquad u'_n \to u' \quad \text{in}\ \mathscr{H}^* \quad \text{as}\ n \to \infty. \tag{4.8}$$

It is obvious that for any $\eta \in \tilde{C}^1(Q)$ and any $\tau_1, \tau_2 \in [T_1, T_2]$

$$-\iint_{Q_{\tau_i,\tau_2}} \omega(u_n)\eta_t\, dt\, dx = \iint_{Q_{\tau_i,\tau_2}} \omega'(u_n)u_{nt}\eta\, dt\, dx - \int_\Omega \omega(u_n)\eta\, dx\Big|_{\tau_1}^{\tau_2}, \tag{4.9}$$

where $Q_{\tau_1,\tau_2} \equiv \omega \times (\tau_1, \tau_2)$. We rewrite (4.9) in the form

$$\int_{\tau_1}^{\tau_2} (u'_n, \omega'(u_n)\eta)\, dt = \int_\Omega \omega(u_n)\,\eta\, dx\Big|_{\tau_1}^{\tau_2} - \int_{\tau_1}^{\tau_2} (\omega(u_n), \eta')\, dt. \tag{4.10}$$

From the estimate

$$|\omega(u_n) - \omega(u)| \leqslant K_1|u_n - u|\big(|u_n| + |u| + K_2\big), \tag{4.11}$$

which is proved in the same way as (4.7), since $u_n \to u$ in $\mathscr{W}$, it follows that $\omega(u_n) \to \omega(u)$ in $C([T_1, T_2]; L^1(\Omega))$ and a fortiori $\omega(u_n) \to \omega(u)$ in $L^1(Q)$. In particular, from this it follows that as $n \to \infty$ the right side of (4.10) tends to the right side of (4.6). We shall prove that the left side of (4.10) tends to the left side of (4.6). Because of the convergence $u_n' \to u'$ in $\mathscr{H}^*$, for this it suffices to show that $\omega'(u_n)\eta \to \omega'(u)\eta$ weakly in $\mathscr{H}$. It follows from the proof of Lemma 4.2 that $\omega'(u_n) \to \omega'(u)$ weakly in $\mathscr{H}$. Then the first term on the right side of the equality

$$A\nabla\{[\omega'(u_n) - \omega'(u)]\eta\} = \{A\nabla[\omega'(u_n) - \omega'(u)]\}\eta + \{A\nabla\eta\}[\omega'(u_n) - \omega'(u)] \tag{4.12}$$

tends to 0 weakly in $L^2(Q)$ as $n \to \infty$. Since the convergence $\omega'(u_n)\eta \to \omega'(u)\eta$ in $L^{l,l_0}(Q)$ is obvious, to complete the proof of weak convergence of $\omega'(u_n)\eta$ to $\omega'(u)\eta$ in $\mathscr{H}$ it suffices to prove that the second term of the right side of (4.12) tends to 0 strongly in $L^2(Q)$. Taking (1.19) into account, we have

$$\|A\nabla\eta[\omega'(u_n) - \omega'(n)]\eta\|_{2,Q} \leqslant \|A\nabla\eta\|_{2\bar{s},2\bar{s}_0,Q}\|\omega'(u_n) - \omega'(u)\|_{\bar{l},\bar{l}_0,Q}. \tag{4.13}$$

Taking into account that $\omega'(u_n) \to \omega'(u)$ in $L^{\bar{l},\bar{l}_0}(Q)$, we conclude that the left side of (4.13) tends to 0 as $n \to \infty$. Lemma 4.3 is proved.

LEMMA 4.4. *Suppose that the function $\omega(u)$ is the same as in Lemma 4.3, and, moreover, that $\omega'(0) = 0$. Then for any function*

$$u \in \mathscr{W}^{0,\Sigma_1} \equiv \{u \in \mathscr{H} : u' \in \mathscr{H}^*\}, \qquad \text{where } \mathscr{H} \equiv \mathscr{H}^{0,\Sigma_1}_{2,(\bar{l},\bar{l}_0)}(A, Q),$$

the function $\omega'(u)$ belongs to $\mathscr{H}^{0,\Sigma_1}_{2,(\bar{l},\bar{l}_0)}(A, Q) \cap C([T_1, T_2]; L^1(\Omega))$, and for any $\tau_1, \tau_2 \in [T_1, T_2]$

$$\int_{\tau_1}^{\tau_2} (u', \omega'(u))\, dt = \int_\Omega \omega(u)\, dx\Big|_{\tau_1}^{\tau_2}. \tag{4.14}$$

PROOF. Lemma 4.4 is proved in the same way as Lemma 4.3, and the proof is even simpler since here $\eta \equiv 1$.

LEMMA 4.5. *Suppose that the function $\omega(u)$ is the same as in Lemma 4.3, and suppose that for the function $u \in \mathscr{W} \equiv \{u \in \mathscr{H} : u' \in \mathscr{H}^*\}$, where $\mathscr{H} \equiv \mathscr{H}_{2,(l,l_0)}(A, Q)$, there exists a sequence $\{u_n\}$, $u_n \in \tilde{C}^1(Q)$, $n = 1, 2, \ldots$, converging to u in $\mathscr{W}$ such that $\omega'(u_n) = 0$ in a neighborhood of the lateral surface $\partial\Omega \times (T_1, T_2)$. Then $\omega'(u) \in \mathscr{H}^{(0)}$, $\omega(u) \in C([T_1, T_2]; L^1(\Omega))$, and for all $\tau_1, \tau_2 \in [T_1, T_2]$ equality* (4.14) *holds.*

PROOF. Lemma 4.5 is proved in exactly the same way as Lemma 4.4.

LEMMA 4.6. *Suppose that the function $\omega(u)$ is the same as in Lemma 4.3, and suppose that for the function $u \in \mathscr{W} \equiv \{u \in \mathscr{H} : u' \in \mathscr{H}^*\}$, where $\mathscr{H} \equiv \mathscr{H}_{2,(l,l_0)}(A, Q)$, there exists a sequence $\{u_n\}$, $u_n \in \tilde{C}^1(Q)$, $n = 1, 2, \ldots$, converging to u in $\mathscr{W}$ such that $\omega'(u_n) = 0$ in an $(n + 1)$-dimensional neighborhood of the set $S \subset \partial\Omega \times (T_1, T_2)$. Let $\eta \in \tilde{C}^1_{0,(\partial\Omega \times (T_1,T_2))\setminus S}(Q)$. Then $\omega'(u)\eta \in \mathscr{H}^{(0)}$, $\omega(u) \in C([T_1, T_2]; L^1(\Omega))$, and for any $\tau_1, \tau_2 \in [T_1, T_2]$ the equality* (4.6) *holds.*

PROOF. Lemma 4.6 is proved in exactly the same way as Lemma 4.3.

The next assertions follow in an obvious way from Lemmas 4.3–4.6.

LEMMA 4.7. *Suppose that conditions* (1.2)–(1.4), (1.11)–(1.13) *and* (1.19) *are satisfied, and that the function* $\omega(u)$ *is the same as in Lemma* 4.3. *Then for any generalized solution* (*local generalized solution*) *of* (1.21) *in the cylinder* Q *for any* $\tau_1, \tau_2 \in [T_1, T_2]$ (*any* $\tau_1, \tau_2 \in (T_1, T_2)$)

$$\int_\Omega \omega(u)\eta\, dx\Big|_{\tau_1}^{\tau_2} - \int_{\tau_1}^{\tau_2} (\omega(u), \eta')\, dt + \iint_{Q_{\tau_1,\tau_2}} [\mathbf{l}'(t, x, u, A\nabla u) \cdot A\nabla u \omega''(u)\eta$$
$$+ \mathbf{l}(t, x, u, A\nabla u) \cdot A\nabla\eta\omega'(u) + l_0'(t, x, u, A\nabla u)\omega'(u)\eta]\, dt\, dx = 0,$$
$$\forall \eta \in \tilde{C}^1_{0, \partial\Omega\times(T_1, T_2)}(Q). \tag{4.15}$$

LEMMA 4.8. *Suppose that conditions* (1.2)–(1.4), (1.11), (1.13), *and* (1.18) *are satisfied, and that the function* $\omega(u)$ *is the same as in Lemma* 4.4. *Then for any generalized solution* u *of problem* (1.21), *for any* $\tau_1, \tau_2 \in [T_1, T_2]$,

$$\int_\Omega \omega(u)\, dx\Big|_{\tau_1}^{\tau_2} + \iint_{Q_{\tau_1,\tau_2}} [\mathbf{l} \cdot A\nabla u \omega''(u) + l_0'\omega'(u)]\, dt\, dx + \int_{\Sigma_3} \sigma U|_{\Sigma_3}\omega'\left(u|_{\Sigma_3}\right) ds = 0, \tag{4.16}$$

where $\omega'(u|_{\Sigma_3}) = \omega'(u)|_{\Sigma_3}$.

LEMMA 4.9. *Suppose that conditions* (1.2)–(1.4), (1.11), *and* (1.13) *are satisfied, and let* u *be a generalized solution of* (1.1) *in the cylinder* Q. *Assume that for the functions* $u(t, x)$ *and* $\omega(u)$ *all the conditions of Lemma* 4.5 *are satisfied. Then for all* $\tau_1, \tau_2 \in [T_1, T_2]$

$$\int_\Omega \omega(u)\, dx\Big|_{\tau_1}^{\tau_2} + \iint_{Q_{\tau_1,\tau_2}} [\mathbf{l}' \cdot A\nabla u \omega''(u) + l_0'\omega'(u)]\, dt\, dx = 0. \tag{4.17}$$

LEMMA 4.10. *Suppose that conditions* (1.2)–(1.4), (1.11)–(1.13), *and* (1.19) *are satisfied, and let* $u(t, x)$ *be a generalized solution of* (1.1) *in the cylinder* Q. *Assume that for the functions* $u(t, x)$, $\omega(u)$, *and* $\eta(t, x)$ *all the conditions of Lemma* 4.6 *hold. Then for all* $\tau_1, \tau_2 \in [T_1, T_2]$ *the equality* (4.15) *is satisfied.*

REMARK 4.1. The function $\omega''(u)$ appearing in identities (4.15)–(4.17) is obviously not defined on the set $\bigcup_{k=1}^N \{(t, x)\colon u(t, x) = u_k\}$, where the u_k, $k = 1, \dots, N$, are the corner points of the function $\omega'(u)$. From the proofs presented above (see Lemmas 4.1–4.6), however, it follows that in these identities $\omega''(u)|_{u=u_k}$, $k = 1, \dots, N$, are understood to be zero (since $\partial\omega'(u(t, x))/\partial x_i = 0$ on the sets $\{(t, x)\colon u(t, x) = u_k\}$, $k = 1, \dots, N$). This is to be kept in mind below.

We now introduce the standard cut-off functions used below. Let the numbers h, h', θ and θ' be such that $0 < h' < h \leqslant 1$ and $0 < \theta' < \theta \leqslant 1$. Suppose that $x_0 \in \mathbb{R}^n$, $t_0 \in \mathbb{R}$ and numbers $\rho > 0$ and $r > 0$ are fixed. We set

$$\xi = \xi\left(|x - x_0|, h', h, \rho\right) = \begin{cases} 1 & \text{for } |x - x_0| \leqslant \rho h', \\ \dfrac{\rho h - |x - x_0|}{\rho h - \rho h'} & \text{for } \rho h' \leqslant |x - x_0| \leqslant \rho h, \\ 0 & \text{for } |x - x_0| \geqslant \rho h, \end{cases} \tag{4.18}$$

$$\eta = \eta(t, t_0, \theta', \theta, r) = \begin{cases} 1 & \text{for } t_0 - (r\theta')^2 \leqslant t, \\ \dfrac{(r\theta)^2 - (t_0 - t)}{(r\theta)^2 - (r\theta')^2} & \text{for } t_0 - (r\theta)^2 \leqslant t < t_0 - (r\theta')^2, \\ 0 & \text{for } t \leqslant t_0 - (r\theta)^2; \end{cases} \tag{4.19}$$

$$\bar{\eta} = \bar{\eta}(t, t_0, \theta', \theta, r) = \begin{cases} 1 & \text{for } t < t_0 + (r\theta')^2, \\ \dfrac{(r\theta)^2 + (t_0 - t)}{(r\theta)^2 - (r\theta')^2} & \text{for } t_0 + (r\theta')^2 \leqslant t \leqslant t_0 + (r\theta)^2, \\ 0 & \text{for } t \geqslant t_0 + (r\theta)^2. \end{cases} \tag{4.20}$$

It is obvious that $\xi \in \tilde{C}^1(\mathbb{R}^n)$, $\eta, \bar{\eta} \in \tilde{C}^1(\mathbb{R})$, and $0 \leqslant \xi \leqslant 1$ in $\mathbb{R}^n$, $\xi = 0$ outside the ball $|x - x_0| \leqslant rh$, $0 \leqslant \eta \leqslant 1$ in $\mathbb{R}$, $\eta = 0$ for $t \leqslant t_0 - (r\theta)^2$, $0 \leqslant \bar{\eta} \leqslant 1$, in $\mathbb{R}$, and $\bar{\eta} = 0$ for $t \geqslant t_0 + (r\theta)^2$.

Thus, if $Q_{(\rho,r)} \equiv K_\rho(x_0) \times [t_0 - r^2, t_0]$ together with its closure is contained in Q, then the function $\Phi \equiv \xi^2\eta$ belongs to the class

$$\tilde{C}^1_{0,(\partial\Omega\times(T_1, T_2))\cup\Omega_{T_1}}(Q).$$

Similarly, if $\overline{Q}_{(\rho,r)} \equiv K_\rho(x_0) \times [t_0, t_0 + r^2]$ together with its closure is contained in Q, then

$$\overline{\Phi} \equiv \xi^2\bar{\eta} \in \tilde{C}^1_{0,(\partial\Omega\times(T_1, T_2))\cup\Omega_{T_2}}(Q).$$

It is obvious also that for the functions (4.18)–(4.20) we have

$$|\nabla\xi| \leqslant \frac{1}{\rho(h - h')}, \quad |\eta'| \leqslant \frac{1}{r^2(\theta^2 - \theta'^2)}, \quad |\bar{\eta}'| \leqslant \frac{1}{r^2(\theta^2 - \theta'^2)}. \tag{4.21}$$

Because of the use of the letter η to denote cut-off functions, below in identities of the form (4.15) and (4.16) we shall use the letter Φ to denote a test function. If the conditions of Lemma 4.7 are satisfied, then for any generalized solution of (1.1) we have by Lemma 4.7

$$\int_\Omega \omega(u)\Phi\, dx\Big|_{\tau_1}^{\tau_2} - \int_{\tau_1}^{\tau_2} (\omega(u), \Phi')\, dt$$
$$+ \iint_{Q_{\tau_1,\tau_2}} \left[\mathbf{l}' \cdot A\nabla u\omega''(u)\Phi + \mathbf{l}' \cdot A\nabla\Phi\omega'(u) + l_0'\omega'(u)\Phi\right] dt\, dx = 0,$$
$$\Phi \in \tilde{C}^1_{0,\partial\Omega\times(T_1,T_2)}(Q), \quad \tau_1, \tau_2 \in [T_1, T_2]. \tag{4.22}$$

Setting $\Phi = \xi^2\eta$ or $\Phi = \overline{\Phi} = \xi^2\bar{\eta}$ in (4.22), we obtain

$$\int_{K_{\rho h}(x_0)} \omega(u)\xi^2\eta\, dx\Big|_{\tau_1}^{\tau_2} - \int_{\tau_1}^{\tau_2}\int_{K_{\rho h}(x_0)} \omega(u)\xi^2\eta_t\, dt\, dx$$
$$+ \int_{\tau_1}^{\tau_2}\int_{K_{\rho h}(x_0)} \left[\mathbf{l}' \cdot A\nabla u\omega''\xi^2\eta + \mathbf{l}' \cdot A\nabla\xi 2\xi\eta\omega' + l_0'\omega'\xi^2\eta\right] dt\, dx = 0, \tag{4.23}$$

where $\tau_1, \tau_2 \in [t_0 - r^2, t_0]$, and

$$\int_{K_{\rho h}(x_0)} \omega(u)\xi^2\eta\, dx\Big|_{\tau_1}^{\tau_2} - \int_{\tau_1}^{\tau_2}\int_{K_{\rho h}(x_0)} \omega(u)\xi^2\eta_t\, dt\, dx$$
$$+ \int_{\tau_1}^{\tau_2}\int_{K_{\rho h}(x_0)} \left[\mathbf{l}'\cdot A\nabla u\omega''\xi^2\eta + \mathbf{l}'\cdot A\nabla\xi 2\xi\bar\eta\omega' + l_0'\omega'\xi^2\bar\eta\right] dt\, dx = 0, \quad (4.24)$$

where $\tau_1, \tau_2 \in [t_0, t_0 + r^2]$. For what follows it is convenient to redefine the function η for $t > t_0$ and the function $\bar\eta$ for $t < t_0$, taking these functions equal to 0 for such values of t. Denoting by $\hat\eta$ either of the functions η and $\bar\eta$, we can write each of these equalities in the unified form

$$\int_{K_{\rho h}(x_0)} \omega(u)\xi^2\hat\eta\, dx\Big|_{\tau_1}^{\tau_2} - \int_{\tau_1}^{\tau_2}\int_{K_{\rho h}(x_0)} \omega(u)\xi^2\hat\eta_t\, dt\, dx$$
$$+ \int_{\tau_1}^{\tau_2}\int_{K_{\rho h}(x_0)} \left[\mathbf{l}'\cdot A\nabla u\omega''\xi^2\hat\eta + \mathbf{l}'\cdot A\nabla\eta 2\xi\hat\eta\omega' + l_0'\omega'\xi^2\hat\eta\right] dt\, dx = 0, \quad (4.25)$$

where $\tau_1, \tau_2 \in [t_0 - r^2, t_0 + r^2]$. It is obvious that in the case $\hat\eta = \eta$ equality (4.25) gives (4.23), while in the case $\hat\eta = \bar\eta$ it gives (4.24). It is also obvious that (4.25) also holds for $\hat\eta \equiv 1$ in $[t_0 - r^2, t_0 + r^2]$.

LEMMA 4.11. *Suppose that* (4.22) *holds for a function*

$$u \in \mathscr{W}(A, Q) \equiv \left\{u \in \mathscr{H}_{2(l,l_0)}(A, Q) \colon u' \in \left(\mathscr{H}_{2(l,l_0)}(A, Q)\right)^*\right\}.$$

Then for any $\alpha > 0$ the transformation

$$\tilde x = (x - x_0)/\alpha, \qquad \tilde t = (t - t_0)/\alpha^2 \quad (4.26)$$

takes $u(t, x)$ into the function $\tilde u(\tilde t, \tilde x) \equiv u(t_0 + \alpha^2\tilde t, x_0 + \alpha\tilde x)$ which belongs to the space $\mathscr{W}(\tilde A, \tilde Q)$, where $\tilde A \equiv \tilde A(x) \equiv A(x_0 + \alpha\tilde x)$ and $\tilde Q$ is the image of Q under the mapping (4.26); *the function $\tilde u$ satisfies the identity*

$$\int_{\tilde\Omega} \omega(\tilde u)\tilde\Phi\, d\tilde x\Big|_{\tau_1}^{\tau_2} - \int_{\tau_1}^{\tau_2}\int_{\tilde\Omega}\omega(\tilde u)\tilde\Phi_{\tilde t}\, d\tilde t\, d\tilde x$$
$$+ \int_{\tau_1}^{\tau_2}\int_{\tilde\Omega}[\tilde{\mathbf{l}}'\cdot\tilde A\tilde\nabla\omega''(\tilde u)\tilde\Phi + \tilde{\mathbf{l}}'\cdot\tilde A\tilde\nabla\tilde\Phi\omega'(\tilde u) + \hat l_0'\tilde\Phi\omega'(\tilde u)]\, dt\, dx = 0,$$
$$\tilde\Phi \in \tilde C^1_{0,\partial\tilde\Omega\times(\tilde T_1, \tilde T_2)}(Q), \quad (4.27)$$

where $\tilde\Omega$ is the image of Ω under the mapping (4.26),

$$\tilde T_1 = T_1\alpha^2 + t_0,\ \tilde T_2 = T_2\alpha^2 + t_0,$$
$$\tilde{\mathbf{l}}'(\tilde t, \tilde x, \tilde u, \tilde p) = \alpha\mathbf{l}'(t_0 + \alpha^2\tilde t, x_0 + \alpha\tilde x, \tilde u, \alpha^{-1}\tilde p),$$
$$\hat l_0'(\tilde t, \tilde x, \tilde u, \tilde p) = \alpha^2 l_0'(t_0 + \alpha^2\tilde t, x_0 + \alpha\tilde x, \tilde u, \alpha^{-1}\tilde p),$$

and the functions $\tilde{\mathrm{l}}'$ *and* $\tilde{l}'_0$ *satisfy the conditions*

$$|\tilde{l}'^i(\tilde{t}, \tilde{x}, \tilde{u}, \tilde{p})| \leqslant \tilde{\mu} \sum_{i=1}^{n} |\tilde{p}_i| + \tilde{\alpha}_1(\tilde{t}, \tilde{x})|\tilde{u}| + \tilde{f}(\tilde{t}, \tilde{x}),$$

$$|\tilde{l}'_0(\tilde{t}, \tilde{x}, \tilde{u}, \tilde{p})| \leqslant \tilde{\alpha}_2(t, x) \sum_{i=1}^{n} |\tilde{p}_i| + \tilde{\alpha}_3(t, x)|\tilde{u}| + \tilde{g}(t, x), \tag{4.28}$$

$$\tilde{\mathrm{l}}(\tilde{t}, \tilde{x}, \tilde{u}, \tilde{p}) \cdot \tilde{p} \geqslant \tilde{\nu}|\tilde{p}|^2 - \tilde{\alpha}_4(t, x)\tilde{u}^2 - \tilde{h}(t, x),$$

where

$$\tilde{\mu} = \mu, \quad \tilde{\nu} = \nu, \quad \tilde{a}_1(\tilde{t}, \tilde{x}) = \alpha a_1(t_0 + \alpha^2\tilde{t}, x_0 + \alpha\tilde{x}),$$
$$\tilde{f}(\tilde{t}, \tilde{x}) = \alpha f(t_0 + \alpha^2\tilde{t}, x_0 + \alpha\tilde{x}), \quad \tilde{a}_2(\tilde{t}, \tilde{x}) = \alpha a_2(t_0 + \alpha^2\tilde{t}, x_0 + \alpha\tilde{x}),$$
$$\tilde{a}_3(\tilde{t}, \tilde{x}) = \alpha^2 a_3(t_0 + \alpha^2\tilde{t}, {}_0 + \alpha\tilde{x}), \quad \tilde{g}(\tilde{t}, \tilde{x}) = \alpha^2 g(t_0 + \alpha^2\tilde{t}, x_0 + \alpha\tilde{x}),$$
$$\tilde{a}_4(\tilde{t}, \tilde{x}) = \alpha^2 a_4(t_0 + \alpha^2\tilde{t}, x_0 + \alpha\tilde{x}), \quad \tilde{h}(\tilde{t}, \tilde{x}) = \alpha^2 h(t_0 + \alpha^2\tilde{t}, x_0 + \alpha\tilde{x}).$$

For any $p \geqslant 1$, $p_0 \geqslant 1$ *and* $D \subset Q$

$$\|\tilde{a}_1^2, \tilde{a}_2^2, \tilde{a}_3, \tilde{a}_4, \tilde{f}^2, \tilde{g}, \tilde{h}\|_{p, p_0, \tilde{D}} = \alpha^{2-n/p-2/p_0}\|a_1^2, a_2^2, a_3, a_4, f^2, g, h\|_{p, p_0, D};$$
$$\|\tilde{a}^{ij}\|_{2\bar{s}, 2\bar{s}_0, \tilde{D}} = \alpha^{-(n/2s+2/s_0)}\|a^{ij}\|_{2\bar{s}, 2\bar{s}_0, D},$$
$$\|\tilde{b}^{ij}\|_{r_i, r_{0i}, \tilde{D}} = \alpha^{-(n/r_i+2/r_{0i})}\|b^{ij}\|_{r_i, r_{0i}, D}, \tag{4.29}$$

where $\tilde{D}$ *is the image of* D *under the mapping* (4.26), $a^{ij}(\tilde{x}) = a^{ij}(x_0 + \alpha\tilde{x})$, $b^{ij}(\tilde{x}) = b^{ij}(x_0 + \alpha\tilde{x})$, $\|a^{ij}\| = A$, *and* $\|b^{ij}\| \equiv B \equiv A^{-1}$.

PROOF. The results of Lemma 4.11 are established in an obvious way by changing variables according to formula (4.26) in the integrals appearing in (4.22) and the norms of the right sides of (4.29).

§5. Local estimates in L^{p, p_0}

Suppose conditions (1.2)–(1.4), (1.11) and (1.12) are satisfied. In Theorem 5.1 we establish conditions on the coefficients in inequalities (1.11) and (1.12) and on the elements of the matrix A ensuring that generalized solutions of (1.1) belong to the space $L^{p, p_0}_{\mathrm{loc}}(Q)$.

THEOREM 5.1. *Suppose that*[2]

$$a_1^2, a_2^2, a_3, a_4 \in L^{\bar{s}, \bar{s}_0}(Q), \quad 1/\bar{s} + 2/\bar{l} = 1, \quad 1/\bar{s}_0 + 2/\bar{l}_0 = 1,$$
$$f^2, h \in L^{d, d_0}(Q), \quad (2 - 2\sigma)/\bar{l} + 1/d = 1, \quad (2 - 2\sigma)/\bar{l}_0 + 1/d_0 = 1, \tag{5.1}$$
$$g \in L^{m, m_0}(Q), \quad (2 - \sigma)/\bar{l} + 1/m = 1, \quad (2 - \sigma)/\bar{l}_0 + 1/m_0 = 1,$$

---

[2] We note that the pairs of indices $\bar{l}$, $\bar{l}_0$ can be different for the different relations in (5.1); it is only important that each such pair satisfy conditions of the form (1.7). However, for brevity we shall assume that the pair $\bar{l}$, $\bar{l}_0$ in (5.1) is the same for all the relations indicated.

where $\sigma \in (0,1]$, *and* $\bar{l}$ *and* $\bar{l}_0$ *are the same as in* (1.7). *Suppose also that*([3])

$$a^{ij} \in L^{2s,2s_0}(Q), \quad i,j=1,\dots,n; \quad \frac{1}{s}+\frac{2\kappa}{\bar{l}}=1, \quad \frac{1}{s_0}+\frac{2\kappa}{\bar{l}_0}=1,$$
$$\kappa \in \left(1, \min\left(\frac{\bar{l}}{2}, \frac{\hat{l}_0}{2\tilde{\alpha}}\right)\right), \quad \text{where } \bar{l}, \bar{l}_0, \hat{l}, \hat{l}_0, \tilde{\alpha} \quad \text{are as in (5.1).} \tag{5.2}$$

Then any local generalized solution u of (1.1) *in the cylinder Q belongs to the space* $L^{\bar{l}/\sigma,\bar{l}_0/\sigma}(Q')$, *where* Q' *is any subdomain of Q such that* $\overline{Q}' \subset \Omega \times (T_1, T_2)$. *For any cylinder*

$$Q_{(\rho)} \equiv Q_{(\rho,\rho^2)} = K_\rho(x_0) \times [t_0 - \rho^2, t_0]$$

which together with its closure is contained in Q

$$\rho^{-(n\sigma/\bar{l}+2\sigma/\bar{l}_0)}\|u\|_{\bar{l}/\sigma,\bar{l}_0/\sigma,Q(\rho/2)} \leqslant c\left(\rho^{-(n/l+2/l_0)}\|u\|_{l,l_0,Q(\rho)}+1\right), \tag{5.3}$$

where $l=\bar{l}/\kappa$, $l_0=\bar{l}_0/\kappa$, $\bar{l}$ *and* $\bar{l}_0$ *satisfy condition* (1.7), *and the constant c depends only on the structure of the equation and on* ρ.

PROOF. Let u be a local generalized solution of (4.1) in Q. Let $\overline{Q}_{(\rho)} \subset Q$. Taking Lemma 4.11 into account, we first assume that $\rho=1$, $t_0=0$ and $x_0=0$. As ω in (4.22) (with $\rho=1$, $t_0=0$ and $x_0=0$) we choose the function

$$\omega(u)=\frac{1}{2}[\varphi(\bar{u})]^2,$$

$$\varphi(\bar{u})=\begin{cases}\bar{u}^q, & 0<\bar{u}\leqslant N,\\ qN^{q-1}\bar{u}-(q-1)N^q, & \bar{u}\geqslant N,\ N>1,\ q\geqslant 1,\\ \bar{u}=(u^2+\varepsilon^2)^{1/2}-\varepsilon, & \end{cases} \tag{5.4}$$

where $\varepsilon>0$. It is obvious that

$$\frac{d\bar{u}}{du}=\frac{u}{\sqrt{u^2+\varepsilon^2}}, \quad \frac{d^2\bar{u}}{du^2}=\frac{\varepsilon}{(u^2+\varepsilon^2)^{3/2}}, \quad \left|\frac{d\bar{u}}{du}\right|\leqslant 1, \quad 0<\frac{d^2\bar{u}}{du^2}<\frac{1}{\bar{u}},$$

and also that

$$\omega'=\varphi\varphi'\frac{d\bar{u}}{du}, \qquad \omega''=\left[\varphi\varphi''+(\varphi')^2\right]\left(\frac{d\bar{u}}{du}\right)^2+\varphi\varphi'\frac{d^21\bar{u}}{du^2}. \tag{5.5}$$

Taking into account that the functions $u\to\bar{u}(\bar{u})$, $\bar{u}\to\varphi(\bar{u})$ and $(\bar{u})\to\varphi'(\bar{u})$ are continuously differentiable and uniformly Lipschitz and that $\bar{u}\to\varphi''(\bar{u})$ is continuous everywhere except at the point $\bar{u}=N$ where it has a discontinuity of first kind, we conclude that for the function $\omega(u)$ defined by (5.4) all the conditions imposed on ω in Lemma 4.3 are satisfied. Therefore, the choice of ω indicated above in (4.22)

([3]) It follows from (5.2) that the indices s and s_0 satisfy the conditions

$$1/s+2/l=1, \qquad 1/s_0+2/l_0=1,$$

while for $l=\bar{l}/\kappa$ and $l_0=\bar{l}_0/\kappa$ all the conditions of (1.9) hold with

$$\alpha=\tilde{\alpha}\left(1+\frac{\kappa-1}{1-\hat{l}/2}\right), \qquad \beta=\frac{2}{\hat{l}_0}\frac{\alpha[\kappa-(1+(\kappa-1)/(1-\hat{l}/2))]}{1-\alpha(1+(\kappa-1)/(1-\hat{l}/2))}.$$

In particular, for these values of α and β the following conditions are satisfied: $\alpha, \beta \in (0,1)$ and $\alpha/\hat{l}_0+(1-\alpha)\beta/2<\frac{1}{2}$.

is legitimate. We then have the equality

$$\frac{1}{2}\int_{K_h}\psi^2(\bar u)\xi^2\eta\,dx\Big|_{t=\tau_1}^{t=\tau_2}-\frac{1}{2}\int_{\tau_1}^{\tau_2}\int_{K_h}\varphi^2(\bar u)\xi^2\eta_t\,dt\,dx$$

$$+\int_{\tau_1}^{\tau_2}\int_{K_h}\left(\mathbf{l}'\cdot A\nabla u\omega''(u)\xi^2\eta+\mathbf{l}'\cdot A\nabla\xi 2\xi\eta\omega'(u)+l_0'\omega'(u)\xi^2\eta\right)dt\,dx=0,$$

$$\tau_1,\tau_2\in[-h^2,0], \tag{5.6}$$

where ξ and η are defined by (4.18) and (4.19), and $K_h\equiv K_h(0)$.

It is obvious that $\omega''>0$ and

$$\varphi'^2+\varphi\varphi''=\begin{cases}((2q-1)/q)\varphi'^2 & \text{for } \bar u<N, \qquad (5.7)\\ \varphi'^2 & \text{for } \bar u\geqslant N. \qquad (5.8)\end{cases}$$

Using (1.11), (5.7), (5.8), and the inequalities

$$|d\bar u/du|\leqslant 1,\qquad |d^2\bar u/du^2|<1/\bar u,$$

setting $v=\varphi(\bar u)$, and taking into account that

$$v_{x_i}=\varphi'(\bar u)(d\bar u/du)u_{x_i},$$

we deduce from (5.6) that

$$\frac{1}{2}\int_{K_h}v^2\xi^2\eta\,dx\Big|_{\tau_1}^{\tau_2}+\nu\int_{\tau_1}^{\tau_2}\int_{K_h}|A\nabla v|^2\xi^2\eta\,dt\,dx$$

$$\leqslant\frac{1}{2}\int_{\tau_1}^{\tau_2}\int_{K_h}v^2\xi^2|\eta_t|dt\,dx$$

$$+\int_{\tau_1}^{\tau_2}\int_{K_h}\Big[(2q-1)qa_4v^2\xi^2\eta+\frac{2q-1}{q}h\varphi'^2\xi^2\eta+2\sqrt{n}\,\mu|A\nabla v|\,|A\nabla\xi|v\xi\eta$$

$$+2qa_1v^2|A\nabla\xi|\xi\eta+2fv|A\nabla\xi|\varphi'\xi\eta+\sqrt{n}\,a_2|A\nabla v|v\xi^2\eta$$

$$+qa_3v^2\xi^2\eta+gv\varphi'\xi^2\eta\Big]\,dt\,dx. \tag{5.9}$$

It is not hard to see that

$$\varphi'(\bar u)\leqslant qv^{1-1/q}. \tag{5.10}$$

Then, taking (5.10) into account and applying the Cauchy inequality, we deduce from (5.9) that

$$\frac{1}{2}\int_{K_h}v^2\xi^2\eta\,dx\Big|_{\tau_1}^{\tau_2}+\frac{\nu}{2}\int_{\tau_1}^{\tau_2}\int_{K_h}|A\nabla v|^2\xi^2\eta\,dt\,dx$$

$$\leqslant\frac{1}{2}\iint_{Q_{(h)}}v^2\xi^2|\eta_t|dt\,dx+2\left(\frac{n\mu^2}{\nu}+1\right)\int_{-1}^{0}\int_{K_h}v^2|A\nabla\xi|^2\eta\,dt\,dx$$

$$+\frac{(n+1)q^2}{\bar\nu}\int_{\tau_1}^{\tau_2}\int_{K_h}(a_1^2+a_2^2+a_3+a_4)v^2\xi^2\eta\,dt\,dx$$

$$+2q^2\iint_{Q_{(h)}}(f^2+h)v^{2-2/q}\xi^2\eta\,dt\,dx+q\iint_{Q_{(h)}}gv^{2-1/q}\xi^2\eta\,dt\,dx, \tag{5.11}$$

where $\bar\nu=\min(\nu,1)$ and $\tau_1,\tau_2\in[-h^2,0]$.

We decompose the segment $[-1, 0]$ into parts by points $t_0 \equiv -1 < t_1 < \cdots < t_{k-1} < t_k < \cdots < t_m \equiv 0$, and we set $Q^k = K_h(0) \times (\tau_{k-1}, \tau_k)$. In (5.11) we set $\tau_1 = t_{k-1}$ and $\tau_2 = t_{k-1} + \theta$, where $\theta \in [0, t_k - t_{k-1}]$. Then for any $k = 1, \ldots, m$

$$\int_{K_h} v^2\xi^2\eta\, dx\Big|^{t=t_{k-1}+\theta} + \int_{t_{k-1}}^{t_k}\int_{K_h} \left|A\nabla\left(v\xi\sqrt{\eta}\right)\right|^2 dt\, dx$$
$$\leqslant \int_{K_h} \left(v\xi\sqrt{\eta}\right)^2 dx\Big|^{t=t_{k-1}} + \frac{2}{\bar{\nu}}(n+1)q^2 \iint_{Q^k} a\left(v\xi\sqrt{\eta}\right)^2 dt\, dx + J, \quad (5.12)$$

where $a \equiv a_1^2 + a_2^2 + a_3 + a_4$, and J is defined by

$$J = \frac{1}{\bar{\nu}} \iint_{Q_{(h)}} v^2\xi^2|\eta_t| dt\, dx + \left[2\left(\frac{n\mu^2}{\nu} + 1\right)\frac{2}{\bar{\nu}} + 1\right] \iint_{Q_{(h)}} v^2|A\nabla\xi|^2\eta\, dt\, dx$$
$$+ \frac{4q^2}{\bar{\nu}} \iint_{Q_{(h)}} (f^2 + h)v^{2-2/q}\xi^2\eta\, dt\, dx + \frac{2q}{\bar{\nu}} \iint_{Q_{(h)}} gv^{2-1/q}\xi^2\eta\, dt\, dx. \quad (5.13)$$

Applying the Hölder inequality and (1.8), we estimate

$$\iint_{Q^k} av^2\, dt\, dx \leqslant \|a\|_{\bar{s},\bar{s}_0,Q^k} \left\|v\xi\sqrt{\eta}\right\|^2_{\bar{l},\bar{l}_0,Q^k} \leqslant c_0^2 b^2 \|a\|_{\bar{s},\bar{s}_0,Q^k} \left\|v\xi\sqrt{\eta}\right\|^2_{\tilde{\mathscr{H}}(Q^k)}, \quad (5.14)$$

where c_0 is the constant in (1.8) for $Q \equiv Q^k$, and it depends only on n, r_i, r_{0i}, $\bar{l}$, $\bar{l}_0$;

$$b \equiv b(Q) \equiv \sup_{i,j=1,\ldots,n} \|b^{ij}\|_{r_i,r_{0i},Q}$$

while $\tilde{\mathscr{H}}(Q^k) \equiv \tilde{\mathscr{H}}_2(A, Q^k)$ (so that $\|v\xi\sqrt{\eta}\|^2_{\tilde{\mathscr{H}}(Q^k)} = \|v\xi\sqrt{\eta}\|^2_{2,\infty,Q^k} + \|A\nabla(v\xi\sqrt{\eta})\|^2_{2,Q^k}$). From (5.12)–(5.14) we now easily obtain

$$\left\|v\xi\sqrt{\eta}\right\|^2_{\tilde{\mathscr{H}}(Q^k)} \leqslant 2\int_\Omega \left(v\xi\sqrt{\eta}\right)^2 dx\Big|^{t=t_{k-1}}$$
$$+ \frac{4}{\bar{\nu}}(n+1)q^2\|a\|_{\bar{s},\bar{s}_0,Q^k}\left\|v\xi\sqrt{\eta}\right\|^2_{Q^k} + 2J, \qquad k = 1, \ldots, m. \quad (5.15)$$

Suppose the lengths of the intervals $[t_{k-1}, t_k]$ are so small that (for fixed q)

$$(4/\bar{\nu})(n+1)q^2\|a\|_{\bar{s},\bar{s}_0,Q^k} \leqslant (1/2), \qquad k = 1, \ldots, m. \quad (5.16)$$

For (5.16) to be satisfied it suffices to require that

$$\|a\|^{\bar{s}_0}_{\bar{s},\bar{s}_0,Q^k} \leqslant \left(\frac{\bar{\nu}}{8(n+1)q^2}\right)^{\bar{s}_0}.$$

It is easy to see that under this condition the number m of points of subdivision of $[-1, 0]$ can be taken to be no more than

$$1 + \left(\frac{8(n+1)q^2}{\bar{\nu}}\right)^{\bar{s}_0} \|a\|^{\bar{s}_0}_{\bar{s},\bar{s}_0 Q_{(h)}}.$$

From (5.15)and (5.16) we then obtain

$$z_k \leqslant 4z_{k-1} + 4J, \qquad k = 1, \ldots, m, \quad (5.17)$$

where we have used the notation $z_k = \|v\xi\sqrt{\eta}\|^2_{\tilde{\mathscr{H}}(Q^k)}$. Taking into account that $v\xi\sqrt{\eta} = 0$ for $t = -1$, we obtain

$$\left\|v\xi\sqrt{\eta}\right\|^2_{\tilde{\mathscr{H}}(Q_h)} \leqslant 4^{m+1}J, \quad (5.18)$$

where

$$m \leqslant 1 + \left(\frac{8(n+1)q^2}{\bar{\nu}} \right)^{\bar{s}_0} \|a\|^{\bar{s}_0}_{\bar{s},\bar{s}_0,Q_{(h)}}.$$

We now estimate the integrals appearing in J. Applying the Hölder inequality and taking condition (5.2) into account, we have

$$\iint_{Q_{(h)}} v^2 |A\nabla\xi|^2 \eta \, dt \, dx \leqslant \|A\nabla\xi\|^2_{2s,2s_0,Q_{(h)}} \|v\|^2_{l,l_0,Q_{(h)}}$$

$$\leqslant \frac{\|A\|^2_{2s,2s_0,Q_{(h)}}}{(h-h')^2} \|v\|^2_{l,l_0 Q_{(h)}},$$

$$\times \iint_{Q_{(h)}} v^2 \xi^2 |\eta_t| \, dt \, dx \frac{1}{h^2 - h'^2} \|v\|^2_{l,l_0,Q_{(h)}} \|\xi\|^2_{2s,2s_0,Q_{(h)}},$$

(5.19)

where $A \equiv (\sum_{i,j=1}^n (a^{ij})^2)^{1/2}$, $l = \bar{l}/\kappa$, $l_0 = \bar{l}_0/\kappa$, and $\bar{l}$, $\bar{l}_0$ and κ are the indices in condition (5.2). We note that $2 < l < \bar{l}$ and $2 < l_0 < \bar{l}_0$ (see the footnote to (5.2)).

Applying the Hölder inequality, the Cauchy inequality, and (1.8), and assuming that

$$f^2, h \in L^{(\bar{l}/(2-2/q))^*, (\bar{l}_0/(2-2/q))^*}(Q), \quad g \in L^{(\bar{l}/(2-1/q))^*, (\bar{l}_0/(2-1/q))^*}(Q), \tag{5.20}$$

where, as always, p^* denotes the index conjugate to p, i.e., $1/p + 1/p^* = 1$, we obtain

$$\frac{4q^2}{\bar{\nu}} \iint_{Q_{(h)}} (f^2 + h) v^{2-2/q} \xi^2 \eta \, dt \, dx$$

$$\leqslant \frac{4q^2}{\bar{\nu}} \|v\xi\sqrt{\eta}\|^{2-2/q}_{\bar{l},\bar{l}_0,Q_{(h)}} \|f^2 + h\|_{(\bar{l}/(2-2/q))^*, (\bar{l}_0/(2-2/q))^*, Q_{(h)}}$$

$$\leqslant \frac{1}{4} \|v\xi\sqrt{\eta}\|^2_{\mathscr{H}(Q_{(h)})} + c\|f^2 + h\|^q_{(\bar{l}/(2-2/q))^*, (\bar{l}_0/(2-2/q))^*, Q_{(h)}}; \tag{5.21}$$

$$\frac{2q}{\bar{\nu}} \iint_{Q_{(h)}} g v^{2-1/q} \xi^2 \eta \, dt \, dx \leqslant \frac{2q}{\bar{\nu}} \|v\xi\sqrt{\eta}\|^{2-1/q} \|g\|_{(\bar{l}/(2-1/q))^*, (\bar{l}_0/(2-1/q))^*, Q_{(h)}}$$

$$\leqslant \frac{1}{4} \|v\xi\sqrt{\eta}\|^2_{\mathscr{H}(Q_{(h)})} + c\|g\|^{2q}_{(\bar{l}/(2-1/q))^*, (\bar{l}_0/(2-1/q))^*, Q_{(h)}}, \tag{5.22}$$

where the constant c in (5.21) depends on n, $\bar{\nu}$, q, r_i, r_{0i}, $\bar{l}$, $\bar{l}_0$ and $b(Q)$. From (5.18), (5.13), (5.19), (5.21), and (5.22) we obtain

$$\|v\xi\sqrt{\eta}\|^2_{\mathscr{H}(Q_{(h)})} \leqslant \frac{c_1}{(h-h')^2} \|v\|^2_{l,l_0,Q_{(h)}} + c_2, \tag{5.23}$$

where c_1 depends on n, ν and $\|A\|_{2s,2s_0,Q_{(h)}}$ while c_2 depends on n, ν, q, r_i, r_{0i}, $l(Q)$, and the norms $\|f^2 + h\|$ and $\|g\|$. Again taking the inequality (1.8) into account, we obtain

$$\|v\|_{\bar{l},\bar{l}_0,Q_{(h')}} \leqslant \frac{c_3}{h-h'} \|v\|_{l,l_0 Q_{(h)}} + c_4, \tag{5.24}$$

where $c_3 = \sqrt{c_1}\, c_0 b$, $c_4 = \sqrt{c_2}\, c_0 b$, c_0 depends on n, r_i, r_{0i}, $\bar{l}$, $\bar{l}_0$ and $\tilde{l}$, $\tilde{l}_0$ is any pair of indices satisfying (1.7) (with $\bar{l}$ replaced by $\tilde{l}$ and $\bar{l}_0$ by $\tilde{l}_0$). Since $\lim_{N\to\infty}\varphi(\bar{u})$-$\bar{u}^q$ for almost all $(t, x) \in Q_{(1)}$, by the Lebesgue theorem we have

$$\lim_{N\to\infty} \|v\|_{l,l_0,Q_{(h)}} = \|\bar{u}^q\|_{l,l_0,Q_{(h)}}, \tag{5.25}$$

provided that

$$\bar{u}^q \in L^{l,l_0}(Q_{(h)}). \tag{5.26}$$

In this case from (5.24) we find by Fatou's lemma that $\bar{u}^q \in L^{\tilde{l},\tilde{l}_0}(Q_{(h')})$ and

$$\|\bar{u}\|_{q\tilde{l},q\tilde{l}_0,Q_{(h')}} \leqslant \left(\frac{c_3}{h-h'}\right)^{1/q} \|\bar{u}\|_{ql,ql_0,Q_{(h)}} + c_4^{1/q}. \tag{5.27}$$

Let M be the least integer such that

$$\sigma^{1/M} < \kappa, \tag{5.28}$$

where σ and κ are the numbers in conditions (5.1) and (5.2). In (5.24) we set

$$h = h_k = 1/2 + 1/2^k, \quad h' = h'_k = h_{k+1}, \quad k = 1, \dots, M, \tag{5.29}$$

where M is the number in (5.28). It is obvious that $h_1 = 1$ and $h'_M > 1/2$. We shall prove successively that for the values

$$q = q_k = \sigma^{-k/M}, \qquad k = 1, \dots, M, \tag{5.30}$$

conditions (5.20) and (5.26) with $h = h_k$ and the corresponding indices $k = 1, \dots, M$ are satisfied. Indeed, since $\sigma \in (0, 1)$, it follows that $\sigma^{k/M} \geqslant \sigma$, $k = 1, \dots, M$. From this it follows easily that the indices conjugate to $\bar{l}/(2 - 2/q_k)$ and $\bar{l}_0/(2 - 2/q_k)$, $k = 1, \dots, M$ (it is obvious that $2/q_k = 2\sigma^{k/M}$) do not exceed the indices conjugate to $\bar{l}/(2 - 2\sigma)$ and $\bar{l}_0(2 - 2\sigma)$ respectively, i.e., the indices $(\bar{l}/(2 - 2\sigma))^*$ and $(\bar{l}_0/(2 - 2\sigma))^*$ in (5.1), while the indices conjugate to $\bar{l}(2 - 1/q_k)$ and $\bar{l}_0/(2 - 1/q_k)$ do not exceed $(\bar{l}/2 - \sigma)^*$ and $(\bar{l}_0/(2 - \sigma))^*$, i.e., the indices m and m_0 in (5.1). Thus, for any $k = 1, \dots, M$ conditions (5.20) are satisfied.

Further, in view of (1.8) the function $\bar{u}^{q_1} \equiv u^{1/\sigma^{1/M}}$ belongs to $L^{l,l_0}(Q_{h_1})$, since $1/\sigma^{1/M} \leqslant \kappa$ and $\bar{u} \in L^{\hat{l},\hat{l}_0}(Q_1)$ for any $\hat{l}$ and $\hat{l}_0$ satisfying a condition of the form (1.7), and, in particular, for $\hat{l} = \bar{l} = l\kappa$ and $\hat{l}_0 = \bar{l}_0 = l_0\kappa$, so that $\bar{u}^\kappa \in L^{l,l_0}(Q_{h_1})$. Thus, for $k = 1$ conditions (5.20) and (5.26) are satisfied for $q = q_1 = 1/\sigma^{1/M}$. From what has been proved it follows that $\bar{u}^{q_1} \in L^{\bar{l},\bar{l}_0}(Q_{h_2})$, i.e., $\bar{u}^{\kappa q_1} \in L^{l,l_0}(Q_{h_2})$. Taking (5.28) into account we conclude that $\bar{u}^{q_2} \equiv u^{1/\sigma^{2/M}} \in L^{l,l_0}(Q_{h_2})$. Repeating these arguments, we successively establish that $\bar{u}^{1/\sigma^{3/M}} \in L^{l,l_0}(Q_{h_3}), \dots,$ $\bar{u}^{1/\sigma} \in L^{l,l_0}(Q_{h_M})$. Repeating this argument once again (with $k = M$), we conclude that $\bar{u}^{1/\sigma} \in L^{\bar{l},\bar{l}_0}(Q_{h_{M+1}})$, and, in particular, $\bar{u}^{1/\sigma} \in L^{\bar{l}/\sigma,\bar{l}_0/\sigma}(Q_{1/2})$, since $h_{M+1} = h'_M > 1/2$. Iterating the estimate (5.27) for $q = q_k$, $h = h_k$ and $h' = h'_k$, $k = 1, \dots, M$, we obtain

$$\|\bar{u}\|_{\bar{l}/\sigma,\bar{l}_0/\sigma,Q_{1/2}} \leqslant c\big(\|\bar{u}\|_{l,l_0,Q_1} + 1\big). \tag{5.31}$$

Suppose now that $\rho > 0$ and $(t_0, x_0) \in Q$ are arbitrary but such that $\overline{Q}_\rho(t_0, x_0) \subset Q$. Applying Lemma 1.1, on the basis of what has been proved we find that $\bar{u} \in L^{\bar{l}/\sigma,\bar{l}_0/\sigma}(Q_{\rho/2})$ and

$$\rho^{-(n\sigma/\bar{l}+2\sigma/\bar{l}_0)}\|\bar{u}\|_{\bar{l}/\sigma,\bar{l}_0/\sigma,Q_{\rho/2}} \leqslant c\big(\rho^{-(n/l+2/l_0)}\|\bar{u}\|_{l,l_0,Q_\rho} + 1\big). \tag{5.32}$$

Letting $\varepsilon \to 0$ in (5.32), we obtain (5.3). It is obvious that the constant c in (5.32) depends only on n, ν, σ, $\bar{s}_0$, ρ, $\|A\|_{2s,2s_0,Q_\rho}$, $\|a\|_{s,s_0,Q_\rho}$, $b(Q_\rho) \equiv \sup_{i,j=1,\dots,n}\|b^{ij}\|_{r_i,r_{0i},Q_\rho}$ and the constant c_0 in inequality (1.8) for the cylinders $Q_{(h)}$, $h \in [1/2, 1]$. Theorem 5.1 is proved.

In Theorem 5.1 the parameter σ is subject to the condition $\sigma \in (0, 1]$. We now consider the case $\sigma = 0$. For this case it is obvious that Theorem 5.1 implies the following result.

THEOREM 5.2. *Suppose that the conditions of Theorem* 5.1 *are satisfied in the case* $\sigma = 0$. *Then any local generalized solution* u *of* (1.1) *in the cylinder* Q *belongs to* $L^{\bar{l}/\sigma,\bar{l}_0/\sigma}(Q')$ *for any* $\bar{l}$ *and* $\bar{l}_0$ *satisfying condition* (1.7), $\sigma > 0$, *and* $\overline{Q}' \subset Q$. *For any cylinder* Q_ρ, $\overline{Q}_\rho \subset Q$,

$$\rho^{-(n\sigma/\bar{l}+2\sigma/\bar{l}_0)}\|u\|_{\bar{l}/\sigma,\bar{l}_0/\sigma,Q_{\rho/2}} \leqslant c\left(\rho^{-(n/l+2/l_0)}\|u\|_{l,l_0,Q_\rho} + 1\right), \tag{5.33}$$

where $l = \bar{l}/\kappa_0$, $l_0 = \bar{l}_0/\kappa$, *and the constant* c *depends only on the structure of the equation* $\bar{l}$, $\bar{l}_0$, σ, *and* ρ.

§6. Global estimates in L^{p,p_0}

Suppose that conditions (1.2)–(1.4) and (1.11) are satisfied; below conditions stronger than (1.13) will be imposed on the coefficients in (1.11). We suppose also that the following condition (which is stronger than (1.18)) is satisfied:

$$\text{for any } u \in \widetilde{\mathscr{H}^{0,\Sigma_1}_{l,l_0;2}}(A, Q) \text{ and } \varepsilon > 0$$

$$\int_{\sigma_3\times(\tau_1,\tau_2)} \lambda u^2\, ds \leqslant \varepsilon\int_{\tau_1}^{\tau_2}\int_\Omega |A\nabla u|^2\, dt\, dx + c\int_{\tau_1}^{\tau_2}\int_\Omega u^2\, dt\, dx, \tag{6.1}$$

where $\tau_1, \tau_2 \in [T_1, T_2]$, and c_ε is a constant not depending on u.

Applying the familiar Sobolev imbedding theorems, we can easily prove that condition (6.1) is certainly satisfied if $\lambda \in L^{\kappa,\infty}(\sigma_3 \times (T_1, T_2))$, where $\kappa > r/(r-2)$ and r is defined by (7.2.9). We assume further that the initial function u_0 in (1.21) belongs to $L^{2/\sigma}(\Omega)$.

THEOREM 6.1 *Suppose that the conditions indicated above and also condition* (6.1) *are satisfied. Then for any* $\bar{l}$ *and* $\bar{l}_0$ *satisfying condition* (1.7) *any generalized solution* u *of problem* (1.21) *belongs to* $L^{\bar{l}/\sigma,\bar{l}_0/\sigma}(Q)$, *and the norm* $\|u\|_{\bar{l}/\sigma,\bar{l}_0/\sigma,Q}$ *is bounded by a quantity depending only on the structure of the equation and the data of problem* (1.21).

PROOF. In view of Lemma 4.8 equality (4.16) holds for any function ω satisfying the conditions presented in Lemma 4.4, and for any $\tau_1, \tau_2 \in [T_1, T_2]$. We choose as ω in (4.16) the function defined by (5.4). It was established in the proof of Theorem 5.1 that this function satisfies all the conditions required of ω in Lemma 4.3. But then it also satisfies all the conditions of Lemma 4.4, since the condition $\omega'(0) = 0$ is obviously satisfied. Therefore,

$$\frac{1}{2}\int_\Omega \varphi^2(\bar{u})\Big|_{t_1}^{t_2} + \iint_{Q_{t_1,t_2}} (\mathbf{l}' \cdot A\nabla u\omega''(u) + l_0'\omega'(u))\, dt\, dx + \int_{\Sigma_3} \lambda u\varphi\varphi'\, ds = 0, \tag{6.2}$$

where $Q_{t_1,t_2} = \Omega \times (t_1, t_2)$. Using (1.11), (5.7), (5.8), (5.10), the inequalities

$$|d\bar{u}/du| \leqslant 1, \qquad 0 < d^2\bar{u}/du^2 < \bar{u}^{-1},$$

and condition (6.1), and setting $v = \varphi(\bar{u})$, we deduce from (6.2) that

$$\frac{1}{2}\int_\Omega v^2\,dx\Big|_{t_1}^{t_2} + \nu \iint\limits_{Q_{t_1,t_2}} |A\nabla v|^2\,dt\,dx$$

$$\leqslant \iint\limits_{Q_{t_1,t_2}} \big[q(2q-1)a_4 v^2 + q(2q-1)hv^{2-2/q}$$

$$+\sqrt{n}\,a_2|A\nabla v|v + qa_3v^2 + qgv^{2-1/q}\big]\,dt\,dx$$

$$+\varepsilon \iint\limits_{Q_{t_1,t_2}} |A\nabla v|^2\,dt\,dx + c_\varepsilon \iint\limits_{Q_{t_1,t_2}} v^2\,dt\,dx. \tag{6.3}$$

Applying the Cauchy inequality and choosing a suitable ε, we obtain

$$\frac{1}{2}\int_\Omega v^2\,dx\Big|_{t_1}^{t_2} + \frac{\nu}{2} \iint\limits_{Q_{t_1,t_2}} |A\nabla v|^2\,dt\,dx$$

$$\leqslant c_1 \iint\limits_{Q_{t_1,t_2}} av^2\,dt\,dx + 2q^2 \iint\limits_{Q_{t_1,t_2}} hv^{2-2/q}\,dt\,dx + q\iint\limits_{Q_{t_1,t_2}} gv^{2-1/q}\,dt\,dx, \tag{6.4}$$

where $a = a_2^2 + a_3 + a_4 + c_\varepsilon$.

Decomposing the segment $[T_1, T_2]$ into sufficiently small parts and arguing exactly as in the derivation of inequality (5.18) in the proof of Theorem 5.1, we obtain

$$\|\bar{u}\|^2_{\tilde{\mathscr{H}}(Q)} \leqslant 4^m J + \|u_0^q\|^2_{2,\Omega}, \tag{6.5}$$

where

$$\tilde{\mathscr{H}}(Q) \equiv \mathscr{H}_2^{\widetilde{0,\Sigma_1}}(A, Q),\ m \leqslant c(n,q,\bar{\nu})\big(\|a\|^{\bar{s}_0}_{\bar{s},\bar{s}_0,Q} + 1\big),$$

$$J = c(q)\bigg(\iint\limits_Q hv^{2-2/q}\,dt\,dx + \iint\limits_Q gv^{2-1/q}\,dt\,dx\bigg).$$

Using estimates of the form (5.21) and (5.22) (with $Q_{(h)}$ replaced by Q), and (1.8), we obtain

$$\|\bar{u}\|^2_{\tilde{l},\tilde{l}_0,Q} \leqslant c\Big(\|h\|^{1/q}_{(\tilde{l}/(2-2/q))^*,(\tilde{l}_0/(2-2/q))^*,Q}$$

$$+ \|g\|^{2/q}_{(\tilde{l}/(2-1/q))^*,(\tilde{l}_0/(2-1/q))^*,Q}\Big) + \|u_0^q\|^2_{2,\Omega}, \tag{6.6}$$

where c depends on n, q, $\bar{\nu}$, $\|a\|^{\bar{s}_0}_{\bar{s},\bar{s}_0,Q}$, $b \equiv \sup_{i,j=1,\dots,n}\|b^{ij}\|_{r_i,r_{0i},Q}$ and the constants in (1.8), and $\tilde{l}$ and $\tilde{l}_0$ satisfy condition (1.7). Of course, in deriving (6.6) it was assumed that the norms of h, g, and u_0^q on the right side of (6.6) are meaningful. In view of (5.1) this is certainly the case if $q = 1/\sigma$. Passing to the limit as $N \to \infty$ in (6.6), we then conclude that $\bar{u}^{1/\sigma} \in L^{\tilde{l},\tilde{l}_0}(Q)$, i.e., $\bar{u} \in L^{\tilde{l}/\sigma,\tilde{l}_0/\sigma}(Q)$, and

$$\|\bar{u}\|_{\tilde{l}/\sigma,\tilde{l}_0/\sigma,Q} \leqslant c\Big(\|h\|^{\sigma/2}_{d,d_0,Q} + \|g\|^{\sigma}_{m,m_0,Q}\Big) + \|u_0\|_{2/\sigma,\Omega}. \tag{6.7}$$

Letting ε tend to 0 in (6.7), we obtain Theorem 6.1.

The next result obviously follows as a corollary from Theorem 6.1.

THEOREM 6.2. *Suppose that the conditions of Theorem* 6.1 *are satisfied with* $\sigma = 0$. *Then any generalized solution of problem* (1.21) *belongs to the space* $L^{\tilde{l}/\sigma,\tilde{l}_0/\sigma}(Q)$ *for any* $\tilde{l}$ *and* $\tilde{l}_0$ *satisfying condition* (1.7) *and any* $\sigma > 0$.

REMARK 6.1. Results analogous to Theorems 6.1 and 6.2 can also be proved for generalized solutions of (1.1) (rather than of problem (1.21) as in Theorems 6.1 and 6.2), assuming in addition to the conditions of these theorems that such solutions are bounded on $(\partial\Omega \times (T_1, T_2)) \cup \Omega_{T_1}$. In §9 we give a proof of an estimate for $\operatorname{ess\,sup}_Q |u|$ for generalized solutions of (1.1) which are bounded on $\partial\Omega \times (T_1, T_2)$. This proof as well as the proof of Theorems 6.1 and 6.2 can be modified without difficulty to obtain estimates of the solutions indicated above in the norms of $L^{p,p_0}(Q)$ as well. We leave this to the reader.

§7. Exponential summability of generalized solutions

THEOREM 7.1. *Suppose that conditions* (1.2)–(1.4), (1.11) *and* (1.12) *are satisfied as well as the condition*

$$(a^{ij})^2, a_1^2, a_2^2, a_3, a_4 \in L^{s,s_0}(Q), \quad 1/s + 2\kappa/\bar{l} = 1, \quad 1/s_0 + 2\kappa/\bar{l}_0 = 1,$$

$$\kappa \in \left(1, \min\left(\frac{\hat{l}}{2}, \frac{\hat{l}_0}{2\bar{\alpha}}\right)\right); \quad f^2, g, h \in L^{\bar{s},\bar{s}_0}(Q), \quad 1/s + 2/\bar{l} = 1, \quad 1/\bar{s}_0 + 2/\bar{l}_0 = 1,$$

$$\text{where } \bar{l}, \bar{l}_0, \hat{l}, \hat{l}_0, \alpha \text{ are as in } (1.7). \tag{7.1}$$

Let u be a local generalized solution of (1.1) *in the cylinder Q. Then for any cylinder $Q_\rho \equiv K_\rho(x_0) \times [t_0 - \rho^2, t_0]$, $\overline{Q}_\rho \subset Q$,*

$$\iint_{Q_{\rho/2}} \exp\{\gamma |u(t,x)|\}\, dt\, dx \leqslant c_1 \exp\left(c_2 \rho^{-(n/l + 2/l_0)} \|u\|_{l,l_0,Q_\rho}\right), \tag{7.2}$$

where $l = \bar{l}/\kappa$ and $l_0 = \bar{l}_0/\kappa$, for some γ, c_1 and c_2 depending only on the structure of the equation and on ρ.

PROOF. Suppose first that $\rho = 1$, $t_0 = 0$ and $x_0 = 0$. We denote by v the same function as in the proof of Theorem 5.1, i.e., $v = \varphi(\bar{u})$, where

$$\varphi(\bar{u}) = \begin{cases} \bar{u}^q, 0 < \bar{u} \leqslant N, N > 1, q \geqslant 1, \bar{u} = (u^2 + \varepsilon^2)^{1/2} - \varepsilon, \varepsilon > 0, \\ qN^{q-1}\bar{u} - (q-1)N^q, \bar{u} \geqslant N. \end{cases} \tag{7.3}$$

It was shown in §5 that (5.11) holds for the function v. The integral $\int_{t_1}^{t_2} \int_{K_h} av^2\xi^2\eta\, dt\, dx$ we here estimate simply by the Hölder inequality:

$$\int_{\tau_1}^{\tau_2} \int_{K_h} av^2\xi^2\eta\, dt\, dx \leqslant \|a\|_{s,s_0,Q_h} \|v\|^2_{\bar{l},\bar{l}_0,Q_h}. \tag{7.4}$$

Estimating the remaining terms on the right side of (5.11) in exactly the same way as in the proof of Theorem 5.1 and applying (1.8) and (1.21), we obtain

$$\|v\xi\sqrt{\eta}\|^2_{\mathscr{H}(Q_h)} \leqslant \frac{c_1 q^2}{(h - h')^2} \|v\|^2_{l,l_0,Q_{(h)}} + c_2 q^q, \tag{7.5}$$

where the constants c_1 and c_2 depend on the same quantities as the analogous constants in (5.23) with the exception of the index q; the dependence on this index is indicated explicitly in (7.5). We note that by (7.1) a condition of the form (5.20), which is used in deriving (7.5), is satisfied for all $q \geqslant 1$. Using (1.8), we deduce from (7.5) that

$$\|v\|_{\bar{l},\bar{l}_0,Q_{h'}} \leqslant \frac{c_3 q}{h - h'} \|v\|_{l,l_0,Q_h} + c_4 q^q. \tag{7.6}$$

Applying the Lebesgue theorem and taking into account that by Theorem 5.2 condition (5.26) is satisfied for any $q \geqslant 1$, by letting N tend to ∞ in (7.6) we obtain

$$\|\bar{u}\|_{ql,ql_0,Q_{(h')}} \leqslant \left(\frac{c_3 q}{h-h'}\right)^{1/q} \|\bar{u}\|_{ql,ql_0,Q_{(h)}} + c_4 q. \tag{7.7}$$

In (7.7) we set

$$q = q_s = \kappa^s, \quad h = h_s = \tfrac{1}{2} + 2^{-(s+1)}, \quad h' = h'_s = h_{s+1}, \quad s = 0, 1, \ldots, \tag{7.8}$$

and we let $p_{(s)} = l\kappa^s$ and $p_{0(s)} = l_0\kappa^2$, $s = 0, 1, \ldots$. From (7.7) it then follows that

$$\|\bar{u}\|_{p_{(s+1)},p_{0(s+1)},Q_{h_{s+1}}} \leqslant (4c_3)^{1/\kappa^s}(2\kappa)^{s/\kappa^s}\|\bar{u}\|_{p_{(s)},p_{0(s)},Q_{h_s}} + c_4\kappa^s, \quad s = 0, 1, \ldots. \tag{7.9}$$

Iterating (7.9), we obtain

$$\|\bar{u}\|_{p_{(s+1)},p_{0(s+1)},Q_{h_{s+1}}} \leqslant c_5\|\bar{u}\|_{l,l_0,Q_1} + c_6\sum_{k=0}^{s}\kappa^k, \quad s = 0, 1, \ldots, \tag{7.10}$$

where

$$c_5 = (4c_3)^{\Sigma_1^\infty \kappa^{-k}}(2\kappa)^{\Sigma_1^\infty k\kappa^{-k}}, \ c_6 = c_4 c_5.$$

We set

$$\hat{p}_2 = \min(p_{(s)}, p_{0(s)}) = \kappa^s \min(l, l_0). \tag{7.11}$$

From (7.11) we then obtain

$$\|\bar{u}\|_{\hat{p}_{s+1},Q_{h_{s+1}}} \leqslant c_7\|\bar{u}\|_{p_{(s+1)},p_{0(s+1)},Q_{h_{s+1}}} \leqslant c_8\|\bar{u}\|_{l,l_0,Q_1} + c_9\hat{p}_{s+1}, \tag{7.12}$$

where $c_7 = c_7(l, l_0)$, $c_8 = c_5c_7$ and $c_9 = c_6c_7/(\kappa-1)\min(l, l_0)$. For any $p \geqslant \min(l, l_0)$ there is an index s such that

$$\hat{p}_s \equiv \min(l, l_0)\kappa^2 \leqslant p < \min(l, l_0)\kappa^{s+1} \equiv \hat{p}_{s+1}.$$

From (7.12) it then follows easily that for any $p \geqslant \min(l, l_0)$

$$\|\bar{u}\|_{p,Q_{1/2}} \leqslant c_8\|u\|_{l,l_0,Q_1} + c_{10}\cdot p, \tag{7.13}$$

where $c_{10} = c_9\kappa$.

Letting ε tend to 0 in (7.13), we obtain

$$\|u\|_{p,Q_{1/2}} \leqslant c_8\|u\|_{l,l_0,Q_1} + c_{10}p, p \geqslant \min(l, l_0). \tag{7.14}$$

From (7.14) it follows easily that for sufficiently small $\gamma > 0$ the series

$$\sum_{m=0}^{\infty} \iint_{Q_{1/2}} \frac{(\gamma|u^m|)^m}{m!} dt\, dx \tag{7.15}$$

converges. Indeed, applying (7.14), we find that

$$\sum_{m=0}^{\infty} \frac{\gamma^m\|u\|_{m,Q_{1/2}}^m}{m!} \leqslant \sum_{m=0}^{\infty} \frac{c_{11}^m\left(\|u\|_{l,l_0,Q_1} + m\right)^m\gamma^m}{m!}, \tag{7.16}$$

where c_{11} depends on c_8 and c_{10}. From (7.16) by means of Stirling's formula we easily obtain

$$\sum_{m=0}^{\infty} \frac{\gamma^m\|u\|_{m,Q_{1/2}}^m}{m!} \leqslant \sum_{m=0}^{\infty} \frac{\left(2c_{11}\gamma\|u\|_{l,l_0,Q_1}\right)^m}{m!} + \sum_{m=0}^{\infty}(2c_{11}\gamma e)^m. \tag{7.17}$$

setting $\gamma = (4c_{11}e)^{-1}$, we obtain an estimate of the right side in (7.17) in terms of $\exp(2e)^{-1}\|u\|_{l,l_0,Q_1} + 2$. Thus, from (7.16) and (7.17) we have

$$\iint_{Q_{1/2}} \exp\{\gamma|u|\}\,dt\,dx = \sum_{m=0}^{\infty} \iint_{Q_{1/2}} \frac{\gamma|u|^m}{m!}\,dt\,dx \leqslant \exp(2e)^{-1}\|u\|_{l,l_0,Q_1} + 2. \tag{7.18}$$

Consequently, (7.2) has been established for $\rho = 1$, $t_0 = 0$ and $x_0 = 0$. The general case easily follows from this by Lemma 4.11. Theorem 7.1 is proved.

THEOREM 7.2. *Suppose that conditions* (1.2)–(1.4), (1.11), (7.1) *and* (6.1) *are satisfied, and let* $u_0 \in L^\infty(\Omega)$. *Then for any generalized solution u of problem* (1.21)

$$\iint_{Q} \exp\{\gamma|u(t,x)|\}\,dt\,dx \leqslant c, \tag{7.19}$$

where γ *and c depend only on the structure of equation* (1.1) *and the data of problem* (1.21).

PROOF. Theorem 7.2 is proved just as Theorem 7.1 on the basis of inequality (6.3) with estimates of the form (6.5) and (6.6), which are valid for all $q \geqslant 1$ taken into account.

REMARK 7.1. An estimate of the form (7.19) can be established under condition (7.1) also for generalized solutions of (1.1) in a cylinder Q which are bounded on $(\partial\Omega \times (T_1, T_2)) \cup \Omega_{T_1}$ (see the proof of Theorem 9.1).

§8. Local boundedness of generalized solutions

THEOREM 8.1. *Suppose that conditions* (1.2)–(1.4), (1.11), *and* (1.12) *are satisfied as well as the condition*

$$(a^{ij})^2, a_1^2, a_2^2, a_3, a_4, f^2, g, h \in L^{s,s_0}(Q), \quad \frac{1}{s} + \frac{2\kappa}{\bar{l}} = 1, \quad \frac{1}{s_0} + \frac{2\kappa}{\bar{l}_0} = 1, \tag{8.1}$$

where κ, $\bar{l}$, *and* $\bar{l}_0$ *are the same indices as in* (5.1).

Then any local generalized solution of (1.1) *in the cylinder Q belongs to the space* $L^\infty(Q')$ (*i.e., has finite* $\operatorname{ess\,sup}_{Q'}|u|$), *where Q' is any subdomain of Q such that* $\overline{Q}' \subset Q$. *For any cylinder* $Q_\rho = K_\rho(x_0) \times [t_0 - \rho^2, t_0]$, $\overline{Q}_\rho \subset Q$,

$$\operatorname*{ess\,sup}_{Q_{\rho/2}} |u| \leqslant c\left(\rho^{-(n/l + 2/l_0)}\|u\|_{l,l_0,Q_\rho} + k_\rho\right), \tag{8.2}$$

where

$$k_\rho = \rho^{2-n/s-2/s_0}\|f^2 + h\|_{s,s_0,Q_\rho} + \rho^{2-n/s-2/s_0}\|g\|_{s,s_0,Q_\rho}, \tag{8.3}$$

and the constant c depends only on the structure of (8.1) *and on* ρ.

PROOF. Let u be a local generalized solution of (1.1) in Q. Let Q_ρ be a cylinder such that $\overline{Q}_\rho \subset Q$. We suppose first that $\rho = 1$, $t_0 = 0$ and $x_0 = 0$. We consider the function

$$\bar{u} = \sqrt{u^2 + (k_1 + \varepsilon)^2}, \qquad \varepsilon > 0, \tag{8.4}$$

where $k_1 = \|f^2 + h\|^{1/2}_{s,s_0,Q_1} + \|g\|_{s,s_0,Q_1}$. It is obvious that

$$\frac{d\bar{u}}{du} = \frac{u}{\sqrt{u^2 + (k_1 + \varepsilon)^2}}, \quad \frac{d^2\bar{u}}{du^2} = \frac{(k_1 + \varepsilon)^2}{\bar{u}\left[\bar{u}^2 + (k_1 + \varepsilon)^2\right]}, \quad \left|\frac{d\bar{u}}{du}\right| \leqslant 1, \; 0 < \frac{d^2\bar{u}}{du^2} < \frac{1}{\bar{u}}.$$

From (8.1) we easily obtain

$$|\mathbf{l}'(t, x, u, p)| \leqslant \mu \sum_{i=1}^{n} |p_i| + \hat{a}_1(t, x)\bar{u},$$

$$|l_0'(t, x, u, p)| \leqslant a_2(t, x) \sum_{i=1}^{n} |p_i| + \hat{a}_3(t, x)\bar{u}, \tag{8.5}$$

$$\mathbf{l}'(t, x, u, p) \cdot p \geqslant \nu|p|^2 - \hat{a}_4(t, x)\bar{u}^2,$$

where ν, μ and a_2 are the same as in (1.11), $\hat{a}_1 = a_1 + k_1^{-1}f$, $\hat{a}_3 = a_3 + k_1^{-1}g$, $\hat{a}_4 = a_4 + k_1^{-2}h$, and $\hat{a}_1^2, \hat{a}_3, \hat{a}_4 \in L^{s,s_0}(Q_1)$, since

$$\|k_1^{-2}f^2\|_{s,s_0,Q_1} \leqslant 1, \qquad \|k_1^{-1}g\|_{s,s_0,Q_1} \leqslant 1, \qquad \|k_1^{-2}h\|_{s,s_0,Q_1} \leqslant 1.$$

It is obvious that the function

$$\omega(u) = \frac{1}{2}[\varphi(\bar{u})]^2, \quad \varphi(\bar{u}) = \begin{cases} \bar{u}^q, 0 < \bar{u} \leqslant N, \\ qN^{q-1} - (q-1)N^q, \bar{u} \geqslant N, \end{cases} \quad N > 1, q \geqslant 1, \tag{8.6}$$

where $\bar{u}$ is defined by (8.4), satisfies all the conditions of Lemma 4.3 regarding the function ω. This is proved in the same way as in the proof of Theorem 5.1 it was proved that the choice of ω according to (5.4) was legitimate. In analogy to the derivation of (5.9) we then establish the inequality

$$\begin{aligned} \frac{1}{2}\int_{K_h} (v\xi\sqrt{\eta})^2\, dx\Big|^{t=\tau} + \nu\int_{-1}^{1}\int_{K_h} |A\nabla v|^2 \xi^2\eta\, dt\, dx \leqslant \frac{1}{2}\int_{-h^2}^{\tau}\int_{K_h} v^2\xi^2|\eta_t|dt\, dx \\ + \int_{-h^2}^{\tau}\int_{K_h} \big(2\sqrt{n}\,\mu|A\nabla v|\,|A\nabla\xi|v\xi\eta + 2q\hat{a}_1 v^2|A\nabla\xi|\xi\eta - \sqrt{n}\,a_2 1|A\nabla v|v\xi^2\eta \\ + q\hat{a}_3 v^2\xi^2\eta + 2q^2\hat{a}_4 v^2\xi^2\eta\big)\, dt\, dx, \qquad \tau \in (-1, 0], \end{aligned} \tag{8.7}$$

where $v = \varphi(\bar{u})$. Applying the Cauchy inequality and (4.21), we derive from (8.7) that

$$\begin{aligned} \frac{1}{2}\int_{K_h} (v\xi\sqrt{\eta})^2\, dx\Big|^{t=\tau} + \frac{\nu}{2}\int_{-h^2}^{\tau}\int_{K_h} |A\nabla v|^2\xi^2\eta\, dt\, dx \\ \leqslant \frac{1}{(h-h')^2}\int_{-h^2}^{0}\int_{K_h}\left[\left(\frac{4n\mu^2}{\nu} + 1\right)\sum_{i,j=1}^{n}(a^{ij})^2 + 1\right]v^2\, dt\, dx \\ + \frac{n+1}{\bar{\nu}}q^2\int_{-h^2}^{0}\int_{K_h} \hat{a}v^2\, dt\, dx, \end{aligned} \tag{8.8}$$

where $\tau \in (-1, 0)$, $\bar{\nu} = \min(1, \nu)$ and $\hat{a} \equiv \hat{a}_1^2 + a_2^2 + a_3 + \hat{a}_4$. Applying the Hölder inequality to the integrals on the right side of (8.8), we obtain

$$\begin{aligned} \frac{1}{2}\int_{K_h} (v\xi\sqrt{\eta})^2\, dx\Big|^{t=\tau} + \frac{\nu}{2}\int_{-h^2}^{\tau}\int_{K_h} |A\nabla v|^2\xi^2\eta\, dt\, dx \\ \leqslant c(n, \nu, \mu)\frac{q^2}{(h-h')^2}\left(\|A\|^2_{2s,2s_0,Q_1} + \|\hat{a}\|_{l,l_0,Q_1}\right)\|v\|^2_{\bar{l},l_0,Q_h}, \end{aligned} \tag{8.9}$$

where

$$A = \left(\sum_{i,j=1}^{n} (a^{ij})^2 \right)^{1/2}, \quad 1/l = \kappa/\tilde{l}, \quad 1/l_0 = \kappa/\tilde{l}_0.$$

From (8.9) it follows easily that

$$\|v\|^2_{2,\infty,Q_{h'}} + \|A\nabla v\|^2_{2,Q_{h'}} \leqslant \tilde{c}(n,\nu,\mu)\left(\|A\|^2_{2s,2s_0,Q_1} + \|\hat{a}\|_{s,s_0,Q_1}\right) \times \frac{q^2}{(h-h')^2}\|v\|^2_{l,l_0,Q_h}. \tag{8.10}$$

Applying (1.8), we deduce from (8.10) that

$$\|v\|_{\tilde{l},\tilde{l}_0,Q_{h'}} \leqslant \mathscr{K}_1 \frac{q}{h-h'}\|v\|_{l,l_0,Q_h}, \tag{8.11}$$

where $\tilde{l}$, $\tilde{l}_0$ is any pair of indices satisfying (1.7), and

$$\mathscr{K}_1 = c\left(\|A\|_{2s,2s_0k,Q_1} + \|\hat{a}\|^{1/2}_{s,s_0,Q_1}\right) \sup_{i,j=1,\dots,n} \|b^{ij}\|_{r_i,r_{0i},Q_1}, \tag{8.12}$$

where the constant c depends only on n, ν, μ, r_i, r_{0i}, $\tilde{l}$, and $\tilde{l}_0$. Since it follows from Theorem 5.2 that $\bar{u}^q \in L^{l,l_0}(Q_n)$ for any q, letting N tend to ∞ in (8.11), we obtain

$$\|\bar{u}\|_{q\tilde{l},q\tilde{l}_0,Q_{h'}} \leqslant \left(\frac{\mathscr{K}_1 q}{h-h'}\right)^{1/q}\|\bar{u}\|_{ql,ql_0,Q_h}. \tag{8.13}$$

We set

$$q = q_s = \kappa^s, \quad p_{(s)} = l\kappa^s, \quad p_{0(s)} = l_0\kappa^s,$$
$$h = h_s = \frac{1}{2} + 2^{-(s+1)}, \quad h' = h'_s = h_{s+1}, \quad s = 0, 1, \dots. \tag{8.14}$$

From (8.13) it then follows that

$$\|\bar{u}\|_{p_{(s+1)},p_{0(s+1)},Q_{h_{s+1}}} \leqslant \left(\mathscr{K}_1\kappa^s 2^{s+2}\right)^{1/\kappa^s}\|\bar{u}\|_{p_{(s)},p_{0(s)},Q_{h_s}}. \tag{8.15}$$

Iterating (8.15), we obtain

$$\|\bar{u}\|_{p_{(s+1)},p_{0(s+1)},Q_{h_{s+1}}} \leqslant (4\mathscr{K}_1)^{\sum_{k=0}^{s}\kappa^{-k}}(2\kappa)^{\sum_{k=0}^{s}k\kappa^{-k}}\|\bar{u}\|_{l,l_0,Q_1}. \tag{8.16}$$

Passing to the limit as $s \to \infty$ in (8.16), we conclude that $\operatorname{ess\,sup}_{Q_{1/2}} \bar{u} < +\infty$, and

$$\operatorname*{ess\,sup}_{Q_{1/2}} \bar{u} \leqslant (4\mathscr{K}_1)^{\sum_{k=0}^{\infty}\kappa^{-k}}(2\kappa)^{\sum_{k=0}^{\infty}k\kappa^{-k}}\|\bar{u}\|_{l,l_0,Q_1}. \tag{8.17}$$

Recalling that $\bar{u} = (u^2 + (k_1+\varepsilon)^2)^{1/2}$, we easily deduce from (8.17) that

$$\operatorname*{ess\,sup}_{Q_{1/2}} |u| \leqslant (4\mathscr{K}_1)^{\sum_0^{\infty}\kappa^{-k}}(2\kappa)^{\sum_0^{\infty}kgk^{-k}}\left(\|u\|_{l,l_0,Q_1} + k_1\right). \tag{8.18}$$

Suppose now that ρ, t_0 and x_0 are arbitrary but such that $\overline{Q_\rho(t_0,x_0)} \subset Q$. In view of Lemma 4.11, from (8.18) we obtain

$$\operatorname*{ess\,sup}_{Q_{\rho/2}} |u| \leqslant (4\mathscr{K}_\rho)^{\sum_0^{\infty}\kappa^{-k}}(2\kappa)^{\sum_0^{\infty}k\kappa^{-k}}\left(\rho^{-n/l-2/l_0}\|u\|_{l,l_0,Q_\rho} + k_\rho\right), \tag{8.19}$$

where k_ρ is defined by (8.3), and $\mathscr{K}_\rho$ has the form

$$\mathscr{K}_\rho = c(n, \nu, \mu, r_i, r_{0i}, \hat{l}, \hat{l}_0)\left(\frac{\|A\|_{2s,2s_0,Q_\rho}}{\rho^{n/2s+1/s_0}} + \frac{\|\hat{a}\|^{1/2}_{s,s_0,Q_\rho}}{\rho^{1-n/2s-1/s_0}}\right)$$
$$\times\left(\sup_{i,j=1,\dots,n}\frac{\|b^{ij}\|_{r_i,r_{0i},Q_\rho}}{\rho^{n/r_i+2/r_{0i}}}\right), \tag{8.20}$$

and it follows from (8.1) that $2 - n/s - 2/s_0 > 0$, so that $k_\rho \to 0$ as $\rho \to 0$. Theorem 8.1 is proved.

§9. Boundedness of generalized solutions of the boundary value problem

THEOREM 9.1. *Suppose conditions* (1.2)–(1.4), (1.11), (8.1), *and* (1.6) *are satisfied, as well as the condition*

$$a_2^2, a_3, a_4, g, h \in L^{s,s_0}(Q), \quad 1/s + 2\kappa/\bar{l} = 1, \quad 1/s_0 + 2\kappa\bar{l}_0 = 1, \tag{9.1}$$

where κ, $\bar{l}$ *and* $\bar{l}_0$ *are the same as in condition* (5.1).

Then any generalized solution u *of problem* (1.21) *has finite* $\operatorname{ess\,sup}_Q|u|$, *and* $\operatorname{ess\,sup}_Q|u| \leqslant c$, *where* c *depends only on the structure of problem* (1.21).

PROOF. Let u be a generalized solution of problem (1.21). In view of Lemma 4.8, for any function ω satisfying the conditions presented in Lemma 4.4 and any $\tau_1, \tau_2 \in [T_1, T_2]$ the equality (1.16) holds. As the function $\omega(u)$ in (4.16) we choose the function defined by

$$\omega(u) = (1/2)[\varphi(\bar{u})]^2, \qquad \bar{u} = \left(u^2 + (k+\varepsilon)^2\right)^{1/2} - k - \varepsilon,$$
$$\varphi(\bar{u}) = \begin{cases} \bar{u}^q, & k+\varepsilon \leqslant \bar{u} \leqslant N, \\ qN^{q-1}\bar{u} - (q-1)N^q, & \bar{u} \geqslant N, \end{cases} \qquad N > (k+\varepsilon), \tag{9.2}$$

where $q \geqslant 1$ and

$$k = \|h\|^{1/2}_{s,s_0,Q} + \|g\|_{s,s_0,Q} + \|u_0\|_{\infty,\Omega}, \qquad \varepsilon > 0. \tag{9.3}$$

Taking into account that

$$\omega'(u) = \varphi(\bar{u})\varphi'(d\bar{u}/du),$$
$$\omega''(u) = \varphi(\bar{u})\varphi'(\bar{u})(d^2\bar{u}/du^2) + \left\{[\varphi'(\bar{u})]^2 + \varphi(\bar{u})\right\}\varphi''(\bar{u})(d\bar{u}/du)^2, \tag{9.4}$$

exactly as in the proof of Theorem 5.1 we establish that $\omega(u)$ satisfies all the conditions of Lemma 4.3. But then all the conditions of Lemma 4.4 are satisfied for this function, because of the obvious equality $\omega'(0) = 0$. Therefore,

$$\frac{1}{2}\int_\Omega [\varphi(\bar{u})]^2\,dx\Big|_{T_1}^{\tau} + \int_{T_1}^{\tau}\int_\Omega [\mathbf{l}' \cdot A\nabla u\omega'' + l_0'\omega']\,dt\,dx + \int_{\Sigma_3}\lambda u\varphi\varphi'\,ds = 0. \tag{9.5}$$

Taking into account inequalities of the form (8.5), formulas (5.7), (5.8), (5.10), the inequalities

$$|d\bar{u}/du| \leqslant 1, \qquad 0 < d^2\bar{u}/du^2 < \bar{u}^{-1},$$

and condition (6.1) and setting $v = \varphi(\bar{u})$, we deduce from (9.5) that

$$\frac{1}{2}\int_\Omega v^2\,dx\Big|^{t=\tau} + \nu\int_{T_1}^{\tau}\int_\Omega |A\nabla v|^2\,dt\,dx$$

$$\leqslant \int_{T_1}^{\tau}\int_\Omega \left(2q^2\hat{a}_4 v^2 + \sqrt{n}\,a_2|A\nabla v|v + q\hat{a}_3 v^2\right)dt\,dx + \varepsilon_1\int_{T_1}^{\tau}\int_\Omega |A\nabla v|^2\,dt\,dx$$

$$+c_{\varepsilon_1}\int_{T_1}^{\tau}\int_\Omega v^2\,dt\,dx + \frac{1}{2}(k+\varepsilon)^{2q}\operatorname{meas}\Omega, \tag{9.6}$$

where $\varepsilon_1 > 0$, and in deriving (9.6) we have taken into account that $v^2 \leqslant (k+\varepsilon)^{2q}$ for $t = T_1$ (see (9.3)). Applying the Cauchy inequality and choosing a suitable $\varepsilon_1 > 0$, we deduce from (9.6) that

$$\frac{1}{2}\int_\Omega v^2\,dx\Big|^{t=\tau} + \frac{\nu}{2}\int_{T_1}^{\tau}\int_\Omega |A\nabla v|^2\,dt\,dx$$

$$\leqslant \int_{T_1}^{T}\int_\Omega\left(2q^2\hat{a}_4 + \frac{na_2^2}{\nu} + q\hat{a}_3 + c_{\nu/4}\right)v^2\,dt\,dx + (k+\varepsilon)^{2q}\operatorname{meas}\Omega. \tag{9.7}$$

From (9.7) we easily obtain

$$\|v\|_{2,\infty,Q}^2 + \|A\nabla v\|_{2,Q}^2 \leqslant \frac{2(n+1)q^2}{\bar{\nu}\nu}\|\hat{a}\|_{s,s_0,Q}\|v\|_{\bar{l},l_0,Q}^2 + (k+\varepsilon)^{2q}\operatorname{meas}\Omega, \tag{9.8}$$

where $\bar{\nu} = \min(\nu, 1)$ and $a = \hat{a}_4 + \hat{a}_3 + a_2^2 + c_{\nu/4}$.

Applying (1.8) and taking into account that

$$(k+\varepsilon)^{2q}\operatorname{meas}\Omega = \frac{1}{T_2 - T_1}\iint_Q (k+\varepsilon)^{2q}\,dt\,dx$$

$$\leqslant \frac{1}{T_2 - T_1}\iint_Q \bar{u}^{2q}\,dt\,dx < (\operatorname{meas}^{1-2/l}\Omega)(T_2 - T_1)^{-2/l_0}\|\bar{u}^q\|^2, \tag{9.9}$$

we deduce from (9.8) that

$$\|v\|_{\tilde{l},\tilde{l}_0,Q}^2 \leqslant c_0^2 b^2\left[\frac{2(n+1)q^2}{\bar{\nu}}\|a\|_{s,s_0,Q}\|v\|_{l,l_0,Q}^2 + (\operatorname{meas}^{1-2/l}\Omega)\right.$$

$$\left.\times(T_2 - T_1)^{-2/l_0}\|\bar{u}^q\|_{l,l_0,Q}^2\right], \tag{9.10}$$

where $\tilde{l}$, $\tilde{l}_0$ is any pair of indices satisfying (1.7), and c_0 and b are the constants in (1.8). We note that in view of Theorem 6.2 the integral $\iint_Q \bar{u}^{2q}\,dt\,dx$ exists for any $q > 1$. Hence, in (9.10) we may pass to the limits as $N \to \infty$. Applying the Lebesgue theorem and Fatou's lemma, we deduce from (9.10) that

$$\|\bar{u}^q\|_{\tilde{l},\tilde{l}_0,Q} \leqslant Kq\|\bar{u}^q\|_{l,l_0,Q}, \tag{9.11}$$

where

$$K = K(n, \nu, \mu, \operatorname{meas}\Omega, T_2 - T_1, r_i, r_{0i}, \tilde{l}, \tilde{l}_0)\Big(\sup_{i,j=1,\dots,n}\|b^{ij}\|_{r_i,r_{0i},Q}\Big)\|a\|_{s,s_0,Q}.$$

It is obvious that (9.11) implies

$$\|\bar{u}\|_{\kappa ql,\kappa ql_0,Q} \leqslant K^{1/q}q^{1/q}\|\bar{u}\|_{l,l_0,Q}, \tag{9.12}$$

where $\kappa > 1$ is the number in (9.1). Setting $q = q_\nu = \kappa^\nu$, $\nu = 0, 1, \ldots$, in (9.12) and letting $p_{(\nu)} = \kappa^\nu l$ and $p_{0(\nu)} = \kappa^\nu l_0$, $\nu = 0, 1, \ldots$, we obtain

$$\|\bar{u}\|_{p_{(\nu+1)}, p_{0(\nu+1)}, Q} \leqslant K^{1/\kappa^\nu} \kappa^{\nu/\kappa^\nu} \|\bar{u}\|_{p_{(\nu)}, p_{0(\nu)}, Q}, \qquad \nu = 0, 1, \ldots \tag{9.13}$$

Iterating (9.13) as $m \to \infty$, we find that

$$\|\bar{u}\|_{p_{(m+1)}, p_{0(m+1)}, Q} \leqslant K^{\sum_0^m \kappa^{-\nu}} \kappa^{\sum_0^m \nu\kappa^{-\nu}} \|\bar{u}\|_{l, l_0, Q}. \tag{9.14}$$

Letting m tend to ∞ in (9.14), we conclude that $\operatorname{ess\,sup}_Q \bar{u} < +\infty$, and we obtain

$$\operatorname*{ess\,sup}_Q |u| \leqslant cK^{\sum_0^\infty \kappa^{-\nu}} \kappa^{\sum_0^\infty \nu\kappa^{-\nu}} \left(\|u\|_{l, l_0, Q} + \|h\|_{s, s_0, Q}^{1/2} + \|g\|_{s, s_0, Q} + \|u_0\|_{\infty, \Omega} \right). \tag{9.15}$$

Letting ε tend to 0 in (9.15) and taking (6.7) with $\sigma = 1$ into account, we obtain

$$\operatorname*{ess\,sup}_Q |u| \leqslant c\left(\|h\|_{s, s_0, Q}^{1/2} + \|g\|_{s, s_0, Q} + \|u_0\|_{\infty, \Omega} \right), \tag{9.16}$$

where c depends on the constant in (9.15) and (6.7) and also on meas Ω and $T_2 - T_1$. Theorem 9.1 is proved.

§10. The maximum principle

In this section it is assumed that all the conditions of §1 related to consideration of global properties of generalized solutions of equation (1.1) are satisfied.

DEFINITION 10.1. We say that a function $u \in \mathscr{W} \equiv \{u \in \mathscr{H}: u' \in \mathscr{H}^*\}$ *does not exceed* a number M on $(\partial\Omega \times (T_1, T_2)) \cup \Omega_{T_1}$ if for each $s > 0$ there exist an $(n+1)$-dimensional neighborhood $\mathfrak{N}_s$ of $(\partial\Omega \times (T_1, T_2)) \cup \Omega_{T_1}$ and a sequence $\{u_s\}$, $u_s \in \tilde{C}^1(Q)$, $s = 1, 2, \ldots$, converging to u in $\mathscr{W}$ such that $u_s \leqslant M - 1/s$, $s = 1, 2, \ldots$, for almost all $(t, x) \in \mathfrak{N}_s$.

It obviously follows from Definition 10.1 that the condition $u \leqslant M$ on

$$(\partial\Omega \times (T_1, T_2)) \cup \Omega_{T_1}$$

for the function $u \in \mathscr{H}$ implies that $u \leqslant M + \varepsilon$ on $(\partial\Omega \times (T_1, T_2)) \cup \Omega_{T_1}$ for all $\varepsilon > 0$.

THEOREM 10.1. *Suppose that the conditions indicated above are satisfied as well as condition* (9.1). *Let u be a generalized solution of* (1.1) *such that $u \leqslant M$ on $(\partial\Omega \times (T_1, T_2)) \cup \Omega_{T_1}$. Then for almost all $(t, x) \in Q$*

$$u(t, x) \leqslant M + ck, \tag{10.1}$$

where c depends only on the structure of the equation, $T_2 - T_1$, and meas Ω, *while the number k is defined by*

$$k = \left(\|a_3\|_{s, s_0, Q} + \|a_4\|_{s, s_0, Q} \right)|M| + \|h\|_{s, s_0, Q}^{1/2} + \|g\|_{s, s_0, Q}. \tag{10.2}$$

PROOF. We suppose first that $M = 0$. We consider the function

$$\omega(u) = (1/2)[\varphi(\bar{u}) - k^q]^2,$$

$$\varphi(\bar{u}) = \begin{cases} \bar{u}^2, & k \leqslant \bar{u} \leqslant N, \\ qN^{q-1}\bar{u} - (q-1)N^q, & \bar{u} \geqslant N, \end{cases} \tag{10.3}$$

$$N > k, \quad q \geqslant 1, \quad \bar{u} = \max(0, u) + k,$$

where $k = \|h\|_{s,s_0,Q}^{1/2} + \|g\|_{s,s_0,Q} + \varepsilon$, $\varepsilon > 0$. It is obvious that the functions $\bar{u} \to \varphi(\bar{u})$ and $\bar{u} \to \varphi'(\bar{u})$ are continuously differentiable and uniformly Lipschitz, while $\bar{u} \to \varphi''(\bar{u})$ is continuous everywhere except at the point $\bar{u} = N$ where it has a discontinuity of first kind. It is also obvious that

$$\omega'(u) = \begin{cases} 0, & u < 0, \\ (\varphi(\bar{u}) - k^q)\varphi'(\bar{u}), & u \geqslant 0; \end{cases} \tag{10.4}$$

$$\omega''(u) = \begin{cases} 0, & u < 0, \\ (\varphi - k^q)\varphi'' + \varphi'^2, & u \geqslant 0, u \neq N, \end{cases}$$

where $\omega'(u)$ is continuous everywhere and $\omega''(u)$ is continuous everywhere except at $u = 0$ and $u = N - k$. Morover, $\omega''(u)$ is bounded on $\mathbb{R}$. From what has been said it then follows that ω satisfies all the conditions of Lemma 4.3 (with $\omega'(u)$ having two corner points $u = 0$ and $u = N - k$). Let u be the generalized solution of (1.1) considered in Theorem 10.1, so that $u \leqslant 0$ on $(\partial\Omega \times (T_1, T_2)) \cup \Omega_{T_1}$. From Definition 10.1 and the form of $\omega'(u)$ it then follows that there exists a sequence $\{u_n\}$, $u_n \in C^1(Q)$, $n = 1, 2, \ldots$, converging to u in $\mathscr{W}$ such that $\omega'(u_n) = 0$ in some neighborhood of $(\partial\Omega \times (T_1, T_2)) \cup \Omega_{T_1}$. From what has been proved it then follows, in particular, that for the functions $u(t, x)$ and $\omega(u)$ all the conditions of Lemma 4.5 are satisfied. Therefore, by Lemma 4.9, for all $\tau \in (T_2, T_2]$

$$\frac{1}{2}\int_\Omega \omega(u)\,dx\Big|^{t=\tau} + \int_{T_1}^{\tau}\int_\Omega (\mathbf{l}' \cdot A\nabla u\omega'' + l_0'\omega')\,dt\,dx = 0, \tau \in (T_2, T_2], \tag{10.5}$$

where $\omega(u)$ is defined by (10.3), and in deriving (10.5) from (4.17) we have also taken into account that, because $u \leqslant 0$ on Ω_{T_1}, (10.3) implies the equality $\omega(u) = 0$ for $t = T_1$.

We now observe that conditions (1.11) imply that on $\{(t, x) \in Q: u > 0\}$

$$\begin{gathered} |l_0'(t, x, u, p)| \leqslant a_2(t, x)\sum_{i=1}^n |p_i| + \hat{a}_3(t, x)\bar{u}, \\ \mathbf{l}'(t, x, u, p) \cdot p \geqslant \nu|p|^2 - \hat{a}_4(t, x)\bar{u}^2, \end{gathered} \tag{10.6}$$

where ν and a_2 are the same as in (1.11), and where $\hat{a}_3 = a_3 + k^{-1}g$, $\hat{a}_4 = a_4 + k^{-1}h$, and $\hat{a}_3, \hat{a}_4 \in L^{s,s_0}(Q)$, $\|k^{-1}g\|_{s,s_0,Q} \leqslant 1$ and $\|k^{-2}h\|_{s,s_0,Q} \leqslant 1$. Taking (10.6), (5.7) and (5.8) into account and setting $v = \varphi(\bar{u})$, we deduce from (10.5) that

$$\begin{aligned} \frac{1}{2}\int_\Omega (v - k^1)^2\,dx\Big|^{t=\tau} &+ \nu\int_{T_1}^{\tau}\int_\Omega |A\nabla v|^2\,dt\,dx \\ &\leqslant \int_{T_1}^{T_2}\int_\Omega \left(2q^2\hat{a}_4 v^2 + \sqrt{n}\,a_2|A\nabla v|v + q\hat{a}_3 v^2\right)dt\,dx, \qquad \tau \in (T_1, T_2). \end{aligned} \tag{10.7}$$

From (10.7), exactly as in the proof of Theorem 9.1 (the part following (9.6)), we obtain

$$\|\bar{u}\|_{p_{(m)}, p_{0(m)}, Q} \leqslant \mathscr{K}^{\sum_0^m \kappa^{-\nu}} \kappa^{\sum_0^m \nu\kappa^{-\nu}} \|\bar{u}\|_{l, l_0, Q}, \qquad m = 1, 2, \ldots, \tag{10.8}$$

where $\mathscr{K}$ has the same form as in (9.11) and $\bar{u} = \max(0, u) + k$. Letting m tend to ∞ in (10.8), we conclude that ess sup $\bar{u} < \infty$ and

$$\operatorname*{ess\,sup}_{Q} \bar{U} \leqslant c\left(\|u_+\|_{l,l_0,Q} + \|h^{1/2}\|_{s,s_0,Q} + \|g\|_{s,s_0,Q}\right), \tag{10.9}$$

where $u_+ = \max(0, u)$ and $c = \mathscr{K}^{\Sigma_0^\infty \kappa^{-\nu}} \kappa^{\Sigma_0^\infty \nu\kappa^{-\nu}}$.

We estimate the norm $\|u_+\|_{l,l_0,Q}$. Since it is now already known that $\bar{u}(t, x)$ has finite ess sup $\bar{u}_Q$ (so that in (10.3) as $\varphi(\bar{u})$ it is possible to take $\varphi(\bar{u}) = \bar{u}^q$ for $\bar{u} \geqslant k$ and $q = 1$), from (10.7) we obtain

$$\frac{1}{2}\int_\Omega u_+^2\,dx\Big|^{t=\tau} + \nu\int_{T_1}^{\tau}\int_\Omega |A\nabla u_+|^2\,dt\,dx$$
$$\leqslant \int_{T_1}^{T_2}\int_\Omega \left(2\hat{a}_4(u_+ + k)^2 + \sqrt{n}\,a_2|A\nabla u_+|\,|u_+ + k| + \hat{a}_3(u_+ + k)^2\right) dt\,dx,$$
$$\tau \in (T_1, T_2]. \tag{10.10}$$

From (10.10) we easily obtain

$$\|u_+\|_{2,\infty,Q_t} + \int_{T_1}^{t}\int_\Omega |A\nabla u_+|^2\,dt\,dx \leqslant c_i\left(|\Omega|k^2 + \|\hat{a}\|_{s,s_0,Q}\|u_+ + k\|^2_{l,l_0,Q_t}\right), \tag{10.11}$$

where $Q_t = \Omega \times (T_1, t)$, and c_1 depends on n and ν. Applying (1.10) with a suitable $\varepsilon > 0$, we deduce from (10.11) that

$$\|u_+(t)\|_{2,\Omega} + \int_{T_1}^{t}\int_\Omega |A\nabla u_+|^2\,dt\,dx \leqslant c_1\left(k^2 + \int_{T_1}^{t}\int_\Omega u_+^2\,dt\,dx\right), \tag{10.12}$$

where $c_2 = c_1 \operatorname{Meas}\Omega$, and c_2 depends on c_1, $\|\hat{a}\|_{s,s_0,Q}$ and other quantities determined by the constants c_0 and b in (1.8). Applying Gronwall's inequality, we now obtain

$$\|u_+\|^2_{2,\infty,Q} + \|A\nabla u_+\|^2_{2,Q} \leqslant c_1 k^2 e^{c_2(T_2 - T_1)}. \tag{10.13}$$

From (1.8) we now find that

$$\|u\|_{l,l_0,Q} \leqslant ck, \tag{10.14}$$

where c depends on c_1, c_2, $T_2 - T_1$, meas Ω, and the indices l and l_0. From (10.9) and (10.14) we obtain

$$u(t, x) \leqslant c\left(\|h\|^{1/2}_{s,s_0,Q} + \|g\|_{s,s_0,Q}\right). \tag{10.15}$$

Thus, Theorem 10.1 has been established in the case $M = 0$. We now eliminate the assumption $M = 0$. Let $\hat{u} = u - M$, $M = \text{const} \neq 0$. It is obvious that $\hat{u} \leqslant 0$ on $(\partial\Omega \times (T_1, T_2)) \cup \Omega_{T_1}$. It is easy to see that $\hat{u}$ is a generalized solution of the equation

$$\hat{u}_t - (d/dx_i)\hat{l}^i(t, x, \hat{u}, \nabla\hat{u}) + \hat{l}_0(t, x, \hat{u}, \nabla\hat{u}) = 0, \tag{10.16}$$

where

$$\hat{\mathbf{l}}(t, x, \hat{u}, \hat{p}) = \mathbf{l}(t, x, \hat{u} + M, \hat{p}),\ \hat{l}_0(t, x, \hat{u}, \hat{p}) = l_0(t, x, \hat{u} + M, \hat{p}),$$

and, (10.16) has the same structure as (1.1); in particular, the conditions

$$\hat{\mathbf{l}}(t, x, \hat{u}, \hat{p})A^*\hat{\mathbf{l}}'(t, x, u, Ap),\ \hat{l}_0(t, x, \hat{u}, \hat{p}) = \hat{l}_0'(t, x, \hat{u}, A\hat{p}), \tag{10.17}$$

are satisfied, where

$$\hat{\mathbf{l}}'(t, x, \hat{u}, \hat{p}) = \mathbf{l}'(t, x, \hat{u} + M, \hat{p}), \hat{l}'_0(t, x, \hat{u}, \hat{p}) = l'_0(t, x, \hat{u} + M, \hat{p}),$$

as well as the inequalities

$$\hat{l}'(t, x, \hat{u}, \hat{p}) \cdot \hat{p} \geqslant \nu \hat{p}^2 - 2a_4\hat{u}^2 - \hat{h},$$

$$|\hat{l}_0(t, x, \hat{u}, \hat{p})| \leqslant a_2 \sum_{i=1}^{n} |\hat{p}_i| + a_3|\hat{u}| + \hat{g}, \tag{10.18}$$

where $\hat{h} = 2a_4M^2 + h$ and $\hat{g} = a_3|M| + g$. From what has been proved we obtain

$$\hat{u}(t, x) \leqslant c'k', \tag{10.19}$$

where $c' = \sqrt{2}\, c$ and

$$k' = \left(\|a_3\|_{s, s_0, Q} + \|a_4\|_{s, s_0, Q}\right)|M| + \|h\|^{1/2}_{s, s_0, Q} + \|g\|_{s, s_0, Q}.$$

Returning to the old variable in (10.19), we obtain (10.1). Theorem 10.1 is proved.

COROLLARY 10.1. *Suppose that all the conditions of Theorem* 10.1 *are satisfied, and let u be a generalized solution of* (1.1) *such that* $m \leqslant u \leqslant M$ *on* $(\partial\Omega \times (T_1, T_2)) \cup \Omega_{T_1}$. *Then for almost all* $(t, x) \in Q$

$$m - c_1k_1 \leqslant u(t, x) \leqslant M + c_2k_2, \tag{10.20}$$

where c_1 *and* c_2 *depend only on the structure of equation* (1.1), $T_2 - T_1$, *and* meas Ω, *while the numbers* k_1 *and* k_2 *are determined by*

$$\begin{aligned} k_1 &= \left(\|a_3\|_{s, s_0, Q} + \|a_4\|_{s, s_0, Q}\right)|m| + \|h\|^{1/2}_{s, s_0, Q} + \|g\|_{s, s_0, Q}, \\ k_2 &= \left(\|a_3\| + \|a_4\|\right)|M| + \|h\|^{1/2} + \|g\|. \end{aligned} \tag{10.21}$$

PROOF. Since the function $\tilde{u} = -u$, where u is a generalized solution of (1.1), is a generalized solution of a completely analogous equation for which exactly the same conditions are satisfied as for (1.1), together with the estimate $u(t, x) \leqslant M + c_1k_1$ proved in Theorem 10.1 we also have the estimate $-u(t, x) \leqslant -m + c_2k_2$. Inequalities (10.20) obviously follow from these estimates. This proves Corollary 10.1.

Bibliography

1. A. D. Aleksandrov, *Investigations on the maximum principle*. I–VI, Izv. Vyssh. Uchebn. Zaved. Matematika **1958**, no. 5 (6), 126–157; **1959**, no. 3 (10), 3–12; **1959**, no. 5 (12), 16–32; **1960**, no. 3 (16), 3–15; **1960**, no. 5 (18), 16–26; **1961**, no. 1 (20), 3–20. (Russian)

2. V. M. Babich, *On the extension of functions*, Uspekhi Mat. Nauk **8** (1953), no. 2 (54), 111–113. (Russian)

3. L. N. Slobodetskiĭ and V. M. Babich, *On boundedness of the Dirichlet integrals*, Dokl. Akad. Nauk SSSR **106** (1956), 604–606. (Russian)

4. I. Ya. Bakel′man, *Hypersurfaces with given mean curvature and quasilinear elliptic equations with strong nonlinearities*, Mat. Sb. **75 (117)** (1968), 604–638; English transl. in Math. USSR Sb. **4** (1968).

5. S. N. Bernstein, *Collected works*. Vol. III: *Differential equations, calculus of variations and geometry* (1903–1947), Izdat. Akad. Nauk SSSR, Moscow, 1960. (Russian)

6. ______, *Limitation des modules des dérivées successives des solutions des équations du type parabolique*, C. R. (Dokl.) Acad. Sci. URSS **18** (1938), 385–389.

7. O. V. Besov et al., *Imbedding theory for classes of differentiable functions of several variables*, Partial Differential Equations (Proc. Sympos. Sixtieth Birthday S. L. Sobolev), "Nauka", Moscow, 1970, pp. 38–63; English transl. in Amer. Math. Soc. Transl. (2) **105** (1976).

8. O. V. Besov, V. P. Il′in and S. M. Nikol′skiĭ, *Integral representations of functions and imbedding theorems*, "Nauka", Moscow, 1975; English transl., Vols. 1, 2, Wiley, 1979.

9. A. V. Bitsadze, *Equations of mixed type*, Izdat. Akad. Nauk SSSR, Moscow, 1959; English transl., Pergamon Press, Oxford, and Macmillan, New York, 1964.

10. M. M. Vaĭnberg, *Variational methods for the study of nonlinear operators*, GITTL, Moscow, 1956; English transl., Holden-Day, San Francisco, Calif., 1964.

11. O. A. Oleĭnik and T. D. Venttsel′, *The first boundary value problem and the Cauchy problem for quasilinear parabolic type equations*, Mat. Sb. **41 (83)** (1957), 105–128. (Russian)

12. M, I. Vishik, *Boundary value problems for elliptic equations degenerating on the boundary of a region*, Mat. Sb. **35 (77)** (1954), 513–568; English transl. in Amer. Math. Soc. Transl. (2) **35** (1964).

13. ______, *Solvability of boundary value problems for quasilinear parabolic equations of higher order*, Mat. Sb. **59 (101)** (1962), 289–325; English transl. in Amer. Math. Soc. Transl. (2) **65** (1967).

14. V. S. Vladimirov, *Generalized functions in mathematical physics*, 2nd ed., "Nauka", Moscow, 1979; English transl., "Mir", Moscow, 1979.

15. A. I. Vol′pert and S. I. Khudyaev, *The Cauchy problem for degenerate second order quasilinear parabolic equations*, Mat. Sb. **78 (120)** (1969), 374–396; English transl. in Math. USSR Sb. **7** (1969).

16. Herbert Gajewski, Konrad Gröger and Klaus Zacharias, *Nichtlineare Operatorgleichungen und Operatordifferentialgleichungen*, Akademie-Verlag, Berlin, 1974.

17. V. P. Glushko, *Classical solutions of boundary value problems for degenerate second order elliptic equations in arbitrary smooth domains*, Teor. Funktsiĭ, Funktsional. Anal. i Prilozhen. Vyp. 5 (1967), 157–176. (Russian)

18. I. I. Danilyuk, *Nonregular boundary value problems in the plane*, "Nauka", Moscow, 1975. (Russian)

19. Yu. A. Dubinskiĭ, *Some integral inequalities and the solvability of degenerate quasilinear elliptic systems of differential equations*, Mat. Sb. **64 (106)** (1964), 458–480; English transl. in Amer. Math. Soc. Transl. (2) 53 (1966).

20. ______, *Quasilinear elliptic-parabolic equations*, Mat. Sb. **77 (119)** (1968), 354–389; English transl. in Math. USSR Sb. **6** (1968).

21. ______, *Nonlinear elliptic and parabolic equations*, Itogi Nauki: Sovremennye Problemy Mat., Vol. 9, VINITI, Moscow, 1976; English transl. in J. Soviet Math. **22** (1979), no. 5.

22. A. V. Ivanov, *On a class of functions*, Trudy Mat. Inst. Steklov. **73** (1964), 159–171. (Russian)

23. ______, *A priori estimates for solutions of linear second order equations of elliptic and parabolic types*, Dokl. Akad. Nauk SSSR **161** (1965), 1270–1273; English transl. in Soviet Math. Dokl. **6** (1965).

24. A. V. Ivanov et al., *Some properties of generalized solutions of second order parabolic equations*, Dokl. Akad. Nauk SSSR **168** (1966), 17–20; English transl. in Soviet Math. Dokl. **7** (1966).

25. A. V. Ivanov, *Harnack's inequality for generalized solutions of quasilinear parabolic equations of second order*, Trudy Mat. Inst. Steklov. **102** (1967), 51–84.

26. A. V. Ivanov, *Local estimates of the maximum of the modulus of the first derivatives of the solutions of quasilinear nonuniformly elliptic equations in divergence form*, Zap. Nauchn. Sem. Leningrad. Otdel. Mat. Inst. Steklov. (LOMI) 7 (1968), 87–125; English transl. in Sem. Math. V. A. Steklov Math. Inst. Leningrad 7 (1968).

27. ______, *Interior estimates of the first derivatives of the solutions of quasilinear nonuniformly elliptic and nonuniformly parabolic equations of general form*, Zap. Nauchn. Sem. Leningrad. Otdel. Mat. Inst. Steklov. (LOMI) **14** (1969), 24–47; English transl. in Sem. Math. V. A. Steklov Math. Inst. Leningrad **14** (1969).

28. ______, *A boundary value problem for degenerate linear second order parabolic equations*, Zap. Nauchn. Sem. Leningrad. Otdel. Mat. Inst. Steklov. (LOMI) **14** (1969), 48–88; English transl. in Sem. Math. V. A. Steklov Math. Inst. Leningrad **14** (1969).

29. ______, *The Dirichlet problem for quasilinear nonuniformly elliptic equations of second order*, Zap. Nauchn. Sem. Leningrad. Otdel. Mat. Inst. Steklov. (LOMI) **19** (1970), 79–94; English transl. in Sem. Math. V. A. Steklov Math. Inst. Leningrad **19** (1970).

30. ______, *Local estimates for the maximum modulus of the first derivatives of solutions of quasilinear nonuniformly elliptic and nonuniformly parabolic equations and systems of general type*, Trudy Mat. Inst. Steklov. **110** (1970), 45–64; English transl. in Proc. Steklov Inst. Math. **110** (1970).

31. ______, *On the Dirichlet problem for quasilinear nonuniformly elliptic equations of second order*, Trudy Mat. Inst. Steklov. **116** (1971), 34–54; English transl. in Proc. Steklov Inst. Math. **116** (1971).

32. ______, *Global a priori estimates for generalized solutions of degenerate parabolic equations of second order*, Vestnik Leningrad. Univ. **1971**, no. 19 (Ser. Mat. Mekh. Astr. vyp. 4), 13–22; English transl. in Vestnik Leningrad Univ. Math. **4** (1977) (1978).

33. ______, *Smoothness of generalized solutions of degenerate parabolic equations of second order*, Trudy Mat. Inst. Steklov. **116** (1971), 55–70; English transl. in Proc. Steklov Inst. Math. **116** (1971).

34. ______, *The structure of a quasilinear second order elliptic equation*, Zap. Nauchn. Sem. Leningrad. Otdel. Mat. Inst. Steklov. (LOMI) **21** (1971), 5–33; English transl. in J. Soviet Math. **1** (1973), no. 4.

35. ______, *On the limit growth of the coefficients of quasilinear elliptic equations*, Zap. Nauchn. Sem. Leningrad. Otdel. Mat. Inst. Steklov. (LOMI) **27** (1972), 73–90; English transl. in J. Soviet Math. **3** (1975), no. 4.

36. ______, *Local estimates of the first derivatives of the solutions of second order quasilinear elliptic equations, and their application to Liouville type theorems*, Zap. Nauchn. Sem. Leningrad. Otdel. Mat. Inst. Steklov. (LOMI) **30** (1972), 40–50; English transl. in J. Soviet Math. **4** (1975), no. 4.

37. ______, *On the solvability of the Dirichlet problem for some classes of second order elliptic systems*, Trudy Mat. Inst. Steklov **125** (1973), 56–87; English transl. in Proc. Steklov Inst. Math. **125** (1973).

38. ______, *The first boundary value problem for quasilinear second order parabolic equations*, Zap. Nauchn. Sem. Leningrad. Otdel. Mat. Inst. Steklov. (LOMI) **38** (1973), 10–32; English transl. in J. Soviet Math. **8** (1977), no. 4.

39. ______, *Theorems of Liouville type for quasilinear elliptic systems*, Trudy Mat. Inst. Steklov. **127** (1975), 5–19; English transl. in Proc. Steklov Inst. Math. **127** (1975).

40. A. V. Ivanov, N. M. Ivochkina and O. A. Ladyzhenskaya, *On a strengthening of results on the solvability of the Dirichlet problem for uniformly elliptic quasilinear equations of second order*, Vestnik Leningrad. Univ. **1975**, no. 1 (Ser. Mat. Mekh. Astr. vyp. 1), 61–72; English transl. in Vestnik Leningrad Univ. Math. **8** (1980).

41. A. V. Ivanov, *A priori estimates for the solutions of nonlinear elliptic equations of second order*, Zap. Nauchn. Sem. Leningrad. Otdel. Mat. Inst. Steklov. (LOMI) **59** (1976), 31–59; English transl. in J. Soviet Math. **10** (1978), no. 2.

42. ______, *A priori estimates of second derivatives of solutions of nonlinear second order equations on the boundary of the domain*, Zap. Nauchn. Sem. Leningrad. Otdel. Mat. Inst. Steklov. (LOMI) **69** (1977), 65–76; English transl. in J. Soviet Math. **10** (1978), no. 1.

43. ______, *Properties of the solutions of nonstrictly and nonuniformly parabolic linear and quasilinear second order equations with measurable coefficients*, Zap. Nauchn. Sem. Leningrad. Otdel. Mat. Inst. Steklov. (LOMI) **69** (1977), 45–64; English transl. in J. Soviet Math. **10** (1978), no. 1.

44. ______, *The first boundary value problem for quasilinear $(A, \vec{b})$-elliptic equations*, Zap. Nauchn. Sem. Leningrad. Otdel. Mat. Inst. Steklov. (LOMI) **84** (1979), 45–88; English transl., to appear in J. Soviet Math.

45. ______, *The first and second boundary value problems for quasilinear $(A, \vec{b})$-elliptic equations*, Zap. Nauchn. Sem. Leningrad. Otdel. Mat. Inst. Steklov. (LOMI) **92** (1979), 171–181; English transl., to appear in J. Soviet Math.

46. A. V. Ivanov, *A general boundary value problem for quasilinear* $(A, \vec{b})$*-elliptic equations*, Zap. Nauchn. Sem. Leningrad. Otdel. Mat. Inst. Steklov. (LOMI) **96** (1980), 57–68; English transl., to appear in J. Soviet Math.

47. A. V. Ivanov and P. Z. Mkrtychyan, *On the solvability of the first boundary value problem for certain classes of degenerating quasilinear elliptic equations of the second order*, Trudy Mat. Inst. Steklov. **147** (1980), 14–39; English transl. in Proc. Steklov Inst. Math. **1981**, no. 2 (147).

48. A. V. Ivanov, *Two-sided and one-sided Liouville theorems for quasilinear elliptic equations of nondivergence type*, Zap. Nauchn. Sem. Leningrad. Otdel. Mat. Inst. Steklov. (LOMI) **100** (1980), 59–80; English transl. in J. Soviet Math. **19** (1982), no. 6.

49. ______, $(A, \vec{0})$*-elliptic equations with weak degeneracy*, Zap. Nauchn. Sem. Leningrad. Otdel. Mat. Inst. Steklov. (LOMI) **112** (1981), 75–84; English transl., to appear in J. Soviet Math.

50. ______, $(A, \vec{0})$*-parabolic equations with weak degeneracy*, Zap. Nauchn. Sem. Leningrad. Otdel. Mat. Inst. Steklov. (LOMI) **111** (1981), 52–62; English transl., to appear in J. Soviet Math.

51. ______, *Boundary value problems for degenerate quasilinear elliptic and parabolic equations of second order*, Dokl. Akad. Nauk SSSR **262** (1982), 15–19; English transl. in Soviet Math. Dokl. **25** (1982).

52. ______, *On the regularity of generalized solutions of the first boundary value problem for degenerate quasilinear elliptic equations of second order*, Zap. Nauchn. Sem. Leningrad. Otdel. Mat. Inst. Steklov. (LOMI) **110** (1981), 30–52; English transl., to appear in J. Soviet Math.

53. ______, *Boundary value problems for degenerate quasilinear elliptic and parabolic equations of second order*, Uspekhi Mat. Nauk **36** (1981), no. 4 (220), 232. (Russian)

54. N. M. Ivochkina and A. P. Oskolkov, *Nonlocal estimates of the first derivatives of solutions of the first boundary value problem for nonuniformly elliptic and nonuniformly parabolic nondivergence equations*, Zap. Nauchn. Sem. Leningrad. Otdel. Mat. Inst. Steklov. (LOMI) **11** (1968), 6–72; English transl. in Sem. Math. V. A. Steklov Math. Inst. Leningrad **11** (1968).

55. ______, *Nonlocal estimates for the first derivatives of solutions of the first boundary value problem for certain classes of nonuniformly elliptic and nonuniformly parabolic equations and systems*, Trudy Mat. Inst. Steklov. **110** (1970), 65–101; English transl. in Proc. Steklov Inst. Math. **110** (1970).

56. A. M. Il'in, *Degenerate elliptic and parabolic equations*, Mat. Sb. **50 (92)** (1960), 443–498. (Russian)

57. V. P. Il'in, *Some inequalities in function spaces and their application to the study of the convergence of variational processes*, Trudy Math. Inst. Steklov. **53** (1959), 64–127; English transl. in Amer. Math. Soc. Transl. (2) **81** (1969).

58. ______, *Approximation of functions in the spaces* $\tilde{W}_p^{(l)}(D)$ *and* $W_p^{(l)}(D)$ *by continuously differentiable functions*, Trudy Mat. Inst. Steklov. **64** (1961), 61–78; English transl. in Amer. Math. Soc. Transl. (2) **44** (1965).

59. A. S. Kalashnikov, *On a nonlinear equation arising in the theory of nonstationary filtration*, Trudy Sem. Petrovsk. Vyp. 4 (1978), 137–146. (Russian)

60. ______, *Quasilinear degenerate parabolic equations with a finite speed of propagation of perturbations*, Partial Differential Equations (Proc. Conf., Novosibirsk, 1978; S. K. Godunov, editor), "Nauka", Novosibirsk, 1980, pp. 80–83. (Russian)

61. M. V. Keldysh, *On certain cases of degeneration of equations of elliptic type on the boundary of a domain*, Dokl. Akad. Nauk SSSR **77** (1951), 181–183. (Russian)

62. A. I. Kozhanov, *The first boundary value problem for some classes of degenerate quasilinear elliptic equations*, Dinamika Sploshnoĭ Sredy Vyp. 17 (1974), 31–42. (Russian)

63. Ī. M. Kolodiĭ, *Qualitative properties of the generalized solutions of elliptic equations*, Ukrain. Mat. Zh. **27** (1975), 320–328; English transl. in Ukrainian Math. J. **27** (1975).

64. V. A. Kondrat'ev, *Boundary value problems for parabolic equations in closed domains*, Trudy Moskov. Mat. Obshch. **15** (1966), 400–451; English transl. in Trans. Moscow Math. Soc. **15** (1966).

65. A. I. Koshelev, *On the boundedness in* L_p *of derivatives of solutions of elliptic differential equations*, Mat. Sb. **38 (80)** (1956), 359–372. (Russian)

66. M. A. Krasnosel'skiĭ, *Topological methods in the theory of nonlinear integral equations*, GITTL, Moscow, 1956; English transl., Macmillan, 1964.

67. S. N. Kruzhkov, *a*) *Some properties of solutions of elliptic equations*, Dokl. Akad. Nauk SSSR **150** (1963), 470–473; English transl. in Soviet Math. Dokl. **4** (1963).

b) *A priori estimates and certain properties of the solutions of elliptic and parabolic equations*, Mat. Sb. **65 (107)** (1964), 522–570; English transl. in Amer. Math. Soc. Transl. (2) **68** (1968).

68. ______, *A priori estimates for generalized solutions of second order elliptic and parabolic equations*, Dokl. Akad. Nauk SSSR **150** (1963), 748–751; English transl. in Soviet Math. Dokl. **4** (1963).

69. ______, *Boundary value problems for degenerate second order elliptic equations*, Mat. Sb. **77 (119)** (1968), 299–334; English transl. in Math. USSR Sb. **6** (1968).

70. S. N. Kruzhkov and I. M. Kolodiĭ, *A priori estimates and Harnack's inequality for generalized solutions of degenerate quasilinear parabolic equations*, Sibirsk. Mat. Zh. **18** (1977), 608–628; English transl. in Siberian Math. J. **18** (1977).

71. N. V. Krylov, *On the theory of degenerate nonlinear elliptic equations*, Dokl. Akad. Nauk SSSR **201** (1971), 1279–1281; English transl. in Soviet Math. Dokl. **12** (1971).

72. ______, *Controlled diffusion type processes*, "Nauka", Moscow, 1977; English transl., Springer-Verlag, 1980.

73. L. D. Kudryavtsev, *On the solution by the variational method of elliptic equations which degenerate on the boundary of the region*, Dokl. Akad. Nauk SSSR **108** (1956), 16–19. (Russian)

74. ______, *Direct and inverse imbedding theorems*, Trudy Mat. Inst. Steklov. **55** (1959); English transl., Amer. Math. Soc., Providence, R. I., 1974.

75. L. P. Kuptsov, *On a function space with pth power integrable first derivatives taken over variable directions*, Trudy Mat. Inst. Steklov. **103** (1968), 96–116; English transl. in Proc. Steklov Inst. Math. **103** (1968).

76. ______, *Some estimates for the solution of degenerate* $(k, 0)$*-elliptic equations*, Mat. Zametki **8** (1970), 625–634; English transl. in Math. Notes **8** (1970).

77. O. A. Ladyzhenskaya and N. N. Ural'tseva, *Quasilinear elliptic equations and variational problems in several independent variables*, Uspekhi Mat. Nauk **16** (1961), no. 1 (97), 19–90; English transl. in Russian Math. Surveys **16** (1961).

78. ______, *a*) *A boundary value problem for linear and quasilinear parabolic equations*, Dokl. Akad. Nauk SSSR **139** (1961), 544–547; English transl. in Soviet Math. Dokl. **2** (1961).

______ *b*) *Boundary value problems for linear and quasilinear parabolic equations*. I, Izv. Akad. Nauk SSSR Ser. Mat. **26** (1962), 5–52; English transl. in Amer. Math. Soc. Transl. (2) **47** (1965).

79. ______, *Linear and quasilinear elliptic equations*, "Nauka", Moscow, 1964; English transl., Academic Press, 1968.

80. O. A. Ladyzhenskaya, V. A. Solonnikov and N. N. Ural'tseva, *Linear and quasilinear equations of parabolic type*, "Nauka", Moscow, 1967; English transl., Amer. Math. Soc., Providence, R. I., 1968.

81. O. A. Ladyzhenskaya and N. N. Ural'tseva, *Total estimates of the first derivatives with respect to x of the solutions of quasilinear elliptic and parabolic equations*, Zap. Nauchn. Sem. Leningrad. Otdel. Mat. Inst. Steklov. (LOMI) **14** (1969), 127–155; English transl. in Sem. Math. V. A. Steklov Math. Inst. Leningrad **14** (1969).

82. O. A. Ladyzhenskaya, *Boundary value problems of mathematical physics*, "Nauka", Moscow, 1973. (Russian)

83. O. A. Ladyzhenskaya and N. N. Ural'tseva, *Linear and quasilinear equations of elliptic type*, 2nd ed., "Nauka", Moscow, 1973; English transl. of 1st ed., Academic Press, 1968.

84. ______, *Estimate of the Hölder norm of solutions of quasilinear second-order elliptic equations of general form*, Zap. Nauchn. Sem. Leningrad. Otdel. Mat. Inst. Steklov. (LOMI) **96** (1980), 161–168; English transl., to appear in J. Soviet Math.

85. ______, *An estimate of the Hölder constant for solutions of quasilinear parabolic equations of nondivergence form*, Uspekhi Mat. Nauk **36** (1981), no. 4 (220), 230. (Russian)

86. E. M. Landis, *Second order equations of elliptic and parabolic type*, "Nauka", Moscow, 1971. (Russian)

87. Lu Ven'-tuan [Lu Wen-tuan], *On imbedding theorems for spaces of functions with partial derivatives summable with different powers*, Vestnik Leningrad. Univ. **1961**, no. 7 (Ser. Mat. Mekh. Astr. vyp. 2), 23–37. (Russian)

88. V. G. Maz'ya, *Some estimates for solutions of second order elliptic equations*, Dokl. Akad. Nauk SSSR **137** (1961), 1057–1059; English transl. in Soviet Math. Dokl. **2** (1961).

89. ______, *The degenerate problem with oblique derivative*, Uspekhi Mat. Nauk **25** (1970), no. 2 (152), 275–276. (Russian)

90. V. P. Mikhaĭlov, *The first boundary value problem for quasielliptic and quasiparabolic equations*, Trudy Mat. Inst. Steklov. **91** (1967), 81–99; English transl. in Proc. Steklov Inst. Math. **91** (1967).

91. S. G. Mikhlin, *Degenerate elliptic equations*, Vestnik Leningrad Univ. **1954**, no. 8 (Ser. Mat. Mekh. Astr. Fiz. Khim. vyp. 3), 19–48. (Russian)

92. P. Z. Mkrtychyan, *Solvability of the first boundary value problem for higher-order quasilinear elliptic equations that are degenerate with respect to the independent variables*, Zap. Nauchn. Sem. Leningrad. Otdel. Mat. Inst. Steklov. (LOMI) **96** (1980), 187–204; English transl., to appear in J. Soviet Math.

93. ______, *On the solvability of the Dirichlet problem for a class of degenerate quasilinear elliptic equations*, Zap. Nauchn. Sem. Leningrad. Otdel. Mat. Inst. Steklov. (LOMI) **102** (1980), 181–189; English transl., to appear in J. Soviet Math.

94. P. Z. Mkrtychyan, *Some properties of generalized solutions of degenerate second-order quasilinear elliptic equations,* Akad. Nauk Armyan. SSR Dokl. **70** (1980), 292–296. (Russian)

95. S. M. Nikol'skiĭ, *The first boundary value problem for a general linear equation*, Dokl. Akad. Nauk SSSR **146** (1962), 767–769; English transl. in Soviet Math. Dokl. **3** (1962).

96. S. M. Nikol'skiĭ, *Approximation of functions of several variables and embedding theorems*, "Nauka", Moscow, 1969; English transl., Springer-Verlag, 1974.

97. O. A. Oleĭnik, A. S. Kalashnikov and Chzhou Yuĭ-lin' [Jou Yuh-lin], *The Cauchy problem and boundary value problems for equations of the type of nonstationary filtration*, Izv. Akad. Nauk SSSR Ser. Mat. **22** (1958), 667–704. (Russian)

98. O. A. Oleĭnik and S. N. Kruzhkov, *Quasilinear second order parabolic equations with several independent variables*, Uspekhi Mat. Nauk **16** (1961), no. 5 (101), 115–155; English transl. in Russian Math. Surveys **16** (1961).

99. O. A. Oleĭnik, *Linear equations of second order with nonnegative characteristic form*, Mat. Sb. **69** **(111)** (1966), 111–140; English transl. in Amer. Math. Soc. Transl. (2) **65** (1967).

100. ______, *Mathematical problems of boundary layer theory*, Uspekhi Mat. Nauk **23** (1968), no. 3 (141), 3–65; English transl. in Russian Math. Surveys **23** (1968).

101. ______, *Formations of the boundary layer in the case of a gradual acceleration*, Sibirsk. Mat. Zh. **9** (1968), 1199–1220; English transl. in Siberian Math. J. **9** (1968).

102. O. A. Oleĭnik and E. V. Radkevich, *Equations of second order with nonnegative characteristic form*, Itogi Nauki: Mat. Anal. 1969, VINITI, Moscow, 1971; English transl., Plenum Press, 1973.

103. A. P. Oskolkov, *Some estimates for nonuniformly elliptic equations and systems*, Trudy Mat. Inst. Steklov. **92** (1966), 203–232; English transl. in Proc. Steklov Inst. Math. **92** (1966).

104. ______, *A priori estimates for first derivatives of solutions of the Dirichlet problem for quasilinear nonuniformly elliptic equations*, Trudy Mat. Inst. Steklov. **102** (1967), 105–127; English transl. in Proc. Steklov Inst. Math. **102** (1967).

105. ______, *A remark on the estimation of the Hölder constant for certain quasilinear nonuniformly elliptic systems*, Zap. Nauchn. Sem. Leningrad. Otdel. Mat. Inst. Steklov. (LOMI) **7** (1968), 178–183; English transl. in Sem. Math. V. A. Steklov Math. Inst. Leningrad **7** (1968).

106. A. V. Pogorelov, *The multidimensional Minkowski problem*, "Nauka", Moscow, 1975; English transl., Wiley, 1978.

107. E. V. Radkevich, *The second boundary value problem for a second order equation with nonnegative characteristic form*, Vestnik Moskov. Univ. Ser. I Mat. Mekh. **1967**, no. 4, 3–11. (Russian)

108. E. S. Sabinina, *On the Cauchy problem for the equation of nonstationary gas filtration in several space variables*, Dokl. Akad. Nauk SSSR **136** (1961), 1034–1037; English transl. in Soviet Math. Dokl. **2** (1961).

109. ______, *A class of degenerate nonlinear parabolic equations*, Dokl. Akad. Nauk SSSR **143** (1962), 794–797; English transl. in Soviet Math. Dokl. **3** (1962).

110. I. V. Skrypnik [Ī. V. Skripnik], *Nonlinear higher order elliptic equations*, "Naukova Dumka", Kiev, 1973. (Russian)

111. ______, *Solvability and properties of solutions of nonlinear elliptic equations*, Itogi Nauki: Sovremennye Problemy Mat., Vol. 9, VINITI, Moscow, 1976, pp. 131–254; English transl. in J. Soviet Math. **12** (1979), no. 5.

112. ______, *Solvability of boundary value problems for nonlinear elliptic equations that are not in divergence form*, Zap. Nauchn. Sem. Leningrad. Otdel. Mat. Inst. Steklov. (LOMI) **84** (1979), 243–251; English transl., to appear in J. Soviet Math.

113. M. M. Smirnov, *Degenerate elliptic and hyperbolic equations*, "Nauka", Moscow, 1966. (Russian)

114. S. L. Sobolev, *Applications of functional analysis in mathematical physics*, Izdat. Leningrad. Gos. Univ., Leningrad, 1950; English transl., Amer. Math. Soc., Providence, R. I., 1963.

115. T. B. Solomyak, *The Dirichlet problem for quasilinear nonuniformly elliptic equations*, Problemy Mat. Anal., Vyp. 1, Izdat. Leningrad. Gos. Univ., Leningrad, 1966, pp. 88–109; English transl. in Problems in Math. Anal., no. 1, Plenum Press, 1968.

116. V. A. Solonnikov, *A priori estimates for certain boundary value problems*, Dokl. Akad. Nauk SSSR **138** (1961), 781–784; English transl. in Soviet Math. Dokl. **2** (1961).

117. A. L. Treskunov, *Sharp estimates in L_p for a certain class of degenerate second order equations of elliptic type*, Zap. Nauchn. Sem. Leningrad. Otdel. Mat. Inst. Steklov. (LOMI) **5** (1967), 232–249; English transl. in Sem. Math. V. A. Steklov Math. Inst. Leningrad **5** (1967).

118. N. N. Ural'tseva, *Degenerate quasilinear elliptic systems*, Zap. Nauchn. Sem. Leningrad. Otdel. Mat. Inst. Steklov. (LOMI) **7** (1968), 184–222; English transl. in Sem. Math. V. A. Steklov Math. Inst. Leningrad **7** (1968).

119. G. M. Fateeva, *On boundary value problems for degenerate quasilinear parabolic equations*, Mat. Sb. **76 (118)** (1968), 537–565; English transl. in Math. USSR Sb. **5** (1968).

120. M. I. Freĭdlin, *The first boundary value problem for degenerate quasilinear parabolic differential equations*, Uspekhi Mat. Nauk **15** (1960), no. 2 (92), 204–206. (Russian)

121. ______, *On the smoothness of solutions of degenerate elliptic equations*, Izv. Akad. Nauk SSSR Ser. Mat. **32** (1968), 1391–1413; English transl. in Math. USSR Izv. **2** (1968).

122. ______, *Existence "in the large" of smooth solutions of degenerate quasilinear equations*, Mat. Sb. **78 (120)** (1969), 332–348; English transl. in Math. USSR Sb. **7** (1969).

123. Avner Friedman, *Partial differential equations of parabolic type*, Prentice-Hall, 1964.

124. D. G. Aronson, *On the Green's function for second order parabolic differential equations with discontinuous coefficients*, Bull. Amer. Math. Soc. **69** (1963), 841–847.

125. D. G. Aronson and James Serrin, *Local behavior of solutions of quasilinear parabolic equations*, Arch. Rational Mech. Anal. **25** (1967), 81–122.

126. ______, *A maximum principle for nonlinear parabolic equations*, Ann. Scuola Novm. Sup. Pisa (3) **21** (1967), 291–305.

127. Serge Bernstein, *Sur la généralisation du problème de Dirichlet*. II, Math. Ann. **69** (1910), 82–136.

128. ______, *Sur les équations du calcul des variations*, Ann. Sci. École Norm. Sup. **29** (1912), 431–485.

129. Elwood Bohn and Lloyd K. Jackson, *The Liouville theorem for a quasi-linear ellliptic partial differential equation*, Trans. Amer. Math. Soc. **104** (1962), 393–397.

130. E. Bombieri, E. DeGiorgi and M. Miranda, *Una maggiorazione a prior relativa alle ipersuperfici minimali non parametriche*, Arch. Rational Mech. Anal. **32** (1969), 255–267.

131. Felix E. Browder, *Nonlinear elliptic boundary value problems*, Bull. Amer. Math. Soc. **69** (1963), 862–874.

132. Shiu-Yuen Cheng and Shing-Tung Yau, *On the regularity of the solution of the n-dimensional Minkowski problem*, Comm. Pure Appl. Math. **29** (1976), 495–516.

133. ______, *On the regularity of the Monge-Ampère equation* $\det(\partial^2 u/\partial x_i \partial x_j) = F(x, u)$, Comm. Pure Appl. Math. **30** (1977), 41–68.

134. Ennio DeGiorgi, *Sulla differenziabilità e l'analiticità delle estremali degli integrali multipli regolari*, Mem. Accad. Sci. Torino Cl. Sci. Fis. Mat. Nat. (3) **3** (1957), 25–43.

135. Jim Douglas, Jr., Todd Dupont and James Serrin, *Uniqueness and comparison theorems for nonlinear elliptic equations in divergence form*, Arch. Rational Mech. Anal. **42** (1971), 157–168.

136. D. E. Edmunds and L. A. Peletier, *Quasilinear parabolic equations*, Ann. Scula Norm. Sup. Pisa (3) **25** (1971), 397–421.

137. Gaetano Fichera, *Sulle equazioni differenziali lineari ellittico-paraboliche del secondo ordine*, Atti Accad. Naz. Lincei Mem. Cl. Sci. Fis. Mat. Nat. Sez. I (8) **5** (1956), 1–30.

138. David Gilbarg, *Boundary value problems for nonlinear elliptic equations in n variables*, Nonlinear Problems (Proc. Sympos., Madison, Wisc., 1962), Univ. of Wisconsin Press, Madison, Wisc., 1963, pp. 151–159.

139. Francesco Guglielmino, *Sulla regolarizzazione delle soluzioni deboli dei problemi al contorno per operatori parabolici*, Ricerche Mat. **12** (1963), 44–66.

140. Erhard Heinz, *Interior gradient estimates for surfaces* $z = f(x, y)$ *with prescribed mean curvature*, J. Differential Geom. **5** (1971), 149–157.

141. J. J. Kohn and L. Nirenberg, *Degenerate elliptic-parabolic equations of second order*, Comm. Pure Appl. Math. **20** (1967), 797–872.

142. O. A. Ladyzhenskaya and N. N. Ural'tseva, *Local estimates for gradients of solutions of non-uniformly elliptic and parabolic equations*, Comm. Pure Appl. Math. **23** (1970), 677–703.

143. Jean Leray and Jacques-Louis Lions, *Quelques résultats de Višik sur les problèmes elliptiques non linéaires par les méthodes de Minty-Browder*, Bull. Soc. Math. France **93** (1965), 97–107.

144. J. L. Lions, *Équations différentielles opérationnelles et problèmes aux limites*, Springer-Verlag, 1961.

145. ______, *Quelques méthodes de resolution des problèmes aux limites non linéaires*, Dunod and Gauthier-Villars, Paris, 1969.

146. George J. Minty, *On a "monotonicity" method for the solution of nonlinear equations in Banach spaces*, Proc. Nat. Acad. Sci. U.S.A. **50** (1963), 1038–1041.

147. Mario Miranda, *Una maggiorzione integrali per le curvature delle ipersuperfici minimali*, Rend. Sem. Mat. Univ. Padova **38** (1967), 91–107.

148. Charles B. Morrey, Jr., *Second order elliptic equations in several variables and Hölder continuity*, Math. Z. **72** (1959/60), 146–164.

149. Jürgen Moser, *A new proof of DeGiorgi's theorem concerning the regularity problem for elliptic differential equations*, Comm. Pure Appl. Mat. **13** (1960), 457–468.

150. ______, *On Harnack's theorem for elliptic differential equations*, Comm. Pure Appl. Math. **14** (1961), 577–591.

151. Jürgen Moser, *A Harnack inequality for parabolic differential equations,* Comm. Pure Appl. Math. **17** (1964), 101–134.

152. ______, *On a pointwise estimate for parabolic differential equations*, Comm. Pure Appl. Math. **24** (1971), 727–740.

153. Zane C. Motteler, *Existence theorems for certain quasi-linear elliptic equations*, Pacific J. Math. **17** (1966), 277–299.

154. M. K. V. Murthy and G. Stampacchia, *Boundary value problems for some degenerate-elliptic operators*, Ann. Mat. Pura Appl. (4) **80** (1968), 1–122.

155. J. Nash, *Continuity of solutions of parabolic and elliptic equations*, Amer. J. Math. **80** (1958), 931–954.

156. Louis Nirenberg, *On nonlinear elliptic partial differential equations and Hölder continuity*, Comm. Pure Appl. Math. **6** (1953), 103–156.

157. L. A. Peletier and J. Serrin, *Gradient bounds and Liouville theorems for quasilinear elliptic equations*, Ann. Scuola Norm. Sup. Pisa Cl. Sci. (4) **5** (1978), 65–104.

158. R. S. Phillips and Leonard Sarason, *Elliptic-parabolic equations of the second order*, J. Math. Mech. **17** (1967/68), 891–917; erratum, ibid. **18** (1968/69), 191.

159. P. A. Raviart, *Sur la résolution des certaines équations paraboliques non linéaires*, J. Functional Anal. **5** (1970), 299–328.

160. Raymond Redheffer, *On the inequality* $\Delta u \geqslant f(u, |\text{grad } u|)$, J. Math. Anal. Appl. **1** (1960), 277–299.

161. Laurent Schwartz, *Théorie des distributions à valeurs vectorielles.* I, Ann. Inst. Fourier Grenoble **7** (1957), 1–141.

162. James Serrin, *Local behavior of solutions of quasi-linear equations*, Acta Math. **111** (1964), 247–302.

163. ______, *The problem of Dirichlet for quasilinear elliptic differential equations with many independent variables*, Philos. Trans. Roy. Soc. London Ser. A **264** (1969), 413–496.

164. ______, *The Dirichlet problem for surfaces of constant mean curvature*, Proc. London Math. Soc. (3) **21** (1970), 361–384.

165. ______, *On the strong maximum principle for quasilinear second order differential inequalities*, J. Functional Anal. **5** (1970), 184–193.

166. ______, *Entire solutions of nonlinear Poisson equations*, Proc. London Math. Soc. (3) **24** (1972), 348–366.

167. ______, *Liouville theorems and gradient bounds for quasilinear elliptic systems*, Arch. Rational Mech. Anal. **66** (1977), 295–310.

168. Guido Stampacchia, *Contributi alla regolarizzazione delle soluzioni dei problemi al contorno per equazioni del secondo ordine ellittiche*, Ann. Scuola Norm. Sup. Pisa (3) **12** (1958), 223–245.

169. ______, *Some limit cases of* L^p*-estimates for solutions of second order elliptic equations*, Comm. Pure Appl. Math. **16** (1963), 505–510.

170. Friedrich Tomi, *Über semilineare elliptische Differentialgleichungen zweiter Ordnung*, Math. Z. **111** (1969), 350–366.

171. Neil S. Trudinger, *Quasilinear elliptic partial differential equations in n variables*, Ph. D. Thesis, Stanford Univ., Stanford, Calif., 1966.

172. ______, *On Harnack type inequalities and their application to quasilinear elliptic equations*, Comm. Pure Appl. Math. **20** (1967), 721–747.

173. ______, *Some existence theorems for quasi-linear, non-uniformly elliptic equations in divergence form*, J. Math. Mech. **18** (1968/69), 909–919.

174. ______, *On the regularity of generalized solutions of linear, non-uniformly elliptic equations*, Arch. Rational Mech. Anal. **42** (1971), 50–62.

175. ______, *Generalized solutions of quasilinear differential inequalities.* I: *Elliptic operators*, Bull. Amer. Math. Soc. **77** (1971), 576–579.

176. ______, *On the analyticity of generalized minimal surfaces*, Bull. Austral. Math. Soc. **5** (1971), 315–320.

177. ______, *Gradient estimates and mean curvature*, Math. Z. **131** (1973), 165–175.

178. ______, *Linear elliptic operators with measurable coefficients*, Ann. Scuola Norm. Sup. Pisa (3) **27** (1973), 265–308.

179. Wolf von Wahl, *Einige neue innere Abschätzungen für nichtlineare elliptische Gleichungen und Systeme*, Math. Z. **134** (1973), 119–128.

180. ______, *Über quasilineare elliptische Differentialgleichungen in der Ebene*, Manuscripta Math. **8** (1973), 59–67.

181. ______, *Über die Hölderstetigkeit der zweiten Ableitungen der Lösungen nichtlinearer elliptische Gleichungen*, Math. Z. **136** (1974), 151–162.

ABCDEFGHIJ–AMS–8987654